AUTOMOTIVE CHASSIS SYSTEMS

Steering, Suspension, Alignment, and Brakes

Second Edition

David A. Coghlan

Centennial College, Scarborough, Ontario

 Delmar Publishers Inc.®

To Anna,
with thanks for her patience and support

For more information address Delmar Publishers Inc.
3 Columbia Circle
Box 15-015
Albany, New York 12212-5015

10 9 8 7 6 5 4 3

Printed in the United States of America
Published simultaneously in Canada
by Nelson Canada,
A division of The Thomson Corporation

ISBN 0-8273-3865-1

Contents

Preface

There are many excellent textbooks that provide a general overview of the various operational systems of a vehicle. There is, however, limited text material available to readers that relates exclusively to automotive suspension, steering, alignment, and brakes. After teaching this subject for a number of years at a community college, I was asked to write a textbook that would deal more completely with these subjects.

The text is divided into twenty-seven chapters. Learning Objectives are stated as performance goals at the beginning of each chapter. At the conclusion of each chapter, students take a Review Test, based on the content of that chapter. This test enables the students to see how well they have understood the material. If there is no need for further review, they can proceed to the Final Test, composed of questions based on the chapter material. It should be noted that the text does not attempt to take the place of a manufacturer's shop manual but is intended to complement a shop manual by providing an expanded view of how suspension, steering, alignment, and brake systems operate and by emphasizing general principles.

This book has been written primarily for students of automotive technology and for practicing mechanics who desire to become more specialized. Readers who master the text will be able to increase their earning potential by expanding and upgrading their knowledge and skills.

ACKNOWLEDGMENTS

The author wishes to express thanks to the following firms for permission to reprint some material and illustrations used in the text: Ammco Tools, Incorporated; Bear Automotive Incorporated; Blackhawk Applied Power of Canada, Limited; Caroliner Company; Chart Industries; Chrysler Corporation; FMC Corporation; Ford Motor Company; General Motors of Canada Limited; Goodyear Canada Incorporated; Hunter Engineering Company; Kansas Jack Incorporated; Monroe Auto Equipment Company; Moog Automotive Canada Limited; Raybestos-Manhattan (Canada) Limited; Rubber Manufacturers Association; Snap-on Tools of Canada Limited; Stewart-Warner Corporation of Canada, Limited; and Wagner Brake Company, Limited.

The author also wants to thank the following people for their valuable assistance during the development of this text: Harry Evans, Blackhawk Applied Power of Canada, Limited; Nancy Logan, Goodyear Canada Incorporated;

Harold Swann, Sonic Automobile Enterprises Limited; and Ray Thompson, Snap-on Tools of Canada Limited.

Finally, thanks are extended to Raymond Oviyach (Triton College, Illinois) for his review of the manuscript and to the staff of Breton Publishers for their guidance and expertise during production.

CHAPTER

Suspension Systems and Their Design

LEARNING OBJECTIVES

After studying this chapter, you should be able to:

- Explain the early development of steering and suspension systems,
- Illustrate and describe the different types of vehicle front and rear suspensions,
- List the names of the various suspension parts,
- Explain the purposes of the major component parts of the various suspension systems,
- Define the terms *jounce* and *rebound*.

INTRODUCTION

The first vehicle invented by human beings was similar in design to a child's sled. It consisted of wooden runners and was pulled by brute strength. As loads became heavier, the effort required to move the skid was exhausting. Since humans are ingenious beings, they found that effort could be reduced greatly by placing the skids on round logs. In time, these wooden rollers gave way to wooden wheels.

The wheel became the basic answer to humanity's transportation problems, but it also introduced problems of its own. To be useful, the wheeled vehicle needed to be maneuverable. The driver needed some method of changing the vehicle's direction. The need to control the direction of travel was even more important than the need for braking. Even today, although our vehicles are heavier, travel at higher speeds, and are required to stop in shorter distances, it is still the steering, not the function of the brakes, that demands the driver's complete attention.

As one considers the steering of the vehicle, it becomes necessary also to consider the subject of suspension. Steering and suspension are related subjects, since the suspension system must absorb road shock that would otherwise be transmitted directly to the steering wheel. Many of the very early vehicles were designed without suspension. Wheels were fastened to wooden axles and fixed as solidly as possible to the frame of the vehicle. As we became more dependent on transportation, we began to think about comfort. Small bumps in the dirt roads were tolerated; larger bumps were avoided if possible.

The vehicles that were designed before the turn of the century had a steering system similar to a child's wagon. The driver steered the horseless carriage by means of a tiller. Although the tiller was simple in design, a loss of directional control and steering stability was encountered when the wheel came into contact with an obstacle.

Redesign of the steering mechanism became necessary. The **front axle*** was no longer allowed to pivot at its center; the axle became stationary, and **kingpins** were attached to the outer ends of the front axle (Figure 1-1). This design allowed each wheel to pivot independently and reduced some of the road shock transmitted to the driver. Directional control and stability remained a problem, however. With increased speeds and more powerful engines, the vehicle could easily go out of control. The design of the kingpin and axle was modified to provide increased steering control. Instead of the kingpin's being placed in a vertical position, the top of the kingpin was tilted in toward the center of the vehicle.

Picture for a moment the time of the Roman era. Julius Caesar is being carried by his slaves to the Roman Senate. These slaves made Caesar's ride comfortable by using the most perfect suspension system: the knees of the slaves. Just as the slaves' knees acted to absorb the bumps in Caesar's ride, so the suspension system of your car absorbs the bumps and bounces on the highway. When a vehicle is being driven, its wheels revolve and meet obstacles in the road. The wheels are subject to **inertia.** A wheel riding over an obstacle moves upward. If the wheel of the vehicle moves up, the body of the vehicle must also move up. Thus, the purposes of a suspension system are:

- To cushion the body of a vehicle from the reactions of a wheel to the irregularities of the road surface,
- To support the total body weight of the vehicle.

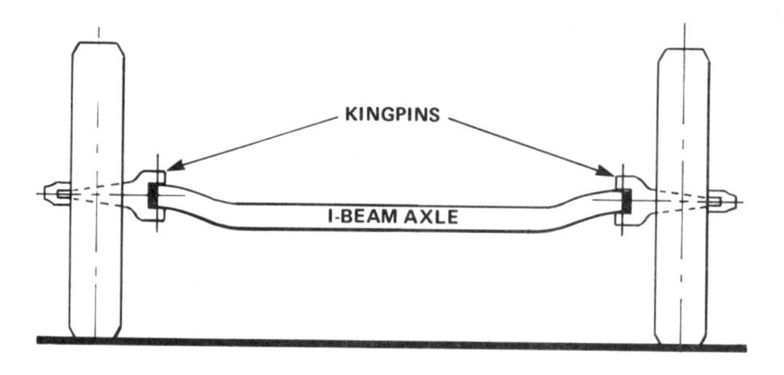

FIGURE 1-1. Simplified Front End Using Kingpins.

In the early years, a number of devices were added to the suspension to reduce the jarring effects of road shock. Heavy leather straps were used to cradle the body from the undesirable reaction of the axles and wheels. Flexible wooden members were probably the first springs until steel became available. The flexibility and strength of tempered steel made it a natural choice for the suspension material for motorized vehicles. **Leaf springs,** which are made of long, thin strips of steel, have been, and still are, most popular. **Coil springs,** made of steel rods in a cylindrical form, with air space between the coils, have also been used by vehicle manufacturers for a number of years. The advantage of a leaf spring is its flexibility to move in a vertical direction. The coil spring, because of its distinct design, will move in both a vertical and a horizontal direction.

Vehicles designed with coil spring suspensions require additional suspension parts to keep the wheels and axles in line with the vehicle body. Until now, two basic designs of front suspension systems have been used. They are the (1) **rigid axle suspension** and the (2) **independent front suspension.** Recently, an air suspension system has replaced the conventional suspension. The new system is air operated and microprocessor controlled to provide automatic front and rear load leveling. Let us examine the rigid axle design.

*Boldfaced words throughout this text are defined in the Glossary at the back of the book.

RIGID AXLE FRONT SUSPENSIONS

Single I-Beam Front Suspension

Figure 1-2 illustrates the single I-beam front suspension. This basic design was developed by pioneering vehicle manufacturers. **Spindle** assemblies that partially pivot are attached by kingpins to the outer ends of the I-beam axle. The steering arm, tie-rod, drag link, intermediate arm, and pitman arm are parts of the **steering linkage** mechanism. These parts are moved by the **steering gear** assembly. Vehicle body weight is supported by heavy-duty leaf springs. One end of the spring is attached to the frame at the support bracket. The opposite end of the spring is connected to the frame by means of a spring shackle. Since the spring must absorb road shock and thus continually lengthen and shorten, the main leaf must be allowed to change its length. This movement is accomplished through the spring shackle, which is a hinge. Leaf springs also keep the I-beam axle in correct **alignment** with the frame. The springs transfer the steering and braking action of the wheels from the axle to the frame. The springs are held to the axle by U bolts. This type of front suspension provides resistance to vertical and twisting forces and thus minimizes tire wear. The suspension also helps to achieve maximum durability and front end stability. This type of axle design is used on medium and heavy trucks.

Front Wheel Drive Axle

Although the *front wheel drive axle* is rigid in design, it uses coil springs to support the body of the vehicle (Figure 1-3). The front driving axle is kept in proper alignment with the vehicle's frame and is prevented from moving forward and backward by use of radius arms. Heavy-duty rubber **bushings** help to absorb undesirable road shock and vibration from the vehicle's frame.

A 2.54 centimeter (1 inch) diameter front track bar also maintains lateral stability and front axle alignment to the body of the vehicle. Road service life of the kingpins and wheel bearings is increased by incorporating the track bar into the front suspension. Double-acting hydraulic shock absorbers control the **oscillation** of the front coil springs and stabilize the ride of the vehicle.

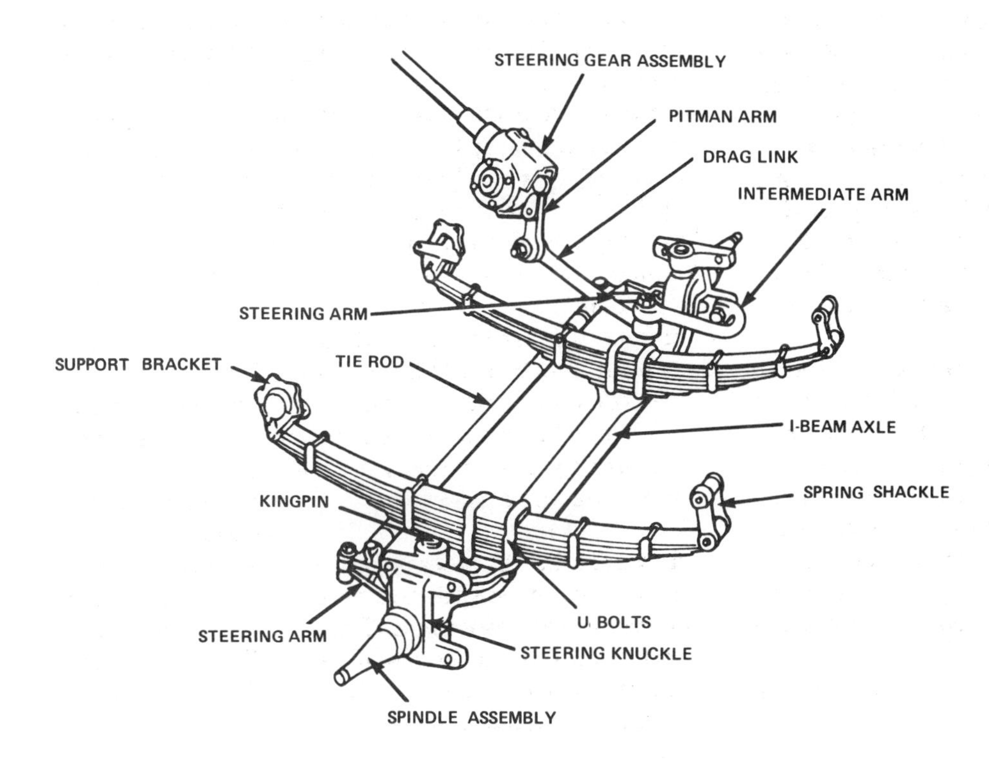

FIGURE 1-2. **Single I-Beam Front Suspension.** (Courtesy of Ford Motor Company)

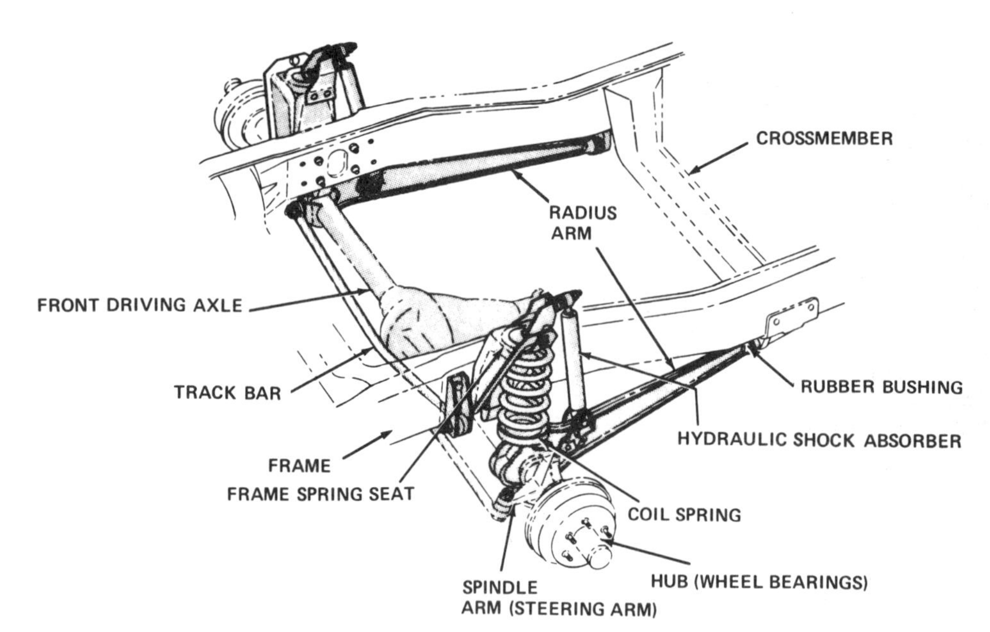

FIGURE 1-3. Front Wheel Drive Axle. (Courtesy of Ford Motor Company)

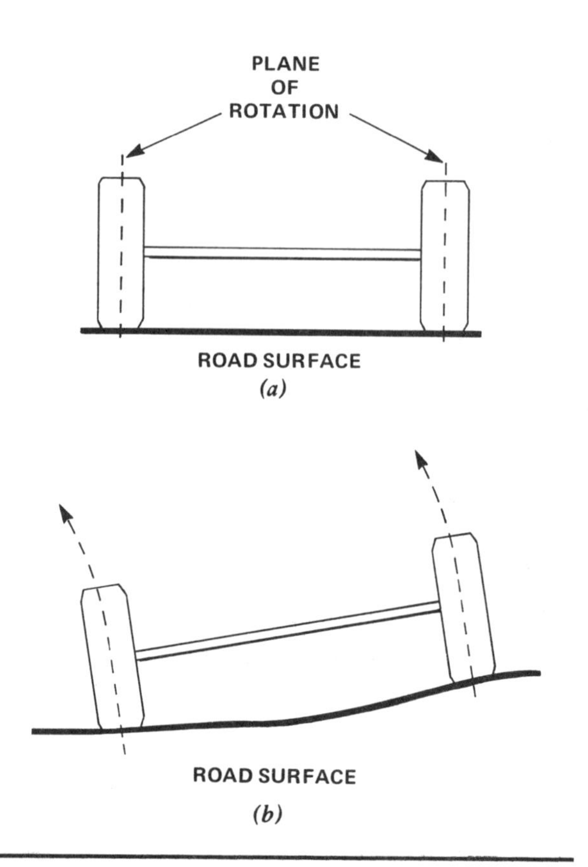

FIGURE 1-4. Plane of Rotation of a Vehicle's Wheels. (a) Flat road surface; (b) uneven road surface. (Courtesy of Hunter Engineering Company)

Pros and Cons of Rigid Axle Suspension. Rigid axles can support considerable vehicle weight, but ride and handling qualities are somewhat diminished. When one front wheel encounters an obstacle or depression on the road surface, one side of the axle is raised or lowered; the opposite wheel must also move from its **plane of rotation** (Figure 1-4). If the suspension or steering linkage systems are subject to wear, some loss of **directional control and stability** will be experienced by the driver.

INDEPENDENT FRONT SUSPENSIONS

The term *independent front suspension* describes a method of supporting the **chassis** on the wheels without the use of rigid axles. With independent suspension, the movements of the two front wheels are not interdependent; one wheel does not force the other wheel to change its plane of rotation.

Long-and-Short-Arm Coil Spring Suspension

Long-and-short-arm coil spring suspension is used by all North American vehicle manufacturers. A coil spring attached to or positioned on the lower control arms supports the weight of the vehicle. This suspension is called the long-and-short-arm sus-

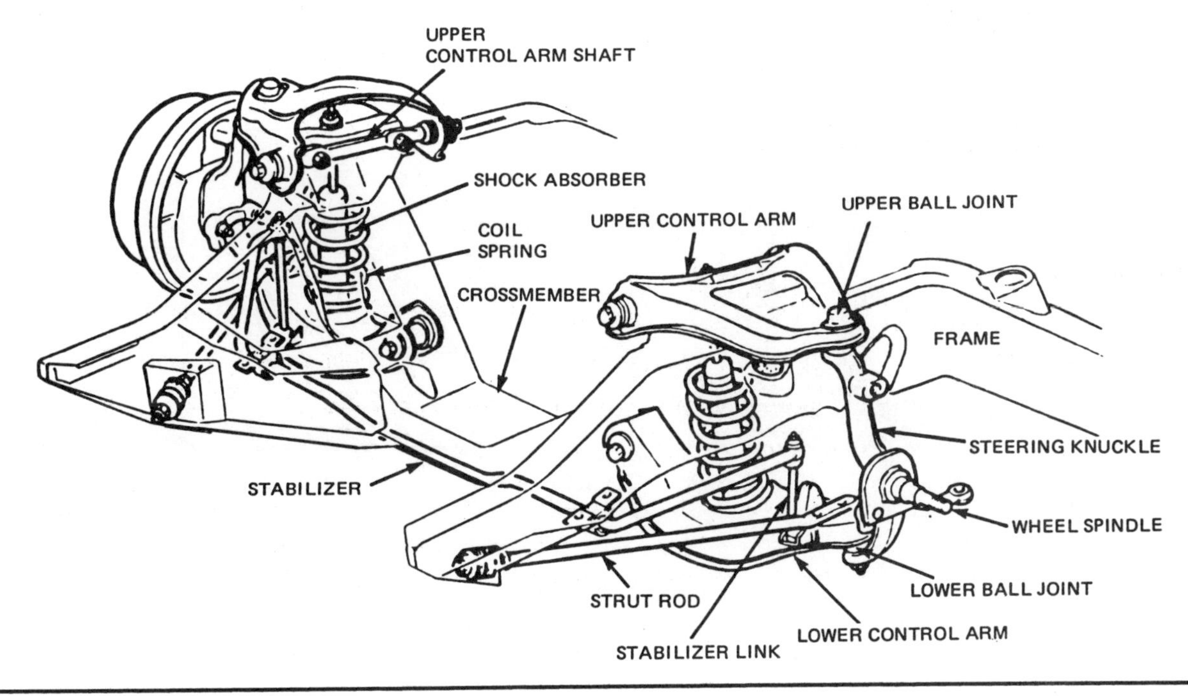

FIGURE 1-5. **Long-and-Short-Arm Suspension.** (Courtesy of General Motors of Canada, Limited)

pension because it is designed with upper and lower **control arms** of unequal lengths (Figure 1-5).

The upper and lower control arms are attached to the frame by control arm shafts. The crossmember serves the same purpose as the I-beam axle discussed previously. The **wheel spindle** and **steering knuckle** are attached to the outer ends of the upper and lower control arms by **ball joints.** The upper and lower ball joints allow the steering knuckle to pivot in much the same manner as the conventional kingpin. Most vehicle manufacturers with independent long-and-short-arm suspension designs use a **stabilizer** or **sway bar.** This control device is attached to the frame and to the lower control arms by stabilizer links. The torsional (twisting) stabilizer bar (made of spring steel) reduces the body roll, thereby reducing the effect of **centrifugal force** when a vehicle rounds a corner. Some vehicles are designed with a **strut rod;** this rod reduces the forward and backward movement of the lower control arm when the **brakes** are applied and when the vehicle is accelerated. The strut rod is located between the lower control arm and the frame of the vehicle.

At first, the reasons for using upper and lower control arms of unequal lengths may not be clear; however, a careful examination of Figure 1-6 will explain the advantage. If the arms were equal in length, the vehicle's **track width** (the distance between the two front tires) would constantly change as the vehicle passed over irregularities in the road. This spreading and retracting action would drag the tires sideways and result in very rapid tire wear from the scuffing and scrubbing action. To avoid this problem, the upper control arm is made shorter, which makes the radius it traces shorter. The result is that the top of the wheel moves faster than the bottom. The change in the vehicle's track width is minimal, and the side scuffing of the front tires is greatly reduced.

Manufacturers use two terms to describe the up-and-down movement of a wheel. When a wheel moves upward in relation to the vehicle, this movement is called **jounce travel.** When a wheel moves downward in relation to the vehicle, this movement is called **rebound travel** (Figure 1-7).

Two Basic Designs. North American vehicle manufacturers use two basic types of front end suspension designs. Some manufacturers use a coil spring attached to or positioned on the lower control arms to support the weight of the vehicle (see Figure 1-5). Other manufacturers support the ve-

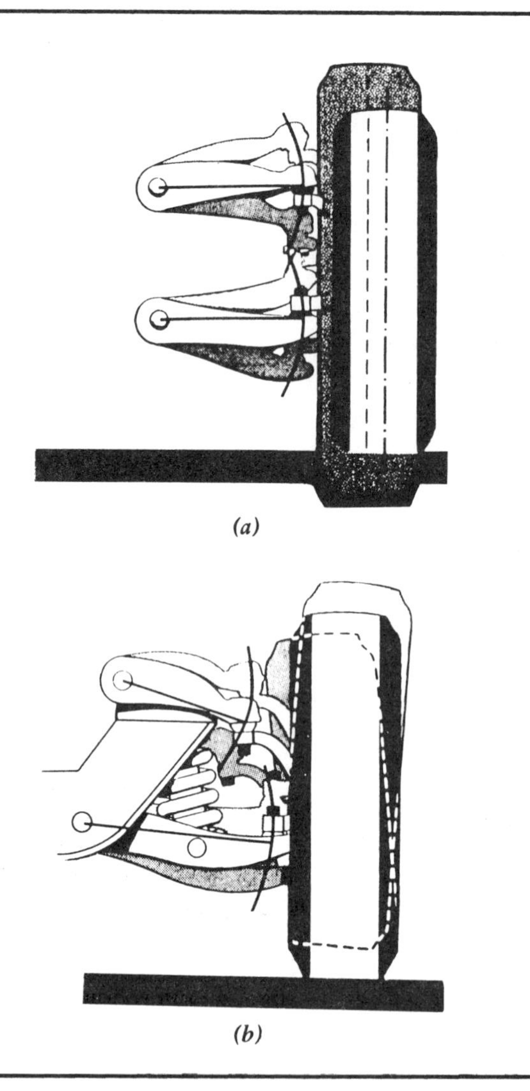

FIGURE 1-6. Control Arm Design. (a) Equal length; (b) unequal length. (Courtesy of Hunter Engineering Company)

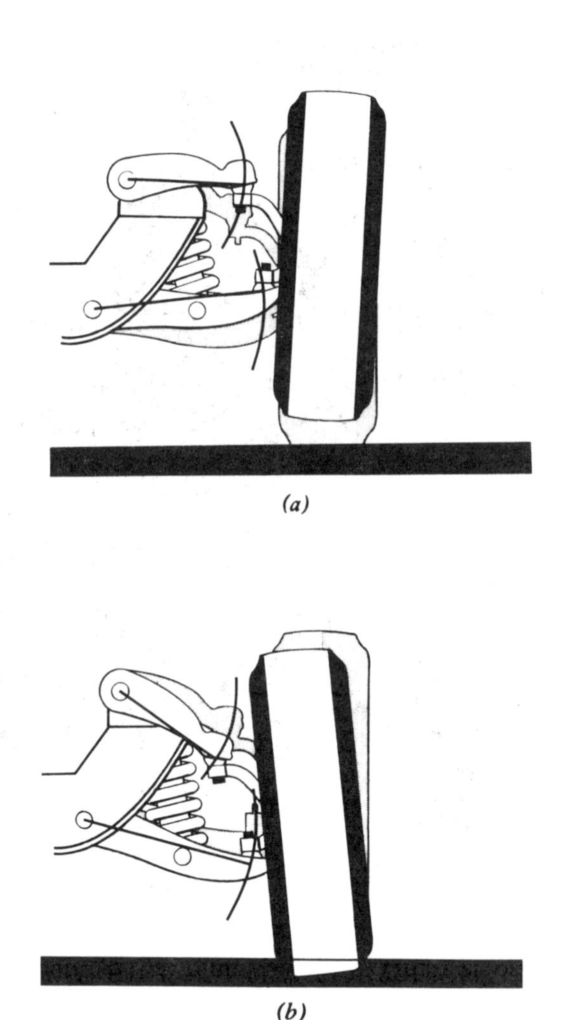

FIGURE 1-7. Types of Up-and-Down Wheel Movement. (a) Jounce travel; (b) rebound travel. (Courtesy of Hunter Engineering Company)

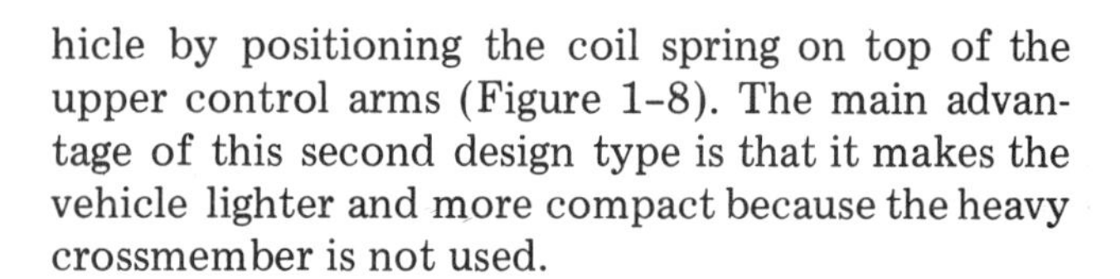
hicle by positioning the coil spring on top of the upper control arms (Figure 1-8). The main advantage of this second design type is that it makes the vehicle lighter and more compact because the heavy crossmember is not used.

Torsion Bars

In 1957, Chrysler Motor Corporation modified its front end design and introduced a slightly different type of front end suspension. **Torsion bars** (rods made of spring steel) were used in place of coil springs. One end of the torsion bar is fixed to the vehicle's frame; the opposite end of the bar is attached to the lower control arm pivot point. Any force that would push the front wheel upward (like a bump in the road) is transmitted through the lower control arm and would therefore cause the torsion bar to twist. When the force is removed, the torsion bar unwinds like a rubber band and lowers the front wheel to its original position (Figure 1-9).

In recent years, Chrysler Motor Corporation has introduced the **transverse torsion bar** front suspension (Figure 1-10). The bars are positioned laterally across the width of the chassis and attached to the control arms. Because of their unique design, the bars eliminate the need for strut rods; thus, the suspension is very compact. Torsion bar suspension allows the front **curb riding height** to be constantly maintained to the manufacturer's specification.

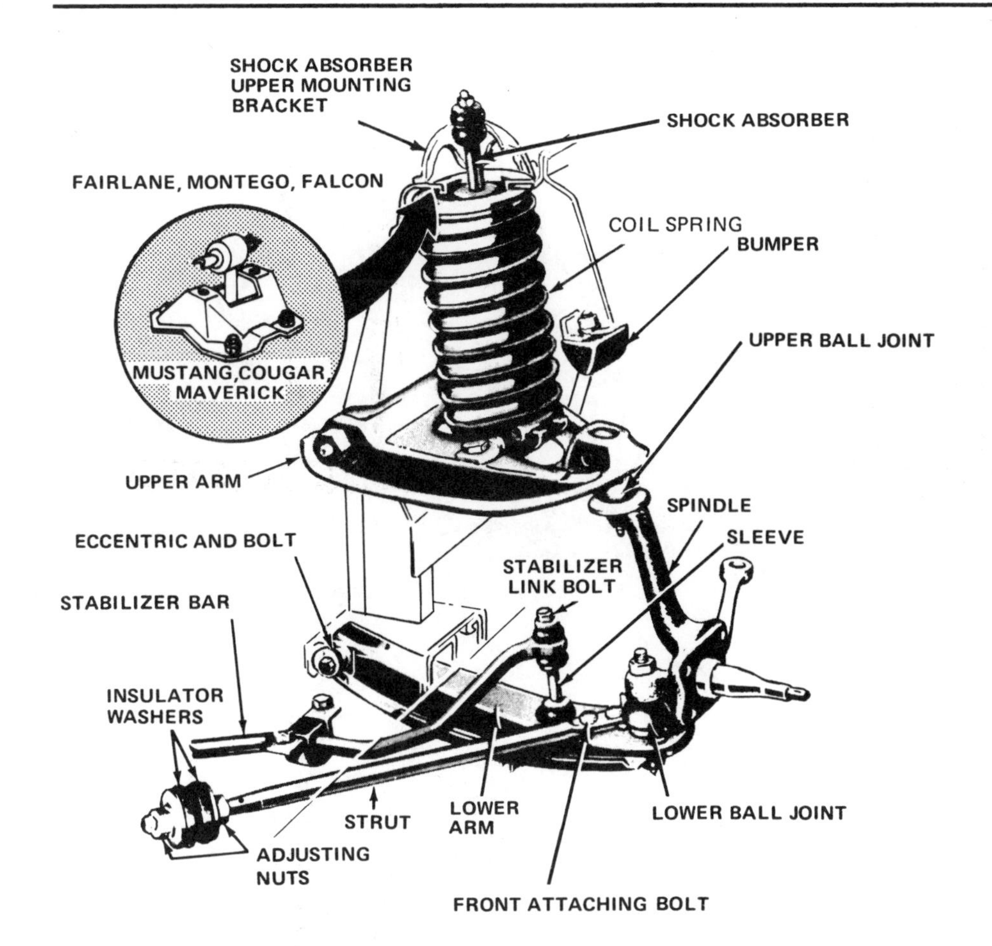

FIGURE 1-8. Suspension with Coil Spring Positioned on Upper Control Arm. (Courtesy of Ford Motor Company)

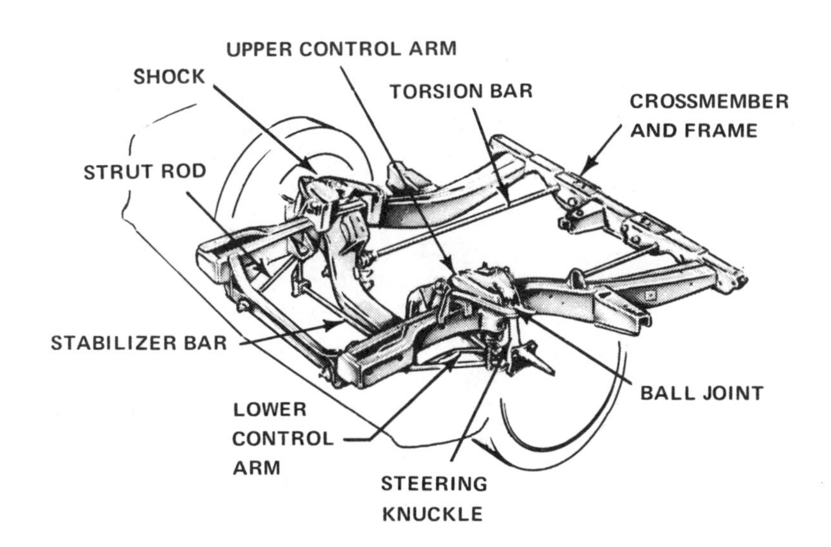

FIGURE 1-9. Torsion Bar Front Suspension. (Courtesy of Chrysler Corporation)

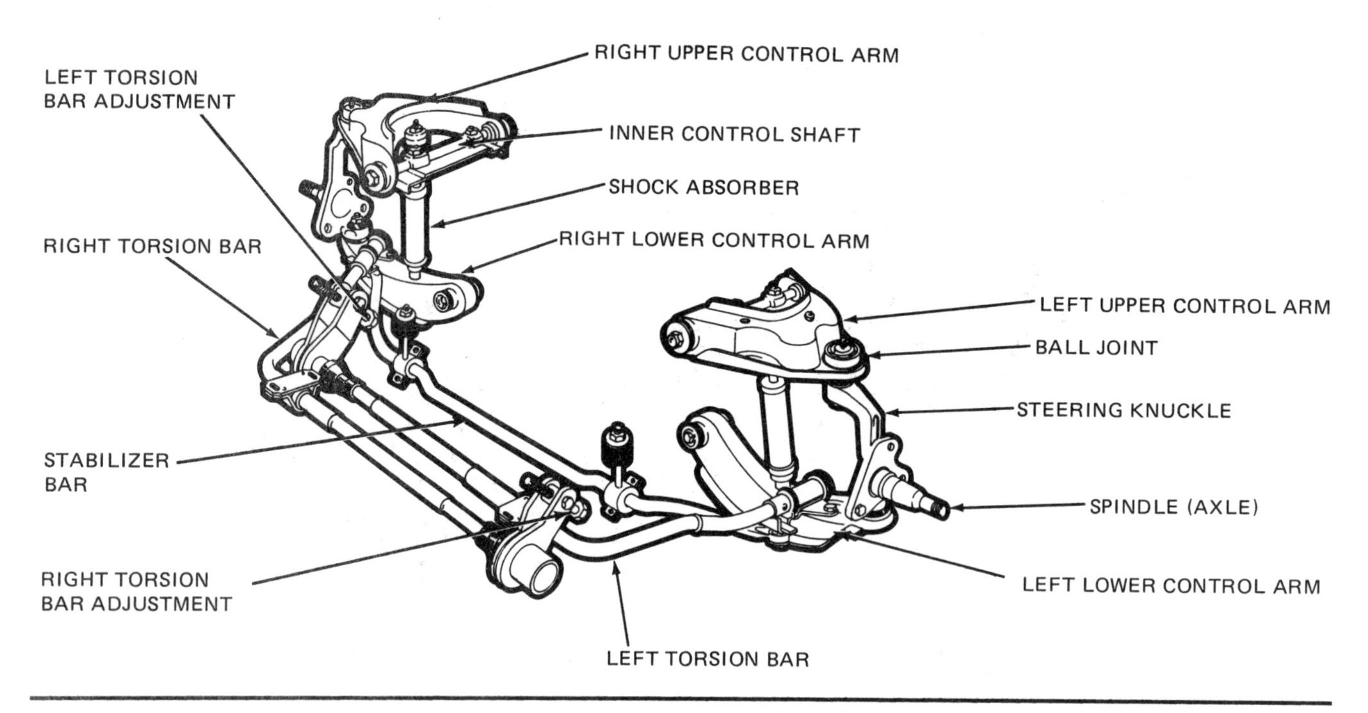

FIGURE 1-10. Transverse Torsion Bar Front Suspension. (Courtesy of Moog-Canada Automotive, Ltd.)

OTHER TYPES OF FRONT END SUSPENSIONS

Twin I-Beam Front Suspension

In recent years, the *twin I-beam front suspension* has become very popular (Figure 1-11). This design provides each front wheel with its own solid I-beam; coil springs absorb road shock and support the vehicle's front weight. Radius arms and coil springs maintain the position of the front wheels in relation to the road surface. The solid design feature of this suspension allows both the caster and camber (two wheel alignment factors we'll discuss later) to be preset by the manufacturer. This suspension uses heavy-duty hydraulic **shock absorbers** to control the oscillation of the front springs and to allow the vehicle to be used for constant off-road use.

MacPherson Strut

Figure 1-12 illustrates another design of front end suspension, one that is used on many domestic and imported vehicles. This type of suspension is called the *MacPherson strut*. The shock absorber, strut, steering knuckle, and spindle are a combined unit supported by a coil spring at the top. An upper mount assembly bolted to the vehicle's body enables the unit to pivot. A ball joint and a lower control arm connect the bottom part of the assembly to the crossmember. Forward and backward movement of the assembly is controlled by a strut rod and bushings. Vehicle body sway is reduced by means of a stabilizer bar.

The strut is the main structural component; it replaces the upper control arm and ball joint found in a conventional long-and-short-arm design. Therefore, the strut must be able to withstand lateral and vertical forces that affect the steering alignment, ride, and handling qualities of the vehicle.

Modified MacPherson Strut. A *modified MacPherson strut* design (Figure 1-13) uses shock struts with coil springs mounted between the lower arms and spring pockets in the crossmember. The springs rest on the lower control arms. Therefore, the front portion of the vehicle's weight is transmitted through the coil springs, control arms, and ball joints to the struts and front wheel spindles. Front strut rods are not needed because the lower control arms are affixed to the crossmember at the front and rear connecting shaft positions.

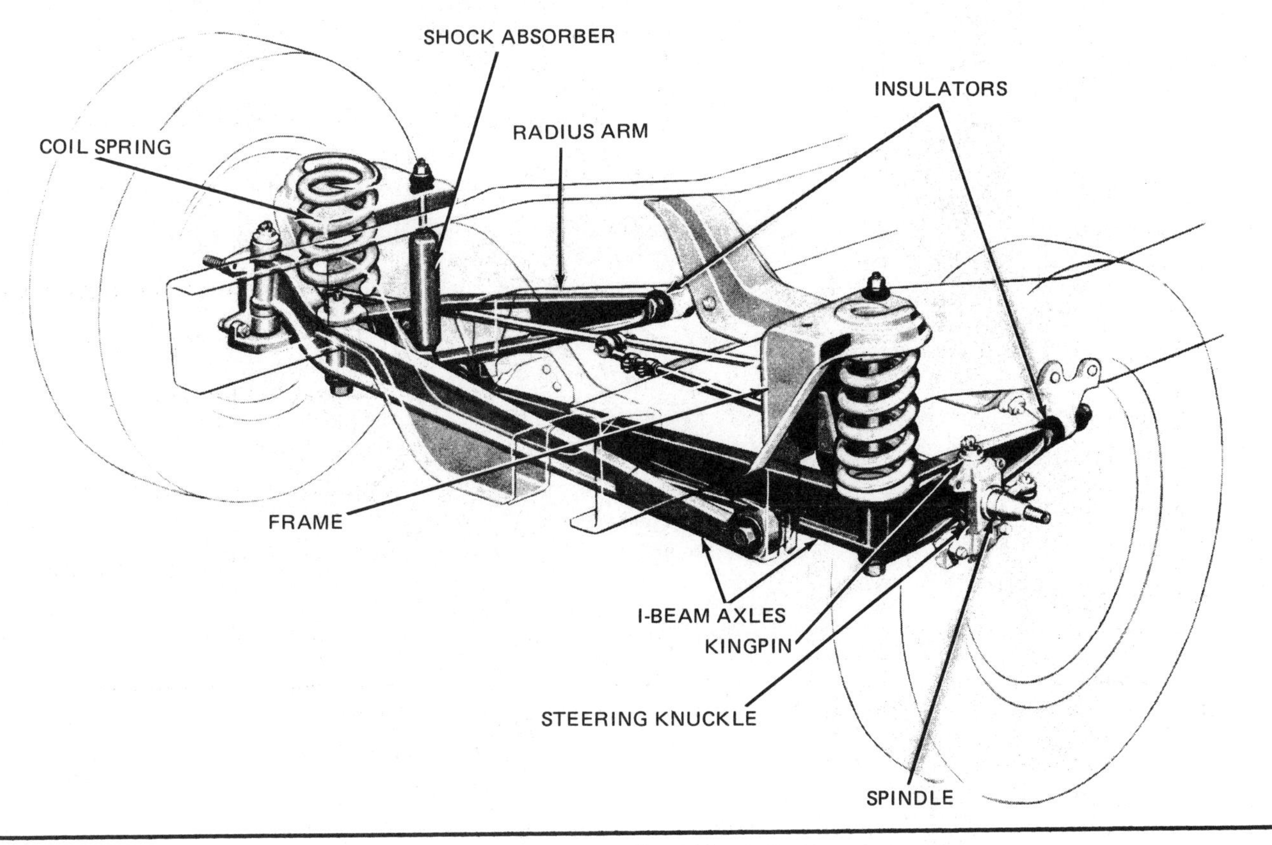

FIGURE 1-11. Twin I-Beam Front Suspension. (Courtesy of Ford Motor Company)

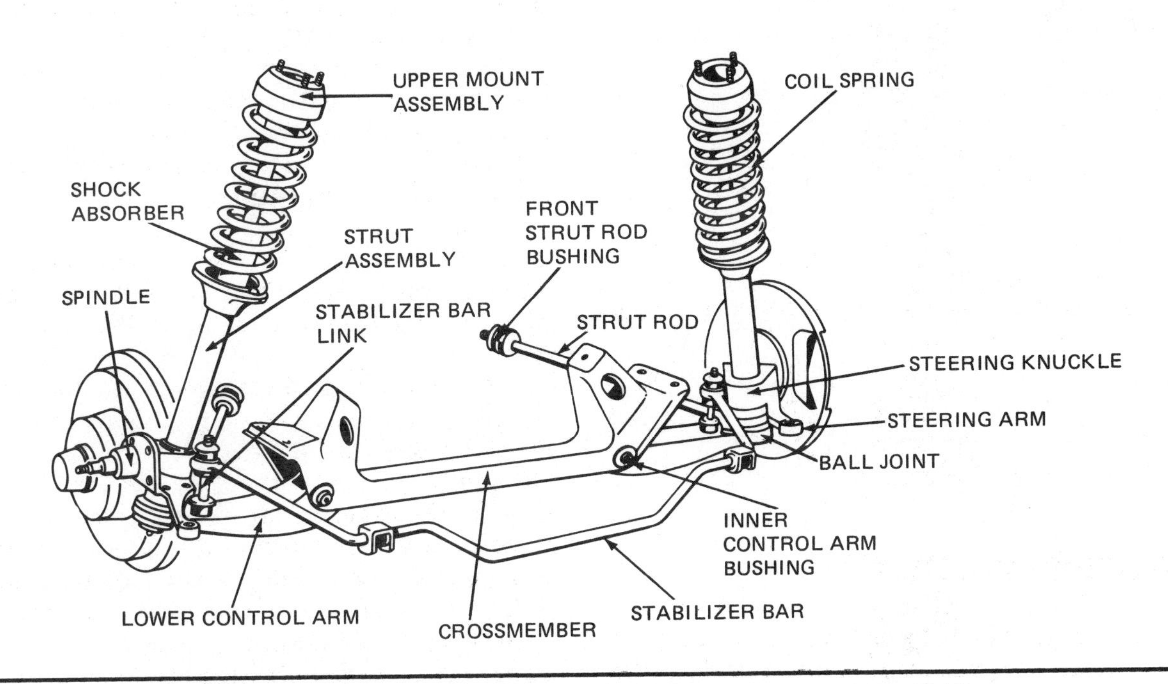

FIGURE 1-12. MacPherson Strut Front Suspension. (Courtesy of Moog-Canada Automotive, Ltd.)

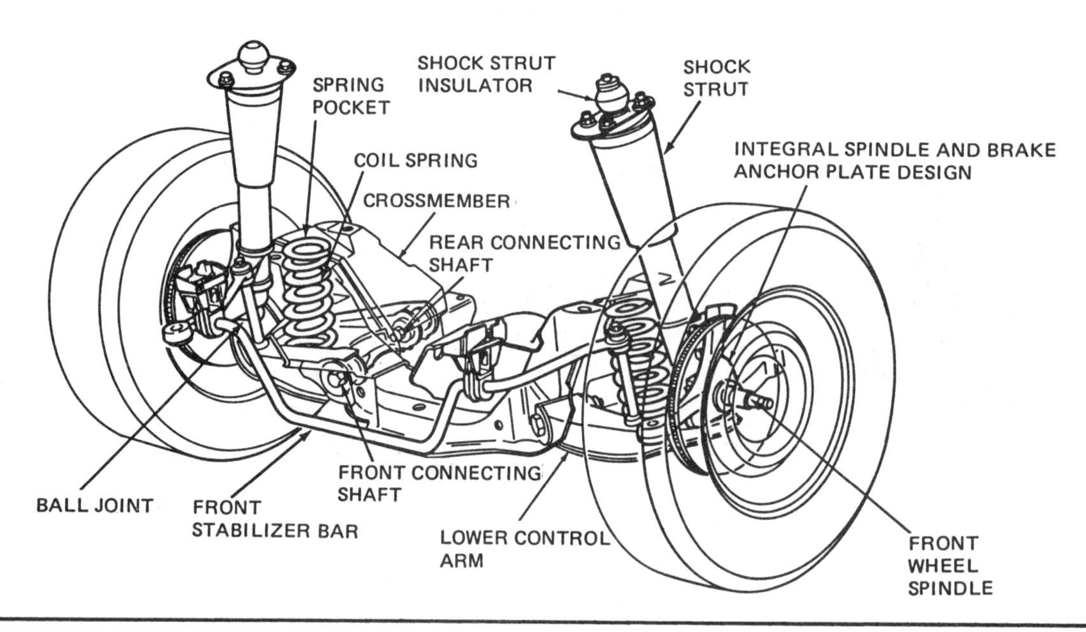

FIGURE 1-13. Modified MacPherson Strut Front Suspension. (Courtesy of Ford Motor Company)

REAR SUSPENSIONS

The *rear suspension* must cushion the ride of the vehicle and support the weight of the vehicle. Although the rear suspension does not actually steer the vehicle, it must keep the rear wheels in line with the body of the vehicle so that the driver can maintain directional control and stability.

Rigid Rear Driving Axle

The most common rear suspension system is the *rigid rear driving axle* design. This design is constructed of a central differential and tubular axle shaft housing. Solid axle shafts are supported near the wheels by axle bearings and are splined to the differential side gears within the gear housing. The axles provide a means of transferring rear wheel driving **torque,** braking, and accelerating forces from the axle housing to the frame of the vehicle.

Hotchkiss Drive

Figure 1-14 illustrates the *Hotchkiss drive* rear axle design. This type of axle design has been used for many years. Two semi-elliptical leaf springs are attached to the axle housing by U bolts. Spring shackles and brackets attach the springs to the body of the vehicle. Spring shackles are flat bars made of steel that allow the springs to change their length in response to road surface and load change conditions. The Hotchkiss drive has spring torque windup. When the vehicle is accelerated or braked, reaction on the torque of the wheels tends to twist the rear axle housing. This twisting action distorts or winds up the springs. (This is what happens when you spin the wheels and leave a patch of rubber on the surface of the road.) The rear of the vehicle drops under severe acceleration forces (Figure 1-15).

Coil Spring Rear Suspension

In *coil spring rear suspension* (Figure 1-16), the rear axle assembly is attached to the frame through a link-type suspension system. Rubber bushed, lower control arms mounted between the axle assembly and the frame of the vehicle maintain the fore and aft relationship of the axle assembly to the chassis and frame. Rubber bushed upper control arms angularly mounted with respect to the centerline of the vehicle control acceleration, braking, and the sideways movement of the axle assembly. The rigid rear axle holds the rear wheels in proper alignment with the body of the vehicle. Some manufacturers use a track bar located between the axle

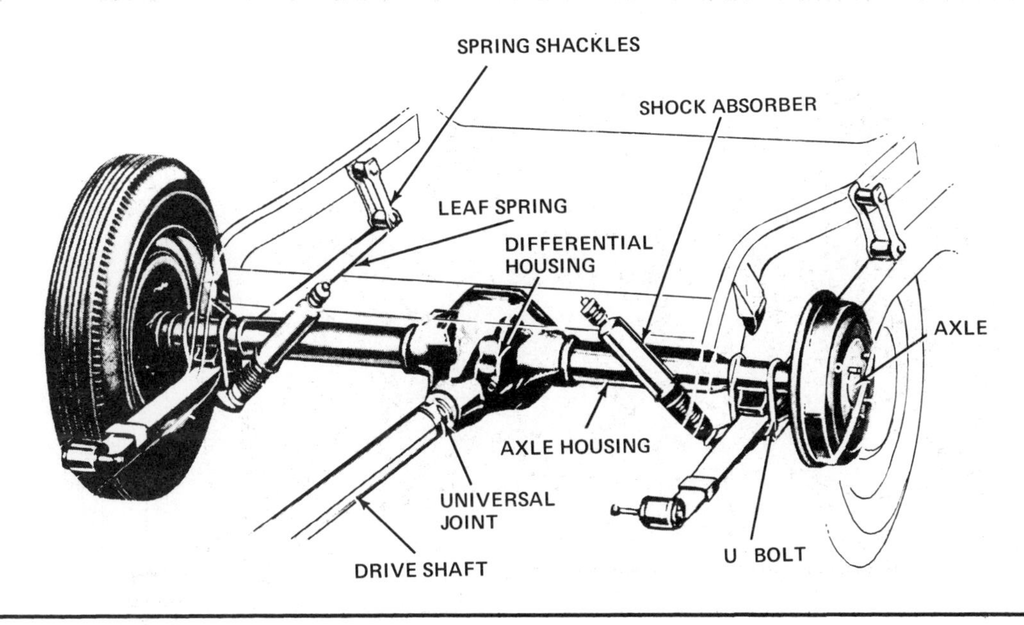

FIGURE 1-14. **Hotchkiss Drive Rear Axle Design.** (Courtesy of Ford Motor Company)

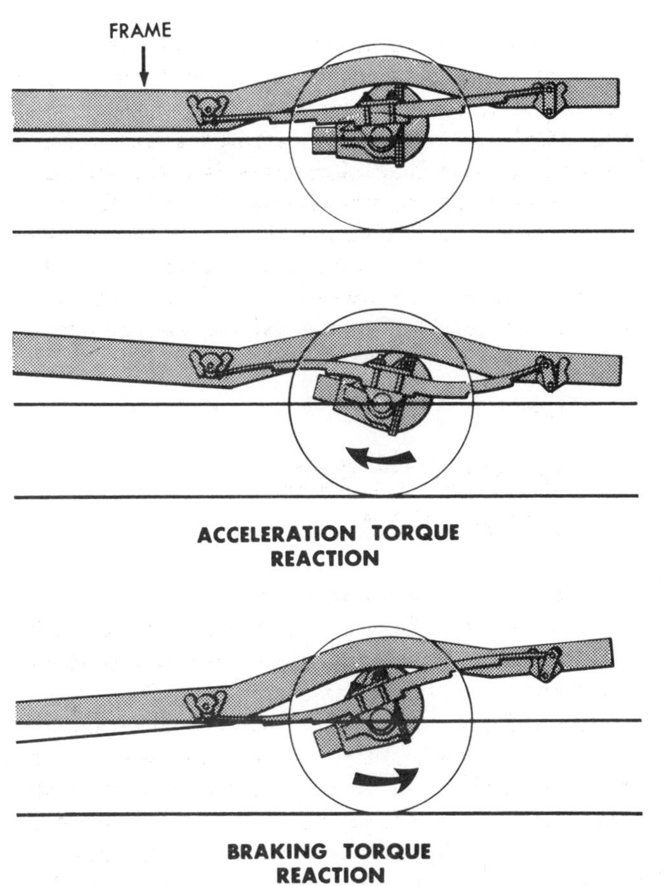

FIGURE 1-15. **Spring Windup Reaction in Hotchkiss Drive Rear Suspension.** (Courtesy of Ford Motor Company)

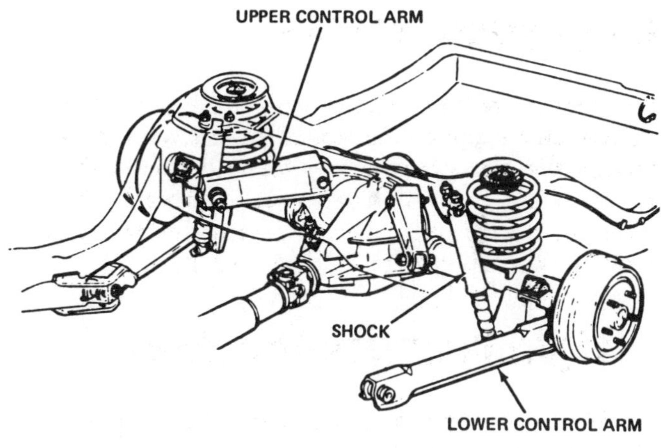

FIGURE 1-16. **Coil Spring Rear Suspension.** (Courtesy of General Motors of Canada, Limited)

housing and the vehicle's frame to keep the rear wheels in line with the vehicle's chassis and body.

The rear coil springs are located between the rear axle housing and the spring seats in the frame of the vehicle. The springs are held in position by the weight of the car and by the shock absorbers, which limit the upward movement of the body of the vehicle. Ride control and body sway are maintained by double-acting shock absorbers mounted at an angle between the axle housing and the frame of the vehicle.

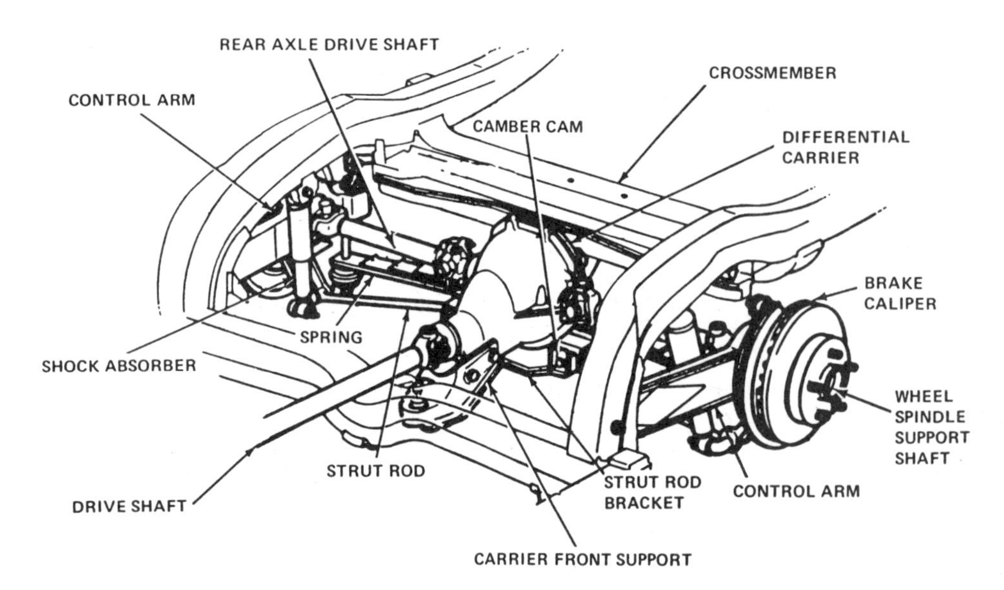

FIGURE 1-17. Independent Rear Suspension. (Courtesy of General Motors of Canada, Limited)

Independent Rear Suspension

In recent years, General Motors Corporation has introduced *independent rear suspension* (Figure 1-17). This design features a transverse multileaf spring mounted on a fixed differential carrier. Each rear wheel is mounted by a three-link independent suspension. These three links are composed of a rear axle drive shaft, a camber control strut rod, and a wheel spindle support shaft.

The advantage of this suspension system includes a reduction of **unsprung weight** and an overall vehicle weight reduction. In addition, **wheel tramp** is greatly reduced, and handling is improved because of the independent action of each rear wheel. The forward and backward motion of the wheel is limited by control arms. The arms also act as mounts for the brake **calipers** and the parking brake assemblies. Because each rear wheel's suspension is independent of the other, one wheel can move vertically without affecting the wheel on the opposite side of the vehicle.

MacPherson Strut Independent Rear Suspension

A true *MacPherson strut independent rear suspension* (Figure 1-18) has, on each side, a shock absorber strut assembly, two parallel control arms, tie-rod, strut rod, forged spindle, and jounce bumper and bracket assembly.

The *shock absorber strut assembly* includes a rubber isolated top mount, upper spring seat, coil spring insulator, coil spring, and lower spring seat. The strut assembly is attached at the top by two studs. The studs retain the top mount of the strut to the inner body side panel. The lower end of the assembly is bolted to the spindle. The two stamped control arms attach to the underbody and spindle with nuts and bolts.

TRUCK REAR SUSPENSIONS

Single Stage Multileaf Suspension

Figure 1-19 illustrates two different types of truck spring assemblies: the conventional single stage progressive spring and the single stage spring with auxiliary. Two sets of leaves are used to suit heavy and light load conditions. The upper leaves soften the ride under light loads. As the load increases, the stiffer lower leaves combine with the leaves of the upper spring to support the weight of the vehicle and its load.

Variable Rate Suspension

Figure 1-20 illustrates a variable rate spring. This design uses radius leaves to link the spring

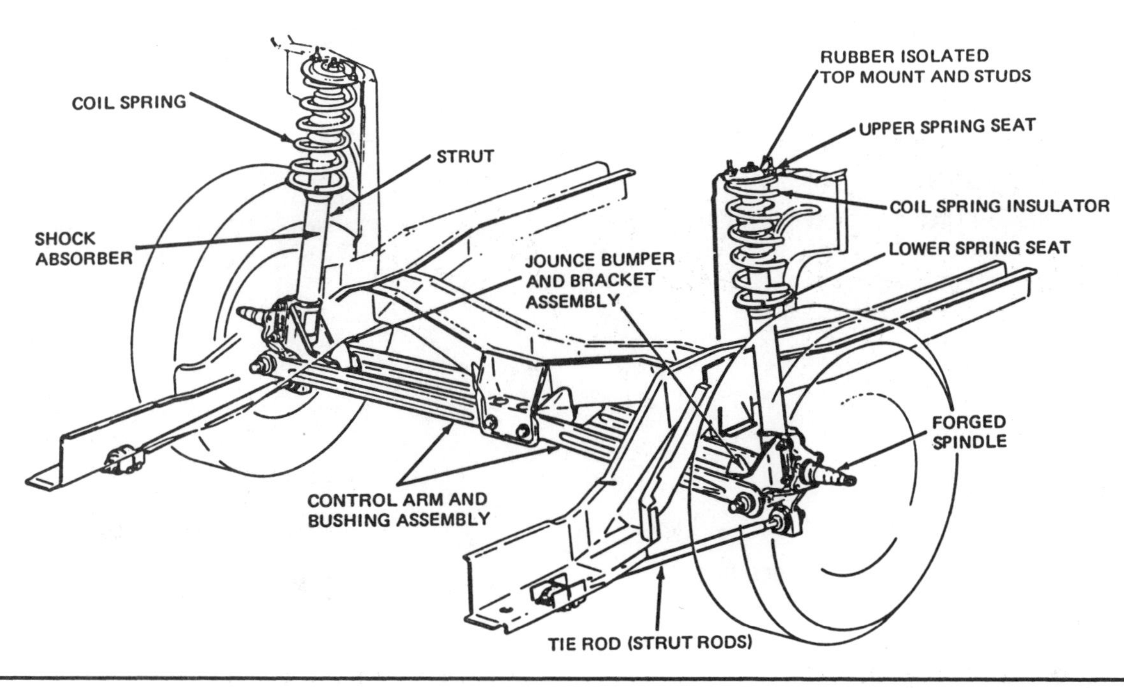

FIGURE 1-18. **MacPherson Strut Independent Rear Suspension.** (Courtesy of Ford Motor Company)

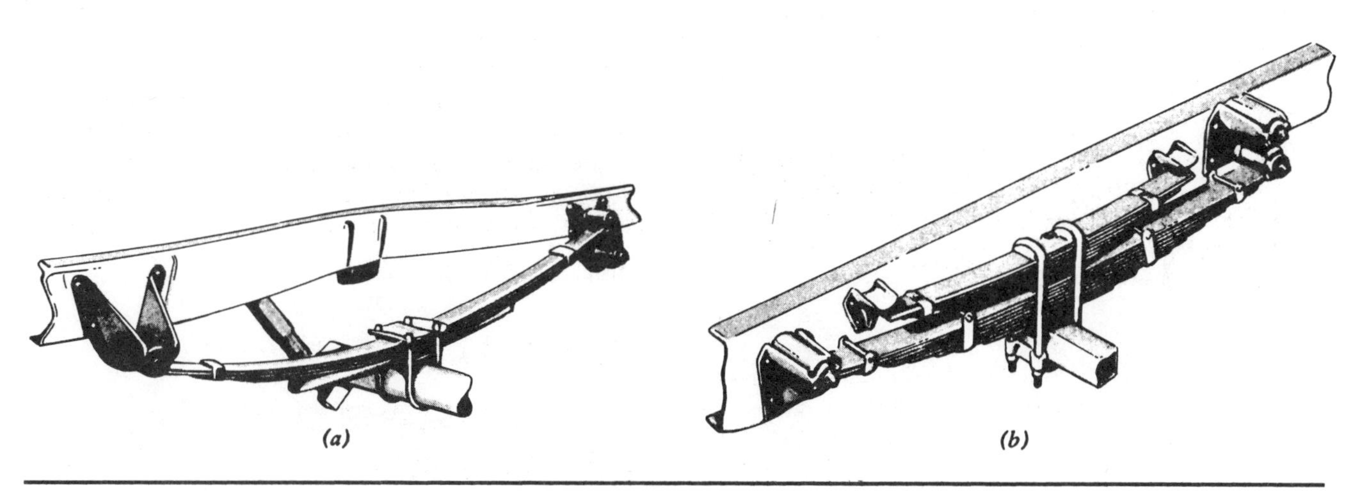

FIGURE 1-19. **Two Types of Truck Spring Assemblies.** (a) Single stage progressive spring; (b) single stage progressive spring with auxiliary. (Courtesy of Ford Motor Company)

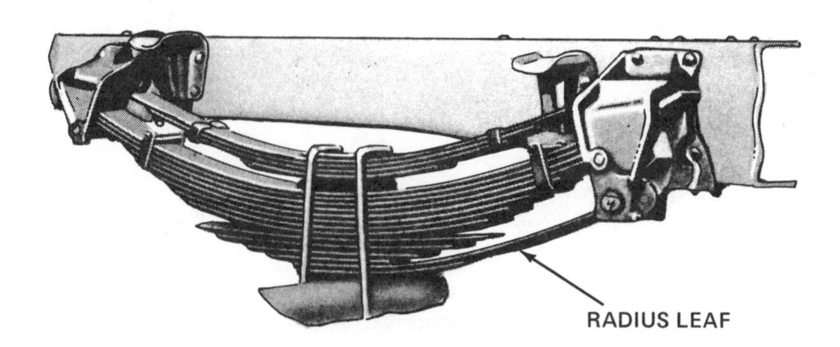

FIGURE 1-20. **Variable Rate Spring.** (Courtesy of Ford Motor Company)

mount on the axle to the forward spring bracket. The radius leaves control accelerating and braking forces, leaving the main spring free to support and cushion the load of the vehicle. **Variable spring rate** is achieved through cam-shaped pads in the mounting brackets. The pads change the length of the main spring and thus its rate as the load changes. The spring becomes longer and stiffer when heavily loaded; it shortens and becomes more flexible under lighter loads.

AIR SPRING SUSPENSION

In the past, some trucks and buses have been designed with *air spring suspension* (Figure 1-21). More recently, Ford Motor Company has introduced air spring suspension for the Mark VII Continental automobile. The prime feature of this type of system is that correct curb riding height is constantly maintained.

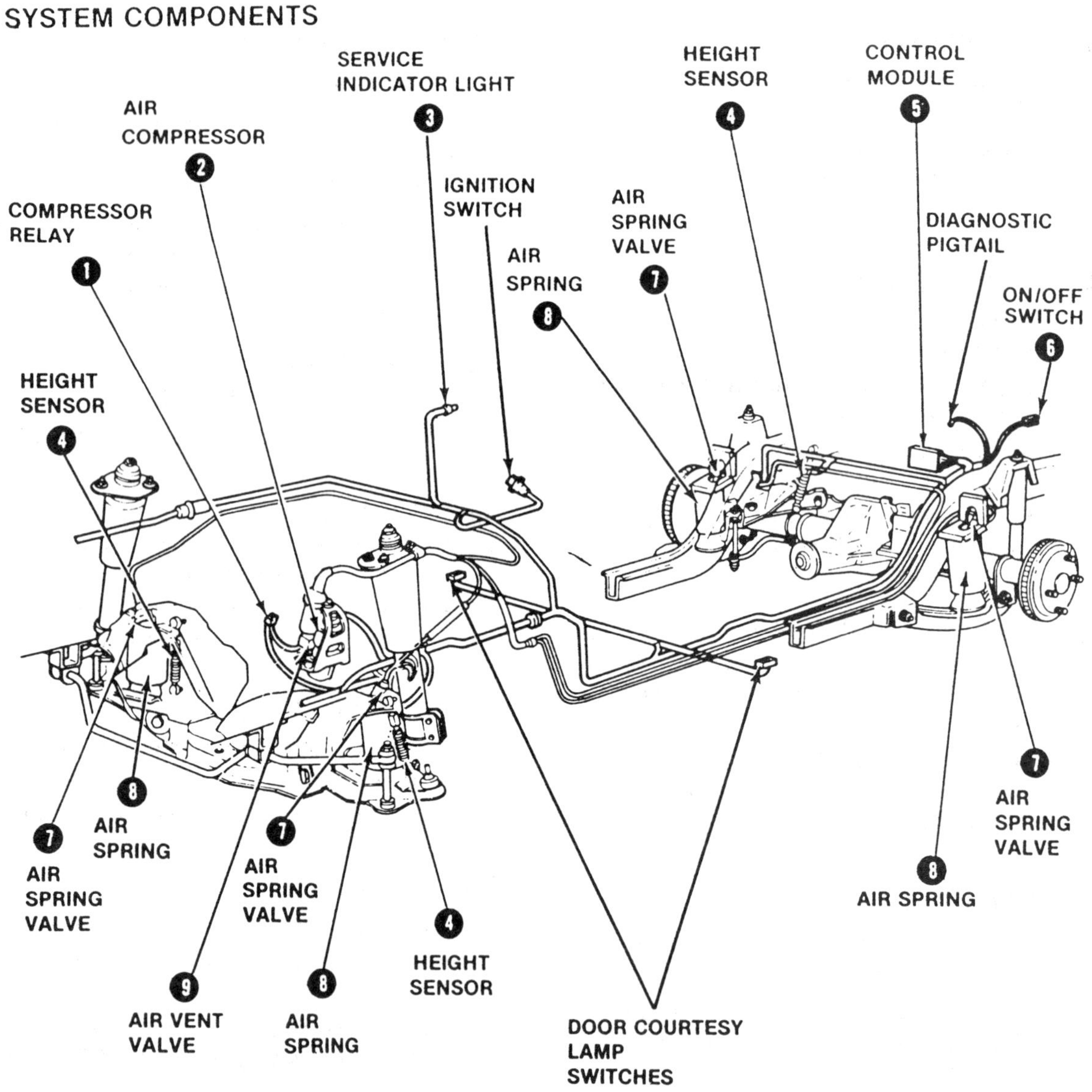

FIGURE 1-21. Overall View of an Air Spring Suspension System. (Courtesy of Ford Motor Company)

Description

Air spring suspension is an air-operated, microprocessor-controlled suspension system. It replaces the conventional coil spring suspension and provides automatic front and rear load leveling.

Four air springs, made of rubber and plastic, support the vehicle load at the front and the rear wheels (Figure 1-22). The front air springs are mounted to a spring pocket in the crossmember and on the lower suspension arms, as on a conventional front spring system. The rear air springs are mounted ahead of the rear axle outside the body subframe side members and on the lower rear suspension arms as on a conventional rear spring system.

A single cylinder, piston-type, electrically operated air compressor, mounted on the left fender apron, supplies the air pressure for operating the system. A regenerative-type dryer is attached to the compressor manifold. All air flow during compression or venting passes through the dryer. A vent solenoid, located on the compressor manifold, controls air exhaustion.

The air compressor, solenoids, height sensors,

1 COMPRESSOR RELAY

- Opens and closes high-amp battery feed circuit to compressor
- Module operates compressor by providing relay ground circuit

2 AIR COMPRESSOR

- Single-cylinder, motor-driven air compressor
- Regenerative air dryer removes moisture from air entering system
- Air exhausted during lowering of vehicle removes moisture from dryer
- Four air lines, one to each air spring, are covered by protective cap

3 SERVICE INDICATOR LIGHT

- Module monitors system operation; when problem is detected, module switches on light
- Light alerts driver to system problem
- During self-diagnosis, light flashes test number and helps isolate system problem

4 HEIGHT SENSOR (3)

- Attached to body and suspension arms
- One rear, two front
- Changes in body height are converted to electronic signal that "tells" module when body is out of trim

5 CONTROL MODULE

- Microprocessor
- Controls system components to maintain constant trim height
- System diagnostic capability
- Alerts driver to possible system problem through service indicator light
- Can fill an empty, service replacement air spring

6 ON/OFF SWITCH

- Cuts battery feed to module
- Disables entire system for tire changing or service

7 AIR SPRING VALVE (4)

- Opened to allow air to enter spring (inflate) or vent from spring (deflate)
- Solenoid-type valve is opened when module provides ground circuit for valve coil

8 AIR SPRING (4)

- Made of rubber and plastic, replaces coil springs
- Inflated or deflated to maintain constant vehicle height

9 AIR VENT VALVE

- Part of air compressor cylinder head
- Opens only when module wants to lower vehicle
- Solenoid-type valve is opened when module provides ground circuit for valve coil

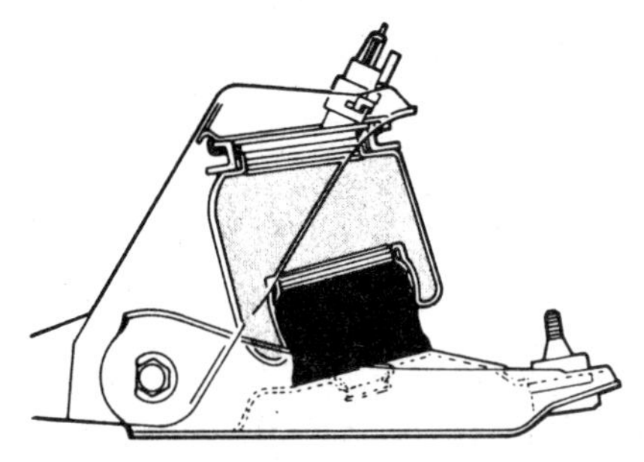

DESIGN

- Air spring is at normal trim height
- Air pressure contained in rubber membrane maintains vehicle height and acts like coil spring
- Air spring valve mounted in end cap opens to allow air to enter and exit spring
- When air is added, vehicle will rise
- When air is removed, vehicle will lower

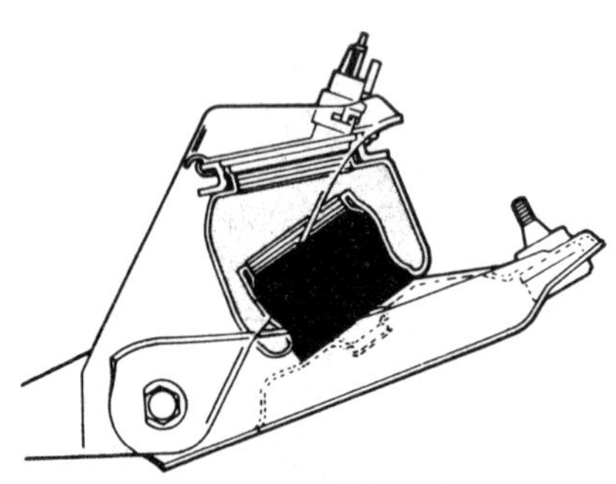

JOUNCE

- When control arm moves upward, piston moves upward into rubber membrane
- As the arm moves upward toward jounce the rate of the air spring increases

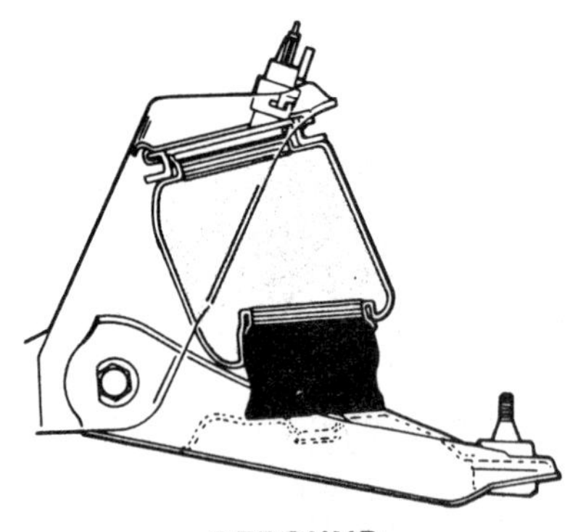

REBOUND

- When control arm moves downward, piston extends outward from rubber membrane
- Rubber membrane unfolds from around piston to allow downward suspension movement

FIGURE 1-22. Air Spring. The air spring, made of rubber and plastic, supports the vehicle load at the front and rear wheels. (Courtesy of Ford Motor Company)

and control module work together to control the air flow to the entire system. All the air-operated parts of the system are connected by nylon tubing. This suspension system uses gas-pressurized front shock struts and rear shock absorbers.

Operation

The air spring suspension leveling system operates by adding or removing air in the air springs to maintain the level of the vehicle at a predetermined front and rear suspension height. The predetermined distance is known as the *vehicle curb riding height.* Curb height is controlled by the height sensors. The distance of the body to ground will change with tire size and inflation pressure.

The height sensors, attached to the body and the suspension arms, lengthen or shorten with suspension travel. Three height sensors are used: one at the left front wheel, one at the right front wheel, and one for the rear suspension.

The system works in the following manner. As weight is added to the vehicle, the body settles under the load. As the body lowers, the height sensors shorten (low out-of-trim) and generate a signal to the control module. The signal activates the air compressor (through a relay) and opens the air spring solenoid valves. As the body rises, the height sensors lengthen. When the preset curb height is reached, the air compressor is turned off, and the solenoid valves are closed by the control module.

A similar action takes place when weight is removed from the vehicle. As weight is removed, the body rises, which causes the height sensors to lengthen (high out-of-trim). This movement generates a signal to the control module. The signal opens the air compressor vent solenoid and the air spring solenoid valves. As the body lowers, the height sensors shorten. When the preset curb height is reached, the air compressor vent solenoid is closed, and the air spring solenoid valves are closed by the control module.

Air required for leveling the vehicle is distributed from the air compressor to each air spring by four nylon air lines. The lines start at the compressor dryer and end at the individual air springs. The dryer is a common pressure manifold for all four air lines so orientation of these lines at the compressor is not required. The air lines are color coded to identify to which air spring they are attached. The dryer contains a dessicant (silica gel) that dries the compressed air before delivering the air to the springs. During venting of any air spring, the previously dried air passes through the dryer to remove moisture from the dessicant (regeneration).

Air required for compression and vent air enters and exits through a common port on the compressor head. Vented air is also controlled by a solenoid valve in the compressor head. Electrical power to operate the air spring suspension system is distributed by the main body wiring harness.

Caution: The compressor relay, compressor vent solenoid, and all air spring solenoids have internal diodes for electrical noise suppression and therefore are polarity sensitive. Care must be taken when servicing these components not to switch the battery feed and ground circuits, or component damage will result.

A microcomputer-based module controls the air compressor motor (through a relay), the vent solenoid, and the four air spring solenoids to provide the air requirements of the springs. The module also provides power and ground to the three digital height sensors and continuously monitors input from the three height sensors and the ignition run/brake on/door open circuits. These inputs are used by the module to make vehicle leveling decisions, which are then carried out by the air spring system components controlled by the module. For service, the module provides a series of diagnostic tests, provides a routine for filling the springs, and operates a system warning lamp.

CONCLUSION

Rear suspension design and operating characteristics are closely related to the front suspension. Spring rates and shock absorber design are closely matched to prevent a harsh ride. The location of the rear axle is determined to a great degree by the vehicle's **center of gravity** and the need to distribute the vehicle's weight as evenly as possible between the front and the rear axles. When the vehicle is driven by the rear wheels, the drive shaft and universal joint angles must also be considered in rear suspension design. This consideration is particularly true with Hotchkiss drive where spring windup can often tip the rear axle excessively.

REVIEW TEST

1. Explain the purpose of a suspension system.
2. List the parts of a single I-beam front suspension system.
3. Briefly list two advantages of a rigid front axle design.
4. Briefly list two disadvantages of a rigid front axle design.
5. A vehicle is equipped with a front wheel drive rigid axle and a coil spring suspension. Explain how the front axle is kept in line with the body of the vehicle.
6. Explain the term *independent front suspension.*
7. Match the names of the following suspension parts with the numbered arrows in Figure 1-23. (Check your answers by reviewing the illustrations in this chapter.) coil spring, lower ball joint, lower control arm, crossmember, shock

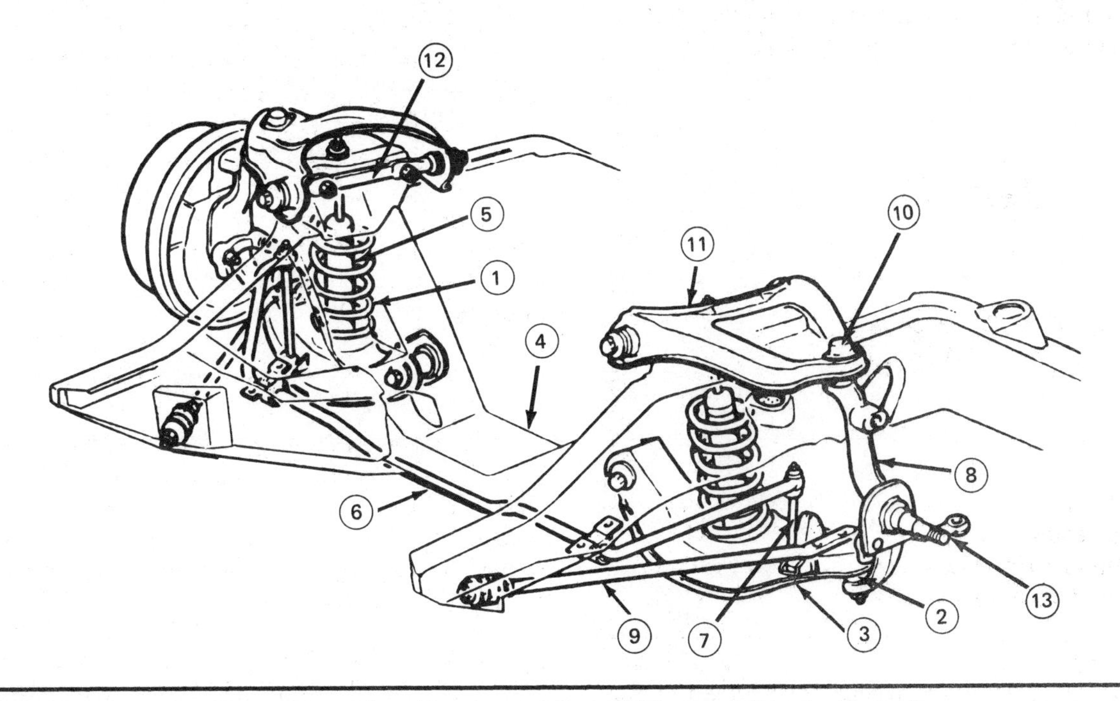

FIGURE 1-23. **Independent Front Suspension.** (Courtesy of General Motors of Canada, Limited)

absorber, stabilizer, stabilizer link, steering knuckle, strut rod, upper ball joint, upper control arm, upper control arm shaft, wheel spindle.

8. Define the terms *jounce travel* and *rebound travel.*
9. Explain the purpose of the following parts: stabilizer bar, strut rod, steering knuckle, and shock absorber.
10. List the parts of a front MacPherson strut suspension.
11. Explain the purpose of a rear leaf spring.
12. A vehicle is equipped with rear coil spring and a rear driving axle. Explain how the axle is kept in line with the body of the vehicle.
13. Define the terms *unsprung weight*, *wheel tramp*, and *center of gravity*.
14. Describe the design of an independent rear suspension system.
15. List the parts of an air spring suspension system.
16. Briefly describe the operation of an air spring suspension system.

FINAL TEST

This examination is multiple choice. Only one answer will be accepted. Carefully read every statement.

1. Mechanic A says that a shock absorber controls the ride of a vehicle. Mechanic B says that a shock absorber controls the riding height of a vehicle. Who is right? (A) Mechanic A, (B) Mechanic B, (C) Both A and B, (D) Neither A nor B.
2. Mechanic A says that all kingpins are placed in a true vertical position. Mechanic B says that all kingpins are tilted inward at the top. Who is right? (A) Mechanic A, (B) Mechanic B, (C) Both A and B, (D) Neither A nor B.
3. Mechanic A says that leaf springs move vertically and change length. Mechanic B says that leaf springs must transmit braking torque to the vehicle's frame. Who is right? (A) Mechanic A, (B) Mechanic B, (C) Both A and B, (D) Neither A nor B.
4. Mechanic A says that a tie-rod is attached to the steering knuckle through the steering arm. Mechanic B says that a pitman arm and an intermediate arm are different names for the same part. Who is right? (A) Mechanic A, (B) Mechanic B, (C) Both A and B, (D) Neither A nor B.
5. Mechanic A says that radius arms are used only on twin I-beam suspension. Mechanic B says that the track rod controls the forward and backward movement of a front driving axle. Who is right? (A) Mechanic A, (B) Mechanic B, (C) Both A and B, (D) Neither A nor B.
6. Mechanic A says that the steering knuckle is located between the upper and lower ball joints. Mechanic B says that a stabilizer bar reduces a vehicle's body sway when turning a corner. Who is right? (A) Mechanic A, (B) Mechanic B, (C) Both A and B, (D) Neither A nor B.
7. Mechanic A says that stabilizer bars do not need to be attached to a vehicle's frame. Mechanic B says that on modern cars, the stabilizer bar and strut perform the same purpose. Who is right? (A) Mechanic A, (B) Mechanic B, (C) Both A and B, (D) Neither A nor B.
8. Mechanic A says that when a vehicle has front control arms that are of unequal length, the wheel track remains relatively constant. Mechanic B says that if a vehicle had equal-length front control arms, the tread of the front tires would be minimal. Who is right? (A) Mechanic A, (B) Mechanic B, (C) Both A and B, (D) Neither A nor B.
9. Mechanic A says that one of the torsion bars is attached to a control arm and the opposite end is attached to the vehicle's frame. Mechanic B says that a torsion bar must twist, support the vehicle's weight, and absorb road shock. Who is right? (A) Mechanic A, (B) Mechanic B, (C) Both A and B, (D) Neither A nor B.
10. Mechanic A says that all vehicles designed with independent front end suspension have upper and lower control arms. Mechanic B says that independent suspension means that both lower control arms are independent of the vehicle's frame. Who is right? (A) Mechanic A, (B) Mechanic B, (C) Both A and B, (D) Neither A nor B.
11. Mechanic A says that it is desirable to reduce the unsprung weight of an automobile. Mechanic B says that it is desirable to reduce wheel tramp, thus improving the vehicle's handling. Who is right? (A) Mechanic A, (B) Mechanic B, (C) Both A and B, (D) Neither A nor B.
12. Mechanic A says that MacPherson strut suspen-

sion has the coil spring at the top of the strut. Mechanic B says that MacPherson strut suspension has a lower ball joint and a control arm. Who is right? (A) Mechanic A, (B) Mechanic B, (C) Both A and B, (D) Neither A nor B.

13. Mechanic A says that an air spring is actually part of the MacPherson strut assembly. Mechanic B says that all air flow during the compression and venting cycles must pass through a dryer. Who is right? (A) Mechanic A, (B) Mechanic B, (C) Both A and B, (D) Neither A nor B.

14. Mechanic A says that air required for leveling the vehicle is distributed from the compressor to each spring through a nylon air line. Mechanic B says that a microcomputer-based module controls the air compressor motor through a relay vent solenoid and four air spring solenoids. Who is right? (A) Mechanic A, (B) Mechanic B, (C) Both A and B, (D) Neither A nor B.

Suspension System Components

LEARNING OBJECTIVES

After studying this chapter, you should be able to:

- Illustrate and describe the basic design of a ball joint,
- Explain the purposes of the weight-carrying and non-weight-carrying ball joints,
- Explain two prime purposes of suspension springs,
- Explain the primary purpose of a hydraulic shock absorber,
- Explain the basic working principle of a hydraulic shock absorber,
- Explain the various devices that may be used to correct an overload condition of a vehicle.

INTRODUCTION

Now that you are familiar with the various designs of the front and rear suspension systems, we can examine the design and operation of the major suspension parts. Ball joints are a pivotable part of a front suspension system. Before we discuss how they are replaced, it is essential to understand the purpose of various types of ball joints. *Note:* Ball joints can never be repaired; they must always be replaced.

Springs do more than just move up and down and support the weight of the vehicle. A spring must also absorb road shock. Shock absorbers, those misnamed devices, do not absorb shock but control the oscillation of a spring, thus providing the vehicle and its passengers with a comfortable and, more important, a safe ride over a rough road.

BALL JOINTS

Ball joints are not a recent addition to the design of a vehicle's independent front end suspension system. They were first introduced by automotive manufacturers in the mid-1950s. Until that time, steering knuckles designed with kingpins allowed the wheel spindle to rotate partially.

Ball joints provided the means for decreasing the number of suspension pivot parts while enabling the front suspension to have complete flexibility and mobility. The other purpose of a ball joint is to connect the upper and lower control arms to the steering knuckle. If you look closely at various makes of vehicles, you will find that ball joints may be riveted, pressed, bolted, or threaded onto the control arms.

Design

A ball joint is essentially what the name implies—a ball, or sphere, inside a cupped socket (Figure 2-1). Your shoulder and hip joints are good examples of a ball-and-socket design. A tapered stud protruding from the ball within the socket allows the steering knuckle to pivot. In contrast to the kingpin (which allows only a circular rotation around an almost vertical axis), the ball joint is free swiveling.

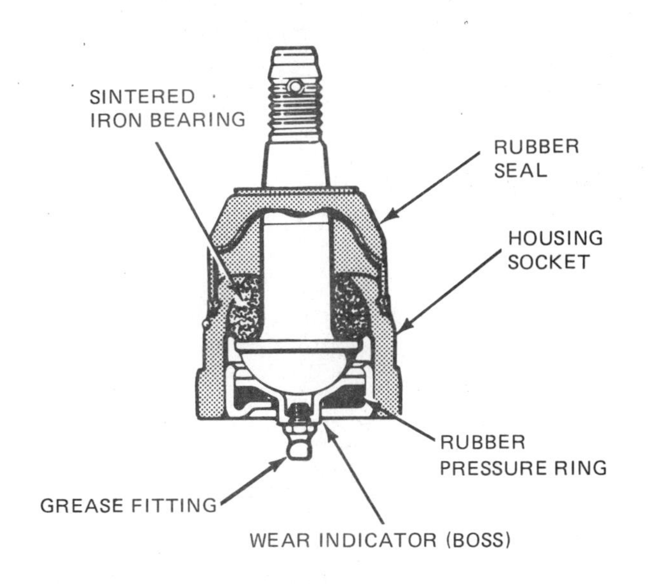

FIGURE 2-1. Internal Design of a Ball Joint. (Courtesy of General Motors of Canada, Limited)

A ball joint stud and rubber seal is about all you can see of a ball joint assembly. The purpose of the rubber seal is to retain lubricating grease. Retention of grease minimizes wear of the internal parts and protects this grease from contamination by water or foreign materials like sand or dirt. If the seal is damaged (torn or pulled off its metal retainer), the ball joint must be replaced. *Note:* Always check and follow the procedures outlined in the manufacturer's shop manual for determining the wear and replacement of the upper and lower ball joints.

Purposes of Ball Joints

Although the upper and lower ball joints allow the steering knuckle to pivot, that is where the similarity stops. Depending on the design of the front suspension system, a ball joint may or may not support a portion of the vehicle's weight.

Long-and-Short-Arm Suspension. Figure 2-2 illustrates a suspension system with the coil spring situated on top of the lower control arm. The arrow through the coil spring indicates the direction of the vehicle's support load. The lower ball joint becomes the weight carrier and is **compression loaded.** The upper ball joint is the follower or guiding ball joint and is **tension loaded.** The tapered stud of the ball joint may be positioned up or down depending on the manufacturer's design.

A suspension system with the coil spring situated on top of the upper control arm is illustrated in Figure 2-3. The arrow through the coil spring indicates the direction of the vehicle's support load. The upper ball joint becomes the weight carrier and is compression loaded. The lower ball joint is the follower or guiding ball joint and is tension loaded.

MacPherson Strut Suspension. In the MacPherson strut front suspension, if the coil spring is located at the top of the strut, the ball joint positioned at the bottom of the strut is a non-weight-carrying ball joint. Vehicle weight is transmitted through the body, upper mount assembly, coil spring, strut, and spindle to the wheel. Thus, the joint allows the strut to pivot; the joint also stabilizes the movement of the structural part.

In the modified MacPherson strut, the coil

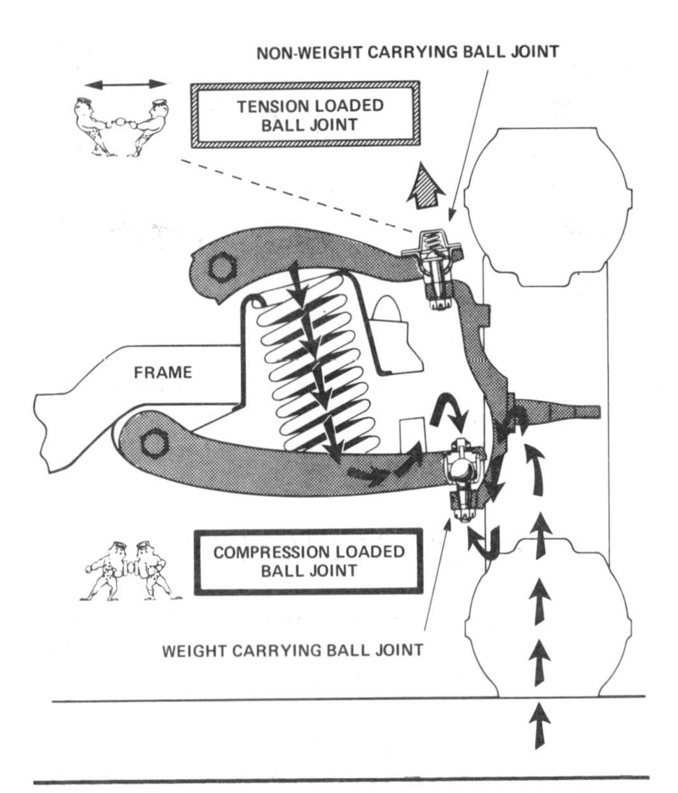

FIGURE 2-2. Ball Joint Placement in Suspension System with Coil Spring Situated on Lower Control Arm. (Courtesy of Moog-Canada Automotive, Ltd.)

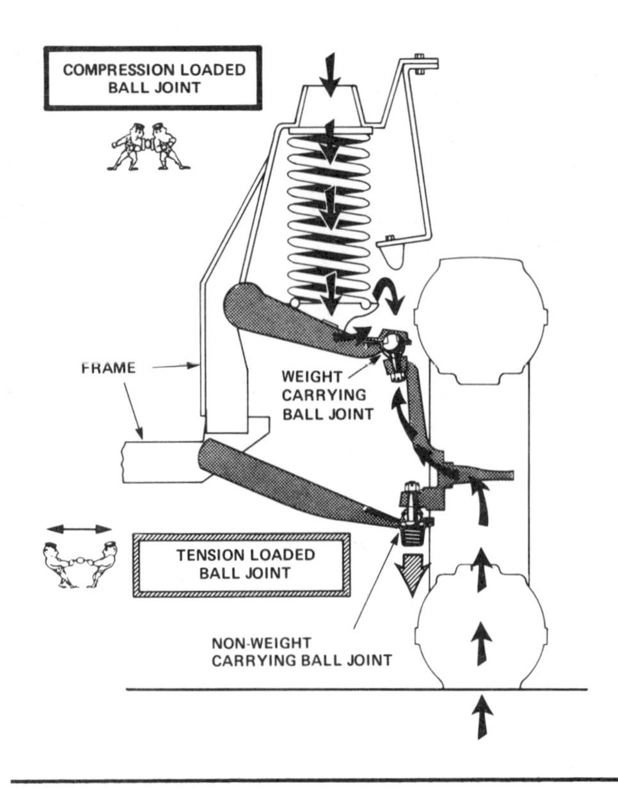

FIGURE 2-3. Ball Joint Placement in Suspension System with Coil Spring Mounted on Upper Control Arm. (Courtesy of Moog-Canada Automotive, Ltd.)

spring is resting on the lower control arm. Therefore, the joint is a weight-carrying ball joint.

FACTS TO REMEMBER

1. Weight-carrying ball joints transmit the load of the vehicle from the control arm to the steering knuckle.
2. Non-weight-carrying ball joints stabilize the position of the steering knuckle and reduce front end wheel alignment factor change.

SPRINGS

Springs are the primary operating parts of a suspension system. They absorb the energy imparted to the wheel by road shock and use that energy to return the wheel and the body of the vehicle to the original position when the bump is past. Springs are without doubt the real shock absorbers in the suspension system, along with the inflated tires. The devices we call shock absorbers are merely spring dampers that control the oscillation and released energy of a spring.

Chapter 1 described the three basic types of springs commonly used in automobile suspension systems. The leaf spring absorbs road shock by flexing (bowing in the middle), the coil spring by twisting and flexing, and the torsion bar by twisting (Figure 2-4).

Springs vary in their resilience, or give. The higher the variable spring rate figure, the stiffer the spring. Current automotive manufacturers use springs that provide a soft, comfortable ride and therefore have a relatively low variable spring rate. Modern automobile suspension systems are engineered to tune out the undesirable vibrations, pitching, and swaying motions that are created as a vehicle is driven over bumps.

Rear Leaf Springs

Although all *rear leaf springs* may appear to be the same, there are slight differences in design. Leaf springs may be symmetrical or asymmetrical. Let's discuss these terms. If the rear axle is located in the

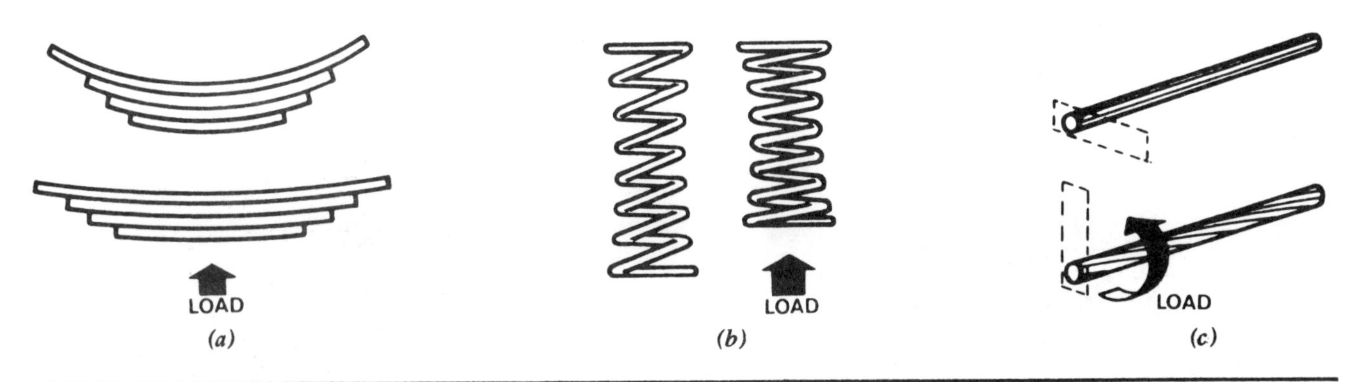

FIGURE 2-4. Three Basic Types of Springs and Their Action under Load. (a) Leaf-spring; (b) coil spring; (c) torsion bar. (Courtesy of Chrysler Corporation)

center of the spring (an equal distance from the two ends of the main leaf), the spring is symmetrical. If the rear axle is not located in the center of the spring, the spring is asymmetrical (Figure 2-5).

The stiff, short portion of the asymmetrical spring tends to resist rear axle acceleration windup far better than a symmetrical rear leaf spring. Comparing rear leaf spring suspension to coil spring rear suspension, the leaf-type spring normally can be mounted closer to the rear wheels, thereby decreasing body sway and increasing the lateral stability of the vehicle.

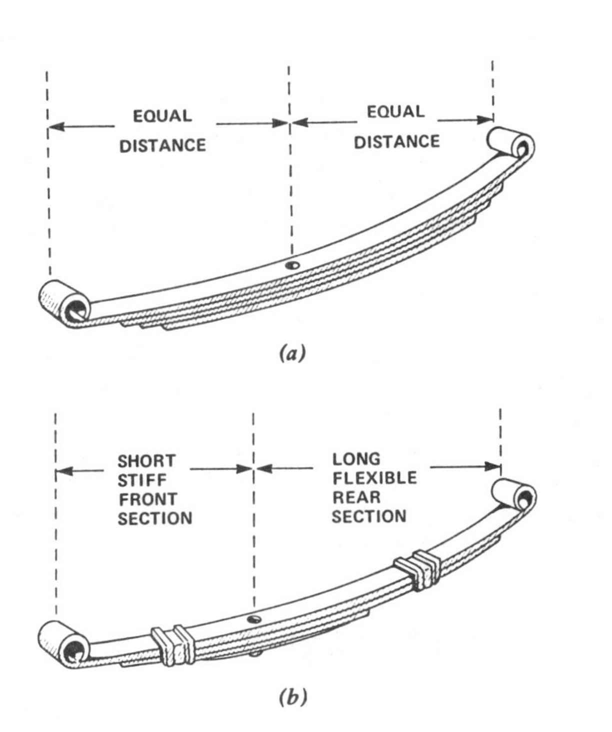

FIGURE 2-5. Rear Leaf Springs. (a) Symmetrical; (b) asymmetrical. (Courtesy of Chrysler Corporation)

A Typical Condition

All springs eventually sag due to the weight they must support. A vehicle that is subject to incorrect curb riding height will have decreased directional control and stability. *Note:* The four springs of a vehicle do not necessarily fail or sag at the same time. If, for example, the right front spring became defective, the right front corner of the vehicle would sag and the left rear corner of the vehicle would be raised.

Always refer to the manufacturer's shop manual for riding height checking procedures and specifications. Defective coil or leaf springs must be replaced. Manufacturers also recommend that front or rear springs be replaced in pairs, with the exception of torsion bars.

A FACT TO REMEMBER

During manufacturing, torsion bars are given a little twist—called *preloading*—in the direction that they will twist when in use. Since the two torsion bars at the front of a vehicle twist in opposite directions, they are made for their respective sides.

CONVENTIONAL SHOCK ABSORBERS

The first hydraulic oil-filled shock absorbers were introduced in the 1920s. They were classified as single acting (control in only one direction). The *direct-acting shock absorber*—a refinement of the earlier design—derives its name from the fact that it is mounted directly between the frame of the vehicle and the control arm or axle. This modern shock is also double acting because it controls motion in both the up and down directions of suspension travel.

Purposes of Shock Absorbers

A shock absorber has three purposes:

1. To dampen the effect of spring oscillation,
2. To control body sway,
3. To reduce the tendency for a tire tread to lift off the road surface.

Without a shock absorber, a deflected spring would vibrate with gradually decreasing oscillation until the energy it has absorbed is dissipated. Therefore, a shock absorber is necessary to dampen the effect of spring oscillation in order to control the ride stabilization of a vehicle. A shock absorber also helps to control body sway and reduces the tendency for the tread of a tire to lift off the surface of the road, a problem often caused by static unbalance. (Static unbalance is discussed in Chapter 6.)

Basic Principle

To understand why a vehicle must be equipped with shock absorbers, hang one end of a thin coil spring (a screen door spring will do fine) on a hook and suspend a weight from its opposite end (Figure 2-6). Now, lift up the weight (deflect the spring) and let it go. What is happening? The thin coil spring alternately expands and compresses, and the weight jiggles up and down. Notice that each movement of the weight becomes smaller and smaller, until eventually it comes to rest in its original position. That is the same reaction an uncontrolled spring has on a vehicle.

Now try the experiment again. This time place the weight in a container filled with very light machine oil. Lift the weight again and release it. You will notice that the weight will come to a halt much sooner.

In effect, this same reaction happens within a shock absorber. The shock absorber uses a specially formulated oily fluid confined in a cylindrical chamber to dampen the oscillation of a spring. A shock absorber is far more efficient than the container of fluid used in our experiment.

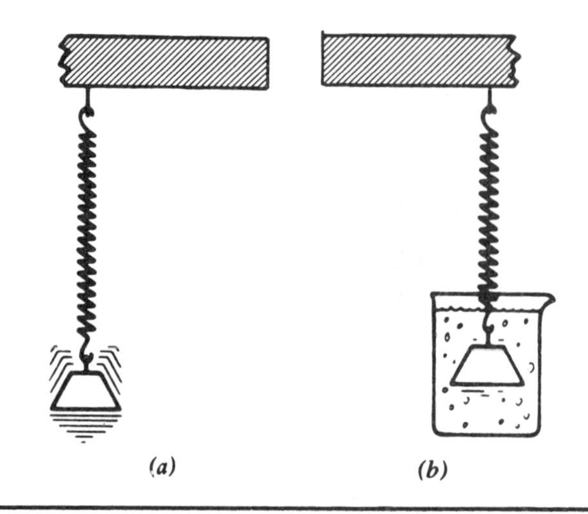

FIGURE 2-6. Simple Experiment Illustrating Basic Working Principle of a Shock Absorber. (a) Undamped spring vibrates; (b) damped spring resists vibration. (Courtesy of Chrysler Corporation)

Design and Operation

Figure 2-7 illustrates a cutaway view of a typical modern shock absorber. The direct-acting shock absorber is a velocity sensitive, hydraulic dampening device—that is, the faster it is moved, the more resistance it has. The amount and type of control in each shock is determined by vehicle engineering specialists. The direct-acting shock absorber works on the principle of fluid displacement through valves. On the wheel jounce, or compression cycle, as well as the wheel rebound, or extension cycle, fluid is compressed through these valves, dampening spring oscillation. This motion energy, through friction, is converted into heat energy, which is then dissipated in the surrounding air.

Compression Cycle. As Figure 2-8 illustrates, during the *compression cycle* of a shock absorber, the piston inside the pressure tube moves down on the compression stroke and displaces hydraulic fluid from chamber B. Some of this hydraulic fluid flows through the piston valving to chamber A, and some of it moves through the base or compression valve to chamber C, which is the reservoir inside the reserve tube. The volume of the piston rod determines the amount of fluid displaced to chamber C. Several valving stages are used in both the piston and base valve to control fluid flow. This operational cycle is called *full-displace valving.*

The velocity of the piston rod and piston affects how the valving stages regulate fluid flow. At slow

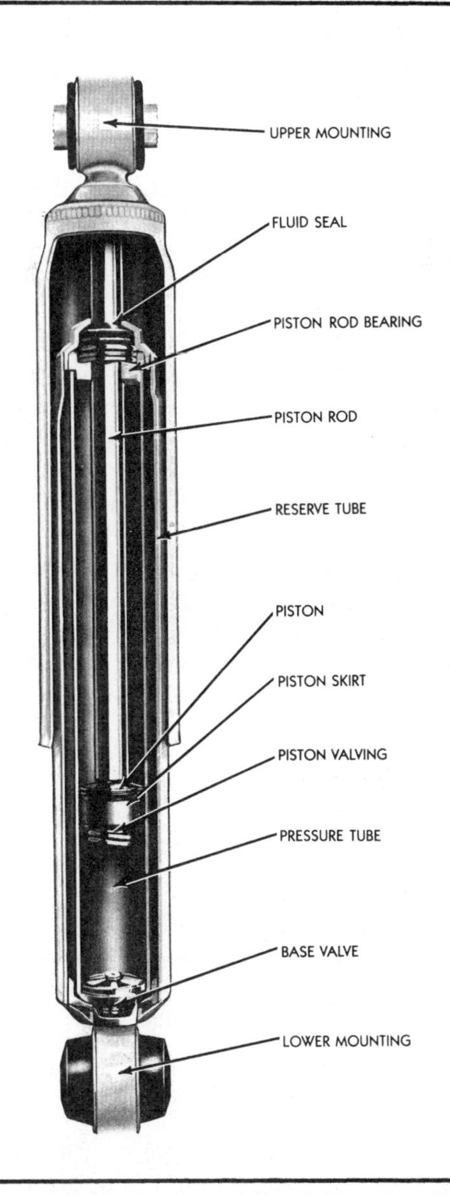

FIGURE 2-7. Cutaway View of Shock Absorber. (Courtesy of Monroe Auto Equipment Company, Division of Tenneco Automotive)

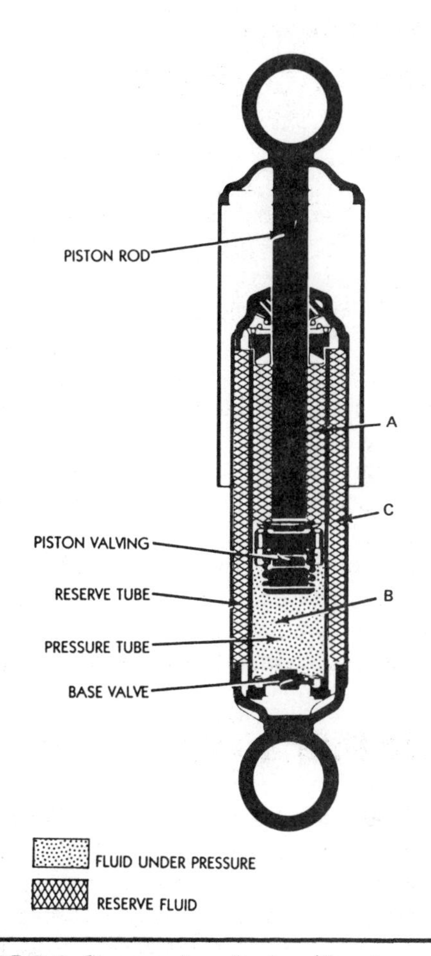

FIGURE 2-8. Compression Cycle. (Courtesy of Monroe Auto Equipment Company, Division of Tenneco Automotive)

piston speeds, the first valving stage, called the *orifice stage*, controls fluid flow. At faster piston speeds, the pressure on the fluid in chamber B becomes higher, and Monroe Auto Equipment Company's *dual rate disc system*, the second and third valving stages, opens. The fourth valving stage, *orifice restriction*, controls the flow at high speeds.

During the four valving stages, the *base* or *compression valve* is also working. As fluid displaced by the piston rod goes through the base valve on its way to the reservoir, three valving stages control its flow. At slow piston rod speeds, fluid flows through an orifice in the valve seat. At faster speeds, the fluid is controlled by discs acting as flat blow-off springs. At very high speeds, fluid flow is controlled by slots in the valve plate. Whatever the piston speed, shock absorber control during the compression cycle is the resistance to movement that results from the hydraulic pressure in chamber B acting against the combined effects of piston and base valving.

Extension Cycle. During the *extension cycle* (Figure 2-9), three stages of piston valving control fluid flow. The stage needed depends on the velocity of the piston rod and piston. As the piston moves

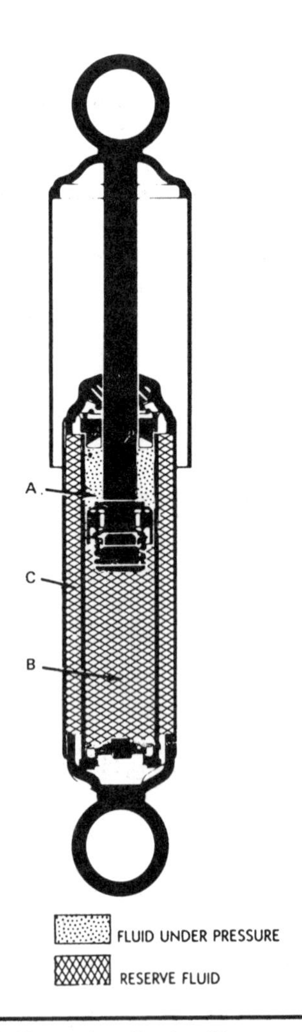

FIGURE 2-9. Extension Cycle. (Courtesy of Monroe Auto Equipment Company, Division of Tenneco Automotive)

up toward the top of the pressure tube, it displaces fluid from chamber A. This displaced fluid then flows through the piston valving into chamber B. At slow piston speeds, the orifice in the piston valve seat regulates the fluid flow. At medium piston speeds, the spring and thickness of the steel discs, which flex at predetermined flow rates or resistance, control fluid flow. At high piston speeds, the control or resistance is determined by the amount of restriction of the inner passages that pass through the piston. Changing pressures force the base valve to unseat, and fluid also flows from chamber C into chamber B. Shock absorber extension control is resistance present as a result of the hydraulic pressure in chamber A acting against the effects of the piston valving.

Shock Ratios

A typical automotive shock absorber will have more extension control than compression control. The reason is that extension controls the heavier, **sprung weight**, or body weight, while compression controls the lighter, unsprung axle, wheel, and tire weight.

The difference in, or **ratio** of, extension and compression resistance can usually be felt by hand. If you were to move a shock absorber with a ratio of 70/30 in and out by hand, you would feel that extending the shock is more difficult than compressing it. The ratio 70/30 means that 70 percent of the total control would be on the extension cycle and 30 percent of the total control would be on the compression cycle. Depending on the need requirement, shock absorber ratios will range from 50/50 to 80/20.

A FACT TO REMEMBER

A conventional shock absorber will not change the curb riding height of a vehicle; the shock is only a spring-dampening device.

STRUT ASSEMBLIES

An increasing number of cars have strut suspensions up front. Strut suspensions do not contain the upper control arm and ball joint found on conventional suspension systems. Strut-equipped cars are lighter and therefore use less gas. Strut suspensions also take up less space in the engine compartment, so there is more room for the transverse mounted engines used in most front wheel drive cars.

Conventional Struts

On most strut assemblies, the coil spring sits on a lower spring seat. The lower seat is part of the strut housing (Figure 2-10). The coil spring supports the weight of the car and maintains proper vehicle height (see Figure 1-12). Within the strut

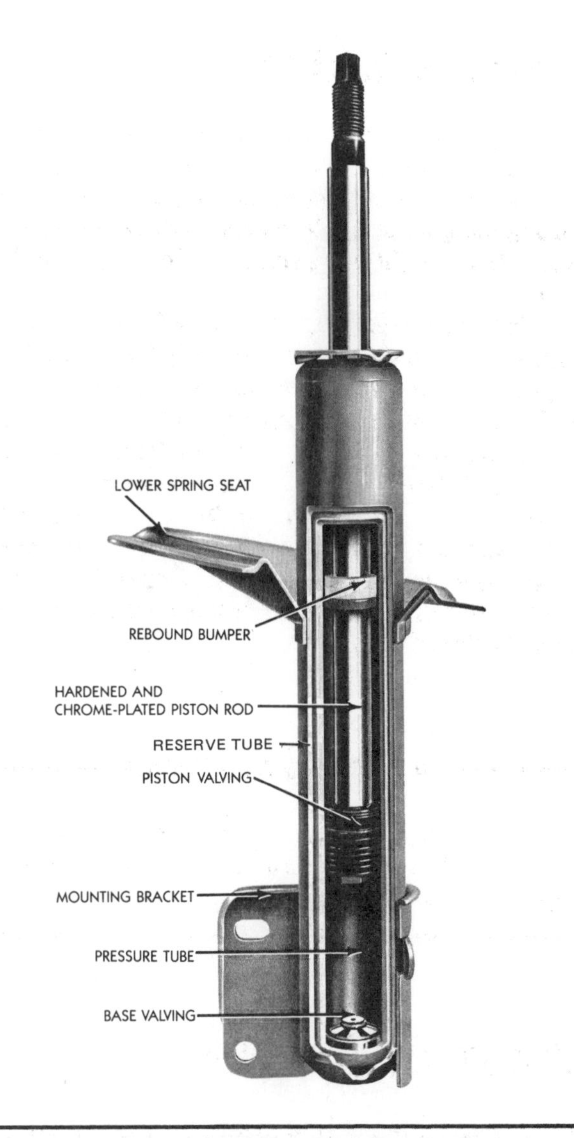

FIGURE 2-10. **Cutaway View of a Conventional Strut.** (Courtesy of Monroe Auto Equipment Company, Division of Tenneco Automotive)

housing, a shock absorber mechanism controls spring movement and dampens excessive vehicle motion by the controlled movement of fluid under pressure. When oil is not under pressure and is continually recirculated, it tends to foam, causing aeration. Adding a charge of nitrogen reduces the foaming because the oil is then under pressure.

Strut units usually pivot with the steering action of the vehicle. Most struts attach at the top by an upper bearing plate to the vehicle body. At the bottom, they attach to the wheel spindle by a mounting bracket that is part of the strut housing. Most conventional strut housings have a lower spring seat and a mounting bracket. On many units, the lower mounting bracket pad allows camber adjustment by either an upper or a lower mounting bolt. A brake line may also attach to the strut housing.

Most U.S.-built cars use sealed units for strut assemblies. The strut housing is factory welded; thus, there is no access to the pressure tube and piston rod assembly. The entire strut housing is discarded and replaced with a new sealed unit when the unit is defective.

Modified Struts

On some strut suspensions, the coil spring is not part of the strut assembly. Instead, the spring is independently located between the lower control arm and the frame (see Figure 1-13).

Strut Cartridges

Most import and a few domestic cars can be serviced with replaceable strut cartridges. A *strut cartridge* is a self-contained unit with a pressure tube and a piston rod assembly. The cartridges are factory sealed and calibrated.

When replacing or repairing struts, it makes good sense to replace both front or both rear shock units to restore a balanced ride. On most vehicles, alignment must be checked and realigned, if necessary, to the manufacturer's specifications after completing a strut job.

REAR SHOCKS

For maximum directional control and stability, the type of rear shock absorber used must be compatible with the type of driving and its load requirements. For example, a heavy load in a trunk, the towing of a travel trailer, or transporting a boat affects the steering, braking, and handling characteristics of a vehicle. The following list provides solutions to this type of load and driving requirement:

1. You may purchase heavy-duty shock absorbers and springs when ordering the vehicle from the dealer.
2. You may install an overload combination

of shock absorbers and variable rated springs for the front and rear of most vehicles (Figure 2-11).

3. You may install rear air adjustable shocks that inflate from 0 to 630 kilopascals (kPa), or 0 to 90 pounds per square inch (psi); that feature all-weather fluid, silicone-lubricated air sleeve boots, and heavy-duty support mountings; and that can be manually operated or automatically adjusted by a load-leveling height control valve (Figure 2-12).
4. You may install air support bags, which are similar to a tire's inner tube, inside the rear coil springs of a vehicle (Figure 2-13).

FIGURE 2-11. Load-Leveler Stabilizing Unit. (Courtesy of Monroe Auto Equipment Company, Division of Tenneco Automotive)

FACTS TO REMEMBER

1. Installing heavy-duty rear shock absorbers with helper springs will no doubt assist the rear suspension in supporting increased weight, but it introduces other problems. Auxiliary overload rear springs will raise the tail end of a vehicle when it is not supporting the increased weight and will affect steering and handling characteristics. Rear tires will need additional inflation pressure—approximately 14 to 28 kPa (2 to 4 psi)—to take care of the additional weight.
2. Many automotive service mechanics have the mistaken idea that shock absorbers, like front or rear springs, should always be replaced in pairs. That is not so. Unlike overworked, sagging springs, the action of a shock absorber does not change with use. Shock absorbers are either good or defective. An individual shock absorber should only be replaced when it has lost some portion of its travel (where it does not resist spring movement) or if it is broken or leaking excessive fluid.

REVIEW TEST

1. Illustrate and describe the design of a ball joint.
2. Explain the purposes of a weight-carrying ball joint.
3. Explain the purposes of a non-weight-carrying ball joint.
4. Define the term *compression-loaded ball joint.*
5. Describe the design of an asymmetrical rear leaf spring.
6. What parts of a vehicle actually absorb road shock?
7. You suspect that a vehicle has sagged springs. How can you verify your suspicions?
8. What is a direct-acting shock?
9. List the three main purposes of a shock absorber.
10. Briefly explain the operation of a shock during the compression and extension cycles.
11. What parts of a vehicle are the sprung weight?
12. What is meant by a shock absorber's ratio?
13. A vehicle has directional control and stability. What does this mean?

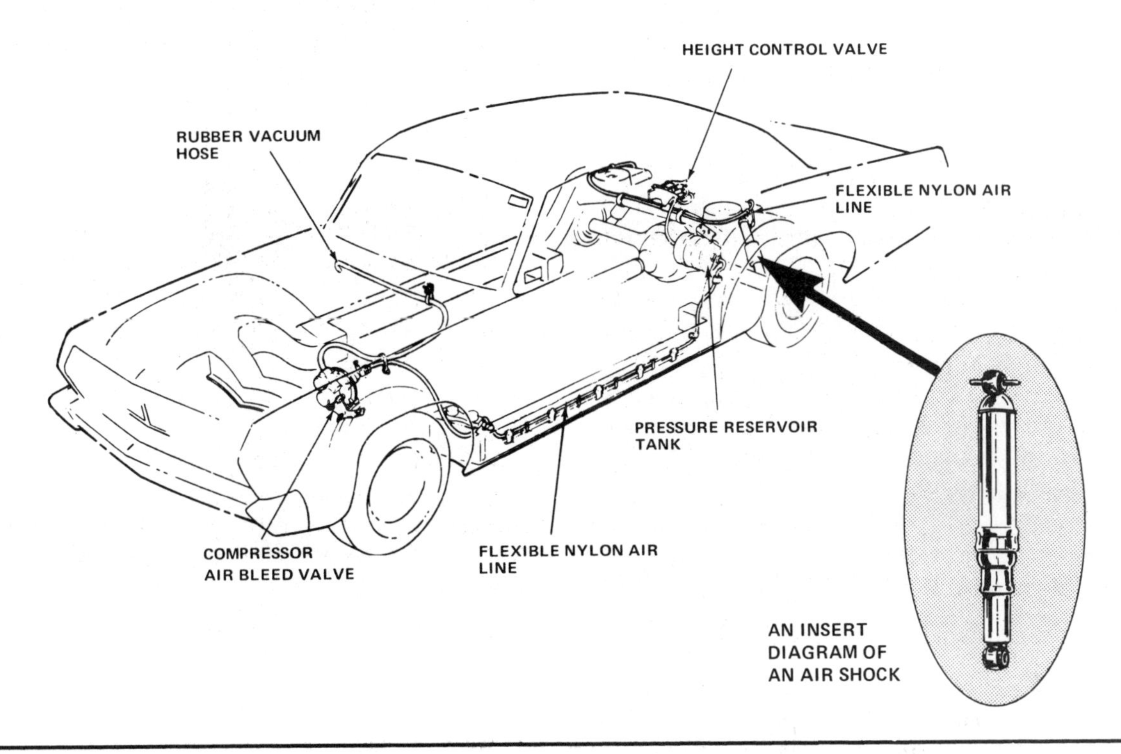

FIGURE 2-12. **Installation of Air Adjustable Shocks.** (Courtesy of Ford Motor Company)

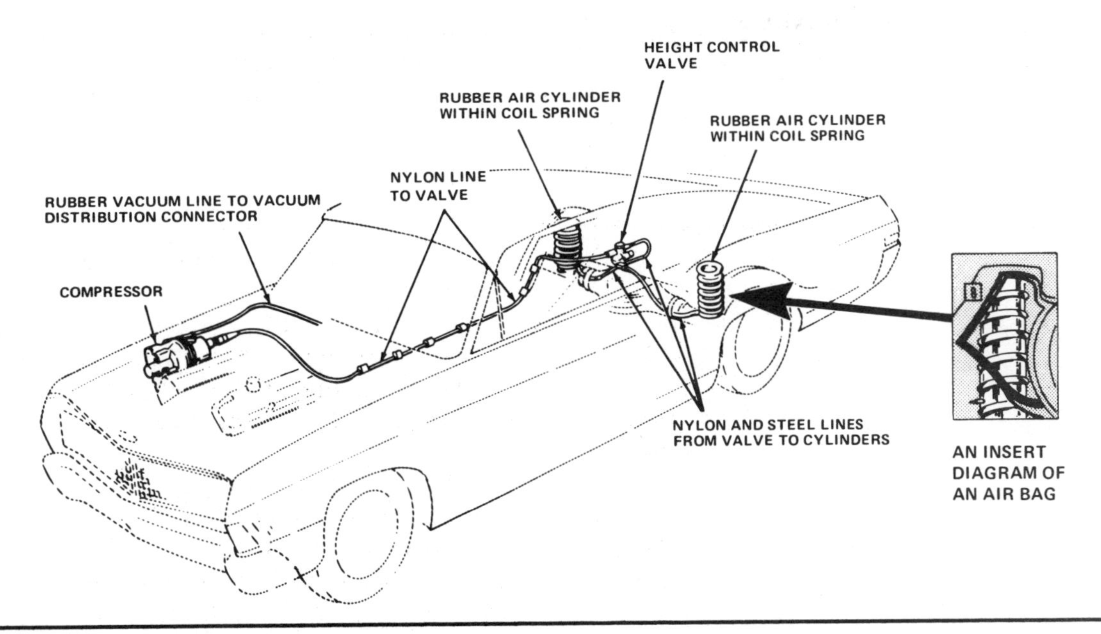

FIGURE 2-13. **Installation of Air Inflatable Bags.**

FINAL TEST

This examination is multiple choice. Only one answer will be accepted. Carefully read every statement.

1. Mechanic A says that ball joints enable the suspension to have complete flexibility and mobility. Mechanic B says that all ball joints do not have the same purpose. Who is right? (A) Mechanic A, (B) Mechanic B, (C) Both A and B, (D) Neither A nor B.
2. Mechanic A says that when a coil spring is resting on the lower control arm, the lower ball joint transmits weight. Mechanic B says that if a coil spring is resting on the lower control arm, the top ball joint stabilizes the position of the steering knuckle. Who is right? (A) Mechanic A, (B) Mechanic B, (C) Both A and B, (D) Neither A nor B.
3. Mechanic A says that a follower or guiding ball joint is always tension loaded. Mechanic B says that a tapered stud of a compression-loaded ball joint is always positioned upward. Who is right? (A) Mechanic A, (B) Mechanic B, (C) Both A and B, (D) Neither A nor B.
4. Mechanic A says that the condition of the left front spring influences the right rear height of a vehicle. Mechanic B says that it is not necessary to replace front coil springs in pairs. Who is right? (A) Mechanic A, (B) Mechanic B, (C) Both A and B, (D) Neither A nor B.
5. Mechanic A says that when a coil spring is resting on the upper control arm, the upper ball joint transmits weight. Mechanic B says that when a coil spring is resting on the upper control arm, the lower ball joint stabilizes the position of the steering knuckle. Who is right? (A) Mechanic A, (B) Mechanic B, (C) Both A and B, (D) Neither A nor B.
6. Mechanic A says that the faster a shock absorber moves, the more resistance it has, thus increasing its dampening effect. Mechanic B says that a defective shock absorber has very little influence on a vehicle's directional control and tire tread life. Who is right? (A) Mechanic A, (B) Mechanic B, (C) Both A and B, (D) Neither A nor B.
7. Mechanic A says that installing auxiliary overload rear springs will not affect the steering and handling characteristics of a vehicle. Mechanic B says that the front or rear shocks do not necessarily need to be replaced in pairs; a shock is either good or defective. Who is right? (A) Mechanic A, (B) Mechanic B, (C) Both A and B, (D) Neither A nor B.
8. Mechanic A says the oil-filled shocks control only spring oscillation. Mechanic B says that oil-filled shocks control the vehicle's curb riding height. Who is right? (A) Mechanic A, (B) Mechanic B, (C) Both A and B, (D) Neither A nor B.

Inspecting and Servicing Suspension Systems

LEARNING OBJECTIVES

After studying this chapter and with the aid of a manufacturer's shop manual, you should be able to:

- Explain how to measure curb riding height,
- Explain how to inspect a vehicle for the causes of incorrect curb riding height,
- Explain how to adjust the front curb riding height of a Chrysler vehicle designed with torsion bar suspension,
- Illustrate and describe how to examine the ball joints of a vehicle,
- Explain the correct procedures for servicing the ball joints,
- Describe how to remove and install press-type ball joints,
- Describe how to examine the kingpins of a vehicle,
- Explain how to inspect and service rubber bushings (noise and vibration insulators),
- Explain how to inspect and test the shock absorbers of a vehicle.

INTRODUCTION

Although vehicle owners may drive their automobiles every day, they are seldom aware of the condition and dependability of their vehicle's suspension system. Statistics also support the fact that many drivers who are involved in accidents are totally unaware of the need for periodic preventive vehicle maintenance. How often should the suspension system be inspected for possible problems? Inspect whenever the manufacturer recommends that the suspension system be lubricated or whenever the driver doubts the dependability or road worthiness of the vehicle.

Every vehicle manufacturer recommends mileage intervals for mechanical checks and services. But regular attention does not guarantee the maintenance of a vehicle. Have you ever had the suspension system of your vehicle inspected? The only thorough method of examining the suspension of a vehicle is by performing a careful, detailed visual inspection of all suspension parts. This chapter deals with this important topic.

The examination of the suspension system may be performed while the vehicle is positioned on the service floor of the garage or on the alignment rack. If you intend to use the alignment rack for the following inspections, position the rack runways to suit the track width of the vehicle. Then position the front wheels on the front turntables of the alignment rack and apply the parking brake.

CURB RIDING HEIGHT

Curb riding height is the distance from the vehicle's frame or lower points of the front and rear suspension to a level surface. When measuring curb riding height, tires must be properly inflated, the fuel tank must be full, and there should be no passengers or load in the luggage compartment. Many wheel alignment technicians have the mistaken idea that the vehicle's riding height can be checked by visual inspection alone. This method is a very inaccurate way to check riding height, however. Relying only on sight will cause a misjudgment of height specifications. When one rear spring is sagged more than the other, a crossed chassis effect results. A sagged right rear spring will produce a serious camber error at the left front wheel. A sagged left rear spring will affect the camber on the right front wheel.

An automotive engineer's first reference point is vehicle curb riding height. Around this height, the control arms move up and down, creating minimal changes. When a sagged spring condition exists, the control arms are working out of their normal geometrical range, creating abnormal suspension changes, thus causing alterations in **camber, caster, toe,** and **turning radius** (Figure 3-1).

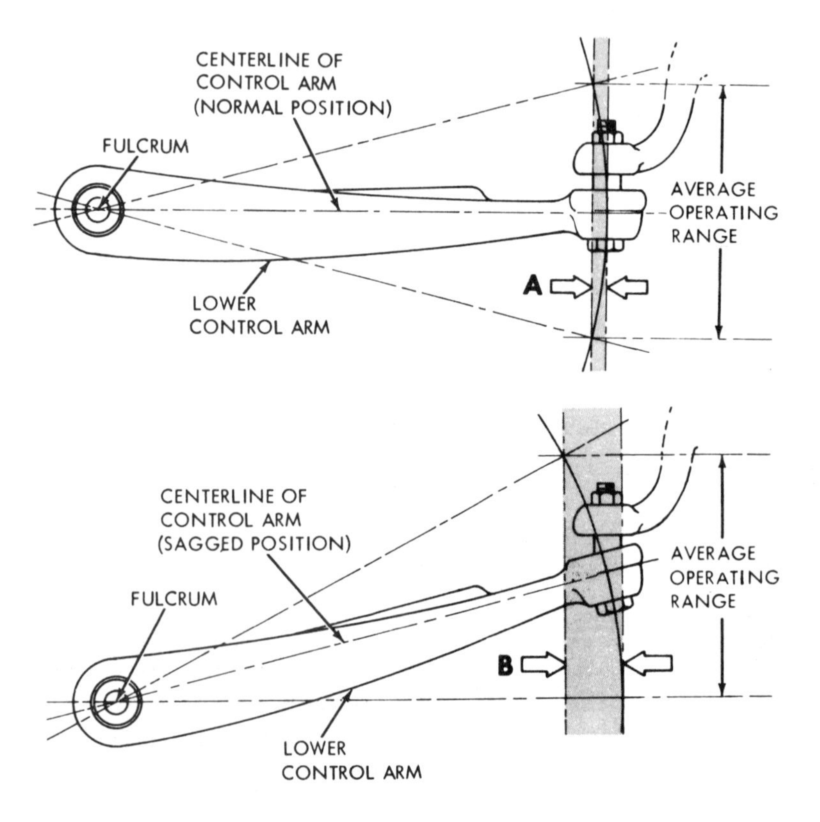

FIGURE 3-1. Side Slip. (When springs are sagged, lateral movement of the wheels is increased enough to cause tires to slide laterally across the pavement, causing tire wear. (Courtesy of Moog-Canada Automotive, Ltd.)

Sagged springs also influence the vehicle's ride characteristics by producing a very rough ride, particularly when rubber strike-out bumpers (see Figure 3-4) are hitting against the frame. Additional stress is also placed on all steering and suspension components, reducing their service life. If sagged springs are the cause of incorrect height, tire treads are subject to increased lateral movement, the wheels will constantly change direction by toeing in and out, and shock absorbers will fail, thus increasing tire tread wear.

When the curb height is correct, alignment angles are normally close to the manufacturer's specifications, and only a small amount of correction is needed. Always measure both the front and the rear of a vehicle at points stated by the manufacturer (Figure 3-2).

The front curb riding height of a Chrysler vehicle designed with torsion bar suspension is adjustable. The height is adjusted by turning the torsion bar adjusting bolts. Side-to-side variation must not exceed 3 millimeters (1/8 inch) (Figure 3-3).

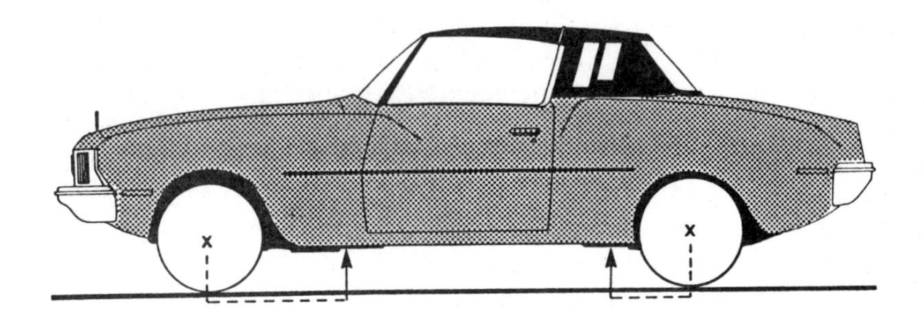

FIGURE 3-2. Points of Measurement for Checking Curb Riding Height.

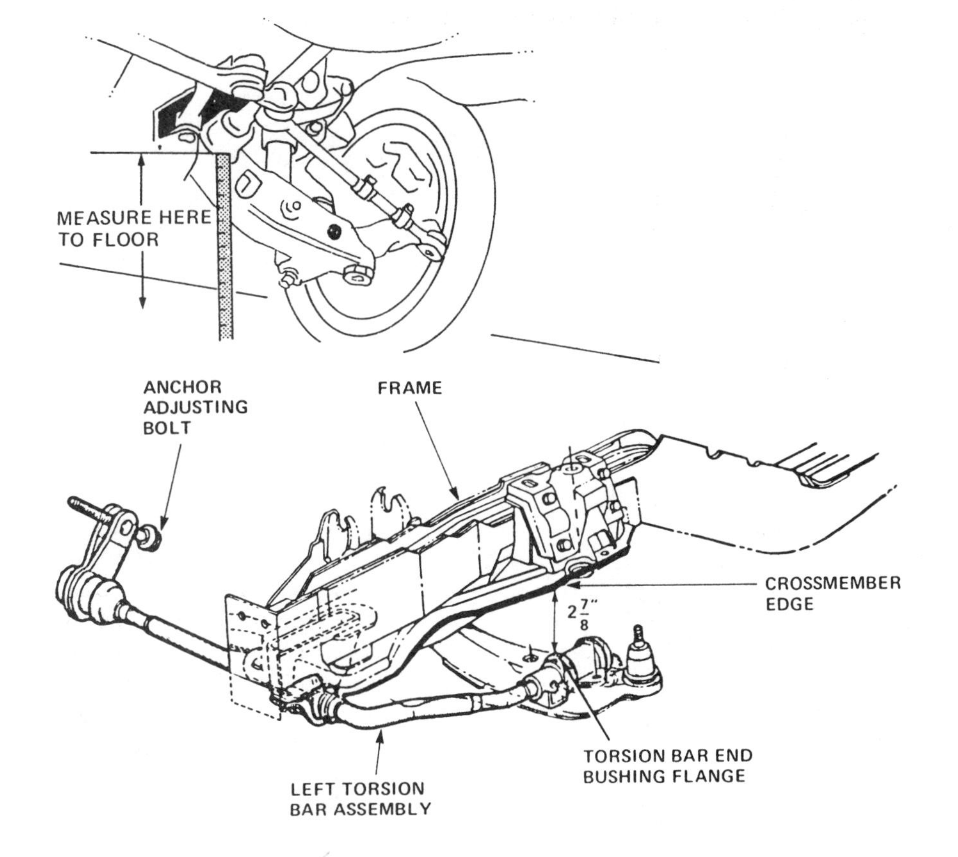

FIGURE 3-3. Measuring Curb Riding Height on Chrysler Vehicles. (Courtesy of Chrysler Corporation)

A FACT TO REMEMBER

If it is necessary to replace the leaf or coil springs of a vehicle, the manufacturer recommends that either the front or rear springs be replaced in pairs.

Caution: When replacing springs, always adopt safe working practices. The vehicle, regardless of its make, must be correctly supported by safety stands. Front coil springs may be removed safely from a vehicle only after the spring has been compressed or when using special coil spring tools. On a rear wheel drive vehicle, a rear coil spring normally may be removed from a vehicle when the rear shock absorber has been removed from its lower mount position.

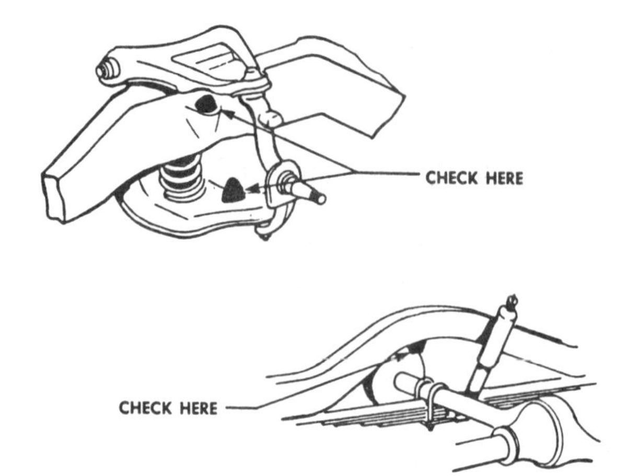

FIGURE 3-4. Checking that Rubber Strike-Out Bumpers Are Not Damaged or Missing. (Courtesy of Moog-Canada Automotive, Ltd.)

A Dangerous Condition

Elevating the chassis above the manufacturer's recommended curb riding height specifications is just as detrimental to overall performance as is a vehicle with sagged springs. Because of design, front end wheel alignment factors change as soon as the vehicle is raised or lowered. Vehicle handling is aggravated by the elevated center of gravity, which increases body roll and wear on the outer edges of the tires. Rear suspensions are affected by chassis heights as well. Many rear suspensions have upper and lower control arms of unequal length. The arms align and position a rear axle and function in a manner similar to the front suspension. Sagged or elevated rear chassis heights, whether the suspension is of leaf spring or coil spring design, can effect changes in front end wheel alignment, load-carrying capacity, drive shaft angles, and vehicle handling. Changing the curb riding height of a vehicle is a dangerous practice that destroys the vehicle's handling characteristics, causing the vehicle to become unstable. Many states and provinces throughout North America have laws prohibiting the alteration of the riding height of a vehicle. Curb riding height will be discussed further in subsequent chapters. The following are signs of incorrect curb riding height:

- Vehicle not sitting level,
- Vehicle not positioned at the manufacturer's curb riding height specifications,
- Missing or damaged rubber strike-out bumpers (Figure 3-4),
- Coil spring clash indicated by a shiny area on coil convolutions,
- Worn shocks, ball joints, and tie-rod ends caused by a change in steering geometry,
- The drive shaft rubbing the vehicle's undercarriage caused by excessive spring compression.

A FACT TO REMEMBER

The only acceptable method of determining chassis riding height is to measure the vehicle as recommended by the manufacturer. If chassis height does not meet specifications, adjust the torsion bars or replace the coil or leaf springs as required. Good judgment should be exercised before replacing a spring when the riding height is only slightly out of adjustment.

BALL JOINT INSPECTION AND SERVICE

Ball joint looseness is critical. The major cause of ball joint failure is pounding and backlash, not necessarily wear due to a lack of lubrication. This key fact was established after many years of study and vehicle testing on the road and in the laboratory. When a wheel is in a jounce or rebound situation, forces are transmitted to the ball joints. The ball joint studs and bearings are slammed back into their original position with tremendous force. This force tends to weaken the metal, causing the housing to fracture because of stress.

Type A Suspension

The proper method of inspecting ball joints is determined by the location of the coil or torsion bar spring. To unload and inspect the lower and upper ball joints in a *type A suspension*, place the jack or lift adapter in a position to free the ball joints from the force exerted by the weight of the vehicle. Place the jack as close as possible to the front wheel under the lower control arm. When the ball joints are properly unloaded, there should be some clearance between the rubber strike-out bumper and the vehicle frame. It is recommended that only one side of the vehicle be raised at a time during the inspection of the ball joints.

The weight-carrying ball joint is attached to the lower control arm. Check the **axial** (vertical) **play** of the lower ball joint by using a pry bar to aid in lifting the wheel assembly (Figure 3-5). A **dial indicator** should be used to calculate accurately the axial play of the joint (Figure 3-6). *Note:* The plunger of the gauge must be truly vertical. Always refer to the manufacturer's shop manual for correct specifications and tolerances. If the axial movement exceeds specifications, replace the worn ball joint.

Weight-carrying ball joints are subject to increased wear because of their function in the suspension system. The non-weight-carrying ball joint is attached to the upper control arm. You will require the assistance of another person to perform this check. Have your assistant grasp the tire at the top and bottom, moving the top ball joint in and out by using the wheel as a lever. This movement will allow you to check visually the **radial** (lateral) **play** of the top joint. If you detect any slack movement in the top ball joint, the part must be replaced (Figure 3-7).

Type B Suspension

To unload and inspect the lower and upper ball joints in a *type B suspension*, place a wooden block or a steel wedge between the vehicle's frame and the underside of the upper control arm before you raise the vehicle (Figure 3-8). The jack or lift adapter must support the vehicle under the crossmember of the front portion of the vehicle's frame (Figure 3-9). **Caution:** Only one side of the vehicle should be raised at a time during the inspection procedure.

FIGURE 3-5. Checking Axial Play of a Ball Joint by Using a Pry Bar. (Courtesy of Moog-Canada Automotive, Ltd.)

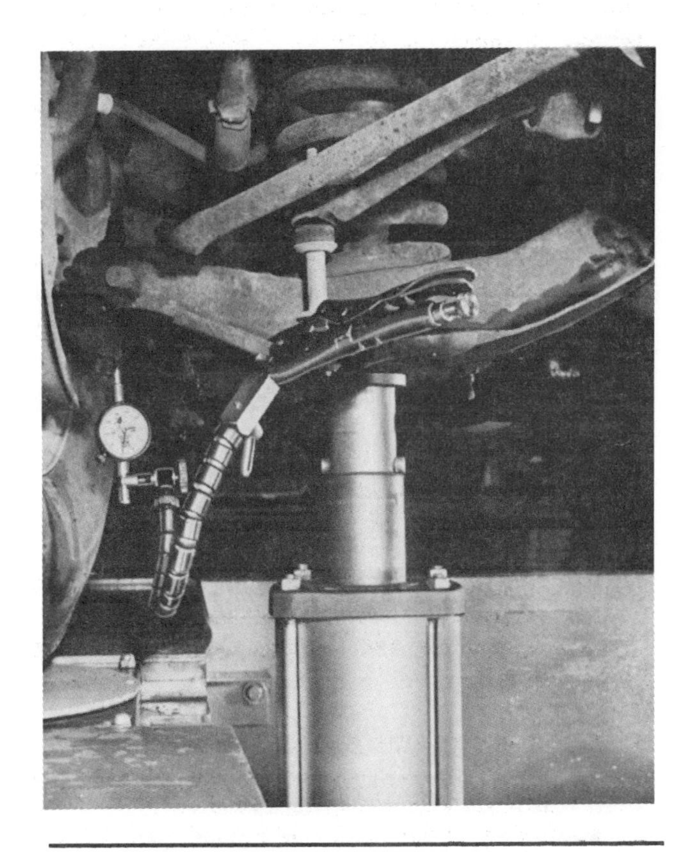

FIGURE 3-6. Jacking Position for Type A Suspension. Measure the movement with a dial indicator. (Courtesy of Moog-Canada Automotive, Ltd.)

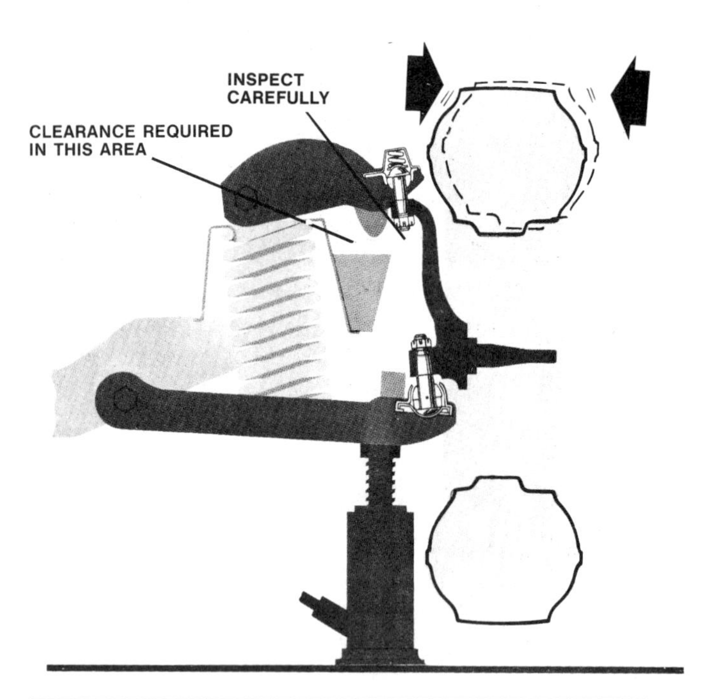

FIGURE 3-7. Inspection of Non-Weight-Carrying Ball Joints. Move the wheel in and out to check for radial movement in the top ball joint. (Courtesy of Moog-Canada Automotive, Ltd.)

The weight-carrying ball joint is attached to the upper control arm. To check the radial play of the upper ball joint, use the procedure described for a type A suspension. Refer to manufacturer's specifications for allowable tolerance. If radial play exceeds specifications, replace the worn part.

The non-weight-carrying ball joint is attached to the lower control arm. Check for axial or radial play in the joint by using the wheel as a lever. If you detect any slack, replace the part.

Wear Indicator Ball Joints

In recent years, vehicle manufacturers have designed the weight-carrying ball joint with a visual *wear indicator.* Wear is indicated by the protrusion of a 12.7 millimeter (1/2 inch) diameter boss (housing) into which the grease fitting is threaded (see Figure 3-10). This round boss (wear indicator) projects 1.27 millimeters (0.05 inch) beyond the surface of the ball joint cover on a new, unworn joint. With normal wear, the lower surface of the boss will retreat slowly inward. When you inspect for wear, the front wheels must support the weight of the vehicle by resting on the garage floor or be positioned on the alignment rack. It is also recom-

FIGURE 3-8A. T-458 Upper Arm Holding Block. This tool is used as a block when the coil spring is above the upper control arm. Its use assists in removing or inspecting the upper and lower ball joints.

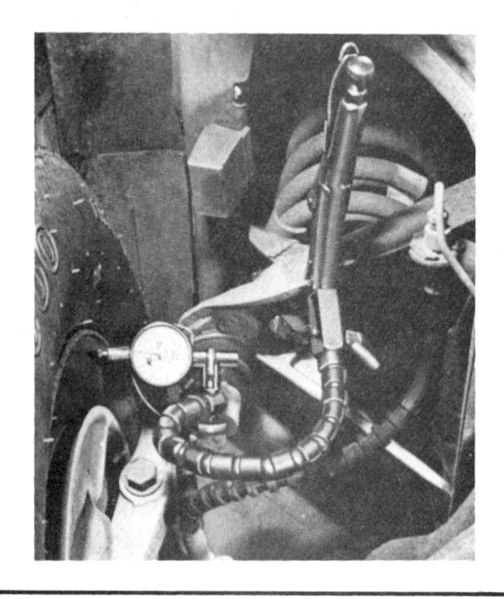

FIGURE 3-8B. Radial Check. Be sure to insert a steel wedge or wooden block where shown while checking a type B suspension. (Courtesy of Moog-Canada Automotive, Ltd.)

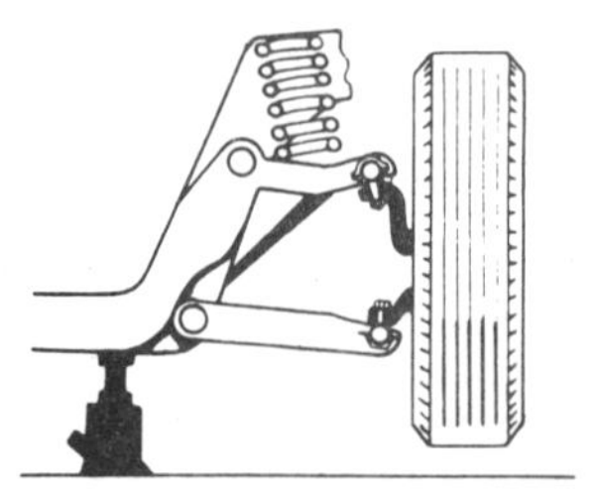

FIGURE 3-9. Jacking Position for a Type B Suspension. (Courtesy of Moog-Canada Automotive, Ltd.)

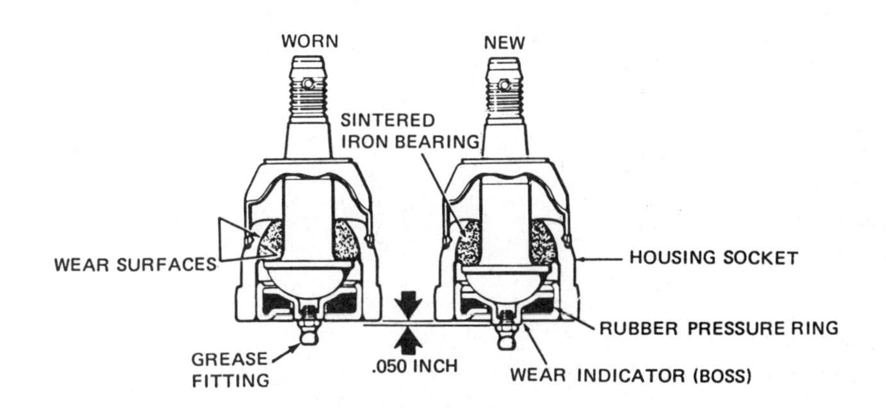

FIGURE 3-10. Determining Ball Joint Wear. When ball joint wear causes wear indicator shoulder to recede within the socket housing, replacement is required. (Courtesy of Moog-Canada Automotive, Ltd.)

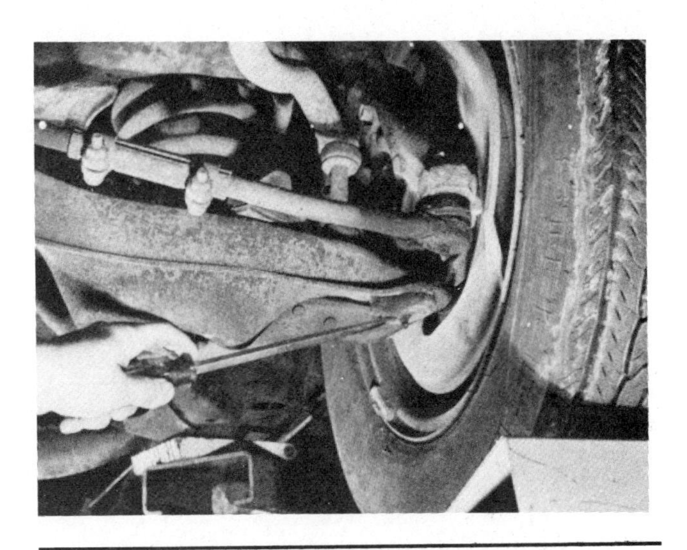

FIGURE 3-11. Wear Indicator Ball Joints. Scrape a screwdriver or nail across the cover. If the boss is below the cover, the ball joint does not need to be replaced. (Courtesy of Moog-Canada Automotive, Ltd.)

mended that the vehicle be at the correct curb riding height. Wipe the grease fitting and boss free of dirt and grease as for a grease job. Observe the bottom of the ball joint by scraping a screwdriver or a fingernail across the cover to remove any existing grease or dirt (Figure 3-10). If the boss is flush or inside the cover surface, replace the ball joint (Figure 3-11).

Not all manufacturers use the same procedures for determining wear in a wear indicator ball joint. For example, on an American Motors Pacer, you must remove the lubrication plug from the lower ball joint. Check ball joint clearance by inserting a stiff piece of wire into the lubrication plug hole until the wire contacts the ball stud. Accurately mark the rod with a knife or scriber where it is aligned with the outer edge of the plug hole. The distance from the ball stud to the outer edge of the plug hole is the ball joint clearance. Measure the distance from the mark to the end of the rod. If the distance measured is 11 millimeters (7/16 inch) or more, the ball joint must be replaced.

On some vehicles with MacPherson strut suspension, ball joint wear is determined by having the front wheels on the floor. Place your fingers on the grease fitting. Then try to shake the fitting from side to side. If there is lateral movement, the ball joint must be replaced.

Replacing Ball Joints

Riveted Ball Joints. Ball joints are often attached to the control arm by *rivets.* Do not use an acetylene torch to remove the head of a rivet, or you may damage the control arm and weaken the metal. Wherever possible, use a center punch to score the head of the rivet. Using a sharp 3 millimeter (1/8 inch) drill bit, drill a small hole through and below the head of the rivet. Then use a sharp chisel to remove the head of the rivet. You may then use an air gun to punch the remaining part of the rivet through the control arm. Always use **tensile bolts and nuts** to attach the ball joint to the control arm. Torque to recommended specifications.

Threaded Ball Joints. Installation of *threaded ball joints* can be made much easier if you use an antifriction lubricant. Engine oil or hypoid rear axle lube should be smeared around the threads of a new ball joint before it is installed in the control arm. The oil will help reduce friction, heat, and scoring of the various parts.

Pressed Ball Joints. Some manufacturers *press*

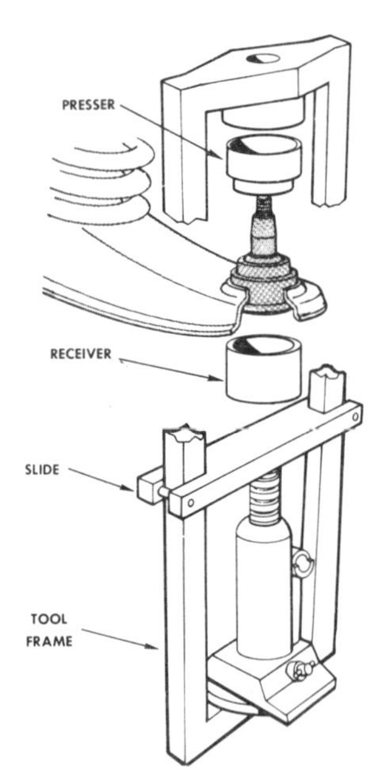

REMOVAL

Under "Removal" section obtain the number of recommended presser and receiver. Seat large end of suggested presser into cup on tool frame. Position presser and frame over ball joint with tapered stud pointing out hole in frame top.

Place selected receiver on slide and raise into position around bottom of ball joint and against underside of control arm. Position jack or ram to be used on frame base and adjust for contact with underside of slide. (Compare to illustration.) Press until ball joint is free of control arm and drops into receiver.

NOTE: Remove grease fitting from all ball joints before pressing out.

INSTALLATION

Under "Installation" section obtain the number of the recommended presser and receiver. Seat suggested receiver into cup on tool frame. Position receiver and frame on control arm over ball joint hole.

Place selected presser and ball joint to be installed on slide and raise into position with ball joint started in lip of control arm. Position jack or ram on frame base and adjust for contact with underside of slide. (Compare with illustration.) Make certain ball joint is straight; then press until shoulder is seated into control arm. (Refer to instruction sheet supplied with ball joint for installation details.)

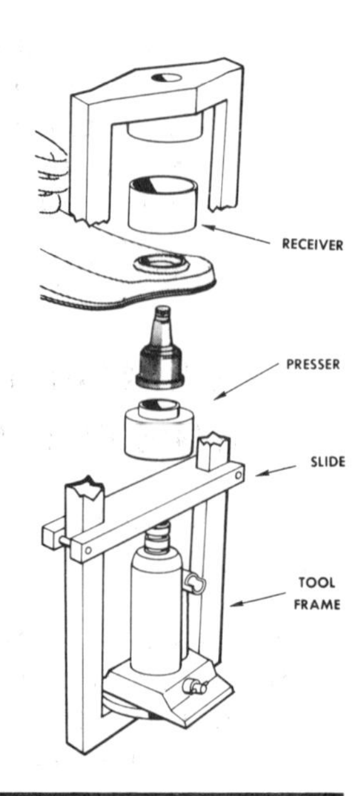

FIGURE 3-12. Removing and Installing Press-Type Ball Joints. (Courtesy of Moog-Canada Automotive, Ltd.)

the ball joints into the control arm. When removing a ball joint from the control arm, always examine the control arm for possible cracks and defects where the ball joint fits the control arm. If a defect is evident, replace the control arm assembly (Figure 3-12).

Lubricating Ball Joints

Ball joints require periodic lubrication. When you examine the vehicle's ball joints, you will observe a hex head lube plug threaded into the housing of each ball joint. The lube plug must be removed and a lube fitting installed into the housing. A manual lube gun designed to deliver lithium-base lubricant at a low pressure of 42 kPa (6 psi) is used to lubricate the ball joints, thereby avoiding damage to the special protective seal. Apply the lubricant slowly; you should not be able to see lube escaping past the seals. When the lubrication is complete, remove the lube fittings and install the lube plugs.

STEERING KNUCKLE INSPECTION

Field research has indicated that out-of-round *steering knuckle* tapered holes that poorly mate with the ball joint are a prime cause of stud breakage. This mismatch allows the stud to fret and work in the taper until the stud fractures. This condition, usually caused by improper installation, can result in complete loss of vehicle control. In order to detect a defective steering knuckle and prevent a broken stud, look for the following signs:

- A stud that is only hand tightened,
- A polished area or a marked tapered area on the stud of the ball joint being removed (Figure 3-13).

Tell-tale warning signs indicate a problem that must receive attention. Before you install a replacement ball joint, you must carefully check the tapered hole in the steering knuckle as follows:

1. Examine for an out-of-round condition.
2. Check that the hole is clean and free of burrs.

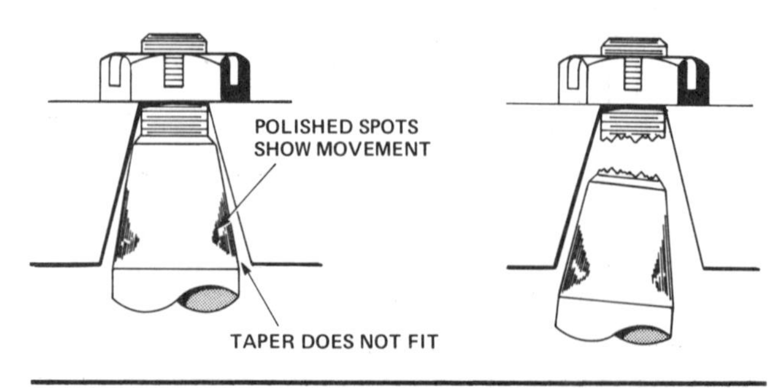

FIGURE 3-13. Signs of a Defective Steering Knuckle. (Courtesy of Moog-Canada Automotive, Ltd.)

3. Press the tapered stud into the hole and check for side movement.
4. You may apply **prussian blue dye** to the stud and rotate it in the tapered hole. (A good hole will show at least 70 percent blue in marking.)

Determining Steering Knuckle Damage

If the vehicle has been involved in a collision, the steering knuckle may be bent. If the damage is severe, the steering knuckle must be replaced. The only accurate way to determine damage to this part is to measure the **steering axis inclination** angle.

Since the camber angle is closely related to the steering inclination angle, the camber angle must first be determined. Then the steering inclination angle is measured. On completion of the measuring process, the figures are compared against the manufacturer's specifications. How this front end factor is measured is discussed in Chapter 18.

Determining Front Spindle Damage

The *front spindle* (or axle) is the part that enables the wheel to be attached to the steering knuckle. Although the part is made of high-quality steel, the spindle may become damaged if the vehicle has been involved in a side collision.

To determine whether the spindle needs to be replaced, attach a runout gauge to where the inner wheel bearing is positioned on the spindle. The gauge is similar to the design of a bridge (Figure 3-14). To check if the spindle is damaged, rotate the entire measuring device around the spindle. If the needle on the dial gauge records more than 0.127 millimeter (0.005 inch) out of round, the centerline (C/L) of the spindle is not straight. The part should be replaced.

KINGPIN INSPECTION

If the vehicle is equipped with *kingpin-type suspension*, place the jack or lift adapter under the axle as close as possible to the wheel. Only one side of the vehicle should be raised at a time during the inspection procedure. Raise the wheel 10 centimeters (4 inches) from the garage floor. Check the torque of the wheel bearing adjusting nut and adjust to the manufacturer's specification. It is recommended that a torque wrench be used for this procedure.

Position a dial gauge indicator clamp on a fixed surface. Then locate the plunger of the gauge on the extreme bottom edge at the inside of the tire. Push the plunger in and out on the bottom of the tire and

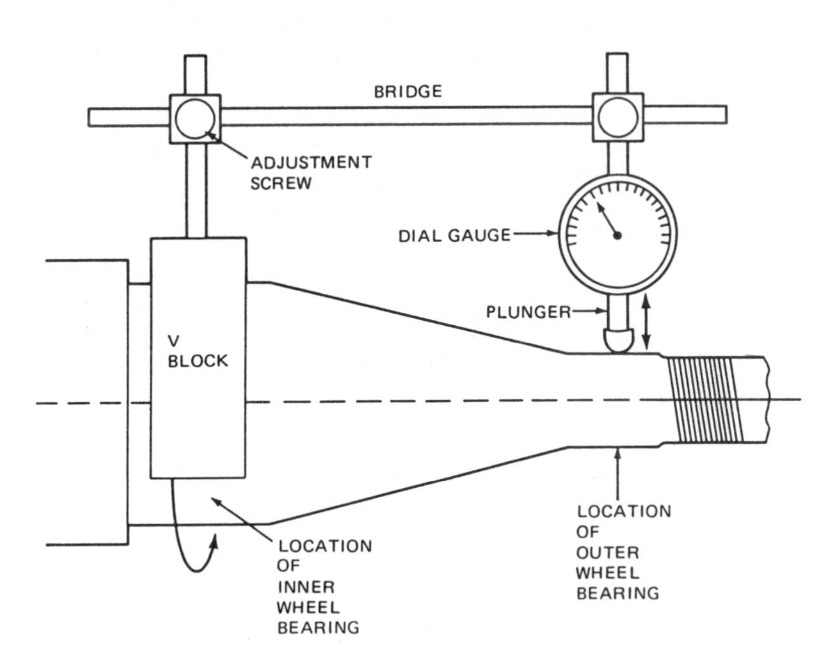

FIGURE 3-14. Spindle Runout Gauge Used to Measure Damage. V block and gauge are rotated around the spindle.

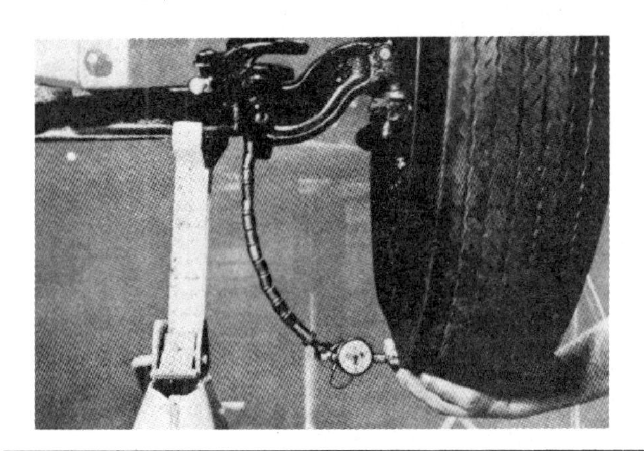

FIGURE 3-15. Checking Kingpin. Affix the gauge to the bottom of the tire. Push in and out on the wheel. (Courtesy of Moog-Canada Automotive, Ltd.)

FIGURE 3-16. Determining Clearance between Steering Knuckle and I-Beam Axle. Insert the feeler gauge between the steering knuckle and the I-beam axle. (Courtesy of Moog-Canada Automotive, Ltd.)

determine the reading on the gauge (Figure 3-15). The kingpins need to be serviced if the readings exceed the following three examples:

- 6.3 millimeters (1/4 inch) for 16 inch wheel diameteı,
- 9.5 millimeters (3/8 inch) for 17 to 18 inch wheels,
- 12.7 millimeters (1/2 inch)for 18 inch wheels and over.

If it is necessary to replace the kingpin and its bushings, the maximum vertical clearance between the steering knuckle and the I-beam axle must not exceed 0.127 millimeter (0.005 inch). This clearance can be determined by placing a **feeler gauge** between the steering kunckle and the I-beam. If the gauge can be inserted, add **shims** until the gauge will not fit (Figure 3-16).

Excessive movement within the kingpin will contribute to steering wheel kickback and loss of directional control and stability. If excessive movement is not corrected, the I-beam axle or steering knuckle will require major repair.

A FACT TO REMEMBER

If the steering knuckles, ball joints, control arms, and kingpins are worn and defective, do not attempt to repair these vital parts. The only method of correction is the replacement of the parts. After the replacement of the suspension parts, it is imperative that the vehicle be aligned since the new parts have less movement and tolerance.

RUBBER BUSHING INSPECTION

Most vehicle manufacturers use *rubber bushings* to insulate unwanted noise and vibration from the vehicle's frame and body. Rubber bushings, like other component parts, eventually require service and replacement. Many vehicle technicians tend to overlook the condition of bushings when inspecting a vehicle's suspension system. Figure 3-17 illustrates some of the locations where rubber bushings are used in a vehicle's suspension system.

When a vehicle is being driven, many forces act on the entire vehicle; among these forces are acceleration, braking, cornering, weight, and road surface deviation. It is vitally important that the vehicle's suspension system not be subjected to any excessive movement that would affect the vehicle's directional control and stability. Since rubber bushings play an important role in the vehicle's handling, there is no point in attempting to align a vehicle if these parts have become fatigued, saturated in oil, out of shape, or loose enough to permit the leaf springs, front and rear suspension arms, shocks, track rods, sway bars, and strut rods to move in an unwanted direction while the vehicle is being driven.

Determining Bushing Condition

Usually, a visual inspection will indicate the condition of bushings. If you are in doubt, one of the following two methods is sure to determine

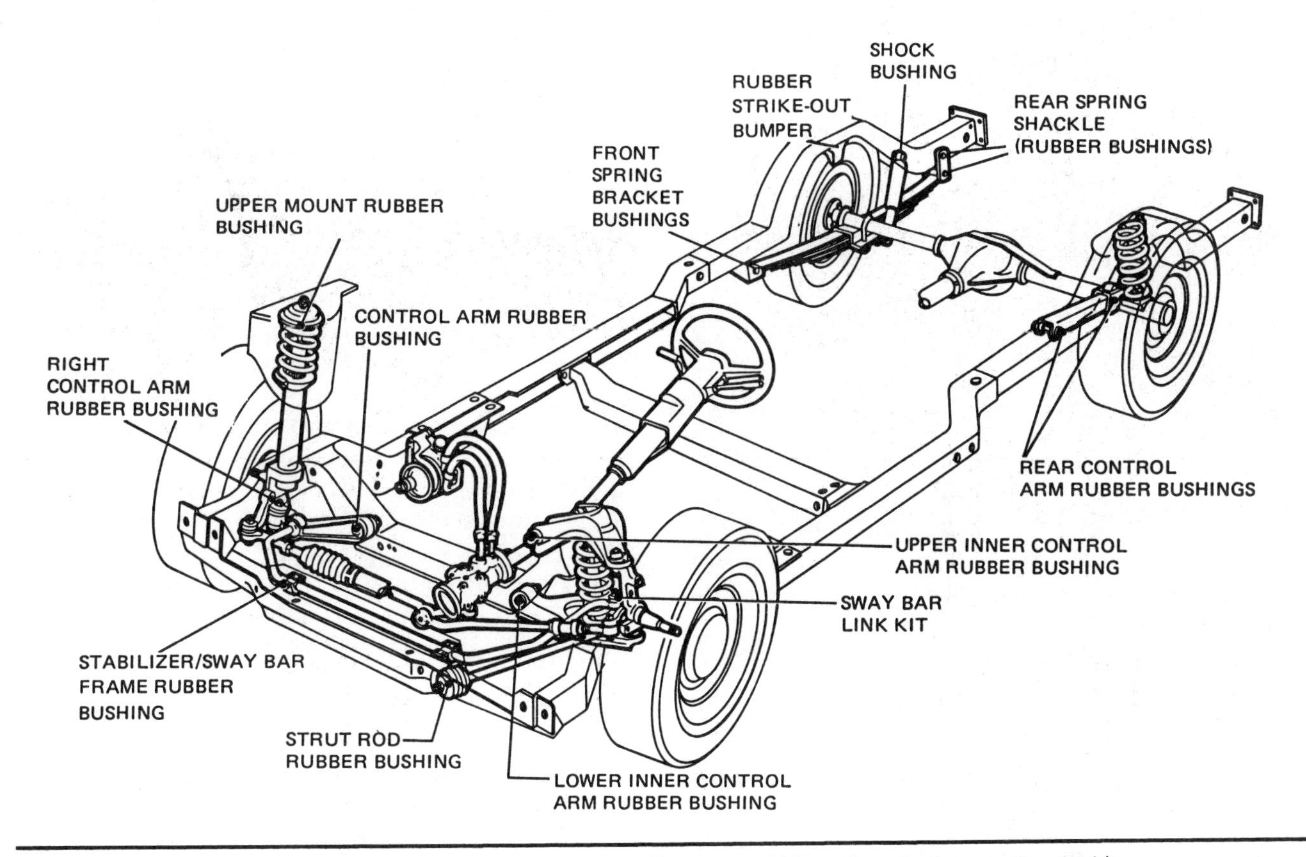

FIGURE 3-17. Rubber Bushing Locations on a Suspension. (Courtesy of Moog-Canada Automotive, Ltd.)

their condition. The first method is to raise the vehicle in a manner that will free the ball joint or control arm from load. Then apply force in several directions. Watch the bushings for evidence of looseness.

For the second method, you must ask another person to sit at the driver's position with his or her foot off the brake pedal. Carefully raise only one side of the vehicle, and place the vehicle on safety stands. If the vehicle is equipped with **disc brakes**, you may need to release the **brake pads** from the brake rotors. By hand, rotate the wheel as fast as possible. At that time, have the other person apply the brake pedal. If the bushings are worn or defective, the quick application of the brake will cause the various suspension parts to move, thus indicating worn parts. If any looseness can be detected or if the rubber is squished out of position, replace the bushings.

Servicing Bushings

Rubber suspension bushings and *steering bushings* must be torqued only when the vehicle is in the normal curb riding height position. Rubber bushings tighten while the vehicle is on a hoist or safety stand, with the wheels hanging free. If the bushings or thrust washers are tightened with the wheels in that position, premature failure and even temporary suspension height changes will result. Other bushings to consider are *track bar bushings*, found in various types of suspension systems, and *front-rear bushings* found in leaf spring suspensions.

When installing rubber suspension bushings mounted in a steel sleeve, always press on the outer sleeve. Do not tighten rubber suspension bushings until all the wheels of a vehicle are resting on the garage floor.

When replacing *strut rod bushings*, always examine the vehicle's front crossmember (where the strut rod enters the crossmember or bracket) for rust and/or possible metal fatigue (Figure 3-18). The strut rod and rubber bushing must stabilize the position of the lower control arm when the vehicle accelerates or brakes. A loose or fatigued bushing will cause the camber, caster, and toe to change constantly, thereby affecting steering control and greatly increasing tire tread wear.

Stabilizer bar rubber bushings and *links* should

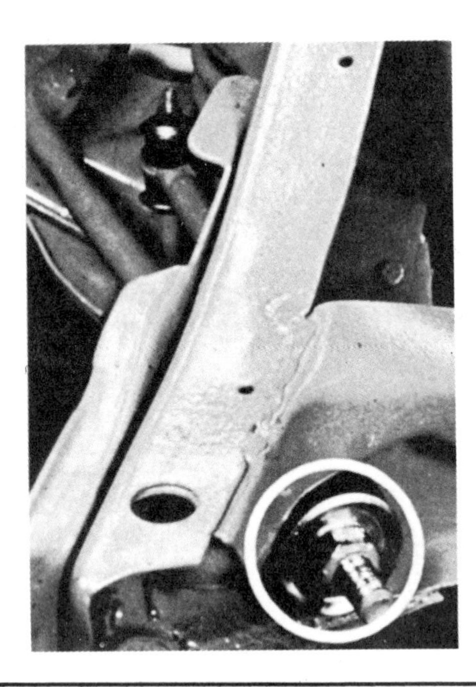

FIGURE 3-18. Examining the Strut Rod Bushings and the Frame. (Courtesy of Moog-Canada Automotive, Ltd.)

also be inspected for deterioration. A broken or nonfunctioning stabilizer bar is cause for rejection when applying for a vehicle inspection sticker in most states and provinces in North America.

SHOCK ABSORBER INSPECTION AND TESTING

Driving a vehicle without safe shock absorbers is tempting fate. If shocks are defective, the springs continue to oscillate, and the vehicle may bounce off the road. Also, the tires will develop flat spots on the tread surfaces. Chapter 2 illustrated and described the purpose and operation of a shock. Now let's examine how to inspect and test a shock absorber. There are five different tests that you should know:

- The bounce test,
- Inspecting a shock's mountings for noise and looseness,
- Visually inspecting for leaking hydraulic fluid,
- Inspecting the shock for ineffective control,
- Inspecting the shock for internal problems.

The first four tests require the vehicle to be on a hoist that supports the wheels or rear axle housing and front control arms.

Conventional Shocks

Bounce Test. Test each front and rear shock absorber by vigorously bouncing each corner of the vehicle and releasing the bumper after the compression cycle. The shock is considered acceptable when, after releasing, it allows one free bounce and then stops the vehicle's vertical movement. This type of test is only a comparison check of side-to-side variation. Do not compare the front or rear shocks since their ratios may be different.

Shock Mountings. If noisy or loose shock mountings are suspected, position the vehicle on a hoist that supports the wheels. Check all mountings for the following conditions:

- Worn or defective grommet bushings,
- Loose mounting nuts,
- Possible interference with other parts,
- Shiny rubber suspension bumpers.

If no apparent defects are noted, proceed to the next tests.

Leaking Hydraulic Fluid. Inspect the front and rear shocks with wheels unsupported to expose the seal cover area at the top of the reservoir. A trace of fluid is not a cause for replacement since the seal permits slight seepage to lubricate the piston rod. If you are in doubt, wipe the top of the lower shock housing; activate the unit, and inspect for leaks again. Fluid seepage will reappear if the shock absorber is leaking. Replace the shock absorber if it is defective.

Ineffective Control. To determine the ability of the shock to control the action of the suspension system, remove the shock from its lower mounting. Then move the shock absorber up and down rapidly. Test for a time lag (loss of control and response) during the operation of the unit. If there is a noticeable loss of response, replace the defective unit.

Internal Problems. Disconnect the shock absorber from the lower mounting and diagnose for the following:

- A grunt or squawk during the cycling period,
- A sharp clicking noise during activation,
- A seizing or binding condition.

If any of these sounds or conditions exist, replace the defective unit.

A PRACTICAL SERVICE TIP

Do not clamp the reservoir or threaded portion of shock absorber in a vise. Damage will result. Always purge (bleed) the shock of air by activating the shock at least five times in a vertical position prior to installation. This action will assist the shock during initial operation.

Special Cases

Some shock absorbers contain a gas-filled cell instead of an air reservoir. These shocks are marked and named Pliacell® or Genetron®. Test this type of shock by removing the unit from the vehicle; invert and activate. Test for a lag condition (loss of control and response). If there is a noticeable loss of response, replace the defective unit.

When an air adjustable shock is being tested, inflate the unit to 630 kPa (90 psi) and apply a solution of soapy water around the valve, air line fittings, lines, and shock absorber. Visually inspect for an air leak. If you detect a leak, repair or replace the defective part. Shock absorber fluid leaks can be detected by visually examining the body or rubber boot for fluid deposits.

Caution: All shocks contain oil. Do not attempt to remove a shock from a vehicle by using an acetylene torch. An explosion could result.

The following are signs of shock absorber wear for vehicles equipped with MacPherson strut suspension:

- Steering problems or sways around corners or curves,
- A dipping or nose-diving condition on quick stops,
- Excessive bouncing or noise after the wheel encounters a bump on the road.

If the shock absorber on the strut is gas filled, the piston rod of the shock should extend from the reservoir when it is held in an upright position. If the rod does not extend, the entire shock unit must be replaced.

REVIEW TEST

1. Explain in detail how to determine if vehicle curb riding height is within the manufacturer's specification.
2. Why is curb riding height so important to sustaining tire tread life?
3. List six signs of incorrect curb riding height.
4. A vehicle is equipped with unequal length front control arm suspension, and the spring is resting on the lower arm. Describe how to diagnose for wear in the upper and lower ball joints.
5. The front coil spring is resting on the upper control arm. Describe how the upper and lower ball joints are diagnosed for wear.
6. What are wear indicator ball joints? How would you determine wear?
7. Describe the correct procedure for replacing a press-type ball joint.
8. The tapered stud of a ball joint has been loose in the steering knuckle. Describe how to determine if the steering knuckle is usable.
9. You suspect that a vehicle's steering knuckle is distorted. Briefly describe how to determine if the part is damaged.
10. Describe how to diagnose for wear in upper inner control arm bushings.
11. A vehicle's strut rod bushings are worn. How will this condition affect the vehicle when the brakes are applied?
12. Describe how to test shock absorbers.
13. What should you do to a shock absorber prior to installation?

FINAL TEST

This examination is multiple choice. Only one answer will be accepted. Carefully read every statement.

1. Mechanic A says that tires must be properly inflated before measuring curb riding height. Mechanic B says that curb riding height is always measured from the height of the front or rear bumpers. Who is right? (A) Mechanic A, (B) Mechanic B, (C) Both A and B, (D) Neither A nor B.
2. Mechanic A says that it is permissible to remove the head of a rivet by using an acetylene torch.

Mechanic B says that safe work practices are for idiots who only try to impress their boss or customer. Who is right? (A) Mechanic A, (B) Mechanic B, (C) Both A and B, (D) Neither A nor B.

3. Mechanic A says that curb riding height measurements are the same for all vehicles. Mechanic B says that curb riding height has no effect on the service life of the steering or suspension parts. Who is right? (A) Mechanic A, (B) Mechanic B, (C) Both A and B, (D) Neither A nor B.
4. Mechanic A says that wheel alignment factors change when curb riding height is raised. Mechanic B says that incorrect curb riding height changes the vehicle's center of gravity. Who is right? (A) Mechanic A, (B) Mechanic B, (C) Both A and B, (D) Neither A nor B.
5. Mechanic A says that a sagged left front spring will affect the camber on the right rear wheel. Mechanic B says that left and right front curb riding height readings must not vary more than 6.3 millimeters (1/4 inch). Who is right? (A) Mechanic A, (B) Mechanic B, (C) Both A and B, (D) Neither A nor B.
6. Mechanic A says that all ball joints must be free from the force exerted by the weight of the vehicle before they can be inspected. Mechanic B says that all wear indicator ball joints are not checked in the same manner. Who is right? (A) Mechanic A, (B) Mechanic B, (C) Both A and B, (D) Neither A nor B.
7. Mechanic A says that axial play of a ball joint is checked by grasping the tire and endeavoring to move the wheel assembly in and out. Mechanic B says that axial play of a ball joint is checked by using a pry bar to aid in lifting the wheel assembly. Who is right? (A) Mechanic A, (B) Mechanic B, (C) Both A and B, (D) Neither A nor B.
8. Mechanic A says that radial play of a ball joint is checked by grasping the tire and endeavoring to move the wheel assembly in and out. Mechanic B says that radial play of a ball joint is checked by using a pry bar to aid in lifting the wheel assembly. Who is right? (A) Mechanic A, (B) Mechanic B, (C) Both A and B, (D) Neither A nor B.
9. Mechanic A says that if you detect slack in a non-weight-carrying ball joint, replace the part. Mechanic B says that when determining ball joint wear, only raise one side of the vehicle prior to inspection. Who is right? (A) Mechanic A, (B) Mechanic B, (C) Both A and B, (D) Neither A nor B.
10. Mechanic A says that ball joints should be lubricated by a high-pressure lube grease gun in order to force the grease under the seal. Mechanic B says that the tapered stud of a ball joint is always positioned upward in order to allow for easy installation into the steering knuckle. Who is right? (A) Mechanic A, (B) Mechanic B, (C) Both A and B, (D) Neither A nor B.
11. Mechanic A says that the maximum clearance between the top of an I-beam axle and the steering knuckle is 0.254 millimeter (0.010 inch). Mechanic B says that if a dial gauge reads 9.5 millimeters (3/8 inch) when measuring lateral movement of a 16 inch diameter wheel (kingpin suspension), the kingpin does not need to be serviced. Who is right? (A) Mechanic A, (B) Mechanic B, (C) Both A and B, (D) Neither A nor B.
12. Mechanic A says that visual inspection is the only way to determine the condition of upper inner control arm bushings. Mechanic B says that a worn strut rod bushing will cause the caster angle to change. Who is right? (A) Mechanic A, (B) Mechanic B, (C) Both A and B, (D) Neither A nor B.
13. Mechanic A says that when installing rubber suspension bushings, always press on the outer steel sleeve. Mechanic B says that rubber suspension bushings must not be tightened until all the wheels are resting on the floor. Who is right? (A) Mechanic A, (B) Mechanic B, (C) Both A and B, (D) Neither A nor B.
14. Mechanic A says that loose or fatigued strut rod bushings will not affect the braking of a vehicle. Mechanic B says that loose or fatigued radius rod bushings will not affect the directional control of a vehicle. Who is right? (A) Mechanic A, (B) Mechanic B, (C) Both A and B, (D) Neither A nor B.
15. Mechanic A says that the only test to determine the condition of a shock is the bounce test. Mechanic B says that testing rear shocks will provide only a comparison check of side-to-side variation. Who is right? (A) Mechanic A, (B) Mechanic B, (C) Both A and B, (D) Neither A nor B.
16. Mechanic A says that a slight trace of fluid at the top of the shock's reservoir is not a reason for shock replacement. Mechanic B says that when a shock piston rod is activated in an up and down motion, there must not be a time lag. Who is right? (A) Mechanic A, (B) Mechanic B, (C) Both A and B, (D) Neither A nor B.

Wheel Bearings

LEARNING OBJECTIVES

After studying this chapter and with the aid of a manufacturer's shop manual, you should be able to:

- Explain the purposes of the front wheel and rear wheel axle bearings,
- Explain how to inspect, diagnose, and service front wheel and rear wheel axle bearings.

INTRODUCTION

Where would humans and transportation be today without the invention of the wheel? Archaeologists suggest that the wheel came into use at least 6000 years ago in Near Eastern countries. The first wheels were made from round wooden logs, with a hole in the center for a wooden axle. The basic design of the wheel remained unchanged for centuries until the period of the Industrial Revolution and the invention of the steam engine. Wooden wheels gave way to wheels made of steel, and transportation began to advance into a new technological era.

Pioneering vehicle manufacturers constructed wheels with wooden spokes, fitted with a band of solid rubber around a steel rim. The rubber band provided the necessary traction. However, it also transmitted road shock because of its inability to flex under the load of the vehicle. The designers went back to their drawing boards. Two brothers, Edward and André Michelin, developed the first detachable cycle tire in 1891. The tire consisted of a separate tube and an outer cover bolted to a rim. This development led to the first pneumatic automotive tire in 1895 and the first commercial truck tire in 1912. These tires were made from woven canvas fabrics held together by rubber binding.

The effect of friction between the threads of the canvas generated excessive heat and caused the tires to have blowouts. To overcome the problem, tires were inflated to increased air pressure. This solution reduced their ability to flex, however, thereby defeating the prime purpose of the pneumatic tire. For example, an early tire designed to carry 395 kilograms (kg) (900 pounds) of weight required a pressure of 420 kPa (60 psi); a similar modern tire would require an inflation of only 210 kPa (30 psi).

To begin our discussion of the wheel, let's start

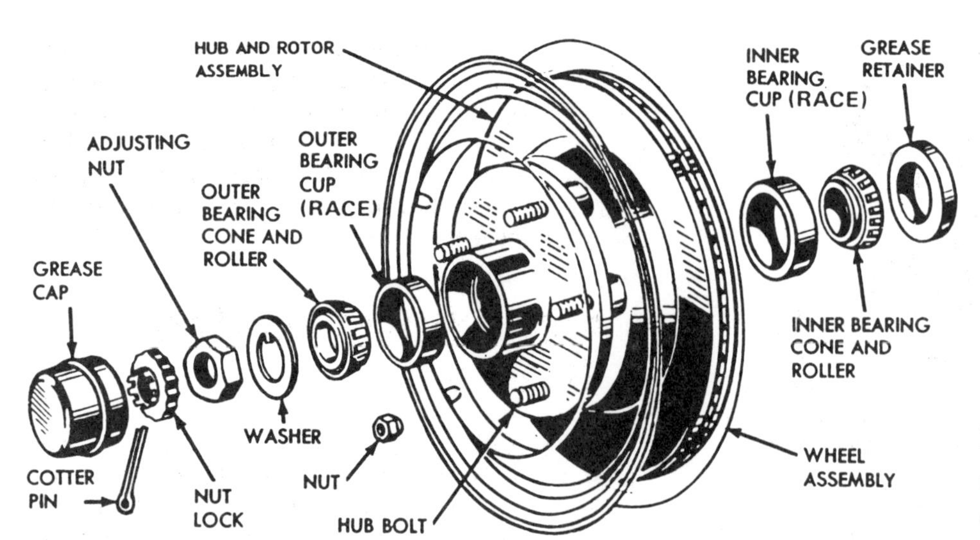

FIGURE 4-1. Assembly of a Wheel. (Courtesy of Ford Motor Company)

at the center of the wheel assembly by first becoming familiar with front wheel bearings on cars with rear differentials. We then discuss wheel hubs and bearings on vehicles with front wheel drive and conclude the chapter with a discussion on rear wheel axle bearings.

FRONT WHEEL BEARINGS

Figure 4-1 shows the assembly of a wheel. The front wheel of a vehicle is bolted to a hub and brake drum or rotor assembly. Two opposed tapered roller bearings and raceways made of case-hardened steel are fitted inside the **hub** and allow the wheel assembly to revolve with a minimum of friction. The grease retainer, recessed in the hub, prevents contamination of the bearings and keeps the lubricating grease away from the brake mechanism. The outer part of the hub is protected from contamination and loss of lubricant by a grease retainer cap.

Figure 4-2 illustrates an important principle of mechanical design. After you have removed the front wheel hub from a vehicle, you will notice the tapered design of the front spindle axle. The center-load line of the wheel is almost in line with the larger, inside wheel bearing. The tapered roller bearings and axle spindle are required to support the radial load of the vehicle, reducing the **shearing** of the spindle. As you examine the illustration, you will observe a center-load line through the kingpin or steering knuckle (ball joint suspension). Both center-load lines are designed to intersect near the road surface in order to reduce the effort required to turn the wheel at the road surface.

Wheel bearings are an integral part of the front wheel assembly. Too few vehicle owners realize the importance and purpose of wheel bearings. Figure 4-3 illustrates the four basic parts of a tapered roller

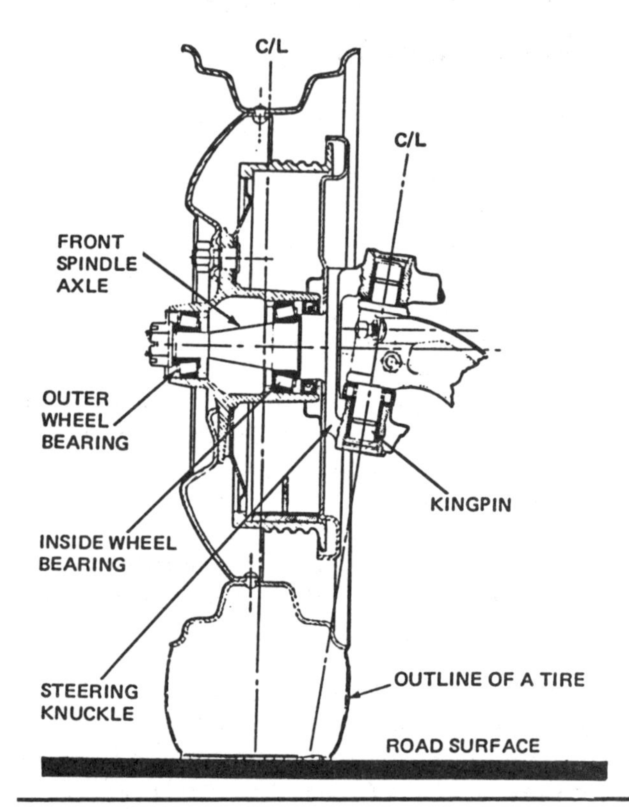

FIGURE 4-2. Loads Directed through Center Lines. The wheel bearings and the kingpin support the vehicle's weight along the center-load lines.

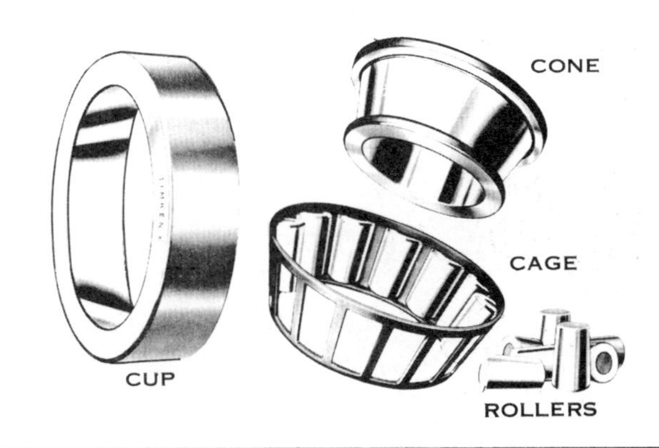

FIGURE 4-3. Major Component Parts of a Wheel Bearing.

bearing: the *cup*, or *outer race;* the *cone*, or *inner race;* the *tapered rollers*, which roll freely between the cup and cone; and the *cage*, which serves as a retainer to maintain the proper spacing between the tapered rollers grouped around the cone. All the parts are assembled into a complete bearing assembly by the manufacturer and are not made to be separated. Bearing part identification numbers are stamped on the cup and cone for reference purposes.

Purposes of Front Wheel Bearings

Front wheel bearings have six purposes:

1. To reduce friction between the revolving wheel hub and spindle,
2. To transmit the **radial** (vertical support) **load** and the **thrust** (lateral, side to side) **load** of a vehicle when the wheel is revolving,
3. To transfer undesirable road shock from the tire to the axle,
4. To keep the front wheels in correct alignment with the vehicle's suspension system regardless of road surface terrain,
5. To transfer the steering action produced by the driver to the wheels of the vehicle,
6. To allow the driver to sense **turning** resistance and road feel in order to control the direction of the vehicle.

Loose or rough front wheel bearings will produce a noise that may be confused with rear axle noises. The difference is that front wheel bearing noise does not change when the vehicle is coasting or accelerating. A light application of the brake while holding the vehicle's speed constant will often cause a bearing noise to diminish. This action tends to reduce some of the weight supported by the bearings. Front wheel bearings may be checked easily for noise by raising the suspected wheel and then spinning it. If the vehicle is equipped with disc brakes, you will need to disengage the brake pads from the rotor. You may also use lateral force to check for excessive looseness.

Front Wheel Bearing Service

Most vehicle manufacturers recommend that front wheel bearings be serviced (lubricated) at intervals of 40,000 to 48,000 kilometers (25,000 to 30,000 miles). If you have been directed by the shop instructor to lubricate the front wheel bearings of a vehicle, it is suggested that you remove the front wheels from the hubs before you remove the brake drum or rotor from the vehicle.

Step-by-Step Procedure

1. Obtain the necessary hand tools, safety stands, vehicle floor jack, and manufacturer's shop manual.

2. Carefully remove the front hubcaps from the wheels. Preloosen the wheel lug bolts or nuts one complete turn. *Note:* Upon removing the hubcaps, you will notice that the front wheels probably have balanced weights attached to the rim. To ensure that you do not disturb the balance of the front wheels, use a piece of crayon or chalk to index each wheel to its hub, as illustrated in Figure 4-4.

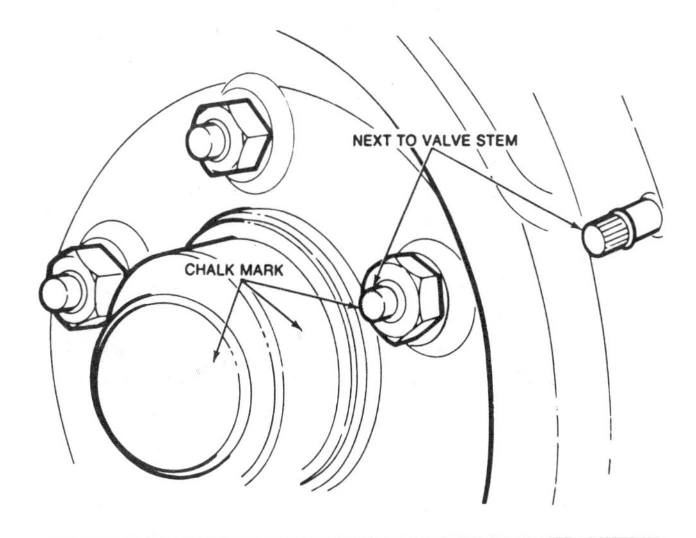

FIGURE 4-4. Indexing the Wheel. Mark each wheel with chalk or crayon to make sure balanced wheels are installed in same position. (Courtesy of Ford Motor Company)

The indexing is done to prevent the wheel from being unbalanced after installation of the wheel, since the wheel assembly may have been balanced on the vehicle. Balancing is discussed in Chapter 7.

3. Place the floor jack under the front cross-member and raise the vehicle. Position the safety stands under the frame of the vehicle (under the front door hinge post). Allow the vehicle to rest on the safety stands. Loosen the wheel lug bolts or nuts, and remove the front wheels. *Note:* If the vehicle is equipped with front disc brakes, the brake calipers will need to be removed from the vehicle before the next step. When the calipers have been detached from the vehicle, wire the assemblies to the frame to prevent damage to the brake flex lines.

4. Remove the grease retainer cap from the hub; use the special pliers if they are available. Remove the cotter pin, nut lock, wheel bearing adjusting nut, and flat washer from the spindle. (See Figure 4-1.) Move the **brake drum** or rotor toward you approximately 7 millimeters (1/4 inch). Push the brake drum or rotor in toward the steering knuckle to free the outer bearing from the hub. Remove the outer bearing. Place the parts in the hubcap.

Caution: Before you remove a brake drum or rotor, put on a respirator so that you do not inhale harmful asbestos dust.

5. Remove the hub from the spindle. Place the hub on the workbench. To protect yourself, vacuum any loose asbestos dust from the many parts of the brake mechanism and from the brake drum or rotor. You may remove the seal and inside bearing from the hub by using a piece of round wooden dowling, 30 centimeters (12 inches) in length, inserted through the outside bearing cup. Place one end of the dowling against the inner wheel bearing; the dowling will be positioned on a slight angle. Gently tap the opposite end of the dowling. Remove the seal and bearing from its recessed position in the hub. Discard the seal (Figure 4-5).

6. Thoroughly clean the bearings with solvent (*not gasoline*); steam or water are not recommended because of the possibility of rust. Dry the bearings with compressed air.

Caution: Compressed air can be dangerous. Direct the flow of air away from you and, if possible, toward the work floor area. Also, be sure to direct the flow of air down through the bearings. Do not allow the bearings to spin or they may fly apart.

7. Clean the lubricant from the inner and outer bearing cups with a solvent (for example, varsol). Inspect the cups for scratches, pits, excessive wear, and other damages. (See the front wheel bearing illustrations in Figure 4-6.)

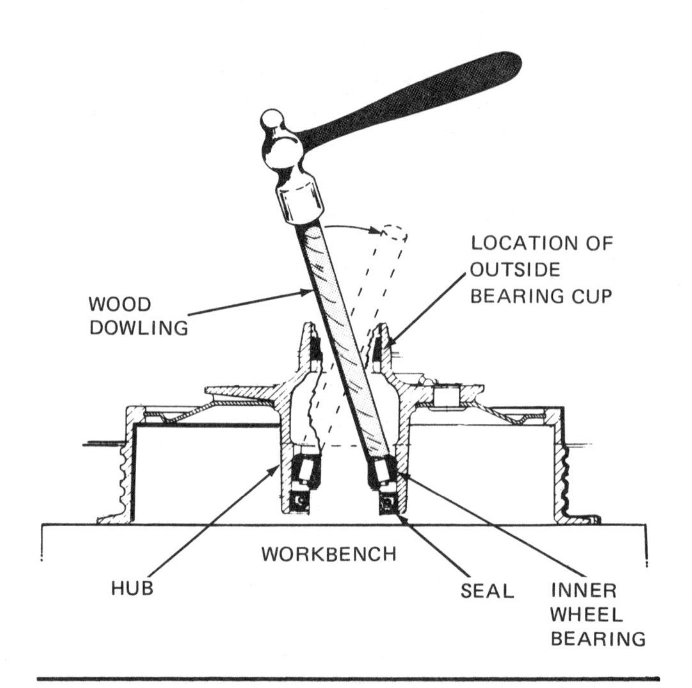

FIGURE 4-5. Removing Wheel Bearing from Hub. As you use a hammer to tap on the end of the wooden dowling, alternate the positions of the dowling to remove the bearing and seal from the hub.

8. If, after close examination, you decide the bearings and cups must be replaced, use the correct tools to remove and install the cups (Figure 4-7). There must be no metal chips under the bearing cups prior to installation.

9. Remove all old grease from the hub and spindle before repacking the bearings. *Note:* Do not immerse the hub, brake drum, or rotor in varsol; immersion will leave a film of oil on the part and will affect the braking of the vehicle. A cloth dampened with varsol will suffice.

10. If a bearing grease packer is not available (Figure 4-8A), the bearing must then be lubricated by hand. Place lubricant in the palm of one hand. Then, holding the bearing in the other hand, scrape the large end of the bearing against your palm, forcing the lubricant completely through the rollers and cage of the bearing (Figure 4-8B). Lubricate the inside cone surfaces with a light film of grease. A 5 millimeter (3/16 inch) layer of grease must also cover the inside of the hub (Figure 4-9).

11. Properly position the inside wheel bearing into the bearing cup. Apply a light film of grease to the lip of the seal and install the new seal.

12. Install the hub, drum, or rotor assembly on the wheel spindle. Keep the hub centered on the spindle to prevent damage to the seal or spindle threads.

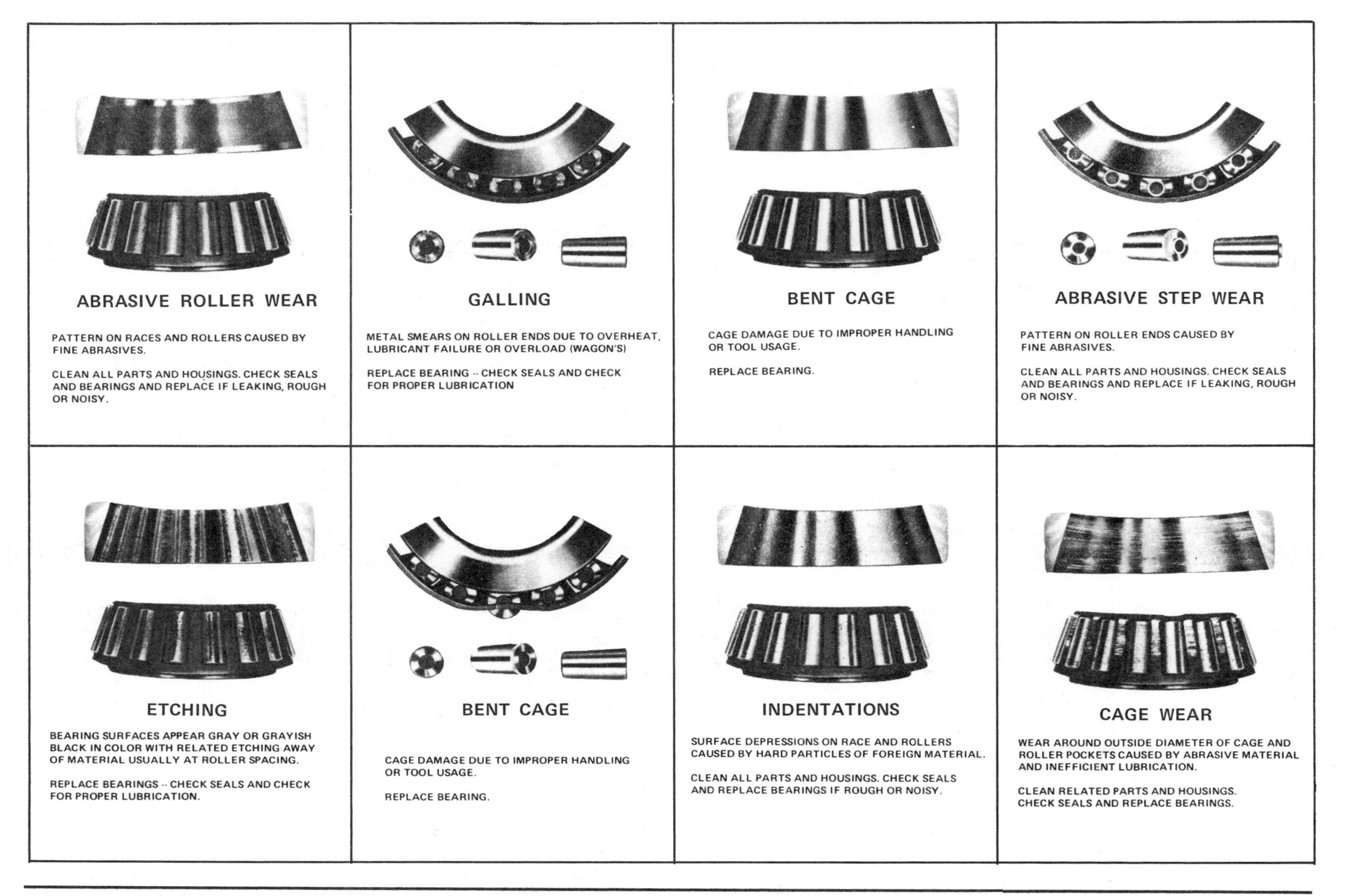

FIGURE 4-6. Types of Wheel Bearing Wear. (Courtesy of General Motors of Canada, Ltd.)

MISALIGNMENT

OUTER RACE MISALIGNMENT DUE TO FOREIGN OBJECT.

CLEAN RELATED PARTS AND REPLACE BEARING. MAKE SURE RACES ARE PROPERLY SEATED.

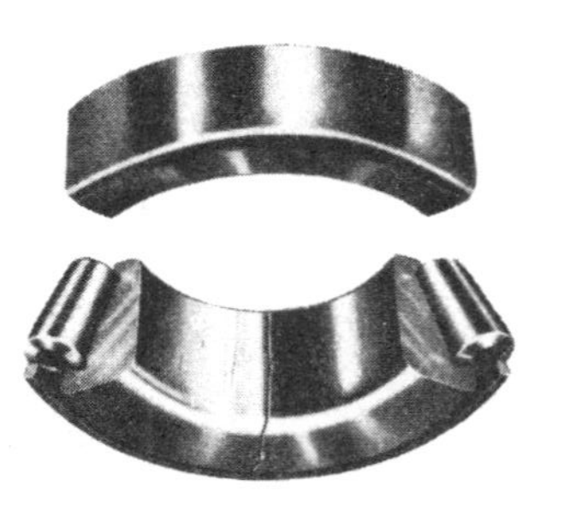

CRACKED INNER RACE

RACE CRACKED DUE TO IMPROPER FIT, COCKING, OR POOR BEARING SEATS.

REPLACE BEARING AND CORRECT BEARING SEATS.

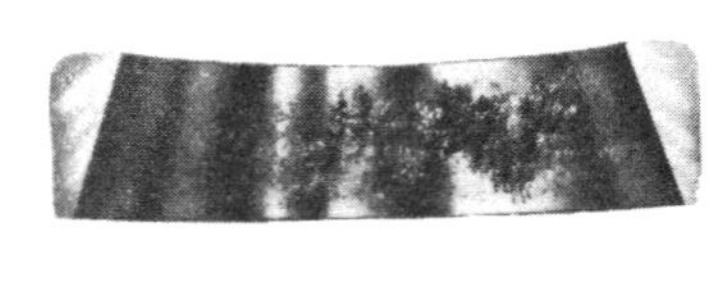

FATIGUE SPALLING

FLAKING OF SURFACE METAL RESULTING FROM FATIGUE.

REPLACE BEARING -- CLEAN ALL RELATED PARTS.

BRINELLING

SURFACE INDENTATIONS IN RACEWAY CAUSED BY ROLLERS EITHER UNDER IMPACT LOADING OR VIBRATION WHILE THE BEARING IS NOT ROTATING.

REPLACE BEARING IF ROUGH OR NOISY.

FRETTAGE

CORROSION SET UP BY SMALL RELATIVE MOVEMENT OF PARTS WITH NO LUBRICATION.

REPLACE BEARING. CLEAN RELATED PARTS. CHECK SEALS AND CHECK FOR PROPER LUBRICATION.

STAIN DISCOLORATION

DISCOLORATION CAN RANGE FROM LIGHT BROWN TO BLACK CAUSED BY INCORRECT LUBRICANT OR MOISTURE.

RE-USE BEARINGS IF STAINS CAN BE REMOVED BY LIGHT POLISHING OR IF NO EVIDENCE OF OVER-HEATING IS OBSERVED.

CHECK SEALS AND RELATED PARTS FOR DAMAGE.

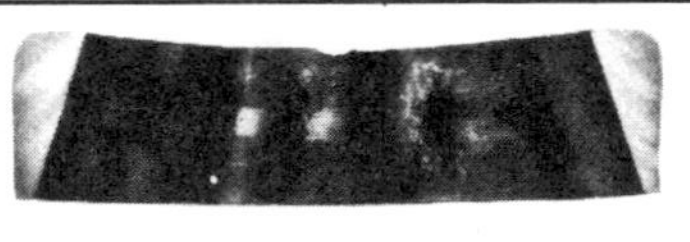

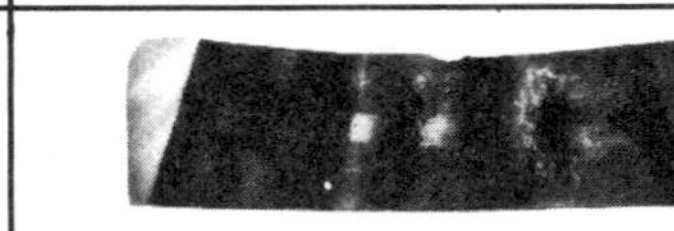

HEAT DISCOLORATION

HEAT DISCOLORATION CAN RANGE FROM FAINT YELLOW TO DARK BLUE RESULTING FROM OVER-LOAD (WAGON'S) OR INCORRECT LUBRICANT.

EXCESSIVE HEAT CAN CAUSE SOFTENING OF RACES OR ROLLERS.

TO CHECK FOR LOSS OF TEMPER ON RACES OR ROLLERS A SIMPLE FILE TEST MAY BE MADE. A FILE DRAWN OVER A TEMPERED PART WILL GRAB AND CUT METAL, WHEREAS, A FILE DRAWN OVER A HARD PART WILL GLIDE READILY WITH NO METAL CUTTING.

REPLACE BEARINGS IF OVER HEATING DAMAGE IS INDICATED. CHECK SEALS AND OTHER PARTS.

SMEARS

SMEARING OF METAL DUE TO SLIPPAGE. SLIPPAGE CAN BE CAUSED BY POOR FITS, LUBRICATION, OVERHEATING, OVERLOADS OR HANDLING DAMAGE.

REPLACE BEARINGS, CLEAN RELATED PARTS AND CHECK FOR PROPER FITS AND LUBRICATION.

FIGURE 4-6. Continued.

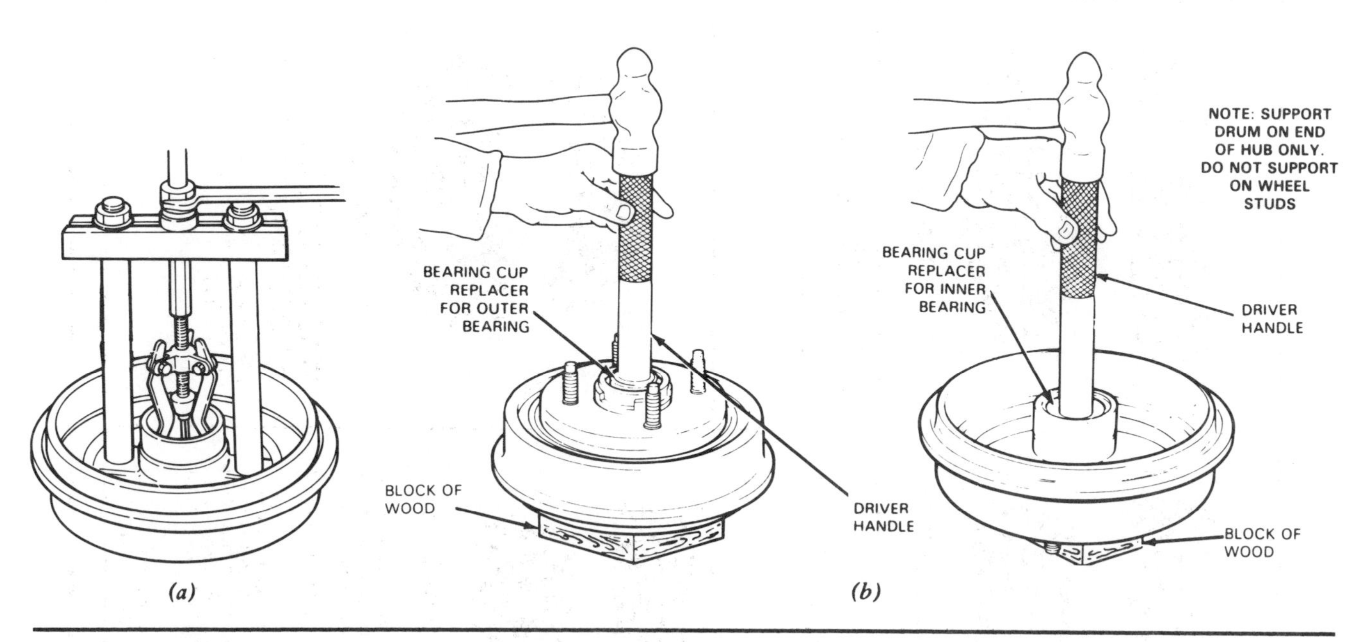

FIGURE 4-7. Replacing Front Wheel Bearing Cups. (a) Removing cups; (b) installing cups. (Courtesy of Ford Motor Company)

FIGURE 4-8A. Model 7150 Bearing Packer. (Courtesy of Ammco Tools, Inc.)

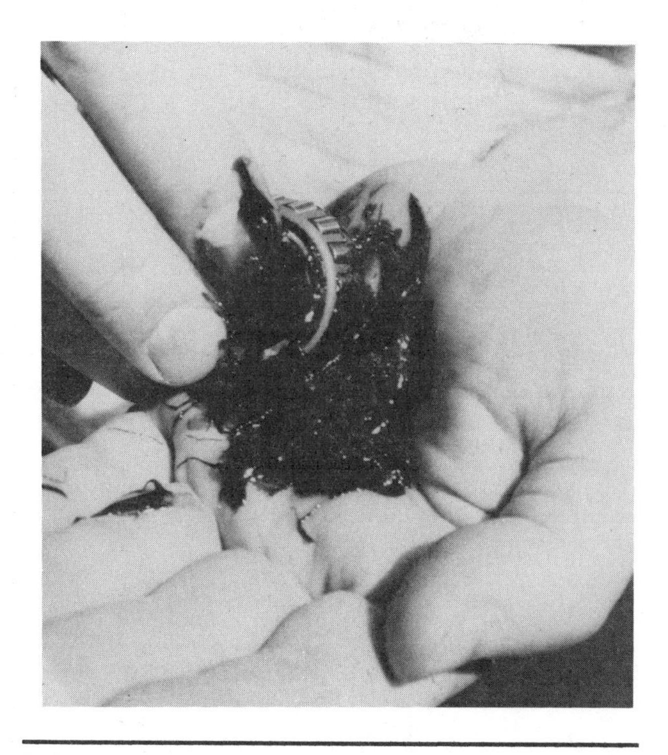

FIGURE 4-8B. Lubricating a Front Wheel Bearing by Hand.

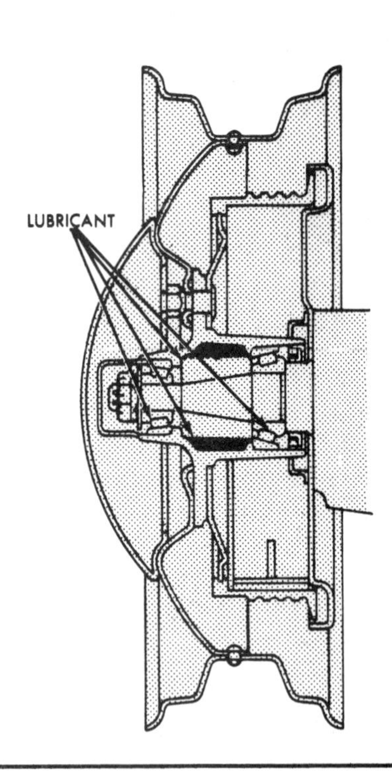

FIGURE 4-9. **Lubrication Sites on Front Wheel Hub.** (Courtesy of Ford Motor Company)

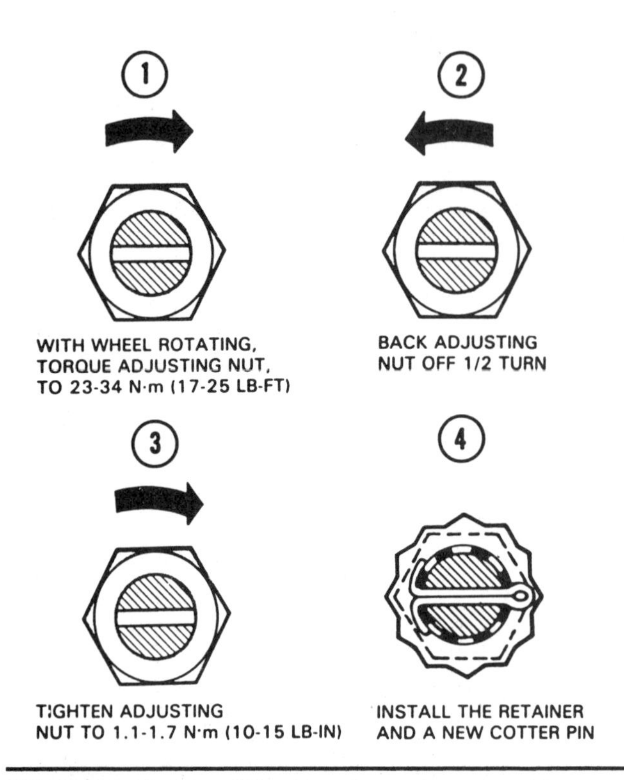

FIGURE 4-10. **Front Wheel Bearing Adjustment.** (Courtesy of Ford Motor Company)

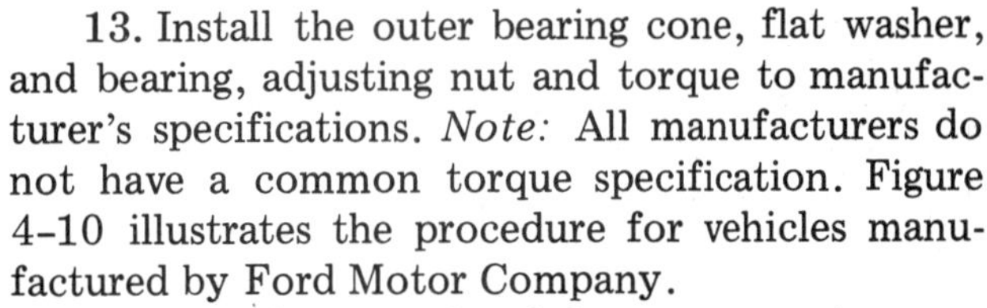

13. Install the outer bearing cone, flat washer, and bearing, adjusting nut and torque to manufacturer's specifications. *Note:* All manufacturers do not have a common torque specification. Figure 4-10 illustrates the procedure for vehicles manufactured by Ford Motor Company.

14. Cover the inside of the grease retainer cap with a light film of grease and install the cap. Mount the caliper according to the manufacturer's instructions. Install the wheel. Tighten the wheel lug bolts or nuts in correct sequence and torque to specifications (this procedure is covered in Chapter 5).

PRACTICAL SERVICE TIPS

1. When servicing and repacking front wheel bearings, use safe working practices.
2. Cleanliness is essential to wheel bearing service.
3. When inspecting, replacing, or repacking bearings, be certain that the inner cones of the bearings are free to creep (turn) on the spindle. The bearings are designed to creep in order to afford a constantly changing load contact between the cones and the rollers. Polishing and applying lubricant to the spindle will permit this movement and prevent rust from forming.
4. Make sure that the wheel bearing cups are secure in the hub. If the cups are not secure, replace the hub.
5. Make all repairs following the manufacturer's recommended procedures.
6. Always tighten wheel lug bolts, nuts, and bearings according to the manufacturer's instructions and torque specifications.

FRONT WHEEL DRIVE HUBS AND BEARINGS

Many front wheel drive cars are designed with permanently sealed front wheel bearings that require no adjustment or lubrication. Each front wheel is bolted to a hub assembly. Inside the steering knuckle's hub are two opposed and tapered roller bearings. Some manufacturers have front wheel

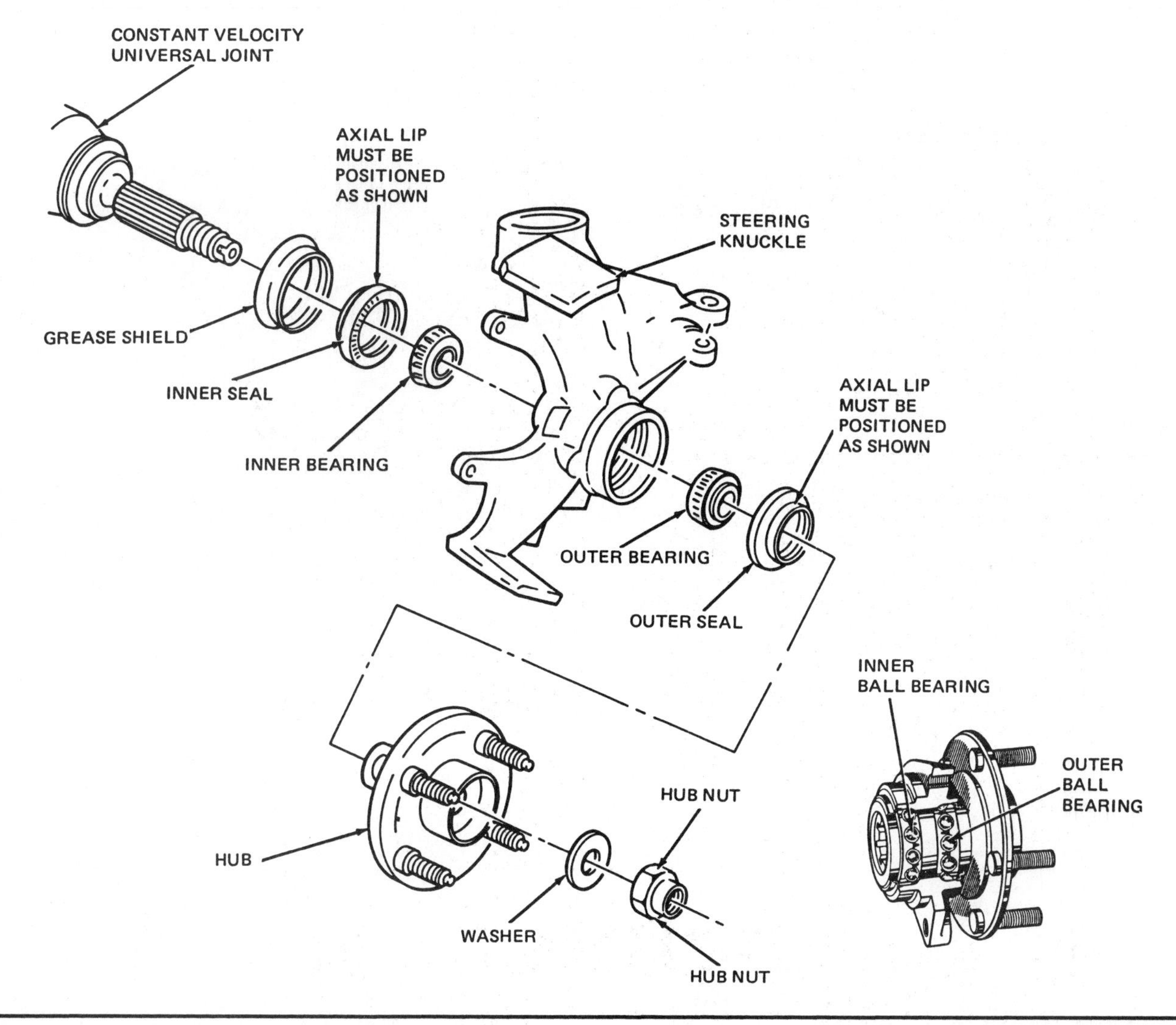

FIGURE 4-11. Front Wheel Drive Wheel Hub and Bearing Assemblies. (a) Wheel hub and bearings. (Courtesy of Ford Motor Company); (b) permanently sealed and lubricated front wheel ball bearings. (Courtesy of General Motors of Canada, Limited)

hubs designed with inner and outer ball bearings (Figure 4-11).

Regardless of the design of the bearings, they have the same purpose as the bearings discussed earlier in this chapter. To enable you to diagnose a suspected defective bearing, study and follow the procedures that are described and shown in the sealed wheel bearing diagnostic check flowchart in Figure 4-12.

If you determine that the wheel bearings need to be replaced, obtain a shop manual for that year and make of car. Follow precisely the removal service and installation. Special tools are often required to remove the hub from and to install the hub to the constant velocity universal joint splined stub shaft (Figure 4-13).

Rear Wheel Bearings on Front Wheel Drive Cars

The rear wheel bearings on many front wheel drive cars require periodic repacking. Lubrication

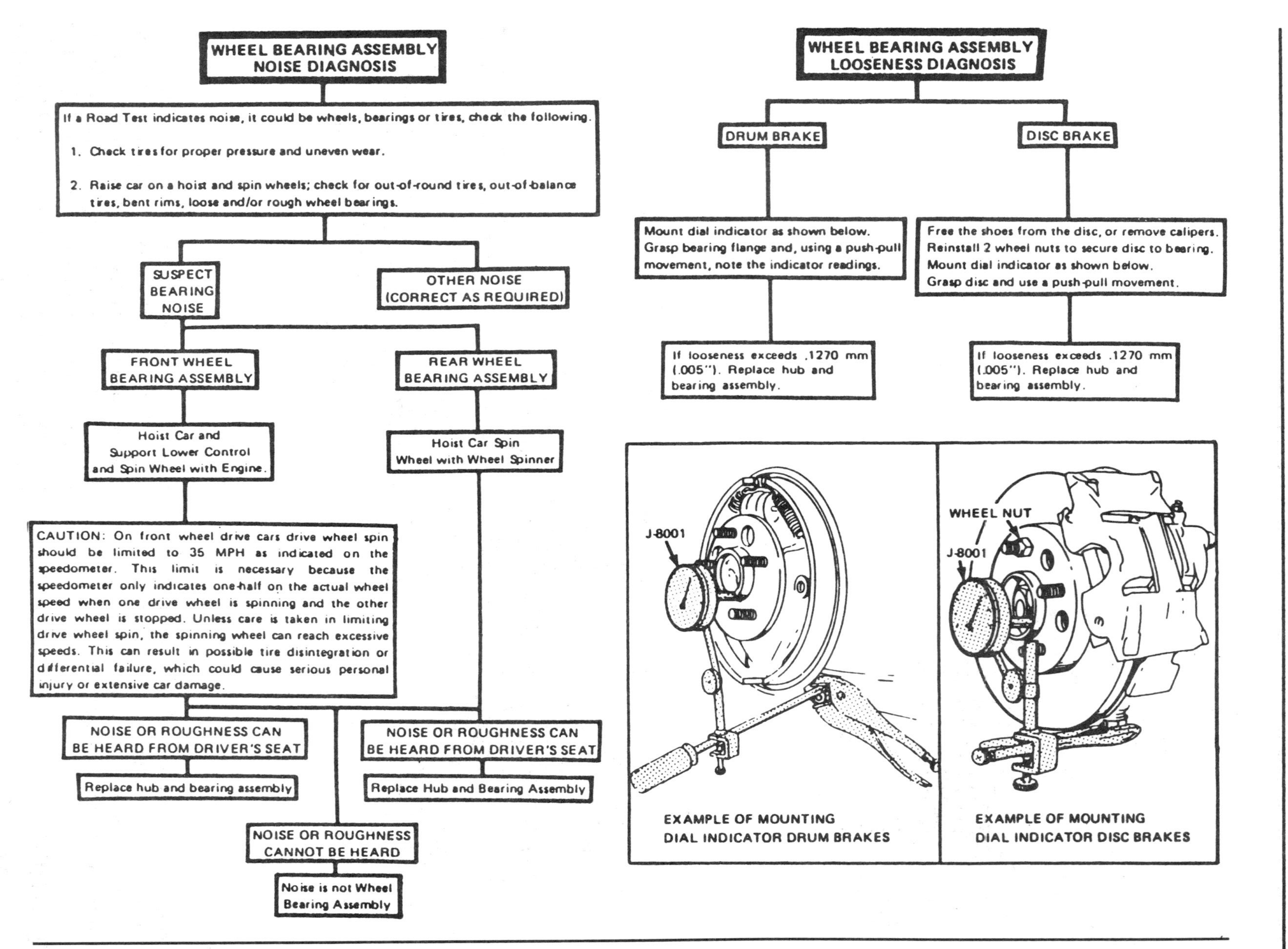

FIGURE 4-12. Sealed Wheel Bearing Diagnosis Chart. (Courtesy of General Motors of Canada, Limited)

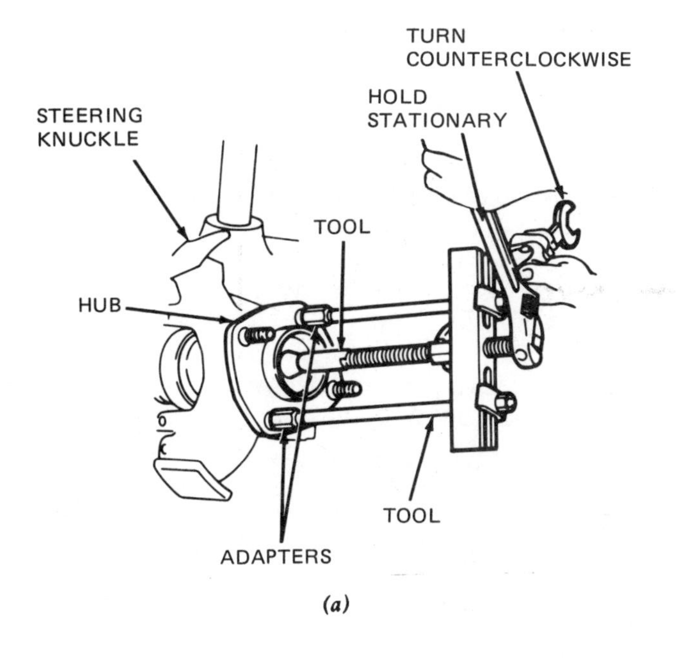

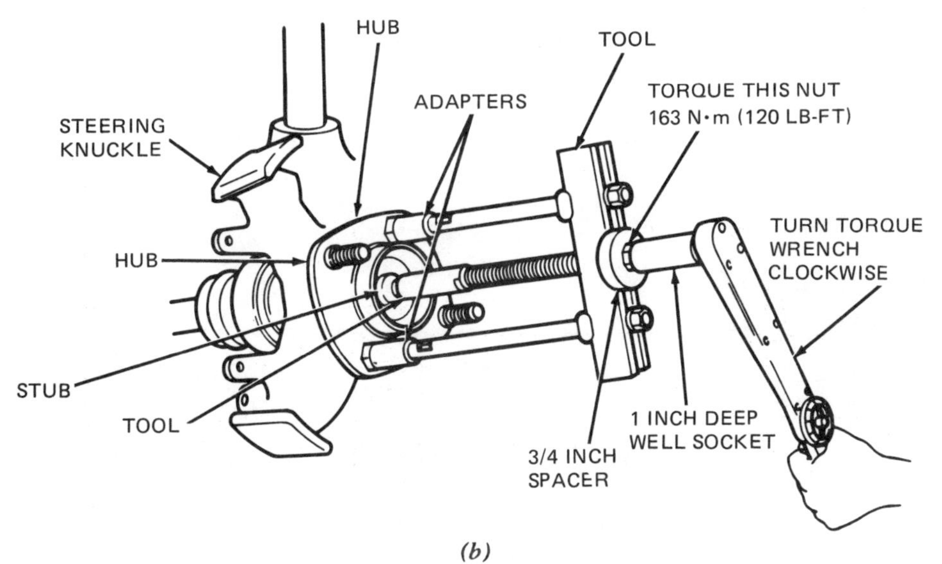

FIGURE 4-13. Replacing Wheel Bearings. (a) Removing hub from constant velocity universal joint stub shaft; (b) installing hub to constant velocity universal joint splined stub shaft. (Courtesy of Ford Motor Company)

and adjustment procedures are similar to those used with front wheel bearings on rear wheel drive cars. Service intervals, lubricants, and adjustment instructions are listed on the service charts.

REAR WHEEL AXLE BEARINGS

Rear wheel axle bearings serve the same purpose as the front wheel bearings. They reduce the revolving friction of the axle shaft and support the weight of the vehicle. Rear wheel axle bearings are not serviced in the same manner as the front wheel bearings, however. Depending on the type of vehicle, the rear wheel axle bearings may be permanently sealed (lubricated when manufactured) or lubricated by gear lube from within the rear axle housing (Figure 4-14). The rear wheel bearings on some trucks with *full-floating axles* require periodic lubrication. Lubricant may be applied through a fitting or by removing and repacking the bearings, depending on

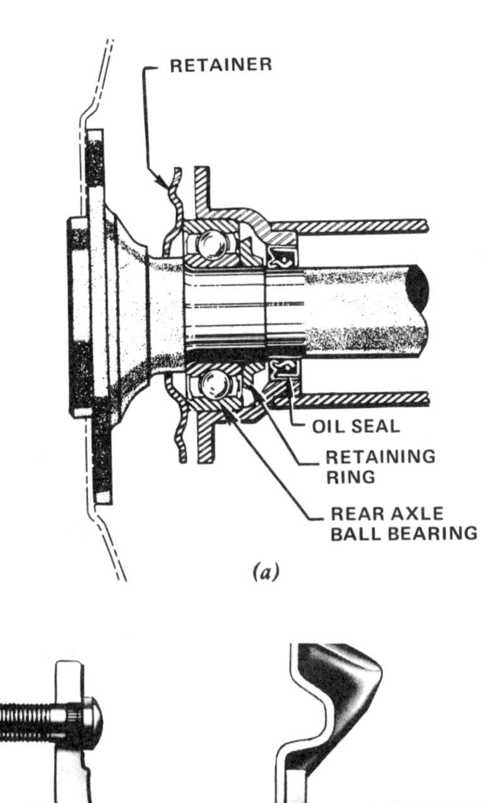

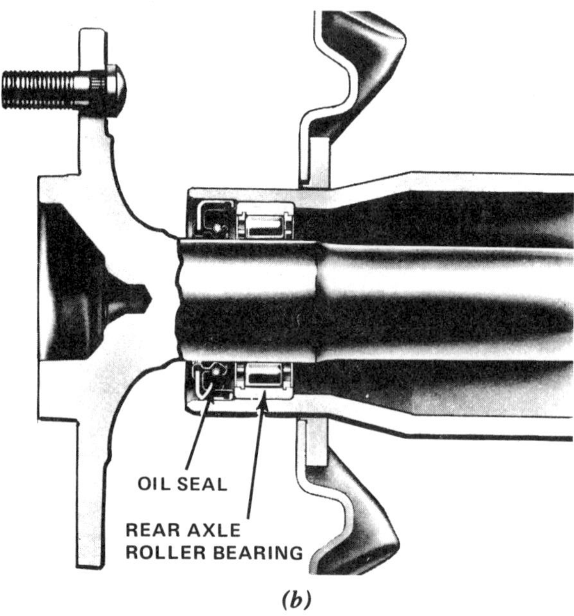

FIGURE 4-14. Two Types of Rear Wheel Axle Bearings. (a) Sealed ball bearing design; (b) roller bearing design. (Courtesy of General Motors of Canada, Limited)

axle design. Service intervals and lubricants are listed on the service charts.

Replacing Rear Wheel Axle Bearings

Rear wheel axle bearings normally require replacement at high vehicle mileage. Bearing failure at low mileage is often caused by the following problems:

- Pulling a trailer with a heavy load,
- Carrying excessive weight in the trunk,
- Overloading the permissible payload of the vehicle.

Failure may result in a partly broken weld of the rear axle housing. Unfortunately, few drivers pay any attention to load restrictions of automobiles or light trucks. A rough rear wheel axle bearing produces a vibration or growl that continues when the vehicle is coasting with the transmission in neutral. Rear wheel axle bearing rollers do not rotate or travel at the same speed as the rear axle. A defective rear axle bearing will knock or click approximately every two revolutions of the rear wheel.

If you suspect that a rear wheel drive axle bearing is defective, raise the vehicle wheels and position the rear axle housing on two safety stands. Release the parking brake. Use the engine and driveline to rotate the rear wheels. Place a long screwdriver or a rod against the axle bearing positions and listen, as with a stethoscope, for bearing noise. If you detect a roughness or a clicking sound, the bearing is probably defective and must be replaced. Follow the procedures described in the shop manual for that model and year of vehicle.

Step-by-Step Procedure

1. If a rear wheel axle bearing is in need of replacement, you may have to partially dismantle the gears inside the crown and pinion housing. Always refer to the manufacturer's shop manual for disassembly, service, and assembly procedures.

2. Inspect the gear lube for signs of gear or bearing metal fatigue. If the gear lube is silvery in color, other serious problems may exist within the gear housing.

3. Carefully clean the bearing and seal area in the housing. Remove any metal particles.

4. Do not use an acetylene torch to remove a pressed lock or bearing from the axle. You will remove the temper (strength and hardness) from the shaft. Instead, drill into or notch the retainer ring in several places with a cold chisel to break the press (Figure 4-15). Use an arbor press or puller to remove the bearing from the axle shaft or housing.

Caution: When removing an axle bearing from an axle shaft and you have the bearing positioned in a press, always wrap a heavy cloth around the bearing. If the bearing shatters during removal, the cloth will prevent the broken pieces of metal from

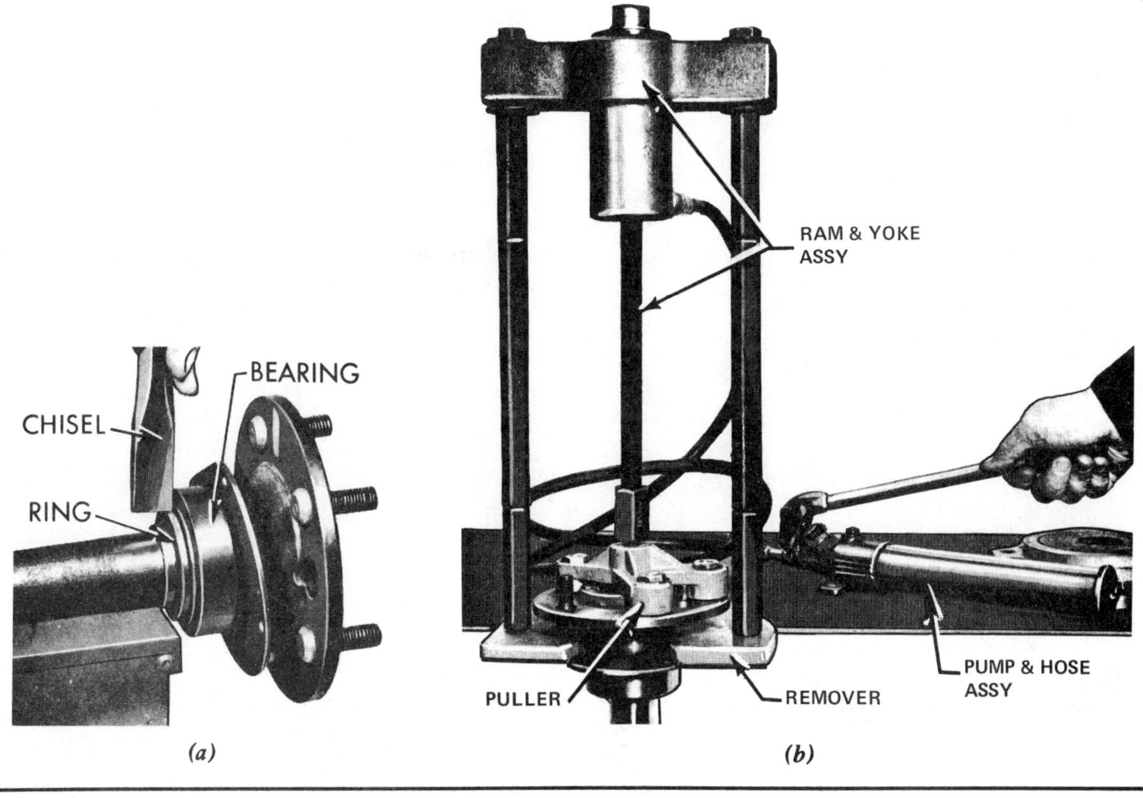

FIGURE 4-15. Removing Lock and Bearing from Axle. (a) Using a cold chisel to notch the retaining ring; (b) removing the rear axle bearing.

scattering and thus prevent injury. Removing a lock ring or bearing from a shaft destroys its press (tightness); neither can be reused.

5. Always install a new seal if it is not part of the new bearing assembly.

6. Care must be taken not to install the seal in a cocked position. Incorrect seal installation will cause loss of gear lubricant and affect the brake mechanism (Figure 4-16).

7. Check the rear axle housing breather vent to ensure that the vent hose or breather is not obstructed. If there is an obstruction, lubricant will be forced past the lip of the seal.

8. When you install the axle, tighten the wheel lug bolts or nuts in the correct sequence and torque to specifications.

REVIEW TEST

1. List six purposes for a wheel bearing.
2. Define the terms *radial load* and *thrust load.*
3. Describe the design of an inside front wheel bearing.
4. Name the major parts of a wheel bearing.
5. Explain how you would diagnose a defective front wheel bearing.
6. Explain why a wheel should be indexed to the hub.

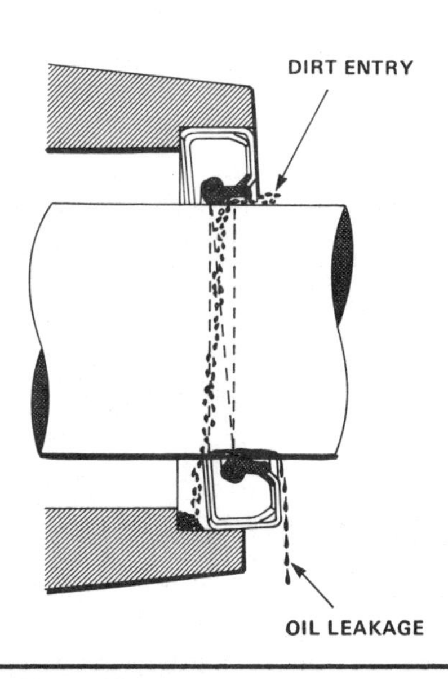

FIGURE 4-16. Results of an Improperly Installed Seal.

7. Describe how to lubricate a wheel bearing.
8. Describe how to torque a wheel bearing.
9. Define the terms *brinelling*, *galling*, *etching*, and *fretting*.
10. Explain how to diagnose a defective front bearing on a front wheel drive automobile.
11. Describe how to determine if a rear wheel drive axle bearing has excessive lateral looseness.
12. Describe how to remove a bearing from a rear axle.

FINAL TEST

This examination is multiple choice. Only one answer will be accepted. Carefully read every statement.

1. Mechanic A says that a wheel bearing must transmit a radial load when a wheel is revolving. Mechanic B says that a wheel bearing must transmit a thrust load when a wheel is revolving. Who is right? (A) Mechanic A, (B) Mechanic B, (C) Both A and B, (D) Neither A nor B.
2. Mechanic A says that the center-load line through a front wheel is very close to the large inside wheel bearing. Mechanic B says that the inside wheel bearing is larger than the outside because it supports a great deal of weight. Who is right? (A) Mechanic A, (B) Mechanic B, (C) Both A and B, (D) Neither A nor B.
3. Mechanic A says that a light application of the brake pedal will often cause a bearing noise to diminish. Mechanic B says that a front wheel bearing noise will be louder when coasting as compared to accelerating. Who is right? (A) Mechanic A, (B) Mechanic B, (C) Both A and B, (D) Neither A nor B.
4. Mechanic A says that gasoline is a suitable solvent to use when washing bearings. Mechanic B says that when the surface of a bearing is subject to galling, the bearing does not need to be replaced. Who is right? (A) Mechanic A, (B) Mechanic B, (C) Both A and B, (D) Neither A nor B.
5. Mechanic A says that an acceptable work practice is to allow a bearing to be spun with compressed air. Mechanic B says that compressed air should be directed only through the bearing. Who is right? (A) Mechanic A, (B) Mechanic B, (C) Both A and B, (D) Neither A nor B.
6. Mechanic A says that when the bearing surface of a cup is subject to brinelling, the bearing has not been rotating. Mechanic B says that when the bearing rollers are yellowish in color, the rollers have not sustained damage. Who is right? (A) Mechanic A, (B) Mechanic B, (C) Both A and B, (D) Neither A nor B.
7. Mechanic A says that when the surface of a bearing is brownish, the lubricant has been contaminated with moisture. Mechanic B says that when the surface of a bearing is subject to galling, the bearing does not need to be replaced. Who is right? (A) Mechanic A, (B) Mechanic B, (C) Both A and B, (D) Neither A nor B.
8. Mechanic A says that all vehicle manufacturers use the same torque specifications for front wheel bearings. Mechanic B says that front wheel bearings must be torqued to the manufacturer's specification for that particular make and model of vehicle. Who is right? (A) Mechanic A, (B) Mechanic B, (C) Both A and B, (D) Neither A nor B.
9. Mechanic A says that the best way to remove old grease from a hub, drum, or rotor is to immerse the part in varsol. Mechanic B says that the best way to lubricate bearings is to pack as much grease as possible into the hub. Who is right? (A) Mechanic A, (B) Mechanic B, (C) Both A and B, (D) Neither A nor B.
10. Mechanic A says that a bearing is correctly lubricated when the outside of the rollers and cage is smeared with grease. Mechanic B says that it is not necessary to apply a light film of grease to the lip of the seal before the hub is installed on the spindle. Who is right? (A) Mechanic A, (B) Mechanic B, (C) Both A and B, (D) Neither A nor B.
11. Mechanic A says that when a bearing adjusting nut is being torqued, the hub, drum, or rotor must be constantly rotating. Mechanic B says that after a bearing adjusting nut has been initially torqued, the nut is tightened further one-half turn. Who is right? (A) Mechanic A, (B) Mechanic B, (C) Both A and B, (D) Neither A nor B.
12. Mechanic A says that vehicles with front wheel drive require the front wheel bearings to be adjusted periodically. Mechanic B says that vehicles with front wheel drive do not require

that the front wheel bearings be lubricated periodically. Who is right? (A) Mechanic A, (B) Mechanic B, (C) Both A and B, (D) Neither A nor B.

13. Mechanic A says that when diagnosing a front wheel bearing noise (front wheel drive) and the lower control arm has been lifted, the speedometer must not exceed a speed of 56 kmh (35 mph). Mechanic B says that when diagnosing a front wheel bearing looseness (front wheel drive) and the dial gauge indicator exceeds 0.1270 millimeter (0.005 inch), replace the hub and bearing assembly. Who is right? (A) Mechanic A, (B) Mechanic B, (C) Both A and B, (D) Neither A nor B.

14. Mechanic A says that when replacing a rear wheel drive axle seal, the lip of the seal must face inward. Mechanic B says that the best way to loosen a rear axle bearing retainer is to notch the ring with a cold chisel. Who is right? (A) Mechanic A, (B) Mechanic B, (C) Both A and B, (D) Neither A nor B.

15. Mechanic A says that if a rear axle seal indicates leakage, always check the vent hose or breather for possible obstruction. Mechanic B says that the rear axle bearings rotate or travel at the same speed as the axle. Who is right? (A) Mechanic A, (B) Mechanic B, (C) Both A and B, (D) Neither A nor B.

Wheel Rims and Tires

LEARNING OBJECTIVES

After studying this chapter, you should be able to:

- Explain the purpose of a wheel rim,
- Illustrate and describe how to determine the dependability of a wheel rim,
- Illustrate and describe how to torque the wheel lug bolts or nuts,
- Identify the various parts of a vehicle's tire and explain the construction of bias ply, bias ply belted, and radial ply tires,
- Identify various tire tread wear conditions,
- Explain how to distinguish road and tire noises,
- Illustrate and describe how to rotate or interchange a vehicle's tires to ensure maximum tread life,
- Explain how to perform tire puncture repair service.

INTRODUCTION

Modern highway speeds require a vehicle to be designed with a low center of gravity. In order to accomplish the design requirement, the wheels of modern vehicles have been decreased in diameter size. Wheels must have sufficient strength and resiliency to carry the weight of the vehicle, transfer driving and braking torque to the tires, and withstand side thrust over a wide range of speeds and road conditions.

This chapter deals primarily with tire tread design, how to distinguish various types of tire tread wear conditions, how to diagnose road and tire noises, how to ensure maximum tread life, and how to perform tire repair service.

WHEEL RIMS

Wheel rims that are used by North American automotive manufacturers are disc wheels with a drop center design (Figure 5-1). The purpose of this design is to prevent the tubeless tire from rolling off the rim when the tire blows out. The center flange support and round rim are joined together by electrical spot welds or rivets. A wheel rim may be attached to the hub or axle by right- or left-hand threaded wheel lug bolts or nuts. Wheel rims must be replaced if they are bent, dented, or heavily rusted, leak air through the welds, have elongated wheel lug holes, or have excessive **lateral** or **radial rim runout**. Wheel rims with excessive runout may cause objectionable vibrations. Wheel runout is measured by placing a dial gauge against the tire or rim of a wheel and taking a reading from the gauge, as illustrated in Figure 5-2.

When a wheel rim and tire are being installed on the hub or rear axle, the wheel lug bolts or nuts should be tightened in the correct sequence to an initial torque specification of 20 newton-meters (15 foot-pounds). Retighten in sequence and torque to a final specification according to the manufacturer's shop manual (Figure 5-3). This procedure will eliminate wheel rim, brake drum, or brake rotor distortion.

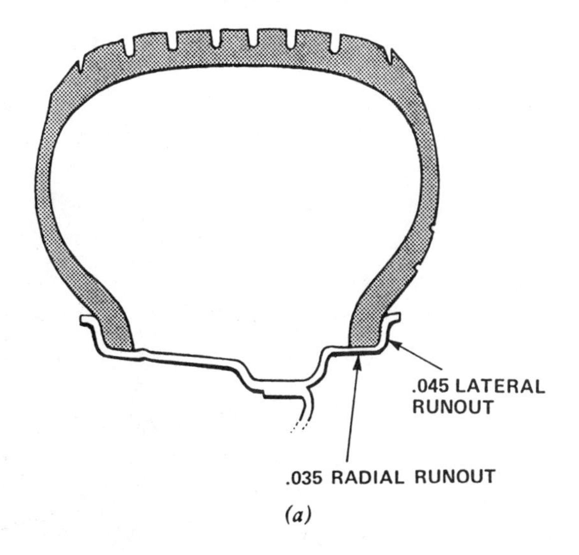

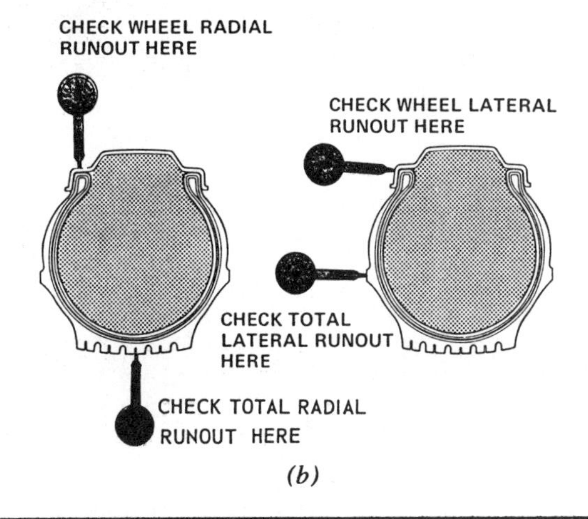

FIGURE 5-2. Wheel Rim Runout. (a) Lateral and radial runout limitations; (b) using a dial gauge to measure runout.

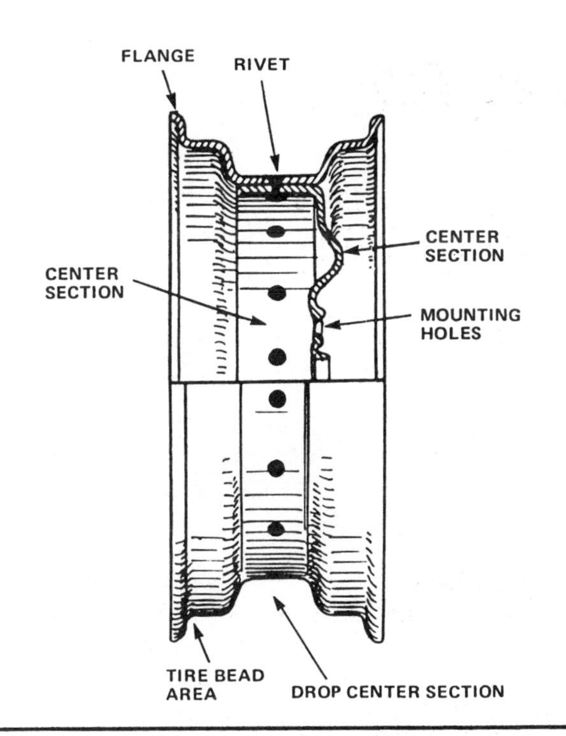

FIGURE 5-1. Typical Drop Center Wheel Rim.

TIRES

Modern tires are a tribute to engineering technology. In fact, tires are made so well that they are taken for granted. Great care is taken by all tire manufacturers because of the vital importance of the safety, steering control, and ride of a vehicle.

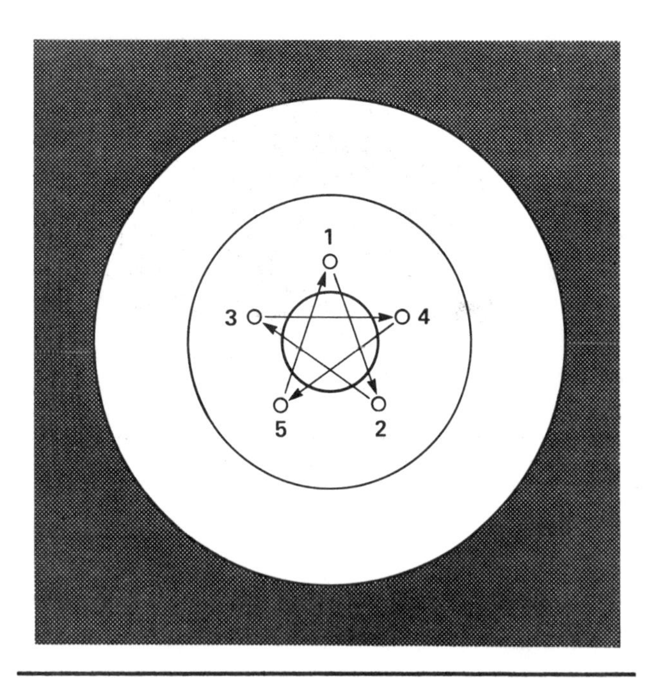

FIGURE 5-3. Proper Tightening Sequence for Lug Nuts or Bolts.

Purposes of Tires

Tires have three purposes:

1. Absorb road shock and provide a comfortable ride by its ability to flex;
2. Provide road adhesion (traction) to control the vehicle's acceleration, steering, and braking requirements;
3. Support the vehicle's weight regardless of load and external forces acting on the vehicle's body.

To realize how these purposes are achieved, it is necessary to understand how a tire is constructed, as well as the basic operation of a rolling wheel on the surface of the road, (which is discussed and illustrated in Chapter 6).

Tire Construction

A tire is made up of five principal parts: the bead, the plies (two or more of which form the carcass), the liner, the treads, and the sidewalls. Let's consider these parts in order.

The tire **bead** (Figure 5-4) is made of several turns of thin, high-tensile steel wires wound into a continuous loop for each bead. The sidewalls of a tire are looped around the beads. Therefore, the entire tire assembly is held on the rim if it becomes deflated.

The *plies*, which together form the carcass, are made of a fabric, usually rayon on original equipment tires, or rayon, nylon, fiberglass, polyester, or kelvar on replacement tires (Figure 5-5). Each ply is a layer of rubber with parallel cords imbedded in its body, called the *matrix*. The plies are wound at their ends around the beads and bonded to the sidewalls. Correct air pressure, exerted evenly against all interior surfaces, produces the tension in the carcass that resists and supports the weight of the vehicle. The inner *liner*, or membrane, provides an airtight seal and is bonded to the tire assembly (Figure 5-6). The *tread* portions of the tire provide traction. They are made of a rubber compound that is highly resistant to abrasive wear. The spaces between the treads permit some distortion of the tire on the road surface, thereby reducing the scrubbing action of the treads (Figure 5-7).

The *sidewalls* are made of a different grade of rubber, designed to help absorb irregularities in the road and to protect the cord plies from damage. The inflation pressure and stiffness of the cord material account for the rigidity of the sidewalls (Figure 5-8).

In general, the more plies a tire has, the stiffer the sidewalls. Therefore, the tire gives the ride less cushioning. Factory-installed two-ply tires have the same strength as four-ply tires because of improved materials and heavier cords with simpler and lighter construction. Six- and eight-ply tires have proportionately greater strength to handle heavier loads.

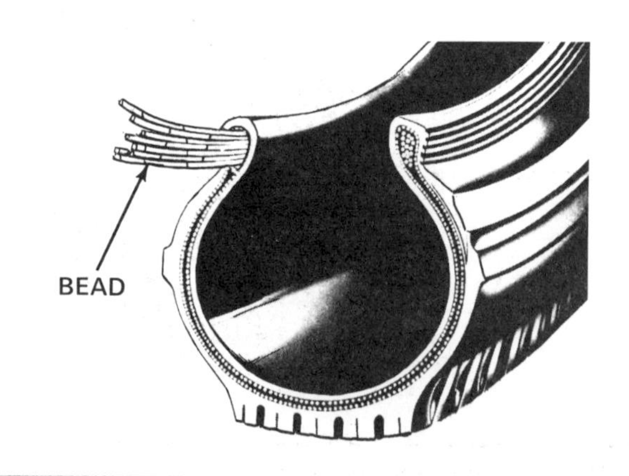

FIGURE 5-4. Tire Bead. (Courtesy of Ford Motor Company)

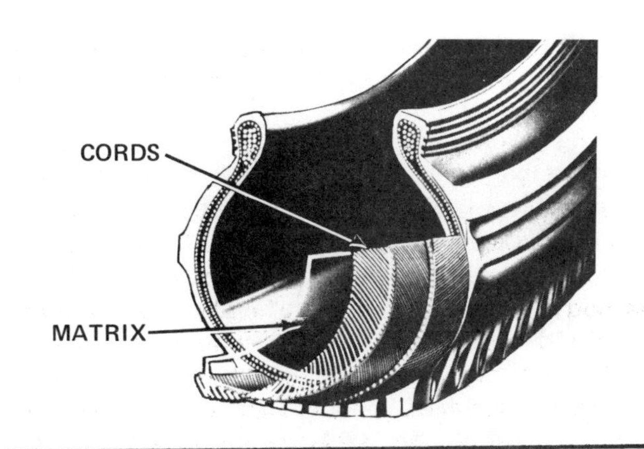

FIGURE 5-5. Tire Plies, Cords, and Matrix. (Courtesy of Ford Motor Company)

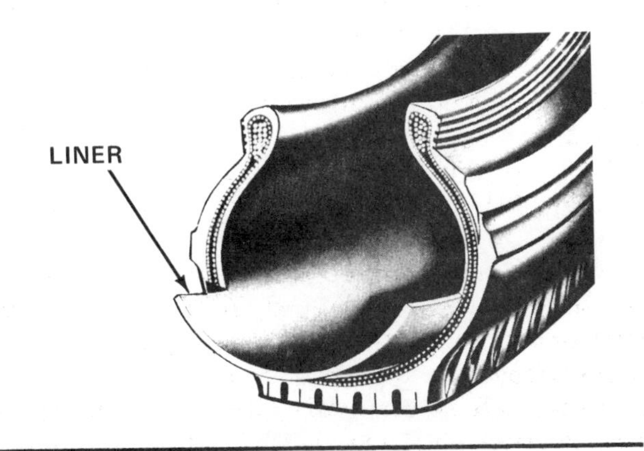

FIGURE 5-6. Tire Liner. (Courtesy of Ford Motor Company)

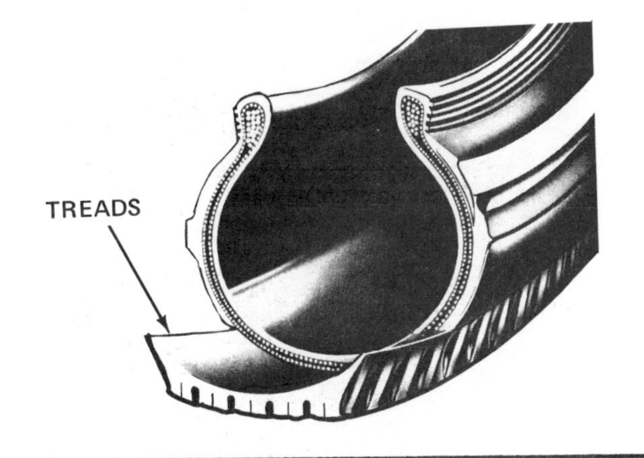

FIGURE 5-7. Tire Tread. (Courtesy of Ford Motor Company)

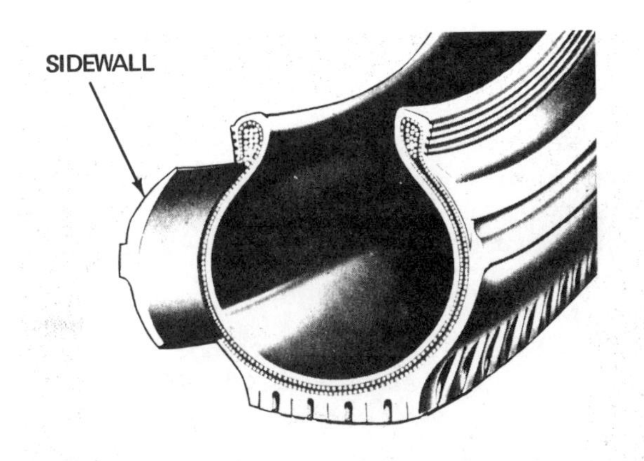

FIGURE 5-8. Tire Sidewalls. (Courtesy of Ford Motor Company)

TYPES OF TIRES

A number of new types of tires, materials, and configurations have been introduced to the motoring public since the late 1970s. Just a few years ago, the average motorist knew little, if anything, about bias ply, bias ply belted, or radial ply tires. In order to select tires properly, you need to be familiar with current tire design technology. Let's examine the three basic designs and types of tires.

Tire Design

Bias Ply. The **bias ply tire** uses the oldest type of pneumatic tire construction (Figure 5-9). *Bias* simply means that the plies are applied in a diagonal, criss-cross design from bead to bead approximately 35 degrees (35°) to the centerline of the tread.

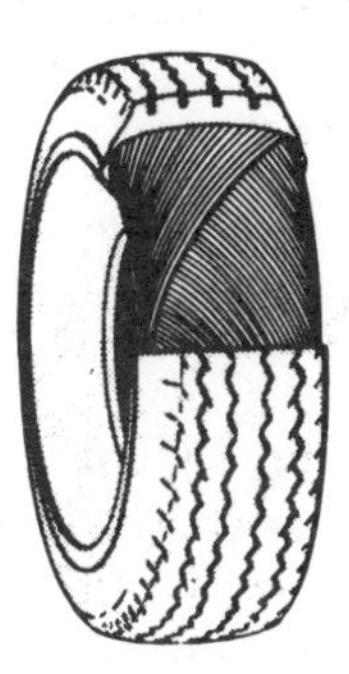

FIGURE 5-9. Bias Ply Tire Construction. Note the alternate plies running in opposite directions.

Bias Ply Belted. The **bias ply belted tire** design is very similar to the bias ply tire. As you study Figure 5-10, you will notice that belts encircle the plies below the tire's tread. This design gives greater rigidity to the tread and flexibility to the sidewall. The belts restrict tread motion during road contact, thereby greatly increasing tread life and reducing squirm.

Radial Ply. Radial ply tires have been used in North America for many years. They offer reduced wear, improved vehicle handling, and overall gas saving performance compared to the two other designs.

The tire's body cords run from bead to bead at an angle of approximately 90° to the centerline of the tire treads. The cord plies are then secured and given added strength and stability by belts. The belts are made of several layers of material, usually polyester, kelvar, or steel (Figure 5-11).

Because of their design, radial tires produce a firm ride that is somewhat harsh at speeds under 64 kilometers (40 miles) per hour but very smooth at highway speeds. All North American vehicles that have been manufactured since 1979 have suspension systems compatible with radial tires. Radial tires have handling characteristics different from conventional tires and must never be used on the same axle with other types of tires. All radial tire manufacturers subscribe to this recommendation.

Tire Coding

Federal legislation requires that all tires carry the following information: manufacturer and tire name, size designation, maximum carrying load at a certain tire pressure and limit, load range, a ten-digit DOT (Department of Transportation) serial number indicating where and when it was made, and the letter *A*, *B*, or *C*, which indicates that the tire has conformed to a uniform tire quality grading system.

Tire Size Designation. Figures 5-12 and 5-13 illustrate a metric tire size designation. The letter *P*

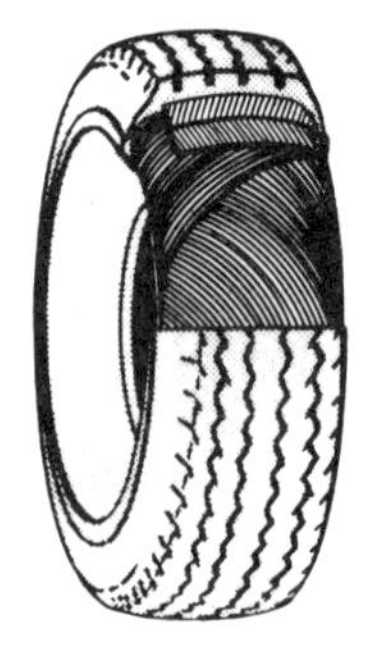

FIGURE 5-10. Bias Ply Belted Tire Construction. Note the alternate plies, plus wrap-around belts.

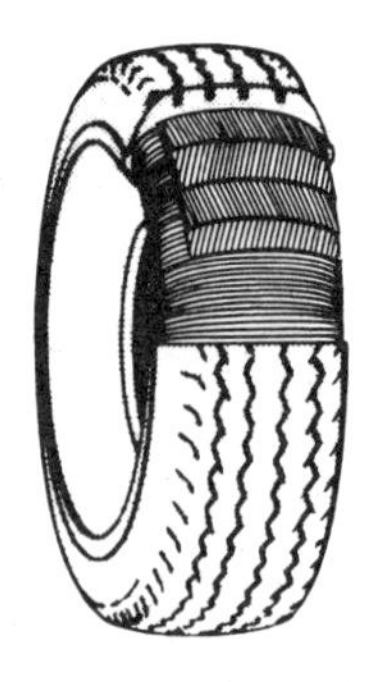

FIGURE 5-11. Radial Ply Tire Construction. Note that the cords run at 90° angles to tire's centerline.

FIGURE 5-12. Tire Sidewall Showing Metric Size Designation. (Courtesy of Goodyear Canada, Inc.)

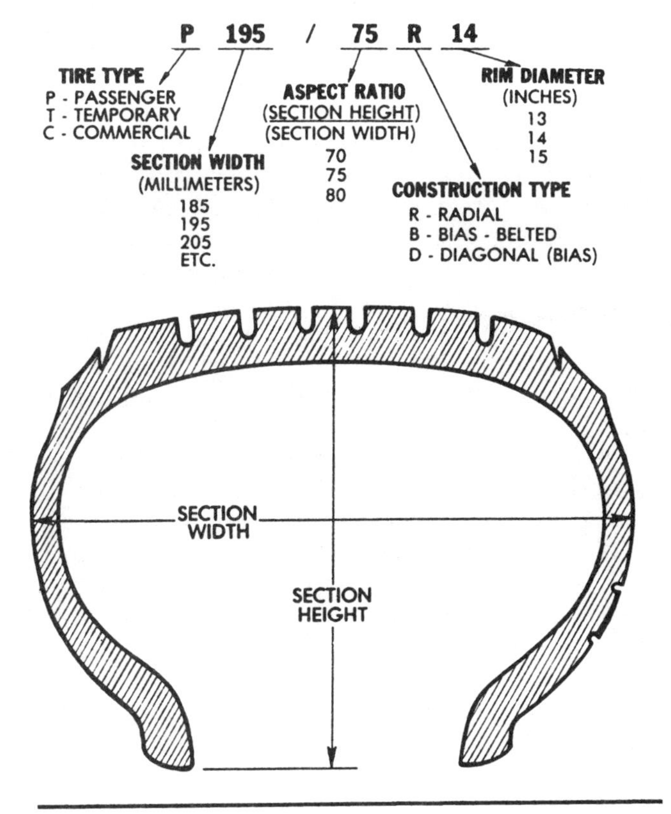

FIGURE 5-13. Metric Tire Size Format. (Courtesy of General Motors of Canada, Limited)

indicates that the tire is for passenger vehicle use. The figure *195*, expressed in millimeters, represents the cross-sectional width of the tire from sidewall to sidewall. Aspect ratio, sometimes referred to as profile height, is 75 percent of the cross-sectional width. The letter *R* designates that the tire is radial construction. The figure *14* is the diameter of the rim expressed in inches. Although not shown in Figure 5-12, load range is also stamped on the sidewall of a tire. For example, a tire rated at load range B will support 635 kilograms (1400 pounds) of weight when inflated to a pressure of 240 kPa (35 psi).

Uniform Tire Quality Grading System. The uniform tire quality grading system requires tire manufacturers to rate their passenger car tires in the areas of tread wear, traction, and temperature resistance. Grade labeling enables consumers to decide what they value most in a tire and then use the ratings as a guide in buying. Each manufacturer chooses its own design, its own materials, and its own way of making tires. The differences are somewhat technical; that is why grade labeling is so useful. It enables the buyer to compare, at a glance, the value of one tire with that of another.

Tread Wear. Tread wear grades enable consumers to compare tire life expectancies. Performance is evaluated on a specific course, approximately 640 kilometers (400 miles) in length, designed to produce tire wear similar to that encountered in general use. A test convoy is set up with one of the cars equipped with course monitoring tires to measure differences in test conditions.

To determine the tread wear grade of the tested tires, the wear rate during 11,587 kilometers (7200 miles) is measured, and the projected mileage to tire wearout on this course is computed. Projected mileage, adjusted for variations in test conditions, is compared to 48,279 kilometers (30,000 miles) on the test course. This calculation arrives at a number or ratio by which various tires may be compared.

Tread wear grades indicate that one tire can be expected to produce more mileage than another; it does not indicate exactly what that mileage will be. For instance, a tire with a grade of 150 can be expected to give 50 percent more kilometers (miles) than a tire rated at 100.

Traction. Traction ratings help consumers determine how quickly one tire will stop on wet pavement in relation to another tested under the same conditions. The tires are tested on wetted asphalt and concrete surfaces. The results are averaged out after ten skid tests on each surface. These testing procedures are carefully controlled. For example, all pavement wetting procedures are standardized to make certain that the same conditions exist for all tires that are tested. After testing is completed, the tire is graded A, B, or C.

Temperature Resistance. Temperature resistance measures how well the tire can stand up to the heat generated during its operation. Temperature tests are conducted on an indoor laboratory test wheel, according to a set of controlled conditions established by the National Highway Traffic Safety Administration.

The tire's temperature rating is determined by how long the tire lasts on the test wheel. Grade C achieves the level of performance that all passenger tires must meet under Motor Vehicle Safety Standard 109. The National Highway Traffic Safety Administration accepts C grade as adequate for driving on all U.S. highways.

Rolling Behavior

To understand a tire's *rolling behavior*, we must realize that a tire actually has not one but two diameters. The *rolling diameter*, measured by the distance of a perpendicular line through the center of the spindle, is always smaller than the *free diameter* (Figure 5-14). The load supported by the tire tends to enlarge the tire's road contact area where the tread touches the road surface. The free diameter is the corresponding distance of a straight line through the center of the spindle of a wheel with the tire properly inflated but carrying no load.

The difference between free diameter and rolling diameter is called *deflection*. The deflection causes the tread to scrub on the road as if the tread were not grooved. This condition may occur when a tire is of bias ply or bias ply belted design but seldom happens when a tire is radial ply construction. Grooves are placed in the tread to compensate for expansion and to allow water to escape from between the tread and the road surface. If water is not expelled, the part of the tread revolving in the road is subject to hydroplaning, and traction is lost.

FACTS TO REMEMBER

1. If tires of different sizes are mixed on a vehicle, it is important to keep the tires either on the front or rear of the vehicle, *not* on the same side.
2. Tires of different construction vary in profile, dimensions, ride, and handling characteristics.
3. Radial tires *must not* be mixed with bias or bias belted tires.

Figure 5-14. A Tire's Rolling and Free Diameters. (Courtesy of Ford Motor Company)

DIAGNOSING TIRE CONDITION

Tires should be inspected regularly for tread wear, cuts, or other damage. Removing nails, small stones, or bits of glass embedded in the tread will help prevent flats or costly tire damage that can lead to failure. Surveys show that while tires are involved in only 1 percent of all highway accidents, in more than half of these cases, the tires are either bald or worn to the cords. Previously damaged or abused tires are additional problems. Tires with any of the wear abuses illustrated in Figure 5-15 are considered unsafe.

Five diagnostic abilities are essential in analyzing problems related to or caused by tires:

- Ability to inspect a tire visually and recognize a tread wear problem,
- Ability to distinguish road and tire noises,
- Ability to detect tire waddle,
- Ability to distinguish a lateral pull condition,
- Ability to determine if a tire is statically and/or dynamically unbalanced (a topic discussed and illustrated in Chapters 6, 7, and 8).

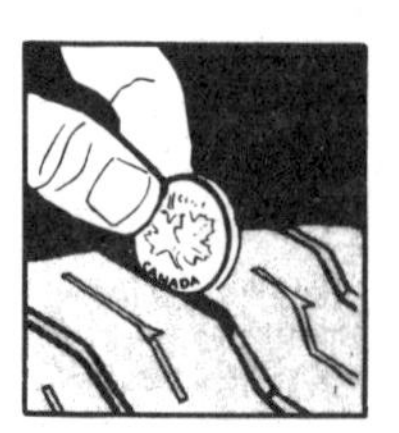

Tread worn below 1/16" depth in center grooves, or where ply cord shows. Measure depth with bottom edge of a penny. If you can see the bottom of the word Canada or United States of America, replace tires.

Tread worn down to the level of tread wear indicators, built into late model tires, which appear as solid bands across the tread surface.

Tread or sidewall cuts, cracks or snags deep enough to expose tire ply cords. Also fabric breaks or temporary blowout patches or "boot" repairs.

Bumps, bulges or knots indicating possible separation of tread or sidewall from tire body. Tire should be removed from the wheel and examined by an expert.

FIGURE 5-15. Tread Wear Abuses. (Courtesy of Rubber Manufacturers Association)

Tire Tread Wear Problems

Underinflation, Overloading, High-Speed Cornering. Tire tread wear caused by **underinflation, overloading, and high-speed cornering** is very noticeable. Underinflation or vehicle overloading decreases the rolling diameter and therefore increases the tire's road contact area. Tires that are subject to these conditions are more difficult to steer and are subject to excessive wear along the outside edges of the tread (Figure 5-16).

The principal problem with underinflation is heat buildup resulting from the flexing of the tire, not road friction. The softer the tire, the more it flexes—and the more it flexes, the higher the temperature. High temperature shortens tire life by accelerating tread wear and weakening body cords, resulting in premature tire failure (Figure 5-17). To remedy this problem:

1. Inflate the tires to the correct pressure.

FIGURE 5-16A. Tire Tread Wear Resulting from Underinflation. (Courtesy of Moog-Canada Automotive, Ltd.)

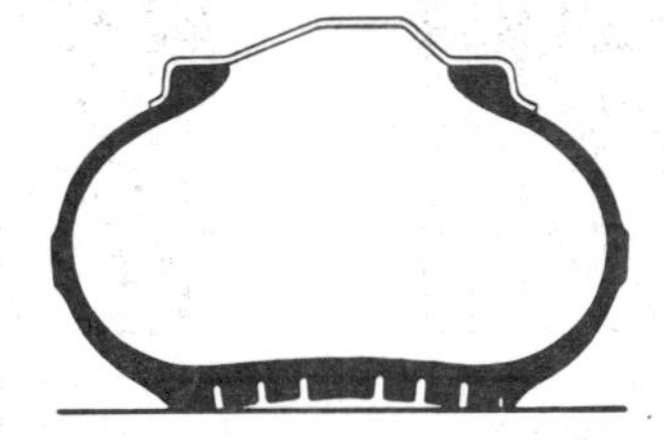

FIGURE 5-16B. Cross-Section of Tire Subject to Underinflation. When a tire has insufficient air pressure to support the load, the central area of the tread rises upward at point of road contact as sidewalls flex. (Courtesy of Moog-Canada Automotive, Ltd.)

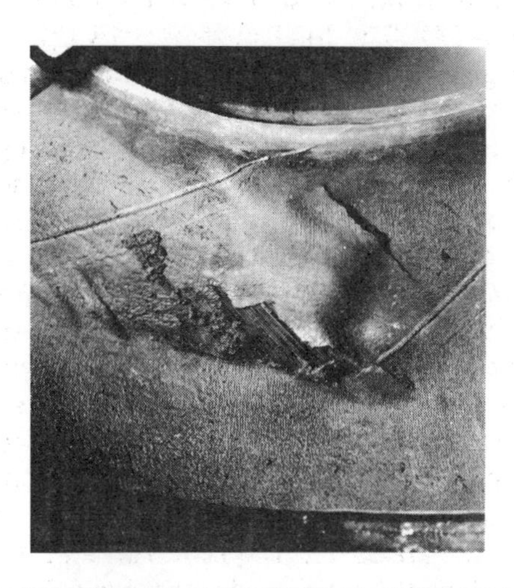

FIGURE 5-17A. Cord Breaks Resulting from Underinflation. (Courtesy of Rubber Manufacturers Association)

FIGURE 5-17B. Fabric Separation Resulting from Underinflation. (Courtesy of Goodyear Canada, Inc.)

2. Decrease the vehicle's load.
3. Suggest to the owner of the vehicle that he or she install tires with a higher load range rating.
4. If the problem is high-speed cornering, recommend that the owner slow down when cornering.

Overinflation. **Overinflation** is just as injurious to a tire as underinflation. Abnormal tread wear caused by overinflation causes the center of the tread to wear since the middle of the tread supports most of the vehicle's weight (Figure 5-18). Overinflation also decreases the tire's ability to absorb road shock by flexing, subjecting the driver to a harsh ride. Under this increased strain, the cords in the tread area eventually snap under impact, causing a break in the tire (Figure 5-19). The tire's ability to provide road adhesion is also decreased since its rolling diameter is increased. The remedy is to decrease the tire pressure to the recommended inflation pressure.

FIGURE 5-18A. Tire Tread Wear Resulting from Overinflation. (Courtesy of Moog-Canada Automotive, Ltd.)

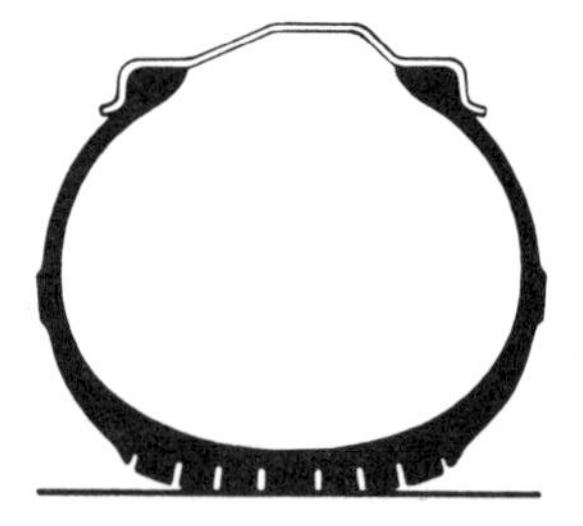

FIGURE 5-18B. Cross-Section of Tire Subject to Overinflation. Overinflation does not permit the tire tread to flatten out at point of road contact, causing the central portion of the tread to wear heavily. (Courtesy of Moog-Canada Automotive, Ltd.)

Incorrect Camber. Tire tread wear on the outside or inside portion of the tread is an indication of incorrect camber (Figure 5-20). (Camber is discussed in detail in Chapter 15.) To remedy this condition, thoroughly examine the suspension for worn or defective parts. Then have the wheel alignment checked and, if necessary, corrected.

Incorrect Toe. Incorrect toe produces a sawtooth or featheredge pattern of tread wear. Before the wear can be seen, it can be felt by rubbing your hand across the surface of the tread. One edge of each tread will feel sharp; the other edge will be round (Figure 5-21). To remedy this condition, thoroughly examine the suspension and steering system for worn or defective parts. Then have the wheel alignment checked and, if necessary, corrected.

Bias Belted Tire Tread Wear Problem

Some bias belted tires have developed a tread wear problem related to their design and construction. Upon inspecting a bias ply belted tire, you may notice that the second inside treads from both edges have developed a tread wear condition (Figure 5-22). This tread wear is neither peculiar nor related to any suspension problem. When the tire's tread is pressed against the road surface, all treads do not support an equal portion of the vehicle's weight. Because of the lighter loads on these two ribs of the tread, they tend to slip, causing a light scrubbing action. This problem can be reduced by rotating the tires at recommended mileage intervals.

FACTS TO REMEMBER

1. Maintaining correct tire inflation pressure is the most important element of good tire care.
2. The vehicle's tire pressure is carefully calculated to provide the vehicle with a satisfactory ride, directional control, steering stability, and maximum tread life.

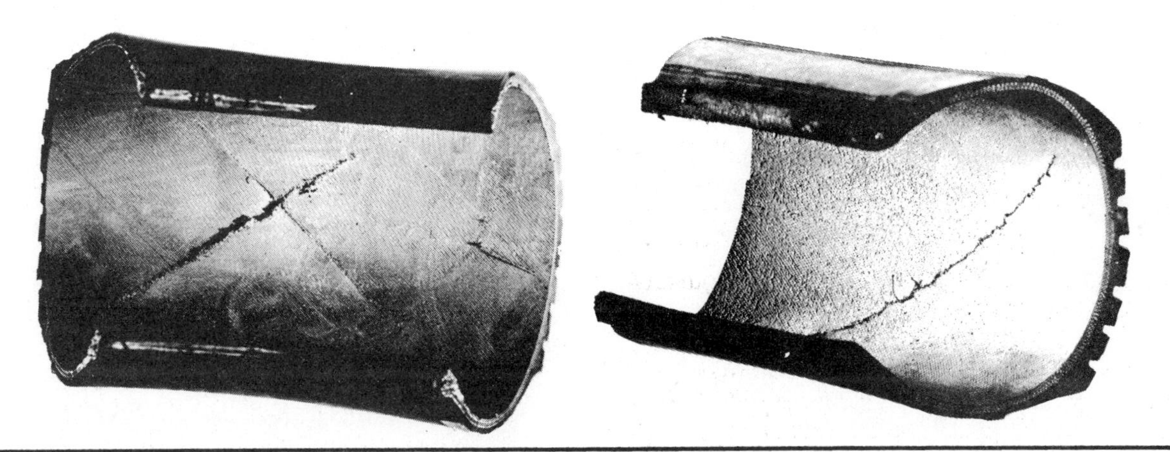

FIGURE 5-19. **Damage to a Tire Caused by Overinflation.** (Courtesy of Rubber Manufacturers Association)

FIGURE 5-20. **Tire Tread Wear Caused by Incorrect Camber.** (Courtesy of Goodyear Canada, Inc.)

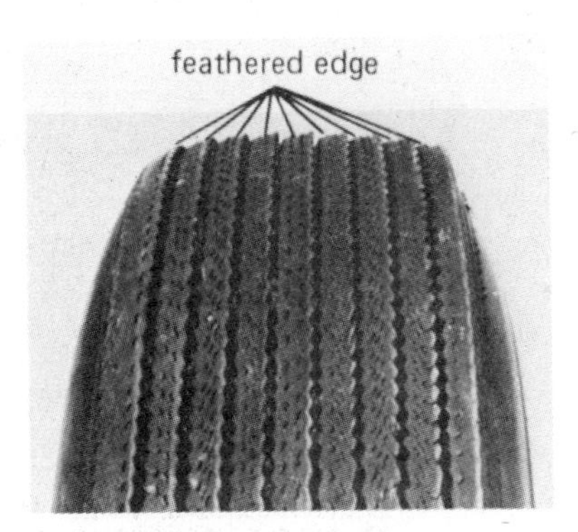

FIGURE 5-21. **Tire Tread Wear Caused by Incorrect Toe.** (Courtesy of General Motors of Canada, Limited)

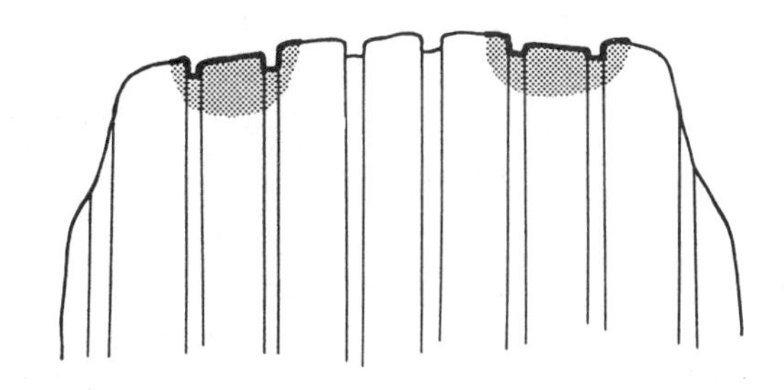

FIGURE 5-22. **Typical Tread Wear Pattern in Bias Ply Belted Tires.**

3. Never bleed air from a hot tire because the pressure is above normal. Tire pressure may be accurately checked after the vehicle has been parked for a period of 3 hours.
4. Tires should be externally examined at regular intervals for tread wear, cuts, or other damage.
5. Camber or toe wear on a tread is a reflection of other problems existing within the vehicle's suspension or steering systems.

Road and Tire Noises

Some road surfaces, such as brick or rough-surfaced concrete, cause a noise that may be mistaken for tire or rear axle noise. Driving on a different type of road (smooth asphalt or dirt) will quickly show if the road surface is the cause of the noise. Road noise is usually the same on acceleration and coast.

Tire noise may easily be mistaken for rear axle noise, even though the offending tires may be located on the front wheels. Uneven tire tread surfaces will produce noise and vibrations that seem to originate elsewhere in the vehicle. This is particularly true with low tire pressure. Tire noise changes with different road surfaces, but rear axle noise will not change. Drive the vehicle at various speeds and note the effects of acceleration and deceleration on the noise. Axle and exhaust noises will show definite variation under these conditions. Tire noise is most pronounced on smooth asphalt roads at speeds between 24 and 64 kilometers (15 and 40 miles) per hour.

Tire Thump and Vehicle Vibration

Tires are the prime cause of vehicle vibration. A tire, although mounted on a round metal rim, is actually elliptical when the tire is supporting the weight of the vehicle. If, in the manufacturing process, the tire becomes excessively elliptical, the **out-of-round** shape of the tire's tread will produce a vibration that is described as *tire thump*. This periodic condition is prominent with each revolution of the tire and is most distinguishable when the vehicle is driven on a smooth asphalt road surface. The vehicle's front suspension and body components (engine hood) are most susceptible to the vibration.

Tire thump may be diagnosed by visually examining the radial runout of the tire's tread. Another method of determining if a tire is out of round is by driving the vehicle over a smooth road surface with the tires inflated to 345 kPa (50 psi) pressure (for test purposes only). Isolate the problem by deflating one tire at a time. Vehicle vibration caused by tire thump will reoccur when you have deflated the offending tire, thereby pinpointing the problem. The following are ways to remedy the problem:

- Substitute a known good tire for a defective tire.
- Remove the offending wheel from the hub or axle and install it at a different stud position.
- Dismount the tire from the rim and remount it 180° from its present position.

Radial Tire Waddle

Tire **waddle** is side-to-side movement at the front and/or rear of the vehicle (Figure 5-23). It is caused by the steel belt's not being straight within the tire and is most noticeable at low speeds of 20 to 50 kilometers (5 to 30 miles) per hour. It may also appear as a ride roughness at speeds of 80 to 110 kilometers (50 to 70 miles) per hour.

It is possible to road test a vehicle to determine on which end of the car the faulty tire is located. If the waddle is on the rear, the rear end of the automobile will shake from side to side, or waddle. The driver will feel as if someone is pushing on the side of the car.

If the faulty tire is on the front, the waddle is more visible. The front sheet metal appears to be moving laterally from side to side, and the driver feels as though he or she is at the pivot point in the car.

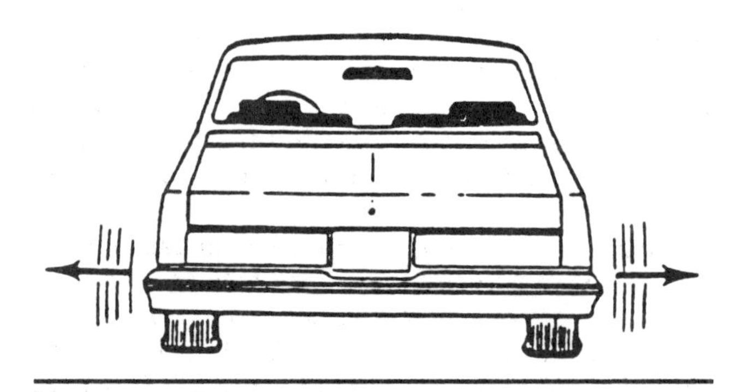

FIGURE 5-23. Radial Tire Waddle.

Steering Pull

There are two types of steering **pull**: (1) a pull that develops when the car is proceeding along a relatively straight road and (2) torque steer, a rare condition that can occur only on front wheel drive vehicles during quick acceleration. Let's discuss some of the reasons for a pull condition and how this condition may be diagnosed and corrected.

Vehicles possess directional control and stability, which means that the vehicle should steer down the road relatively straight without the need for constant steering correction. The most probable causes for loss of steering control are incorrect tire inflation, a defective suspension spring, incorrect wheel base measurement, and more often than not, incorrectly adjusted wheel alignment factors (discussed in detail in Chapters 14 to 20).

Torque steering can influence the steering wheel to turn left or right. The torque steering behavior relates to the dual torque and steering requirements that are applied to the drive axle of front wheel drive vehicles. Torque steer can be solved most often without tire replacement by moving tires to alternate axle positions. Torque steer is often dominated by vehicle causes. Tires that meet all manufacturing tolerances may still generate some level of torque steer.

To road test a vehicle and verify the condition, a moderate constant-speed test on a straight, level roadway is preferred. To verify torque steer, a moderately hard acceleration from a dead stop on a straight, level road to a speed of 80 kilometers (50 miles) per hour is recommended. The pull flowchart in Figure 5-24 can be used to assist in diagnosis.

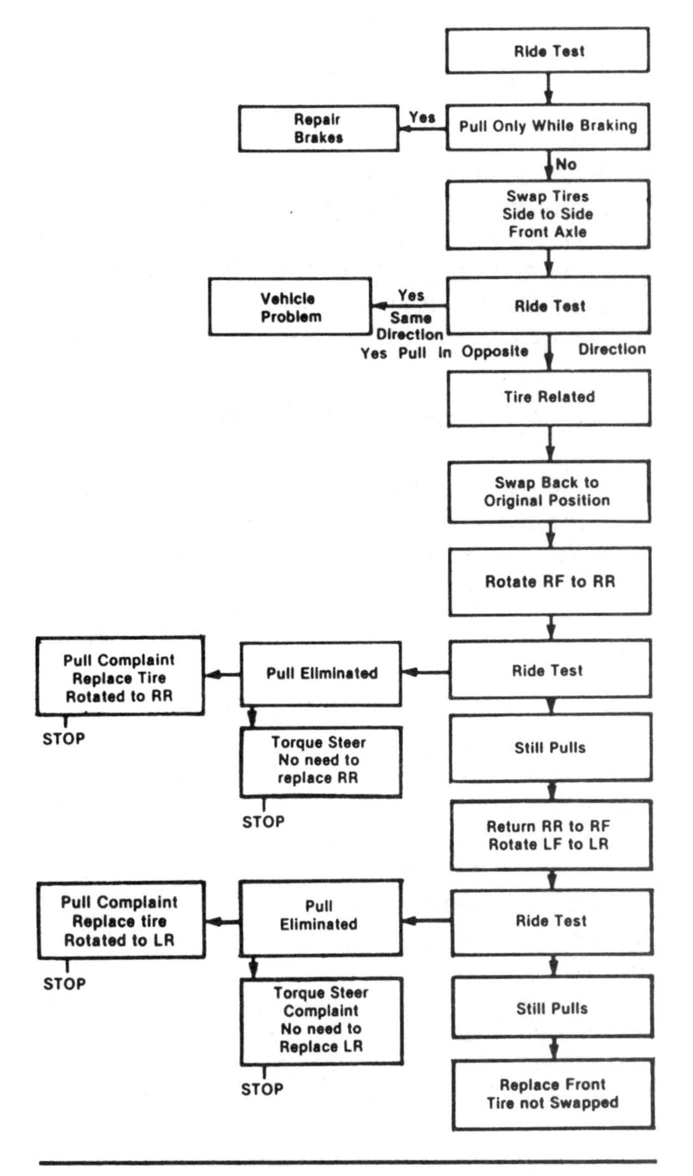

FIGURE 5-24. Steering Pull Diagnostic Chart. (Courtesy of Goodyear Canada, Inc.)

SERVICING TIRES

Tire Rotation

Vehicle manufacturers state in their service manuals that if a tire is bias ply, it may be *rotated*—that is, moved from the left front wheel to the right rear wheel. When a tire has been moved in this manner, the wheel will rotate in the opposite direction. Rotating tires may be done to extend tire tread life.

When a vehicle is equipped with radial tires, vehicle manufacturers recommend that tires not be rotated because they would be forced to revolve in the opposite direction. Instead, they should be *interchanged*—that is, moved from the left rear to the left front or from the right rear to the right front. For a number of reasons, a vehicle's front radial tires tend to wear at a faster rate in the shoulder tread area. This tread wear situation requires regular interchanging of a vehicle's radial tires.

Manufacturers recommend that the tires of a vehicle be rotated or interchanged at the first 12,000 kilometers (7,500 miles) and then at least every 24,000 kilometers (15,000 miles). Figure 5-25 shows two plans for rotating or interchanging a vehicle's tires; one for four tires and one for five tires, depending on whether they are bias ply or radial ply. *Note:* Use the four-wheel plan if the vehicle is equipped with a space-saver spare tire.

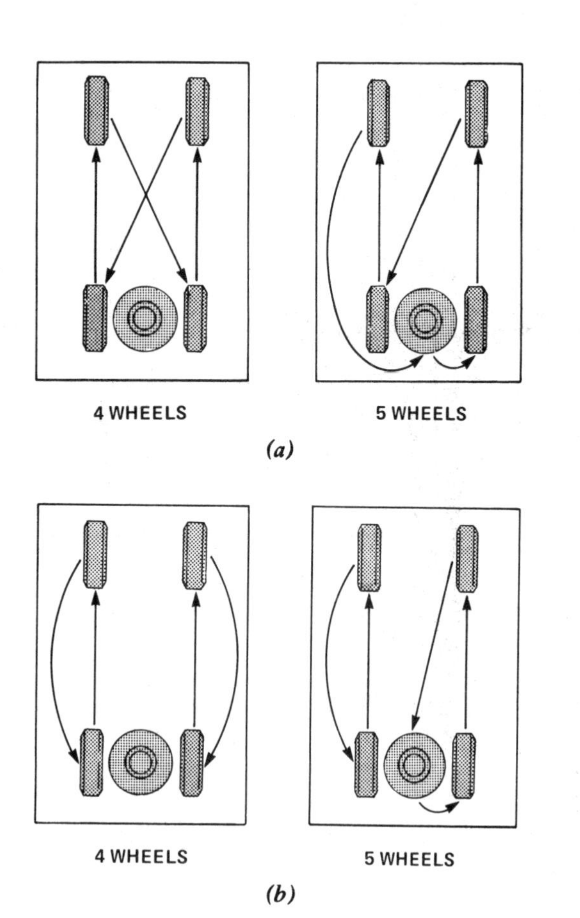

FIGURE 5-25. Tire Maintenance. (a) Rotation patterns for bias ply tires; (b) interchange patterns for radial ply tires.

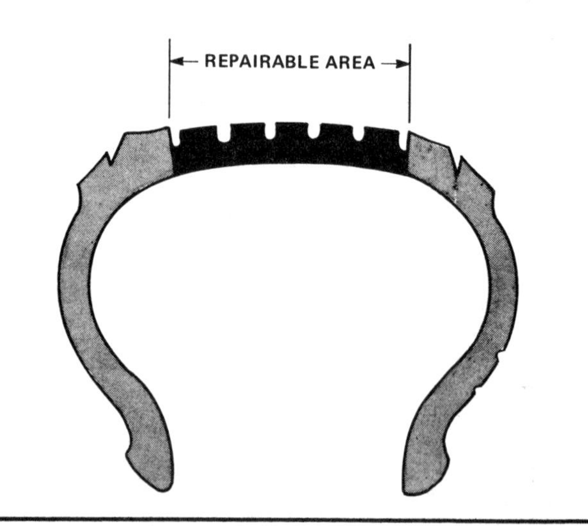

FIGURE 5-26. Repairable Tread Area.

Tire Repair

Tires with the following types of damage *cannot* be safely repaired:

- Ply separation,
- Broken or damaged bead wires,
- Loose or broken cords,
- Tread separation,
- Cracks which extend into the tire fabric,
- Sidewall punctures,
- Tires with visible tread-wear indicators.

Tires with punctures in the tread area up to 6.3 millimeters (0.250 inch) in diameter can be permanently repaired, so long as the damage is in the repairable tread area shown in Figure 5-26. Externally applied plugs, blowout patches, and aerosol-type sealants should be considered only as emergency repair measures. Tires with these types of repairs should not be driven over 80 kilometers (50 miles) per hour or for more than a distance of 160 kilometers (100 miles) before permanent tire tread repair is made.

To permanently repair a tire puncture, the tire must first be removed from the wheel rim and then the step-by-step procedures illustrated in Figure 5-27 followed. As the figure shows, the tire should be inspected and properly prepared before patching the inside of the tire liner.

Caution: When using a demounting and mounting tire changing machine, always follow the manufacturer's instructions. Tire rim bead seats should be cleaned with a wire brush or coarse steel wool to remove dried lubricants, old rubber, and light rust. Allowing air pressure to build within the assembly in an attempt to seat the bead is a dangerous practice. Inflation beyond 275 kPa (40 psi) pressure may break the bead or even the rim with an explosive force, possibly causing serious injury to the person inflating the tire. Injuries caused by unsafe work practices or unawareness of safe work practices could include severed fingers, broken arms, and severe facial lacerations. Use an extension gauge with a clip-on chuck so air pressure buildup can be closely watched and so you can stand well back from the wheel assembly during the seating process (Figure 5-28).

A. INSPECTION
Before repairing, mark injury, then remove nail or other puncturing object from tire. Before deflating, apply soap solution to the tire to determine if air loss is from one or more than one puncture. Unseat the beads and apply approved bead lubricant. Then remove tire from wheel carefully to avoid further damage to the tire, particularly to the bead, and place on spreader.

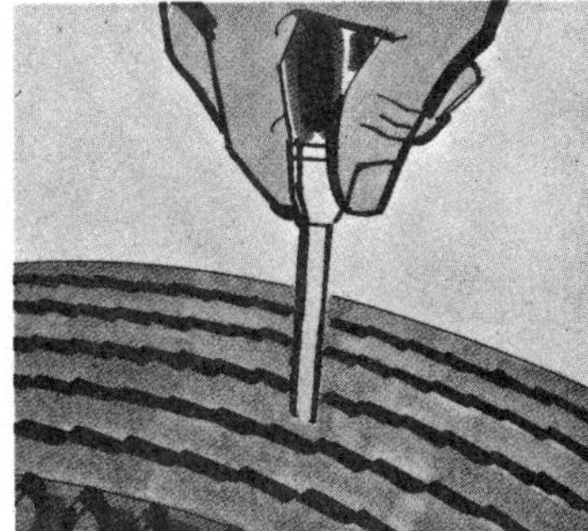

B. PROBING
Probe puncture with blunt, smooth surface awl or other hand probing tool to determine size and direction of injury, making sure no foreign material is left in the injury. Injuries exceeding 1/4" should not be repaired.

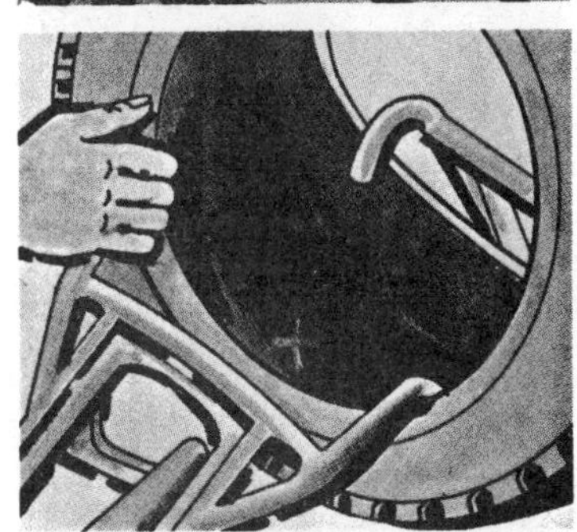

C. INTERNAL EXAMINATION
Spread the beads, marking the puncture with tire crayon. Inspect for evidence of other damage, e.g. in the bead area.

D. SELECT PROPER REPAIR MATERIAL
Center the patch over injury and outline with crayon, an area larger than the patch.

Repair materials must be selected which are recommended for the construction type (i.e. radial, belted-bias, bias) of the tire to be repaired.

E. CLEANING
Clean punctured area thoroughly with a pre-buff chemical cleaner,* covering the outlined area. This serves to remove dirt, mold lubricants, etc., also keeps buffing tools clean. Chemical cleaning is not a substitute for mechanical buffing. Make certain that no loose or frayed wire ends protrude through the liner.
*DO NOT USE GASOLINE.

F. BUFFING
Buff cleaned area thoroughly to a flat, smooth velvet surface (RMA No. 1 buffed texture for chemical vulcanizing repairs, or to RMA No. 3 buffed texture for uncured repairs),* taking care not to gouge liner or expose casing fabric. Remove dust from buffing with a vacuum cleaner. Chemical cleaning is not a substitute for mechanical buffing.

G. PUNCTURE PREPARATION
Ream puncture channel with a fine reamer from inside in a clockwise direction to prepare injury. Any indication of loose steel wires must be removed to prevent damage to the repair.

(Figure continues on next page.)

FIGURE 5-27. Step-by-Step Procedure for Tire Repair. (Courtesy of Rubber Manufacturers Association)

H. AFTER COMPLETING BASIC PREPARATION, FINISH REPAIR BY SELECTING ONE OF THE FOLLOWING REPAIR METHODS:

CHEMICAL VULCANIZING REPAIRS

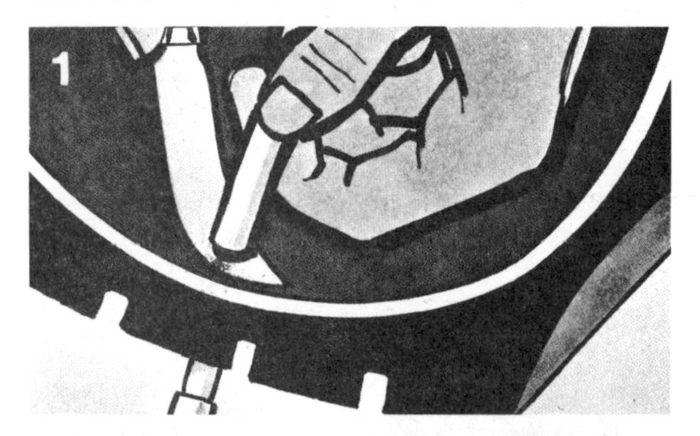

1: FILL INJURY—Cement puncture channel and fill injury from the inside with contour conforming material. Cut or buff material flush with innerliner. Follow repair material manufacturer's recommendations. It is very important to fill the injury to prevent rust of the steel wires or deterioration of fabric.

2: CEMENTING—Always use chemical vulcanizing cement recommended by the repair material manufacturer and apply a thin, even coating to the prepared and buffed surface. IMPORTANT: CEMENT MUST BE ALLOWED TO DRY THOROUGHLY. Keep dirt and other impurities from contaminating the cement remaining in the can.

3: PATCH APPLICATION—Tire must be in a relaxed position when the repair patch is being installed. (DO NOT SPREAD THE BEADS EXCESSIVELY.) Remove backing from suitable patch and center over injury. Stitch patch down thoroughly with stitching tool, working from center out.

HOT VULCANIZING REPAIRS

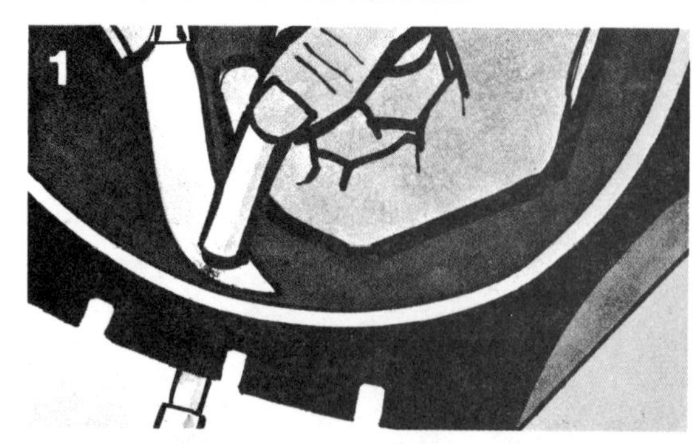

1: FILL INJURY—Cement puncture channel and fill injury from the inside with contour conforming material. Cut or buff material flush with innerliner. Follow repair material manufacturer's recommendations. It is very important to fill the injury to prevent rust of the steel wires or deterioration of fabric.

2: CEMENTING—Always use cement recommended by the repair material manufacturer. Apply a thin, even coating to the prepared and buffed surface. IMPORTANT: CEMENT MUST BE ALLOWED TO DRY THOROUGHLY. Keep dirt and other impurities from contaminating the cement remaining in the can.

3: PATCH APPLICATION—Apply hot vulcanizing and cure according to repair material manufacturer's recommendations.

FIGURE 5-27. Continued.

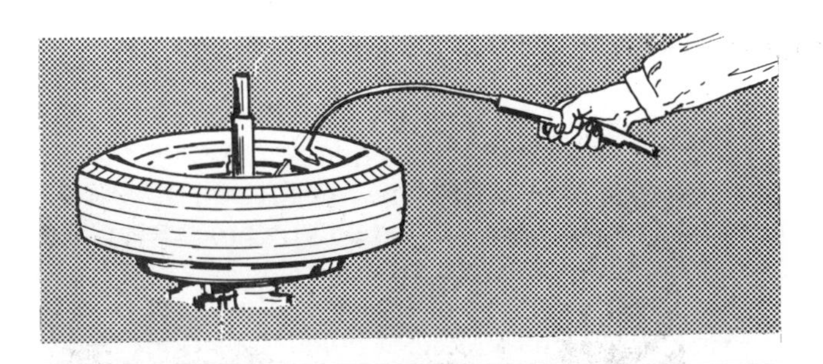

FIGURE 5-28. Using an Extension Gauge to Watch Air Pressure Buildup while Seating a Tire.

PRACTICAL SERVICE TIPS

1. It is recommended that radial tires be interchanged, not rotated.
2. Tires may be repaired only within the repairable tread area.
3. After a tire has been repaired, tire manufacturers recommend the wheel be statically and dynamically balanced.

REVIEW TEST

1. Describe how to measure the lateral and radial runout of a tire and rim assembly.
2. List the three basic purposes of a tire.
3. List the five major parts of a tire. Explain the purpose of each part.
4. Briefly explain the difference between a bias ply tire and a radial ply tire.
5. List the information found on the sidewall of a tire.
6. On the sidewall of a tire is the following marking: C 205/80 R 15. Explain what each letter and number represents.
7. What can cause wear on the outside edges of the tire tread?
8. A tire's tread has sustained wear on the inside. What are some of the probable causes? What remedies would you recommend to the vehicle's owner?
9. The owner of a vehicle is complaining of a vibration at 80 kilometers (50 miles) per hour. Explain in detail how you would diagnose the cause of the problem.
10. Describe how to repair a radial tire.

FINAL TEST

This examination is multiple choice. Only one answer will be accepted. Carefully read every statement.

1. Mechanic A says that the purpose of the drop-center part of the rim is to prevent the tire from rolling off the rim if the tire became deflated. Mechanic B says that lateral runout is measured in the center of the tire tread and that radial runout is measured against the sidewall. Who is right? (A) Mechanic A, (B) Mechanic B, (C) Both A and B, (D) Neither A nor B.
2. Mechanic A says that the purpose of a tire is to absorb road shock. Mechanic B says that the purpose of a tire is to provide road adhesion. Who is right? (A) Mechanic A, (B) Mechanic B, (C) Both A and B, (D) Neither A nor B.
3. Mechanic A says that the purpose of a tire's bead is to prevent the tire from leaving the rim if the tire becomes deflated. Mechanic B says that the purpose of a tire's liner is to provide an airtight seal and prevent leakage. Who is right? (A) Mechanic A, (B) Mechanic B, (C) Both A and B, (D) Neither A nor B.
4. Mechanic A says that when a tire is radial construction, the plies of the tire are designed on a bias. Mechanic B says that when a tire is bias ply construction, the plies of the tire are perpendicular to the beads. Who is right? (A) Mechanic A, (B) Mechanic B, (C) Both A and B, (D) Neither A nor B.
5. Mechanic A says that the cross-sectional width of a tire is actually the distance across the face of the tread. Mechanic B says that the profile height of a tire is a percentage of the tread width. Who is right? (A) Mechanic A, (B) Mechanic B, (C) Both A and B, (D) Neither A nor B.
6. Mechanic A says that the only information on the sidewall of a tire is the serial number, make, and size of the tire. Mechanic B says that Federal regulations require that the sidewall of a tire contain at least eight pieces of information. Who is right? (A) Mechanic A, (B) Mechanic B, (C) Both A and B, (D) Neither A nor B.
7. Mechanic A says that incorrect camber will cause a tire's tread to wear on the outside edge.

Mechanic B says that a tire subject to overloading will develop a featheredge wear pattern. Who is right? (A) Mechanic A, (B) Mechanic B, (C) Both A and B, (D) Neither A nor B.

8. Mechanic A says that a tire with less air pressure than is recommended will operate at a lower temperature and have more traction. Mechanic B says that a tire with less air pressure than is recommended will be subject to accelerated tread wear and weakened body cords. Who is right? (A) Mechanic A, (B) Mechanic B, (C) Both A and B, (D) Neither A nor B.
9. Mechanic A says that a defective tire will produce a noise only when the vehicle is accelerating on a rough road. Mechanic B says that the recommended way to detect a tire thump condition is to inflate each tire to 345 kPa (50 psi). Who is right? (A) Mechanic A, (B) Mechanic B, (C) Both A and B, (D) Neither A nor B.
10. Mechanic A says that when you are attempting to seat the bead of a tire, it is safe to use at least 345 kPa (50 psi) of pressure. Mechanic B says that it is safe to repair a radial tire as long as the hole is not larger than 9.5 millimeters (0.375 inch) in diameter. Who is right? (A) Mechanic A, (B) Mechanic B, (C) Both A and B, (D) Neither A nor B.

Static Wheel Balance

LEARNING OBJECTIVES

After studying this chapter, you should be able to:

- Illustrate and describe static wheel balance,
- Illustrate and describe static wheel unbalance,
- Illustrate and describe the principle and operation of a revolving wheel on the surface of the road,
- Explain the term *centrifugal force* as it applies to static wheel unbalance,
- Explain how static wheel unbalance induces destructive forces as the vehicle accelerates in speed.

INTRODUCTION

During the early years of automotive design and production, vehicles were constructed with rigid I-beam front axles and leaf spring suspension. Because of the design of the suspension and the relatively low speeds of vehicles, balancing the wheels was not considered critical to maintaining maximum tire life and vehicle performance.

People have always tried to improve whatever they make. The ride and control of a vehicle is no exception. Since the introduction of independent coil spring suspension and the increase in highway speeds, it has become essential that the wheels of a vehicle be accurately balanced.

Correct wheel balance is vital to:

- Extended tire tread life,
- Extended service suspension life,
- Safe handling performance,
- Lessened driver fatigue.

To understand how wheel balance affects the performance of a vehicle and is an integral factor in maintaining wheel alignment service life, let's start by becoming acquainted with some basic terms and definitions.

STATIC WHEEL BALANCE

Static wheel balance means that a tire and wheel assembly is evenly weighted around the center of its axis. To help you understand the definition, consider the following examples. You have used a floor jack to raise the left front wheel 2 inches off the surface of the garage floor. The brakes are not dragging, and you are able to rotate the wheel. You rotate the wheel 120° (one-third of a revolution) from point A to point B. You stop the wheel so it does not move. You again rotate the left front wheel another 120° in the same direction from point B to point C. You stop the wheel, and it stays in that position (Figure 6-1). The tire and wheel assembly is statically balanced.

To illustrate the definition further, imagine dividing a tire and wheel assembly exactly in two and placing both sections on a scale. If both sections weigh exactly the same (that is, both sides of the scale balance), then the tire and wheel assembly is statically balanced.

STATIC WHEEL UNBALANCE

Static wheel unbalance means that a tire and wheel assembly is unevenly weighted around the center of its axis. To help you understand this definition,

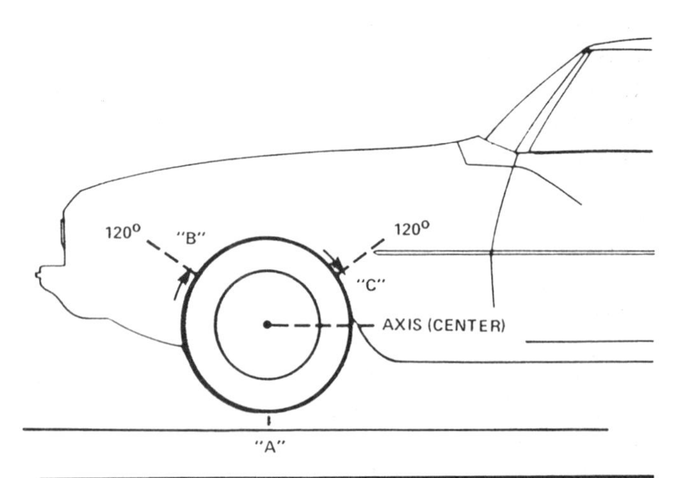

FIGURE 6-1. Static Wheel Balance. If a wheel is statically balanced, gravity will not force it to rotate from its rest position.

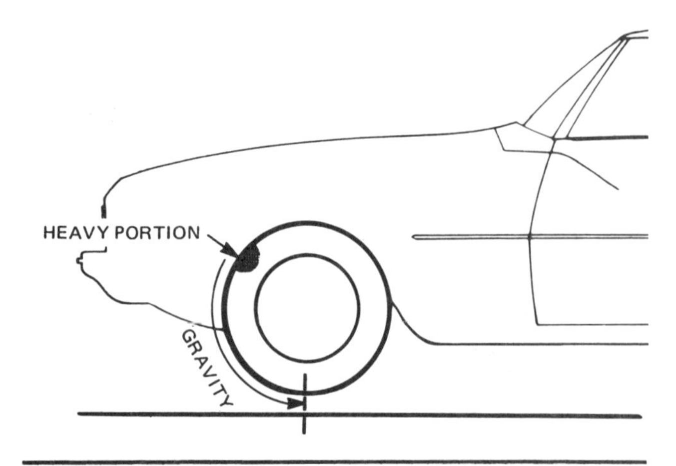

FIGURE 6-2. Static Wheel Unbalance. If a wheel is statically unbalanced, gravity will force it to rotate when the heavy portion of the wheel is not at the closest point to the ground.

consider the following examples. Again, using a floor jack, you raise the left front wheel 2 inches off the surface of the garage floor. You slowly rotate the left front wheel 120°, and then stop the wheel. This time you notice that the wheel rotates in the opposite direction and eventually stops at its original position. Gravity has caused a heavy portion of the tire to return the tire to the original position (Figure 6-2). The tire and wheel assembly is statically unbalanced.

To illustrate the definition further, again imagine dividing the tire and wheel assembly exactly in two and placing both sections on a scale. One section tilts the scale because it is heavier than the opposite section. This tire and wheel assembly is statically unbalanced.

As you study and apply the theory of static wheel balance, you will realize that a wheel is basically a lever. The light portion of a lever or wheel is obviously 180° from the heavy position (Figure 6-3).

PURPOSES OF WHEEL BALANCE

A tire and wheel assembly (based on the average wheel size and highway speed) will revolve approximately 775 to 900 revolutions per minute. A tire and wheel assembly that is statically unbalanced becomes an uncontrolled mass of weight in motion. Since tires and wheels are the chief causes of vehicle

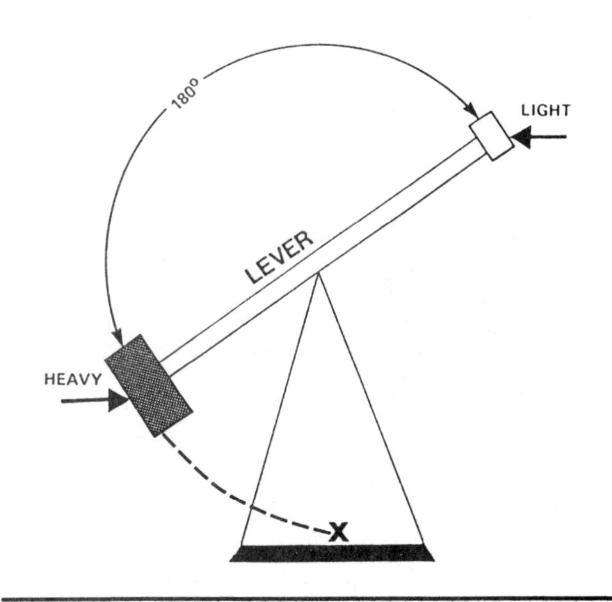

FIGURE 6-3. Unbalanced Wheel.

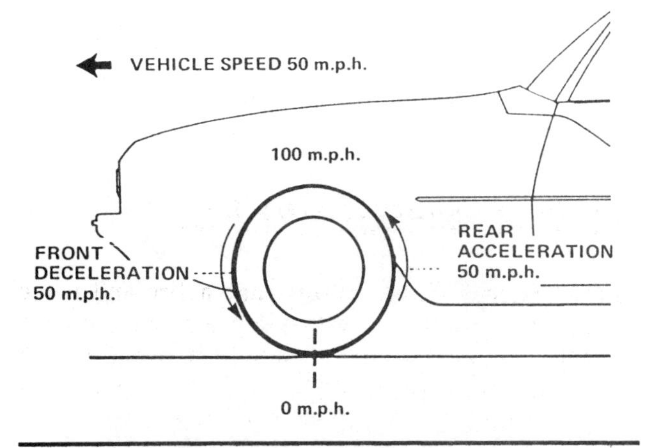

FIGURE 6-4. Wheel Assembly in Motion. The speed of a wheel at different points on its surface also varies.

vibration, it is an absolute necessity that the wheels of a vehicle be accurately balanced. A tire and wheel assembly that is balanced will provide:

1. Increased tire tread life,
2. Increased suspension life,
3. Increased body and chassis life,
4. Increased directional control and stability,
5. Reduced driver fatigue.

In order to achieve these five purposes, the wheel assembly must roll smoothly on the road surface.

BASIC PRINCIPLE OF A REVOLVING WHEEL

To realize what happens when a tire and wheel assembly revolves on the road surface, it is necessary to understand the operation of a wheel assembly in motion. Carefully study Figure 6-4 and these related facts:

1. The vehicle has accelerated to a speed of 80 km/h (50 mph).
2. The bottom of the tire tread (the portion that contacts the surface of the road) momentarily stands still in relation to the road surface regardless of the speed of the vehicle. Figure 6-4 illustrates a side view of the left front wheel.
3. The top of the tire travels at twice the speed of the vehicle and in the same direction—in this case, 160 km/h (100 mph).
4. The back portion of the tire tread travels up and forward and accelerates at a rate that will bring it from 0 km/h (0 mph) to twice the speed of the vehicle in one-half a revolution.
5. The front portion of the tire tread travels down and forward and decelerates at a rate that will bring it from twice the speed of the vehicle to 0 km/h (0 mph) in one-half a revolution.

At this point, you may find it somewhat difficult to accept the fact that the bottom portion of a tire tread stops for a split second. Consider the following illustrations. Figure 6-5 illustrates the operation of a bulldozer. The bottom portion of the track contacts the ground until the large drive gear picks up that portion of track. The bottom portion of the vehicle's track must remain stopped in relation to the ground in order to allow the bottom of the track to gain traction. If the bottom of the track were to slip, the bulldozer would not move. At point A, the track is not moving in relation to the ground. This process is called *static friction.* At point B, the top of the track is moving at twice the speed of the vehicle.

To illustrate this point further, imagine that there has been a light snowfall. You have moved your automobile out of the garage and have stopped it in order to get out and close the garage door. As you walk back to the garage, you notice the imprint left by the tires.

The tire treads left their imprint because they were rolling on the surface of the snow. A tire must momentarily stop in relation to the surface of the snow in order to leave an imprint and allow the

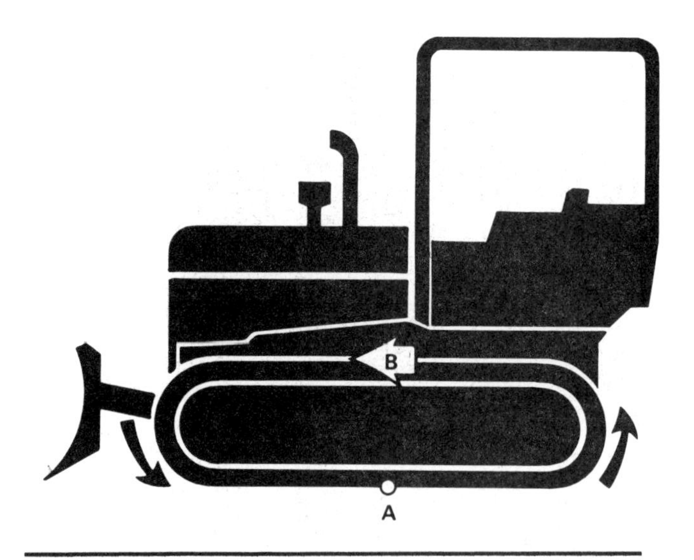

FIGURE 6-5. An Illustration of Traction. A bulldozer's track must momentarily stop when it contacts the ground in order to gain traction.

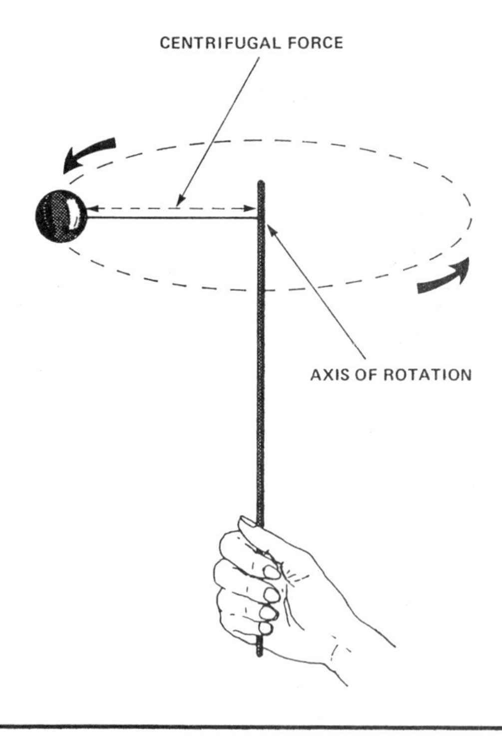

FIGURE 6-6. Centrifugal Force. Centrifugal force will tend to pull an object away from its axis of rotation.

tread to gain traction. If the tire tread were to slip on the surface of the snow, the tread would not leave an imprint, and the vehicle would not move.

Although they differ in shape, a bulldozer's track and an automobile's tire have much in common while in motion. The bottom portion that contacts the ground or road surface must momentarily stop. This momentary stop allows the bulldozer's track and the tire's tread to gain traction. Without traction, the vehicles could not move.

Effects of Static Wheel Unbalance

You have been given a definition of the term *static wheel unbalance.* In order to understand how static wheel unbalance influences a revolving wheel, it is necessary to understand the term *centrifugal force.* Centrifugal force is the force that tends to pull a rotating mass away from its axis (center) of rotation.

Figure 6-6 illustrates the action of centrifugal force on a weight connected by a string to a rotating shaft. As the speed of the shaft's rotation increases, the weight will tend to pull away from its axis of rotation.

Illustrated in Figure 6-7 is the left front wheel of a vehicle. The vehicle has accelerated to a speed of 80 km/h (50 mph). The speed of the top of the

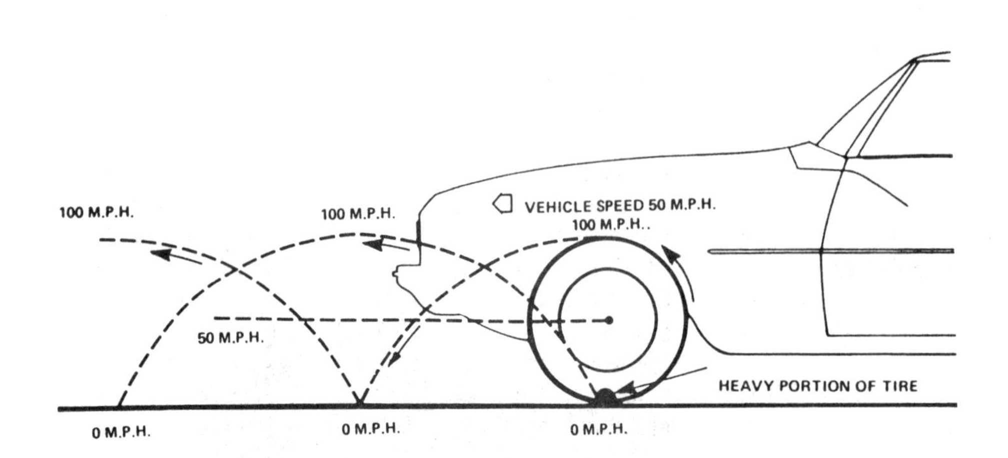

FIGURE 6-7. Normal Rotation Pattern of a Tire along a Road Surface at 80 km/h (50 mph).

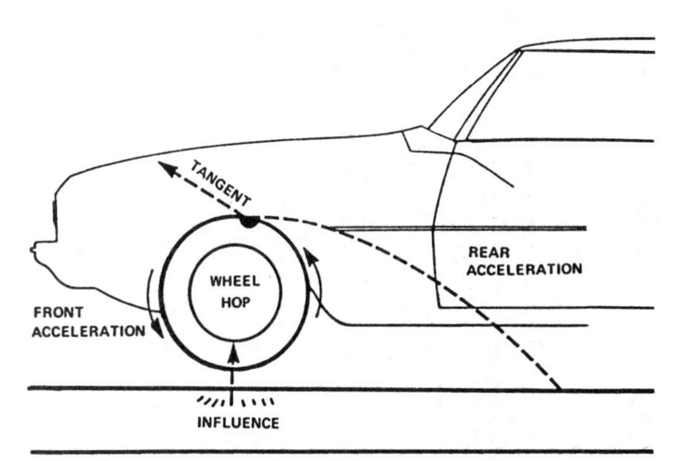

FIGURE 6-8. Wheel Hop Caused by Action of Centrifugal Force on Heavy Portion of Tire.

wheel is 160 km/h (100 mph). Notice the position of the heavy portion of the tire.

Figure 6-8 illustrates the wheel having revolved almost one-half a revolution. The heavy portion of the tire has accelerated from 0 km/h (0 mph) to almost 160 km/h (100 mph). The heavy portion of the wheel assembly, influenced by centrifugal force, endeavors to move on a tangent line away from the center of the wheel. The effect of this extra force causes the wheel to lift vertically from the road surface. This action is called **wheel hop or wheel tramp.** The vertical lift of the wheel allows the wheel assembly to spin momentarily. As the wheel spins, the heavy portion of the wheel assembly decelerates, causing the tire tread to scuff as it contacts the road surface. The scuffing causes the tire to develop flat spots (*cupping*) on the tread surface (Figure 6-9).

Additional Effects of Static Wheel Unbalance

Since a statically unbalanced wheel is subject to vertical influence, the erratic motion of the wheel is then transmitted to the vehicle's suspension and absorbed by the chassis and body. This condition causes the component parts of the vehicle to vibrate, thereby decreasing the life of the tires, suspension, steering, and body.

Table 6-1 provides proof of the destructive forces resulting from static wheel unbalance. The table shows how the destructive force increases at various speeds according to the number of ounces the wheel are out of balance.

The tire and wheel assembly of a vehicle requires only 28 grams (1 ounce) of weight to balance it

FIGURE 6-9A. Tire Tread Wear Caused by an Unbalanced Condition. (Courtesy of Moog-Canada Automotive, Ltd.)

FIGURE 6-9B. Tire Tread Wear Caused by Static Wheel Unbalance and Wheel Bearing Looseness. (Courtesy of Moog-Canada Automotive, Ltd.)

statically at 0 km/h (0 mph). Now the vehicle accelerates to a speed of 32 km/h (20 mph). Due to the effect of centrifugal force, the heavy portion of the tire has effectively increased its force to 0.39 kilograms (0.86 pound). At 112 km/h (70 mph), the effect of centrifugal force has increased the pounding force of the 28 grams (1 ounce) of weight to an astounding figure of 4.78 kilograms (10.53 pounds).

TABLE 6-1. Pounding Force at Various Speeds, by Number of Grams the Wheels Are Out of Balance

Grams (Ounces) Out of Balance	Pounding Force in Kilograms (Pounds)					
	32 km/h (20 mph)	48 km/h (30 mph)	64 km/h (40 mph)	80 km/h (50 mph)	97 km/h (60 mph)	112 km/h (70 mph)
			CARS			
28 g (1 oz)	0.39 kg (0.86 lb)	0.87 kg (1.93 lb)	1.56 kg (3.44 lb)	2.42 kg (5.34 lb)	3.51 kg (7.73 lb)	4.78 kg (10.53 lb)
57 g (2 oz)	0.70 kg (1.72 lb)	1.75 kg (3.86 lb)	3.12 kg (6.88 lb)	4.87 kg (10.74 lb)	7.01 kg (15.46 lb)	9.55 kg (21.06 lb)
85 g (3 oz)	1.17 kg (2.58 lb)	2.63 kg (5.79 lb)	4.68 kg (10.32 lb)	7.31 kg (16.11 lb)	10.52 kg (23.19 lb)	14.33 kg (31.59 lb)
113 g (4 oz)	1.56 kg (3.44 lb)	3.50 kg (7.72 lb)	6.24 kg (13.76 lb)	9.74 kg (21.48 lb)	14.02 kg (30.92 lb)	19.10 kg (42.12 lb)
142 g (5 oz)	1.95 kg (4.30 lb)	4.38 kg (9.65 lb)	7.80 kg (17.20 lb)	12.18 kg (26.85 lb)	17.53 kg (38.65 lb)	23.88 kg (52.65 lb)
170 g (6 oz)	2.34 kg (5.16 lb)	5.25 kg (11.58 lb)	9.36 kg (20.64 lb)	14.61 kg (32.22 lb)	21.04 kg (46.38 lb)	28.66 kg (63.18 lb)
			TRUCKS			
170 g (6 oz)	1.77 kg (3.90 lb)	4.00 kg (8.82 lb)	7.13 kg (15.72 lb)	11.07 kg (24.42 lb)	16.08 kg (35.46 lb)	
227 g (8 oz)	2.36 kg (5.20 lb)	5.34 kg (11.76 lb)	9.51 kg (20.96 lb)	14.77 kg (32.56 lb)	21.45 kg (47.28 lb)	
283 g (10 oz)	2.95 kg (6.50 lb)	6.68 kg (14.70 lb)	11.88 kg (26.20 lb)	18.46 kg (40.70 lb)	26.81 kg (59.10 lb)	
340 g (12 oz)	3.54 kg (7.80 lb)	8.00 kg (17.64 lb)	14.26 kg (31.44 lb)	22.15 kg (48.84 lb)	32.17 kg (70.92 lb)	
454 g (16 oz)	4.72 kg (10.40 lb)	10.67 kg (23.52 lb)	19.01 kg (41.92 lb)	29.54 kg (65.12 lb)	42.89 kg (94.56 lb)	
567 g (20 oz)	5.10 kg (13.00 lb)	13.33 kg (29.40 lb)	23.77 kg (52.40 lb)	36.92 kg (81.40 lb)	53.61 kg (118.20 lb)	

FACTS TO REMEMBER

1. Wheel balance is an integral part of wheel alignment service life.
2. Static wheel unbalance induces vehicle vibration and decreases the suspension life of a vehicle.
3. Static wheel unbalance causes a tire and wheel assembly to lift off the road surface.
4. As the speed of the vehicle increases, the effects due to the heavy portion of the wheel acted on by centrifugal force are greatly multiplied.

REVIEW TEST

1. Define the term *static wheel balance.*
2. Define the term *static wheel unbalance.*
3. List five reasons why a wheel should be balanced.
4. Describe the principle and operation of a wheel as it rolls on the road.
5. Define the term *centrifugal force.* Draw a picture illustrating this concept.
6. Explain why a tire will develop flat spots on the surface of the tread when the wheel is statically unbalanced.

FINAL TEST

This examination is multiple choice. Only one answer will be accepted. Carefully read every statement.

1. Mechanic A says that static wheel unbalance causes wheel hop or wheel tramp. Mechanic B says that static wheel unbalance decreases tire tread life. Who is right? (A) Mechanic A, (B) Mechanic B, (C) Both A and B, (D) Neither A nor B.
2. Mechanic A says that a revolving wheel must possess static friction to achieve traction. Mechanic B says that the tread of a revolving wheel does not stop in relation to the road surface. Who is right? (A) Mechanic A, (B) Mechanic B, (C) Both A and B, (D) Neither A nor B.
3. Mechanic A says that if a tire and wheel assembly is statically balanced, it has even weight around its axis. Mechanic B says that when a wheel revolves, the tire tread is subject to acceleration and deceleration forces. Who is right? (A) Mechanic A, (B) Mechanic B, (C) Both A and B, (D) Neither A nor B.
4. Mechanic A says that wheel hop is the horizontal movement of the tire and wheel assembly. Mechanic B says that the top of a tire travels at twice the speed of the vehicle. Who is right? (A) Mechanic A, (B) Mechanic B, (C) Both A and B, (D) Neither A nor B.
5. Mechanic A says that the effects of static unbalance remain the same regardless of vehicle speed. Mechanic B says that the effects of static unbalance are most critical at a speed of 50 km/h (30 mph). Who is right? (A) Mechanic A, (B) Mechanic B, (C) Both A and B, (D) Neither A nor B.
6. Mechanic A says that wheel balance and wheel alignment are an integral part of alignment service. Mechanic B says that wheel balance has very little influence on a vehicle's directional control and stability. Who is right? (A) Mechanic A, (B) Mechanic B, (C) Both A and B, (D) Neither A nor B.
7. Mechanic A says that the light portion of a wheel is always 120° from the heavy portion. Mechanic B says that a wheel is essentially a lever; the heavy and light portions are opposite in position. Who is right? (A) Mechanic A, (B) Mechanic B, (C) Both A and B, (D) Neither A nor B.
8. Mechanic A says that static wheel unbalance produces baldness on the inside half of the tire's tread. Mechanic B says that static wheel unbalance produces feathering across the face of the tire's tread. Who is right? (A) Mechanic A, (B) Mechanic B, (C) Both A and B, (D) Neither A nor B.
9. Mechanic A says that static wheel balance increases the suspension life of a vehicle. Mechanic B says that static wheel balance increases a vehicle's body and chassis life. Who is right? (A) Mechanic A, (B) Mechanic B, (C) Both A and B, (D) Neither A nor B.

Balancing a Wheel

LEARNING OBJECTIVES

After studying this chapter and with the aid of the necessary tools and equipment, you should be able to:

- Demonstrate how to statically balance a wheel of an automobile by using a bubble wheel balancer,
- Describe the use of the other portable types of wheel balancing equipment.

INTRODUCTION

This chapter illustrates and describes how to use portable wheel balancing equipment. The equipment is not difficult to use. With some instruction and practice, you will be able to balance wheels accurately and quickly. If you are a student in an automotive technical school or college or serving an apprenticeship in a business, ask your instructor or employer if you may be allowed to use the equipment to learn how to statically or **kinetically balance** a wheel.

When you balance a wheel, you will need the following tools and equipment:

- Static wheel balancer,
- Automotive floor jack,
- Vehicle safety stands,
- Wire cleaning brush,
- Wheel weight pliers,
- Tire and wheel runout gauge,
- Tool box,
- Newton-meter/foot-pound torque wrench,
- Tire pressure gauge,
- Shop service manual.

PRELIMINARY CHECKS

A vehicle with new tires should be driven at least 160 kilometers (100 miles) before you attempt to balance the wheels. This driving will allow the tires to take on their correct shape.

Since tires are the main cause of vehicle vibration, considerable time can be saved by performing the following preliminary checks:

- Lateral tire/rim runout,
- Radial tire/rim runout,

- Front wheel bearing adjustment,
- Tire tread and sidewall inspection,
- Tire inflation,
- Objects inside tire.

To make these checks, the vehicle must be raised. Place a floor jack under the front crossmember, and lower the control arm or rear axle housing so that a wheel is raised approximately 12 centimeters (5 inches) from the floor.

Checking Lateral Runout. The lateral runout of the wheel assembly is determined by positioning a dial gauge indicator against the sidewall of the front tire and then slowly rotating the wheel (Figure 7-1). If the manufacturer's specifications are exceeded, you should then determine the lateral runout of the wheel rim. If the rim's runout specifications are exceeded, either the tire or wheel rim must be replaced. When the manufacturer's tire/rim runout specifications are exceeded, the vehicle will probably vibrate.

Checking Tire Inflation. Proper tire inflation is often overlooked prior to wheel balancing. Keep in mind that the tire can maintain its correct shape and support the vehicle's weight only if the tire is correctly inflated. A higher tire inflation pressure than is recommended will cause a harsh ride, tire bruising, carcass damage, and rapid wear at the center of the tread. Low tire pressure will produce tire squash, hard steering, damage to the rim, high temperatures, and rapid wear on the outer edges of the tire. These conditions will adversely affect vehicle handling and possibly lead to a sudden failure that could result in a loss of vehicle control without warning.

Check the air pressure in the tire and inflate it to the manufacturer's specification. A tire placard is often located on the vehicle's left front door edge or on the inside of the vehicle's glove box. The tire placard specifies the maximum load-carrying capacity for the vehicle, the correct tire size, and the cold tire pressures for both front and rear tires.

Checking for Objects inside Tire. If an object is inside a tire, the tire's heavy spot will be constantly changing, and the tire therefore will be impossible to balance. Before removing the wheel from the hub or axle, slowly rotate the wheel and listen for an object revolving inside the tire. If there is an object, demount the tire, remove the object, and reinstall the tire.

Checking Radial Runout. The radial (elliptical) runout of the wheel assembly is determined by positioning a dial gauge indicator against the tread surface of the tire and then slowly rotating the wheel (Figure 7-2). If the manufacturer's specifications are exceeded, it is then necessary to dismount the tire from the rim and determine the radial runout of the rim. If rim radial runout specifications are exceeded, the tire or wheel rim must be replaced.

FIGURE 7-1. Measuring Lateral Runout. (Courtesy of FMC Corporation, Automotive Service Equipment Division)

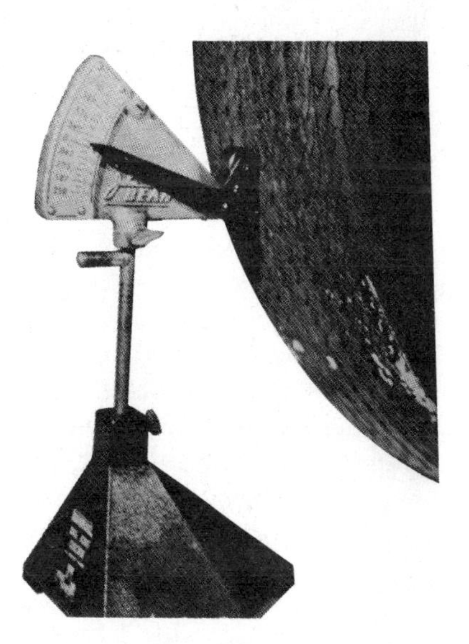

FIGURE 7-2. Measuring Radial Runout. (Courtesy of FMC Corporation, Automotive Service Equipment Division)

Checking Front Wheel Bearing Adjustment. Since North American vehicles use tapered roller bearings, it is a good procedure to determine the preload of the bearings. Obtain the front bearing torque specifications from the manufacturer's shop manual and adjust as per instructions. If wheel bearings are not torqued properly, the wheel will shake laterally, and this condition will contribute to an unbalanced wheel.

Inspecting Tire Tread. There is no point in balancing a wheel if the tire's tread is worn beyond specified limitations. If the tire's tread wear bars are visible, the tire must be replaced (see Figure 5-12). The sidewalls of a tire also must be examined for bulges, blisters (both indicating ply separation), fabric cracks, and cuts. If any of these conditions is present, the tire must be replaced.

OFF-THE-CAR STATIC BALANCING

If after careful examination, you determine that the tire is acceptable for balance, apply the vehicle's parking brake, which enables you to loosen the wheel lug bolts or nuts only two revolutions. Raise the vehicle and position it on safety stands. Remove the wheel from the hub or axle. Use the wheel weight pliers to remove any existing weights from the wheel rim. Clean the inside of the rim by using the wire cleaning brush, and remove all rocks and gravel from the tire tread.

Bubble Wheel Balancing Equipment

Position the static wheel balancer on a flat surface of the workshop floor. Use the adjustable legs at the base of the equipment to center the spirit level. The small triangular mark on the face of the spirit level indicates the position for the valve stem (Figure 7-3). Align the valve stem with the triangular mark. Do not disturb the equipment as you position the wheel assembly on the balancer.

Once you have positioned the wheel assembly on the balancer, push the release lever down and allow the wheel to settle. The spirit level will indicate where the wheel assembly is light in weight. The area of the wheel assembly requiring weight will be located 180° from the heavy portion of the wheel assembly (Figure 7-4).

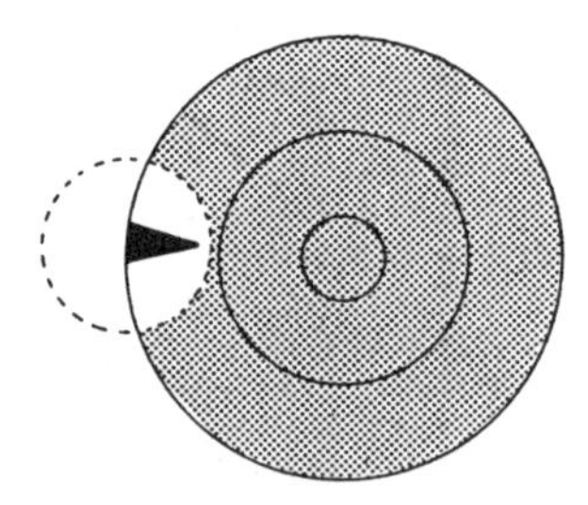

FIGURE 7-3. **Spirit Level of a Static Wheel Balancer.** The triangular mark indicates the position for the valve stem.

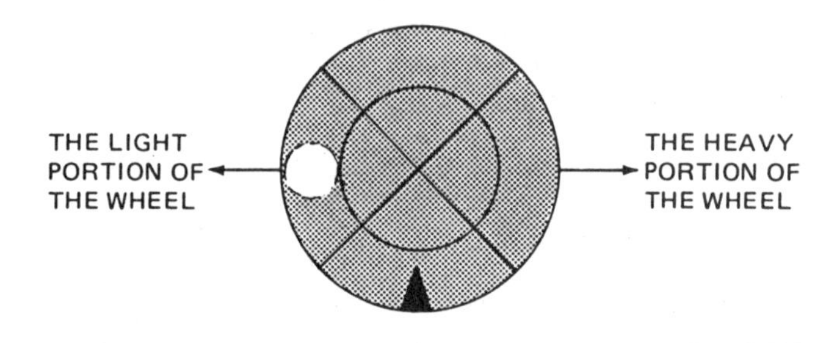

FIGURE 7-4. **Spirit Level Indicating the Light Side of Wheel.** (Courtesy of the Bada Company)

Two static wheel balancing methods use bubble wheel balancing equipment. Some manufacturers of this equipment recommend the following procedure.

Step-by-Step Procedure

1. Place sufficient weight at the tire's bead (in line with the bubble in the spirit level) to balance the wheel.
2. Chalk mark the position of the required weight on the sidewall of the tire from the outside bead to the inside bead.
3. Divide the total amount of weight by 2.
4. Carefully remove the wheel from the wheel balancer.
5. Use the wheel weight pliers and attach half the required weight to the inside of the wheel rim and bead at the chalk mark. *Note:* If the wheel rim is made of magnesium, you will require a **stick-on lead weight.** The weight must be attached to the inside and middle of the wheel rim (Figure 7-5).
6. Carefully reposition the wheel onto the balancer (don't forget to align the valve stem with the mark). Push the lever down.
7. Position the other half of the required weight at the chalk mark and determine if the wheel is still balanced. Increase or decrease the weight in order to balance the wheel assembly. The bubble will be po-

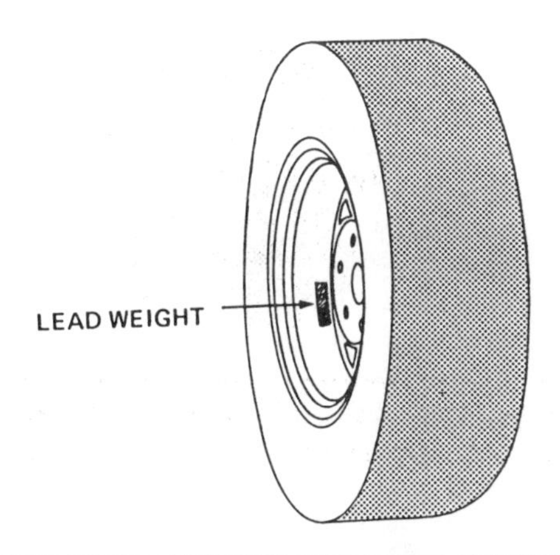

FIGURE 7-5. Position of Stick-On Lead Weight Used for Magnesium Alloy Rim.

sitioned in the middle of the spirit level when the wheel assembly is statically balanced.

Static wheel balancers do not have the capability to detect the exact heavy spot of an unbalanced wheel assembly. Therefore, it is necessary to divide the required weight on each side of the wheel when statically balancing it.

8. Attach the weight to the wheel rim (Figure 7-6).

9. Remove the wheel from the balancer. Install the wheel on the hub or axle. Torque the wheel nuts to the manufacturer's specification.

The second method of statically balancing a wheel requires a greater degree of skill by the service technician.

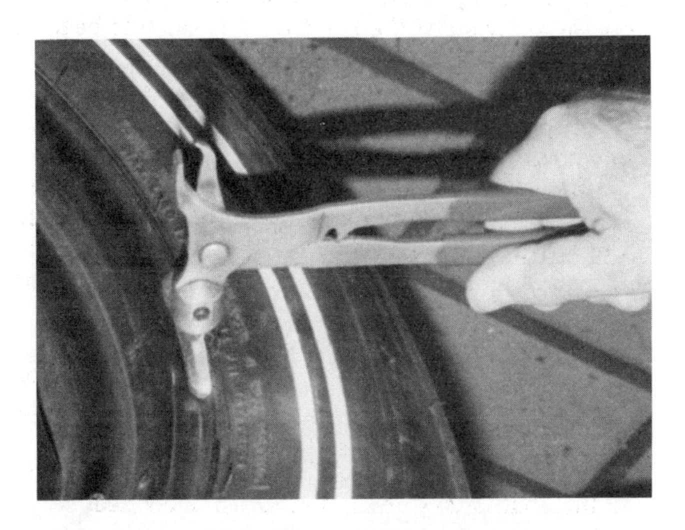

FIGURE 7-6. Attaching a Weight to the Wheel Rim.

Step-by-Step Procedure

1. Place the unbalanced wheel on the balancer and allow the wheel to settle.
2. Locate the light portion of the wheel assembly. Chalk mark the sidewall of the tire.
3. Obtain four equal weights. Move the weights in pairs approximately 50° to the left and right of the chalk mark.
4. Increase or decrease the four weights until the spirit level indicates that the wheel assembly is statically balanced. Chalk mark the position of the required weights on the sidewall of the tire from the outside bead to the inside bead.
5. Remove the wheel from the balancer. Attach two weights at the chalk mark positions on the inside of the tire rim.
6. Position the wheel on the balancer. You may have to shift the position of the weights slightly on the outside of the tire in order to statically balance the wheel.
7. Use the wheel weight pliers and attach the weights to the wheel rim.
8. Check the balance of the wheel assembly after the wheel weights have been installed.
9. Install the wheel on the hub or axle. Torque the wheel nuts to the manufacturer's specification.

This method of balancing a wheel reduces the effect of the heavy portion of the wheel assembly when the wheel is in motion, increasing tire tread life.

FACTS TO REMEMBER

1. Before balancing a wheel, always check its lateral and radial runout.
2. Never attempt to balance a tire that has been subject to excessive tread wear.
3. Make it a rule to always determine the torque of the front wheel bearings when balancing a wheel.
4. Always clean the inside of a wheel rim and remove the existing wheel weights prior to balancing the wheel assembly.

Vertical Static Balancer. Figure 7-7 illustrates a *vertical static wheel balancer.* A shaft, positioned through the rim, supports the wheel on nonfriction rollers. Gravity pulls the heavy portion of the wheel to the bottom. A calibrated scale and a positioning arm indicate the amount and location of the required weight.

FIGURE 7-7. Marking Weight Position with Vertical Static Balancer. (Courtesy of Snap-on Tools of Canada Ltd.)

ON-THE-CAR STATIC BALANCING

On-the-car balancing, referred to as *kinetic balancing*, is a variation of static balancing. Both off-the-car and on-the-car methods correct wheel hop or wheel tramp, but on-the-car balancing is more precise. There are two types of on-the-car balancers: mechanical and electronic.

Mechanical Balancers

A mechanical type on-the-car balancer uses a mechanical adapter that attaches to the wheel rim with expanding clamps (Figure 7-8). The wheel disc cover is removed, and the mechanical balancer is locked to the wheel. A wheel spinner, held against the wheel, rotates the wheel.

The balance adapter (the center piece of the equipment) has two green knobs and two red knobs. Start the wheel spinning, and increase the speed of the wheel only fast enough to produce the vibration of the front bumper. While the wheel is being spun, you briefly hold the four knobs. Inside the adapter are two weights that are moved by manipulating these knobs. The weights counteract the vibration of the wheels; they also pinpoint, as indicated by a gauge, the position and the amount of the required weight. Watch the vehicle's bumper and continue to adjust the knobs until the vibration stops. Slowly stop the wheel's spinning by applying the balancer's brake plate against the tire's tread. Divide the total amount of weight required by 2, and attach half the required weight to the inside of the wheel rim at the indicated position. Set the scale of the adapter to zero and spin the wheel again. Adjust the knobs to eliminate the vibration. Attach the required weight to the indicated position on the outside of the tire rim.

If the vehicle has rear wheel drive, the wheel

FIGURE 7-8. On-the-Car Wheel Balancer. (Courtesy of Hunter Engineering Company)

spinner is not required since the wheel is driven by the engine, driveline, and differential. Some vehicles are equipped with conventional and nonslip differentials. These terms are discussed later in this chapter.

Electronic Balancers

This type of equipment utilizes a wheel spinner, a strobe light, and a sensor to balance a wheel. When you are using this equipment to balance a front wheel on a rear wheel drive vehicle, only the wheel that is being balanced is raised, approximately 12 centimeters (5 inches) from the floor.

After performing all the preliminary checks and procedures, connect the line cord to an electrical outlet. The electronic sensor is positioned under the control arm close to the lower ball joint. A small magnet, which is part of the sensor mechanism, must be in contact with the control arm. Using a lock screw, adjust the vertical rod of the sensor to the correct height (Figure 7-9A). With a piece of chalk, mark a reference line on the outside of the tire's sidewall (Figure 7-9B). Spin the wheel with the spinner only fast enough to produce the vibration of the front bumper. As the vibration becomes more noticeable, the electronic strobe light will flash because a heavy portion of the tire, moving to the bottom of the wheel, activates the sensor, and the strobe flashes. At this point, remove the spinner from against the revolving wheel and allow the wheel to spin freely using its own kinetic energy.

On the top face of the strobe light is a graduated scale. As the wheel begins to slow, the needle on the

FIGURE 7-9A. Adjusting the Height of an Electronic On-the-Car Balancer. (Courtesy of Stewart-Warner Corporation of Canada, Limited)

FIGURE 7-9B. Marking a Chalk Reference Line Prior to Activating the Electronic Sensor. (Courtesy of Stewart-Warner Corporation of Canada, Limited)

scale will move to a lower position and will then move back up the scale. Now, shine the strobe light on the sidewall of the tire and observe the position of the chalk mark. Also observe where the needle is on the scale and the amount of unbalance. Each graduated line on the face of the scale is equal to 7 grams (¼ ounce). Slow the wheel down gradually to stop it, using the machine's brake plate. Turn the tire so the chalk mark is positioned to where the strobe light illuminated the mark. The heavy portion is now at the bottom of the tire, and the correcting weight will be attached 180° opposite (the top of the tire).

Imagine that the side of the tire is the face of a clock. Mentally divide the tire into parts, with the top being 12 o'clock (Figure 7-10). Start the wheel spinning again. If some vibration is still present, the strobe light will be shone on the tire to locate the position of the weight. If the weight is at the 6 o'clock position, you have attached too much weight. If the weight is at the 12 o'clock position, you have not added enough weight. If the weight is at the 3 or 9 o'clock position, you may improve the slight unbalanced condition by shifting the weight approximately 2.5 centimeters (1 inch) toward the 12 o'clock position. Spin the wheel again. If the needle of the scale is in the green position on the graduated scale, the wheel is balanced.

Front Wheel Drive Vehicles. Many cars now being driven in North America are front wheel drive. Since this type of vehicle has constant velocity universal joints, the suspension must not be allowed to hang free during wheel balancing. The floor jack must support the vehicle under the lower control arm as close to the wheel as possible to prevent damage to the universal joints and drive shaft.

If this procedure cannot be followed, remove the wheel and balance it on an off-vehicle balancer. This procedure is discussed in Chapter 8.

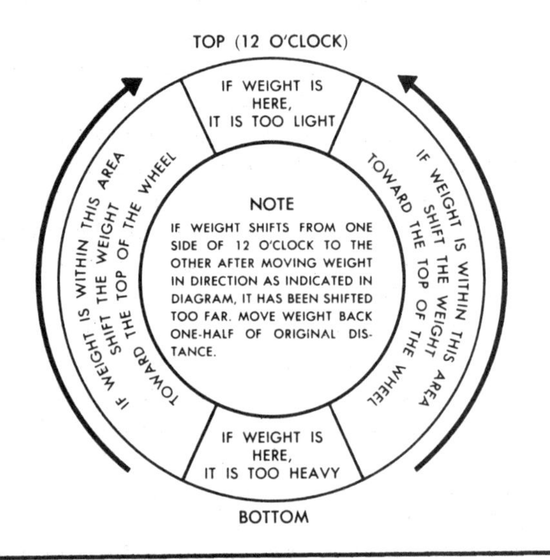

FIGURE 7-10. Guide for Action to Take if Wheel Remains Unbalanced. (Courtesy of Stewart-Warner Corporation of Canada, Limited)

REAR WHEEL BALANCING

To determine if a vehicle has a **conventional** or a **nonslip differential**, raise the rear of the vehicle so that both rear wheels are off the garage floor. With the automatic transmission in park or the manual transmission in gear, attempt to rotate one wheel by hand. If the wheel can be rotated, the rear axle is conventional; if it cannot be rotated, the rear axle has a nonslip differential. *Note:* The parking brake must not be applied.

Conventional Differential

When using an on-the-car wheel balancer to balance a rear wheel on a vehicle equipped with a conventional differential, raise only the wheel to be balanced off the floor. Then attach the balance adapter to the wheel. Use the same balancing procedure as described for the front wheels; however, do not allow the speedometer to exceed a speed of 56 km/h (35 mph). Since the opposite wheel is stationary and resting on the floor, the speed of the wheel being balanced is 112 km/h (70 mph).

Nonslip Differential

When balancing the rear wheel on a vehicle equipped with a nonslip differential with an on-the-car wheel balancer, lift the rear of the vehicle and place safety stands under the frame. The wheel not being balanced must be removed from the axle. Leave the brake drum on the axle and torque the wheel nuts. Use a floor jack and lift the axle housing to remove the acute angle from the drive shaft, but still support the vehicle by using the safety stands. Balance the opposite wheel. When that wheel has been balanced, do not remove it from the axle; correctly install and balance the opposite wheel. Spin the wheel only fast enough to produce a vibration, and be sure not to exceed speeds over 90 km/h (55 mph). The speed shown on the speedometer is the actual speed at which the wheel is revolving. *Note:*

Front wheel drive cars are not equipped with a nonslip transaxle (differential).

5. Always read the manufacturer's operating instructions before using the equipment.
6. Develop and practice safe working habits.

FACTS TO REMEMBER

1. Wheel unbalance causes vehicle vibration and decreases the service life of steering and suspension parts.
2. Wheel balance is not a substitute for excessive lateral and radial tire/rim runout.
3. Wheel balance cannot correct worn tire treads, damaged sidewalls, loose wheel bearings, or an object inside the tire.
4. A wheel assembly may be statically balanced but dynamically unbalanced. This subject is discussed in Chapter 8.

REVIEW TEST

1. List all items that should be checked prior to balancing a wheel.
2. Describe in detail how to check each item.
3. Describe how to statically balance a wheel using two methods.
4. Explain how to determine if a vehicle has a conventional or a nonslip differential.
5. Describe how to kinetically balance a front wheel on a vehicle that has a rear driving axle.

FINAL TEST

This examination is multiple choice. Only one answer will be accepted. Carefully read every statement.

1. Mechanic A says that a machine that statically balances a wheel indicates only a general area of unbalance. Mechanic B says that the heavy portion of a wheel is always on the outside of the tire tread. Who is right? (A) Mechanic A, (B) Mechanic B, (C) Both A and B, (D) Neither A nor B.
2. Mechanic A says that radial runout is the elliptical movement of a revolving wheel. Mechanic B says that lateral runout is the side-to-side movement of a revolving wheel. Who is right? (A) Mechanic A, (B) Mechanic B, (C) Both A and B, (D) Neither A nor B.
3. Mechanic A says that lateral runout is checked by positioning a dial gauge indicator against the side of a wheel. Mechanic B says that radial runout is checked by positioning a dial gauge indicator in the center of the tire tread. Who is right? (A) Mechanic A, (B) Mechanic B, (C) Both A and B, (D) Neither A nor B.
4. Mechanic A says that excessive radial runout may be corrected by static balance. Mechanic B says that excessive radial runout cannot be corrected by static balance. Who is right? (A) Mechanic A, (B) Mechanic B, (C) Both A and B, (D) Neither A nor B.
5. Mechanic A says that manufacturer's specifications for lateral runout are only a general guide and may be exceeded. Mechanic B says that when the manufacturer's tire/rim runout specifications are exceeded, this condition will probably produce vehicle vibration. Who is right? (A) Mechanic A, (B) Mechanic B, (C) Both A and B, (D) Neither A nor B.
6. Mechanic A says that a loose front wheel bearing will contribute to a problem caused by an unbalanced wheel. Mechanic B says that a tire that is overinflated will contribute to a problem caused by an unbalanced wheel. Who is right? (A) Mechanic A, (B) Mechanic B, (C) Both A and B, (D) Neither A nor B.
7. Mechanic A says that it is possible to make a tire perfectly round after sustaining tread wear by kinetically balancing a wheel. Mechanic B says that wheel nuts and wheel bearings do not need to be torqued to manufacturer's specifications since specs are only a general guide. Who is right? (A) Mechanic A, (B) Mechanic B, (C) Both A and B, (D) Neither A nor B.
8. Mechanic A says that it is impossible for a wheel to produce a vibration that is subject to excessive lateral runout. Mechanic B says that it is impossible for a wheel to produce vibration when it is kinetically unbalanced. Who is right? (A) Mechanic A, (B) Mechanic B, (C) Both A and B, (D) Neither A nor B.
9. Mechanic A says that when balancing a rear wheel on a vehicle with a nonslip differential, the true speed of the wheel is shown on the speedometer. Mechanic B says that when bal-

ancing a rear wheel on a vehicle with a nonslip differential, the speedometer is indicating half the speed of the wheel. Who is right? (A) Mechanic A, (B) Mechanic B, (C) Both A and B, (D) Neither A nor B.

10. Mechanic A says that an unbalanced wheel affects the service life of the steering and suspension parts. Mechanic B says that balancing a wheel is no substitute for tire sidewall problems or an object inside the tire. Who is right? (A) Mechanic A, (B) Mechanic B, (C) Both A and B, (D) Neither A nor B.
11. Mechanic A says that when balancing a wheel on a front wheel drive vehicle, the jack is placed under the vehicle's front crossmember. Mechanic B says that all front wheel drive cars are equipped with a nonslip transaxle (differential). Who is right? (A) Mechanic A, (B) Mechanic B, (C) Both A and B, (D) Neither A nor B.
12. Mechanic A says that when using an electronic car balancer, the reading on the scale is taken while the spinner is turning the wheel. Mechanic B says that when using an electronic on-the-car balancer, the correcting weight is always placed at the 6 o'clock position. Who is right? (A) Mechanic A, (B) Mechanic B, (C) Both A and B, (D) Neither A nor B.

Dynamic Wheel Balance

LEARNING OBJECTIVES

After studying this chapter, you should be able to:

- Illustrate and describe dynamic wheel balance,
- Illustrate and describe dynamic wheel unbalance,
- Illustrate and describe how dynamic wheel unbalance induces a wheel to shake laterally as it revolves on the surface of the road,
- Explain how dynamic wheel unbalance affects tire tread life and the vehicle's suspension and produces driver fatigue,
- Illustrate and describe how dynamic wheel unbalance is corrected.

INTRODUCTION

Wheel balance is divided into two distinct categories: static balance (Chapter 6) and dynamic balance. When the entire wheel assembly is in static balance, the weight of the assembly is evenly distributed around the axis (center) of the wheel. However, it is possible for a wheel to be in perfect static balance and at the same time be dynamically unbalanced.

Dynamic unbalance produces lateral forces that influence the wheel assembly to wobble or shake from side to side. This condition drastically reduces tire tread life expectancy and causes premature failure of all suspension and steering parts. Dynamically balanced wheels contribute to the smooth ride and good handling characteristics of a vehicle. This chapter will help you to understand the importance of dynamic wheel balance.

DYNAMIC WHEEL BALANCE

The meaning of the term **dynamic wheel balance** can be explained by dividing a wheel assembly into four equal sections. Figure 8-1 illustrates a top sectional view of a dynamically balanced wheel. If sections A and C (the two outside sections) are equal in weight and if sections B and D (the two inside sections) are equal in weight, the wheel assembly is dynamically balanced. To understand the meaning of the term **dynamic wheel unbalance**, again divide the wheel assembly into four equal sections. Figure 8-2 illustrates a top view of a dynamically unbalanced wheel. Section D is the heaviest section. Sections A and C (the two outside sections) are equal in weight; however, since sections B and D (the two inside sections) are not equal in weight, the wheel assembly is dynamically unbalanced.

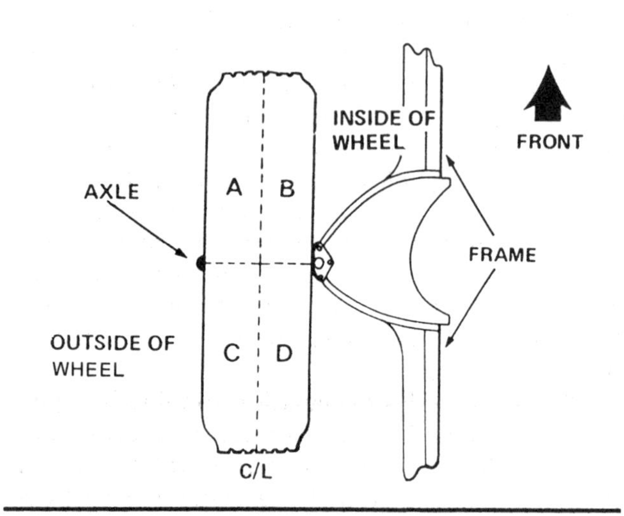

FIGURE 8-1. Top Sectional View of a Dynamically Balanced Wheel.

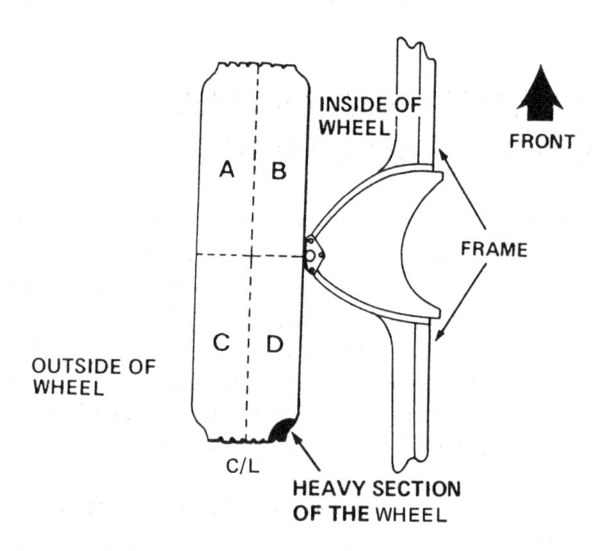

FIGURE 8-2. Top Sectional View of a Dynamically Unbalanced Wheel.

EFFECTS OF DYNAMIC WHEEL UNBALANCE

Lateral Wheel Movement

Dynamic wheel unbalance induces *lateral* (side to side) wheel movement. To understand how this happens, remember that centrifugal force is the force that tends to pull a rotating mass away from its *axis* (center) of rotation (see Figure 6-6). Now let's apply the definition to a dynamically unbalanced wheel using the following example.

Figure 8-3 shows a heavy spot on a tire on a dynamically unbalanced left front wheel. When the wheel rotates, centrifugal force will endeavor to influence the heavy portion of the tire to move to the centerline of the wheel, thereby forming a 90° angle to the axle shaft. The hub and wheel assembly are attached to the front axle shaft, which is made of tempered metal and will not bend. The shaft and steering knuckle are attached to the upper and lower ball joints. The ball joints allow the steering knuckle to pivot.

To see the action of centrifugal force, the vehicle must be accelerated to a speed of 100 km/hr (60 mph), and you must take a stop-action, X-ray photo of the left front wheel. The photo would show the heavy spot has, in fact, moved to the centerline of

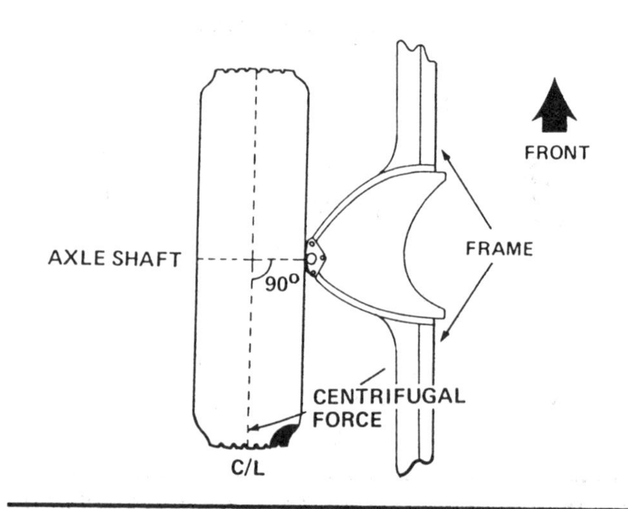

FIGURE 8-3. Action of Centrifugal Force on Heavy Portion of a Dynamically Unbalanced Wheel.

the wheel, thus forcing the rear of the wheel outward (Figure 8-4).

If you were to take a second stop-action photo of the rotating wheel when the heavy spot has revolved 180° (one-half revolution), you would see from the photo that the front of the wheel is now forced outward and the rear of the wheel inward (Figure 8-5). Dynamic wheel imbalance, due to the effect of centrifugal force, induces the wheel assembly to move laterally (shake) from side to side.

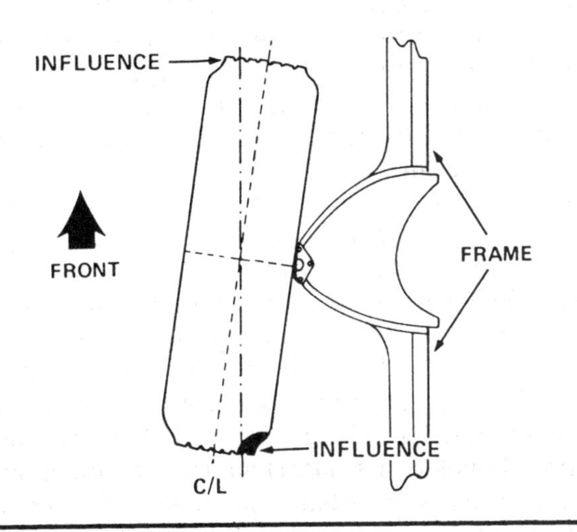

FIGURE 8-4. Outward Movement of the Rear Portion of a Rotating Wheel due to Influence of Centrifugal Force and the Position of the Unbalanced Weight.

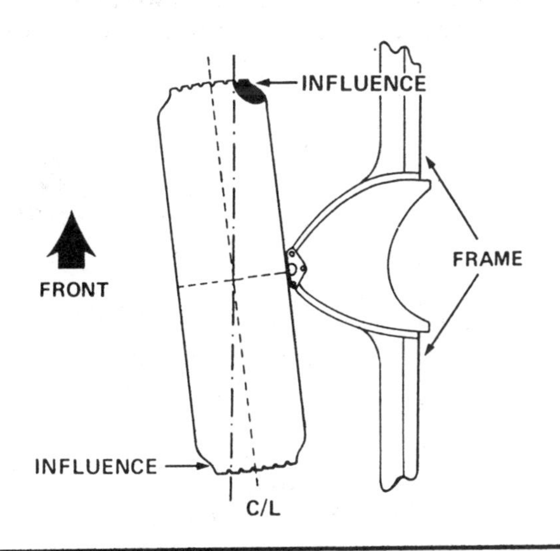

FIGURE 8-5. Outward Movement of the Front Portion of a Rotating Wheel due to Influence of Centrifugal Force and the Position of the Unbalanced Weight.

Vehicle Handling Characteristics

A dynamically unbalanced wheel produces lateral wheel movement. The action of the wheel causes the suspension, steering, and body of the vehicle to react. This condition is reflected in the oscillation of the steering wheel at medium and high speeds. The driver of a vehicle may be exceptionally strong and hold the steering wheel with a viselike grip; however, even that person cannot prevent the steering wheel from oscillating. This condition produces driver fatigue and affects the directional control and stability of the vehicle.

Tire Tread Life. Tire tread life is also reduced by the lateral movement of a dynamically unbalanced wheel. Chapter 6 stated that the bottom of the tire's tread (the portion contacting the surface of the road) momentarily stops in relation to the road surface regardless of the speed of the vehicle in order to allow the tread to gain the necessary traction. If a revolving tire is subject to lateral movement, the tire's tread cannot momentarily stand still and is forced to pivot (scuff) at the road surface. This additional movement of the tire's tread increases tread wear.

To illustrate this explanation, obtain a lead pencil with an eraser. Turn the pencil upside down; hold the pencil in a vertical position and rotate the pencil. The result will be that the eraser will wear due to the constant rotation and friction. The same result applies to the tread of a tire; the more the tread is subjected to movement, the greater the tread wear.

CORRECTING DYNAMIC WHEEL UNBALANCE

To correct dynamic wheel unbalance, the service technician may balance the wheel assembly by using a wheel balancer that is designed to balance a wheel both statically and dynamically when the wheel assembly has been removed from the vehicle (Figure 8-6). This type of equipment is designed to detect the light portion of the wheel and to indicate the amount of weight required to balance the wheel dynamically.

Two important facts form the basis for understanding wheel balance:

1. When a service technician dynamically balances a wheel, he or she attaches lead weights

FIGURE 8-6. Combination Static and Dynamic Wheel Balancer. (Courtesy of FMC Corporation, Automotive Service Equipment Division)

to the wheel rim to counteract the lateral influence of the heavy portion of the wheel assembly.

2. A wheel assembly that is accurately balanced must be statically and dynamically balanced.

Before proceeding further, let's consider an example problem. Figure 8-7 illustrates a top view of a wheel that is unbalanced. Determine where, at the rim, you would attach weight to correctly balance the wheel.

If your answer was position Z, you have missed the whole point of the chapter. Go back to the beginning. If your answer was position W, you did not understand important fact 2. By placing a weight at position W, you have balanced the wheel dynamically but not statically. Sections C and D are heavier than sections A and B. If your answer was position X, you did not understand important fact 2. By placing a weight at position X, you have balanced the wheel statically but not dynamically. Sections A and C are not equal in weight; sections B and D are not equal in weight.

If your answer was position Y, you were right. Placing a correct weight at position Y will counter-

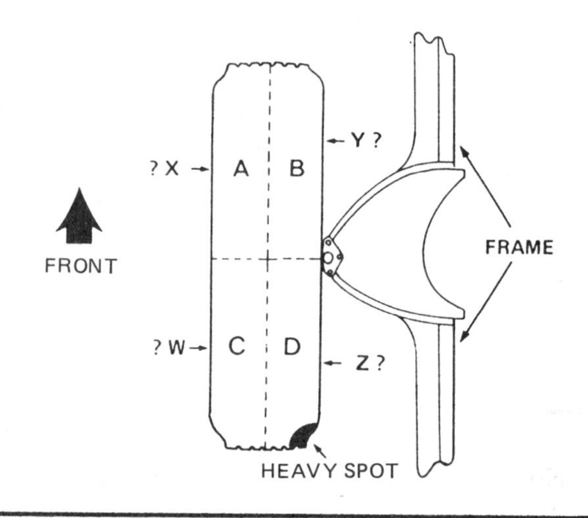

FIGURE 8-7. Example Unbalanced Wheel for Determining the Position of Added Weight to Dynamically Balance the Wheel.

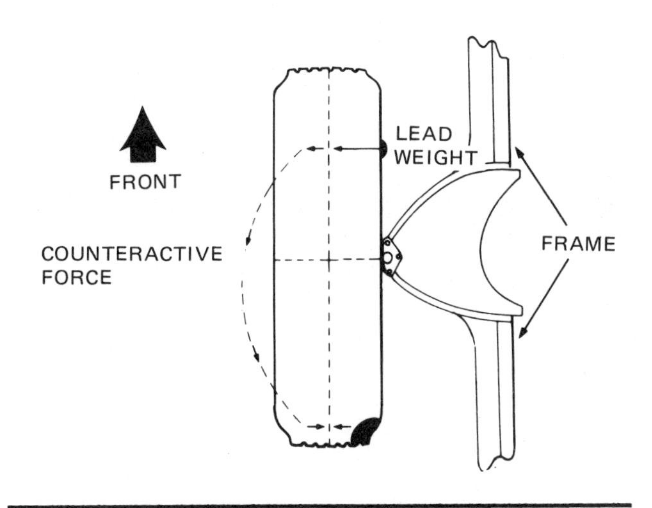

FIGURE 8-8. Correct Position of the Balancing Weight.

act the lateral force of the heavy portion of the tire. *Note:* The influence of a lead weight attached to the rim will always attempt to move (due to centrifugal force) to the centerline position of the wheel (Figure 8-8).

Electronic On-the-Car Balancer

The electronic on-the-car wheel balancer is designed to balance the wheel statically and dynamically. The electronic balancer balances the entire wheel assembly, including the brake rotor or brake drum. An electronic sensor placed under the suspension or against the backing plate will detect any lateral or vertical movement and transmit a signal to the strobe unit while the wheel is spinning.

Regardless of the type of equipment in your shop, correcting dynamic wheel unbalance is simply a matter of adding sufficient weight to the light section of a wheel to counteract the forces on the heavier section. The manufacturer's operating instructions should always be followed when using any wheel balancing equipment.

FACTS TO REMEMBER

1. Normal tire wear will cause wheels to become unbalanced.
2. Vehicle owners should have the wheels checked for correct balance at intervals of 16,000 kilometers (10,000 miles).
3. A wheel assembly must be statically and dynamically balanced if the wheel is to be balanced accurately.

REVIEW TEST

1. Define and illustrate the term *dynamic wheel balance.*
2. Define and illustrate the term *dynamic wheel unbalance.*
3. Describe and illustrate how dynamic wheel unbalance influences a revolving wheel.
4. Explain how dynamic wheel unbalance decreases tire tread life.

Note: Use Figure 8-9 to answer questions 5 through 8.

5. By placing a lead wheel weight at position 1, the wheel assembly would be __ balanced and __ unbalanced.
6. By placing a lead wheel weight at position 2, the wheel assembly would be __ balanced but __ unbalanced.
7. By placing lead wheel weights at positions 2 and 3, the wheel assembly would be __ balanced but __ unbalanced.
8. By placing a lead wheel weight at position 3, the wheel assembly would be __ and ____ ____.

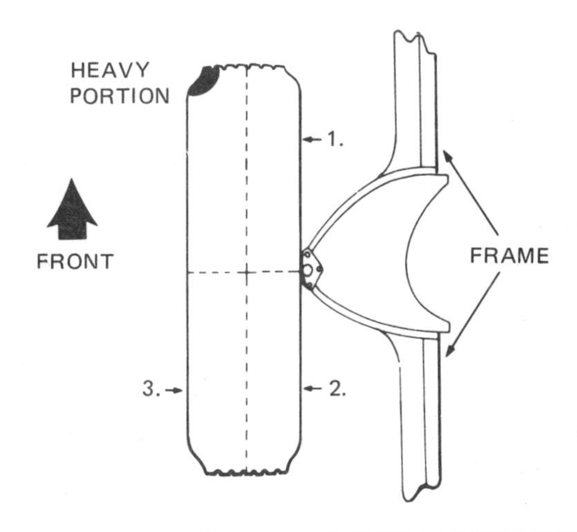

FIGURE 8-9. Illustration for review questions 5 through 8.

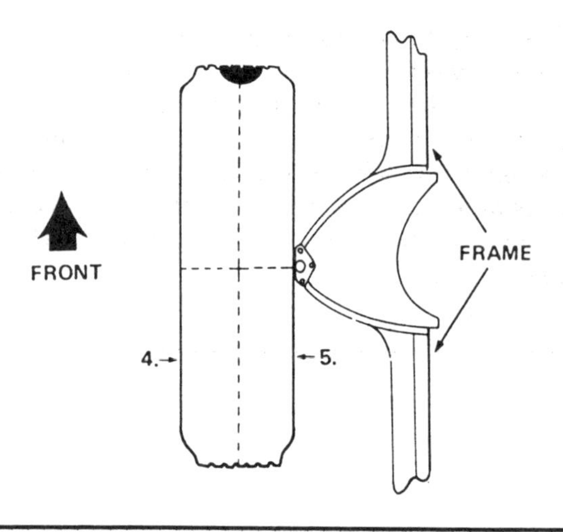

FIGURE 8-10. Illustration for review questions 9 and 10.

9. By placing a lead wheel weight at position 5, the wheel assembly would be _____ but _____ _____.
10. By placing lead wheel weights at positions 4 and 5, the wheel assembly would be ____ and ____ ____.

Note: Use Figure 8-10 to answer questions 9 and 10.

FINAL TEST

This examination is multiple choice. Only one answer will be accepted. Carefully read every statement.

1. Mechanic A says that dynamic unbalance produces lateral forces that influence the wheel assembly to shake laterally. Mechanic B says that a wheel assembly is dynamically balanced when there is no centrifugal influence to move weight to the centerline of the wheel assembly. Who is right? (A) Mechanic A, (B) Mechanic B, (C) Both A and B, (D) Neither A nor B.
2. Mechanic A says that if a wheel assembly is statically balanced, it is also dynamically balanced. Mechanic B says that it is possible for a wheel assembly to be dynamically balanced and be statically unbalanced at the same time. Who is right? (A) Mechanic A, (B) Mechanic B, (C) Both A and B, (D) Neither A nor B.
3. Mechanic A says that dynamic wheel unbalance will be reflected in the oscillation of the steering wheel. Mechanic B says that dynamic wheel unbalance is considered a low speed unbalance condition. Who is right? (A) Mechanic A, (B) Mechanic B, (C) Both A and B, (D) Neither A nor B.
4. Mechanic A says that correcting dynamic wheel unbalance is simply a matter of adding sufficient weight to the light section of the wheel. Mechanic B says that it is not necessary to clean the inside of the wheel rim prior to dynamic balancing. Who is right? (A) Mechanic A, (B) Mechanic B, (C) Both A and B, (D) Neither A nor B.
5. Mechanic A says that the excessive lateral wheel runout will produce steering wheel oscillation. Mechanic B says that tire tread cupping is produced by dynamic wheel unbalance. Who is right? (A) Mechanic A, (B) Mechanic B, (C) Both A and B, (D) Neither A nor B.
6 Mechanic A says that the rear wheels of a vehicle are more susceptible than the front wheels to dynamic unbalance. Mechanic B says that the rear wheels of a vehicle are never susceptible to dynamic unbalance because the rear wheels do not provide steering or directional stability. Who is right? (A) Mechanic A, (B) Mechanic B, (C) Both A and B, (D) Neither A nor B.
7. Mechanic A says that the best way to remember the term *dynamic balance* is to divide the wheel into four sections. Mechanic B says that the wheel is dynamically balanced when the two outside sections are equal in weight and the two inside sections are equal in weight. Who is right? (A) Mechanic A, (B) Mechanic B, (C) Both A and B, (D) Neither A nor B.
8. Mechanic A says that as a wheel revolves, the tire is subject to acceleration, deceleration, and lateral forces. Mechanic B says that centrifugal and lateral forces seldom produce a noticeable unbalanced wheel condition at lower speeds. Who is right? (A) Mechanic A, (B) Mechanic B, (C) Both A and B, (D) Neither A nor B.
9. Mechanic A says that a front wheel is more sus-

ceptible to an unbalanced condition because of the steering and suspension systems. Mechanic B says that although a brake rotor or drum is attached to a front wheel, the rotor or drum's unbalanced condition cannot influence a revolving unbalanced wheel. Who is right? (A) Mechanic A, (B) Mechanic B, (C) Both A and B, (D) Neither A nor B.

10. Mechanic A says that if the heavy portion is located in the center of the tire's tread, the corrective weight is placed only on the outside of the rim. Mechanic B says that if the heavy spot is located in the center of the tire's tread, the corrective weight is placed on the inside and outside of the rim. Who is right? (A) Mechanic A, (B) Mechanic B, (C) Both A and B, (D) Neither A nor B.

Manual Steering

LEARNING OBJECTIVES

After studying this chapter and with the aid of a manufacturer's shop manual and the necessary tools and equipment, you should be able to:

- Explain the purpose of the energy-absorbing steering column;
- Identify the parts and describe the internal operation of the rack and pinion steering gearbox;
- Describe how to diagnose for wear and looseness, disassemble, inspect, and reassemble the rack and pinion steering mechanism;
- Identify the parts and describe the internal operation of the recirculating ball manual steering gearbox;
- Describe how to disassemble, diagnose, inspect, and reassemble a recirculating ball manual steering gearbox;
- Describe the main design features of two other types of manual steering gearbox mechanisms;
- Explain the meaning and advantages of parallelogram steering linkage design;
- Explain how to perform a detailed inspection of a vehicle's steering linkage mechanism;
- Describe how to remove and replace steering linkage parts.

INTRODUCTION

The steering system of a modern vehicle is designed to provide the driver with maximum directional control, vehicle maneuverability, and steering sensitivity, as well as to reduce the effort required to turn the front wheels of the vehicle. To achieve these requirements, the design engineer must calculate the weight supported by the front wheels and the total allowable safe steering versatility of the vehicle; the vehicle's steering gear and steering linkage make this possible. The rotary motion of the steering wheel is transferred through the steering column to the steering gear mechanism.

When the front wheels are turned from the full-left to the full-right steering position, they are normally required to rotate through an arc of slightly more than 60° of rotation. Based on the weight of the vehicle's front end, design calculations, and the **mechanical advantage** (M.A.) (ratio) of the vehicle's steering gear, the driver will be required to rotate the steering wheel three to six complete turns from a full-left to a full-right steering position. The total number of turns that the steering wheel rotates compared to the total number of degrees the front wheels turn is known as the **overall steering ratio** (O.S.R.). As the overall steering ratio decreases, the sensitivity and maneuverability of the vehicle's steering system increases and the effort required to turn the steering wheel is greater. Hydraulic power steering ratios will be less than manual steering ratios because of the advantage of hydraulic assistance.

This chapter discusses the purpose of the energy-absorbing steering column, rack and pinion steering, the recirculating ball manual steering gearbox, and other types of steering gear mechanisms. We will conclude with coverage of the steering linkage and how it is serviced.

ENERGY-ABSORBING STEERING COLUMN

In recent years, North American vehicle manufacturers have introduced the **energy-absorbing steering column.** The outer mast jacket, shift tube, and **steering shaft** are designed to collapse under severe front end impact conditions. When an automobile is involved in a frontal collision, the primary force (forward movement of the vehicle) is suddenly halted, while the secondary force (the driver) continues in a forward direction. The secondary impact occurs when the driver is thrust forward into the steering column (Figure 9-1). Upon secondary impact, the outer mast jacket, shift tube, and steering shaft compress by shearing several plastic pins (Figure 9-2).

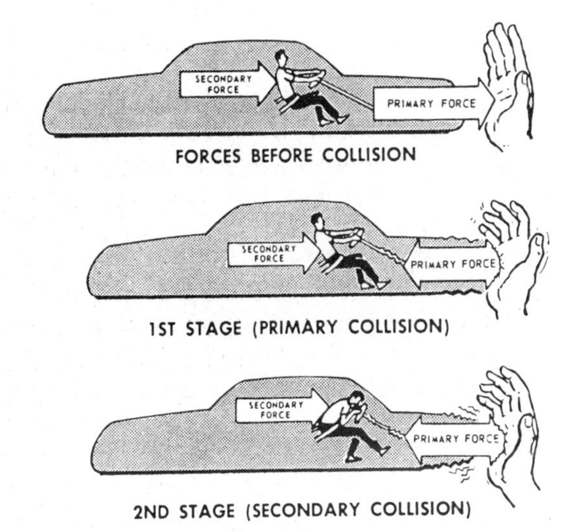

FIGURE 9-1. **Reaction Forces during a Collision.** (Courtesy of General Motors of Canada, Limited)

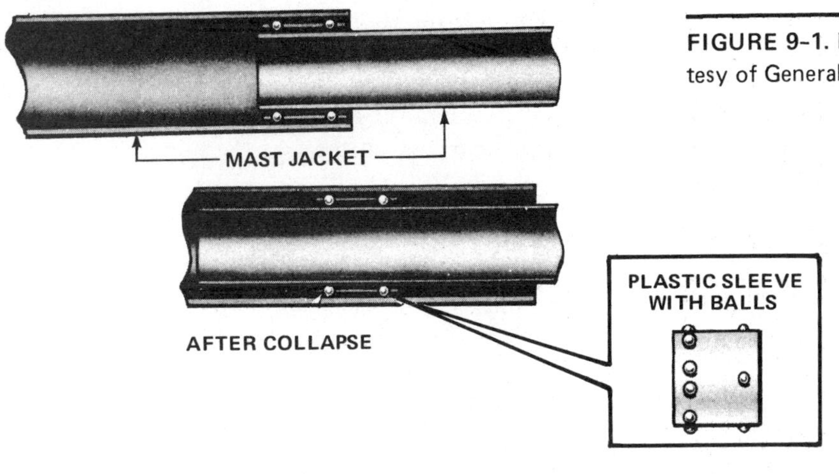

FIGURE 9-2. **Steering Column Jacket Design.** (Courtesy of General Motors of Canada, Limited)

When you suspect damage to a vehicle's steering column, a detailed visual inspection is required for any distortion or the compression of parts in the steering column. Defective steering column parts, because of their importance, cannot be restored; they must be replaced. *Note:* When disassembling and reassembling the steering column, use only the designated screws, bolts, and nuts, and tighten these parts to the manufacturer's specified torque. This service precaution will ensure the proper operation and absorbing action of the steering column. If you are required to perform this type of repair, always follow the instructions specified in the manufacturer's shop manual.

FACTS TO REMEMBER

1. A detailed visual inspection is called for in all cases where damage is evident or whenever the vehicle is being repaired due to a front end collision.
2. If damage is evident, the affected parts must be replaced.
3. Always refer to the manufacturer's shop manual for steering column diagnostic procedures.

RACK AND PINION STEERING

Design and Operation

Rack and pinion steering is not a new engineering concept. It was originally used by import vehicle manufacturers. In the early 1970s, it was adapted to a limited number of domestically built lightweight automobiles. The unique features of this type of steering mechanism, because of its design and low steering gear ratio (8:1 to 10:1), provides the driver with easy turning effort and fast steering response. This mechanism can be quite sensitive to rough road conditions.

Figure 9–3 illustrates rack and pinion steering design. The cast aluminum external ribbed gear housing is fastened to the vehicle (normally on the front main crossmember) at two or three rubber-insulator-mounted positions. The steering shaft is connected to the input pinion shaft and gear by either a flexible rubber coupling or a universal joint. The rubber insulators and flexible rubber coupling reduce the transfer of noise and dampen vibration.

The machined input pinion gear meshes with the rack gear, which moves laterally inside the gear housing assembly. The movement of the rack gear is transmitted through ball socket housings to the left and right tie-rods. The ball housings are threaded into the rack gear and then locked to the gear by retaining pins. The internal gear mechanism is protected from contaminants by flexible rubber bellow seals secured in position by metal clamps. Tie-rod ends threaded onto the tie-rods articulate motion (move from side to side in a circular motion) to the steering arms and knuckles. The lubricant inside the gear housing is a fluid-type grease.

Diagnosing Problems

Problems generally associated with the malfunction of this type of gear mechanism can be discussed under two headings: excessive play experienced at the steering wheel and noise or roughness inside the steering gear.

Excessive Play Experienced at the Steering Wheel. Rack and pinion steering has fewer external operating pivotal parts than a parallelogram steering linkage mechanism and, therefore, less ability to isolate and dampen road vibrations. The slightest additional slack or movement in or between any of the components will greatly exaggerate looseness and produce kickback at the steering wheel. The problem is to determine if the gear and steering mechanism is operating correctly.

With the front wheels raised off the floor and assuming that the front suspension and steering linkages are in good condition and the gear in proper adjustment, there should be no more than 9.5 millimeters (0.375 inch) of free movement at the rim of the steering wheel. If there is more movement than suggested, a further examination of the steering mechanism is required.

Step-by-Step Procedure

Note: The following procedures are best conducted when the entire vehicle is supported by a hoist. Start at one side of the vehicle and work across to the opposite side.

1. Check the tires and rims for any signs of damage, wear, lateral runout, and radial runout. Correct any of these conditions.

2. Following manufacturer's procedures and specifications and using a dial gauge, check for ball joint looseness. If specifications are exceeded, ball joints must be replaced.

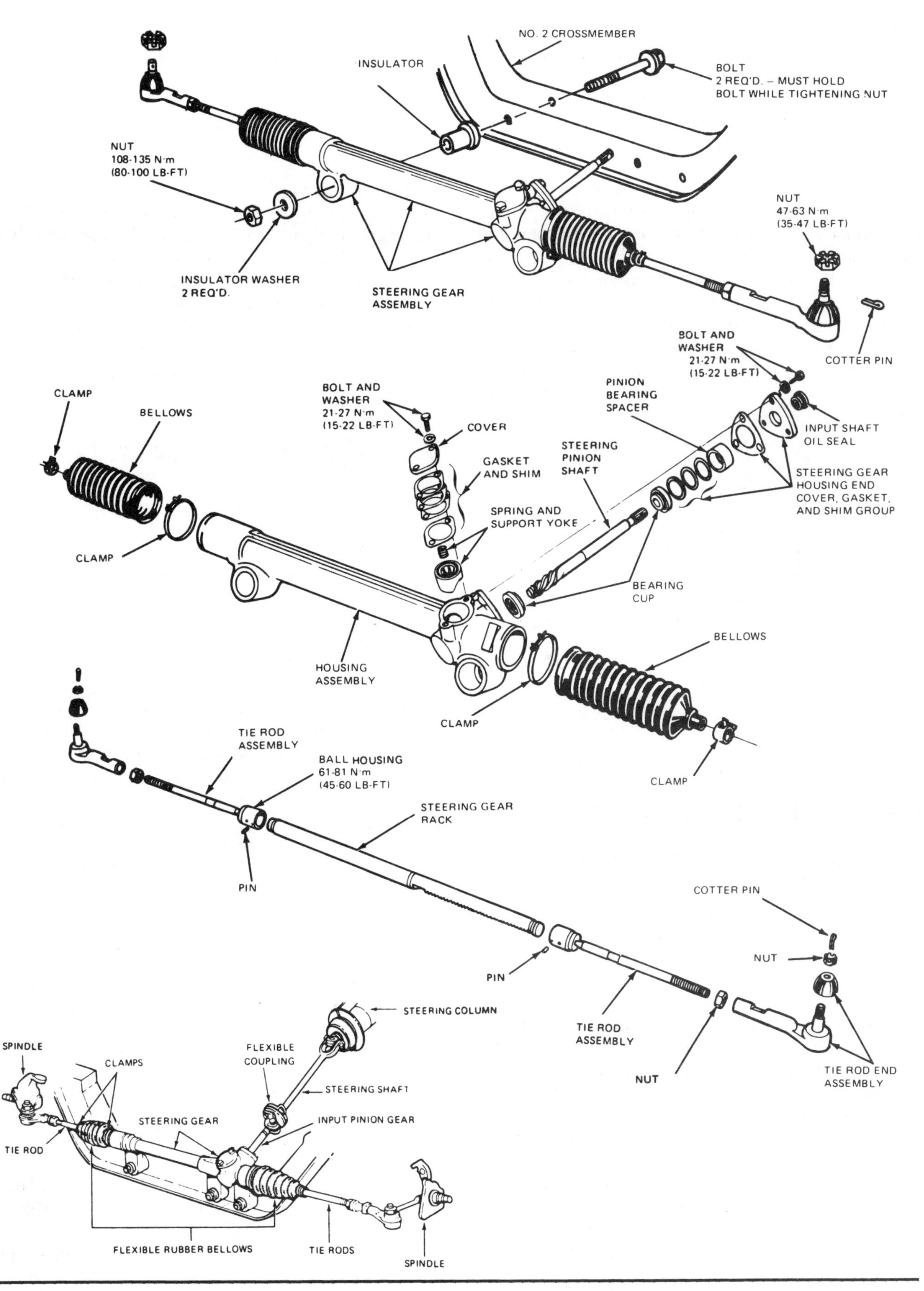

FIGURE 9-3. Rack and Pinion Steering Design. (Courtesy of Ford Motor Company)

3. Using your hands, try to move the outer tie-rod ends up and down. If the part has the slightest perceptible slack, the part should be replaced (Figure 9-4).

4. Visually examine the bellow seals for cracks, splits, and fluid leaks. If any of these conditions are present, the seals must be replaced.

5. To examine the inner tie-rod sockets, squeeze the bellows boot until the inner sockets can be felt. Push and pull on the tire. If looseness exists, the part must be serviced. It is recommended that the gear assembly be removed from the vehicle (Figure 9-5).

6. Carefully examine the rubber insulators for deterioration or looseness by pushing and pulling on the tire. Also, determine if the rubbers are saturated in oil. If these problems are present, the bushings must be replaced (Figure 9-6).

7. While the vehicle is raised, inspect the flex-type steering coupling or universal joint for wear or looseness. If play is found, the defective part must be replaced. During this stage of inspection, grab the pinion shaft and try to move it in and out of the gearbox. If movement exists, internal service of the gear housing is required.

8. By turning the left tire outward, you can grasp the rack. Try to move the rack up and down. If there is movement, follow the manufacturer's instructions for adjusting the rack to pinion gear preload.

Noise or Roughness inside the Steering Gear. With the front wheels supporting the weight and the vehicle stationary, there should be no audible knock in the steering gear when the steering wheel is rotated from stop to stop. If you hear a severe knock, the problem may be a damaged gear face, a damaged upper and/or lower pinion shaft bearing, or an incorrect rack preload. If any of these conditions exists, the gear housing must be removed from the vehicle and the unit disassembled for inspection and repair.

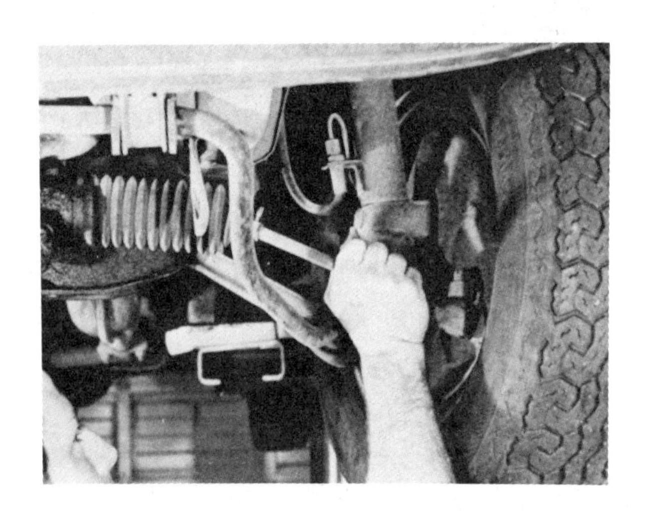

FIGURE 9-4. Diagnosing Wear in Outer Tie-Rod End. (Courtesy of Moog-Canada Automotive, Ltd.)

FIGURE 9-5. Checking for Looseness in Inner Tie-Rod Socket. (Courtesy of Moog-Canada Automotive, Ltd.)

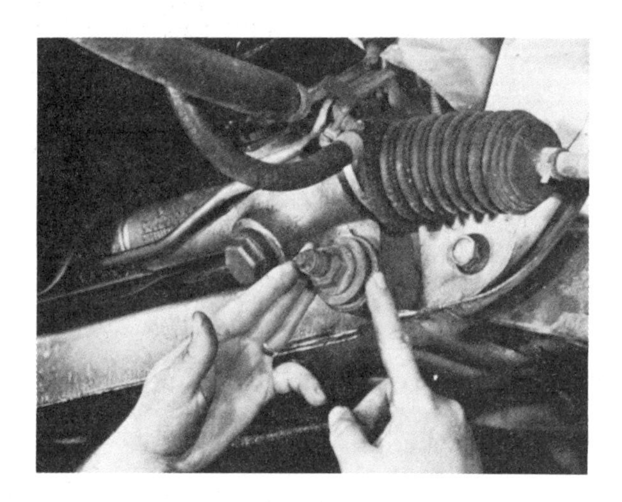

FIGURE 9-6. Examining Insulator Bushings. (Courtesy of Moog-Canada Automotive, Ltd.)

A FACT TO REMEMBER

A faint knock produced by the steering gear while driving on an extremely rough road is not uncommon. According to vehicle manufacturers and because of the design of the steering gear, a slight noise will not affect the operation of the gear assembly.

Servicing Rack and Pinion Steering

Separating the Flex Coupling from the Steering Shaft. Determine if the vehicle manufacturer connects the pinion shaft to the steering shaft by using a flex coupling or a universal joint. If a flex coupling is used, index the two halves of the coupling by using, preferably, a crayon mark. This procedure will enable you to reassemble the coupling so that the steering wheel is in its proper rotational position. If a universal joint has been used, obtain a manufacturer's shop manual. Determine how to separate the universal joint to avoid damage during the disassembly procedures.

Removing the Bolt from the Insulator. It is not uncommon for the steel mounting bolt to become seized from inside the inner sleeve of the rubber insulator. If you encounter this situation, unthread the nut from the bolt so that the nut is flush with the end of the bolt. Obtain an air hammer. With the flat adapter, rattle (push) on the nut end of the bolt. To encourage the bolt to slide out of the sleeve, use the appropriate-sized box end wrench fitted to the head of the bolt, and turn the handle of the wrench.

Holding the Steering Gear Housing. To prevent damage to the housing and if the tool fixture is available, mount the assembly as shown in Figure 9-7.

Removing the Tie-Rod from the Rack. There are several different types of tie-rods. Therefore, the methods used to remove these tie-rods also vary. One thing in common to all tie-rods is that they are threaded onto the rack. Loosen or remove the retaining mechanism from the inner tie-rod ball socket housing, either by drilling out the solid pin or by removing the hollow roll pin (Figure 9-8).

To prevent damage to the rack teeth, affix an appropriate-sized wrench to the rack and ball housing, as illustrated in Figure 9-9. Then, position the correct-sized spanner to the ball housing. Hold the wrench and turn the spanner counterclockwise to unthread the tie-rod from the rack gear.

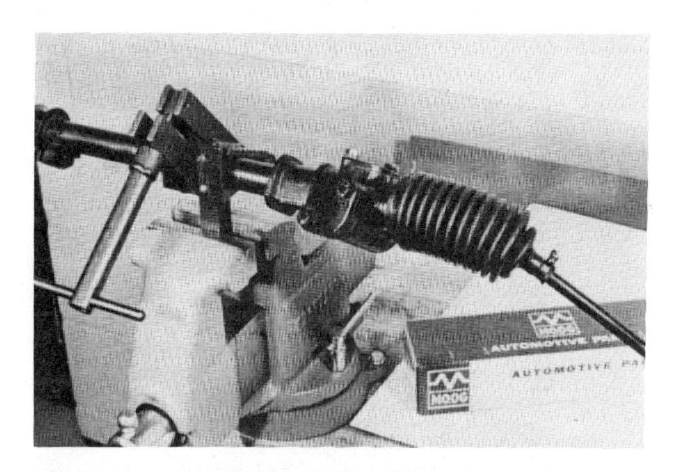

FIGURE 9-7. Holding the Housing in a Fixture. (Courtesy of Moog-Canada Automotive, Ltd.)

FIGURE 9-8A. Drilling Out the Solid Pin. (Courtesy of Moog-Canada Automotive, Ltd.)

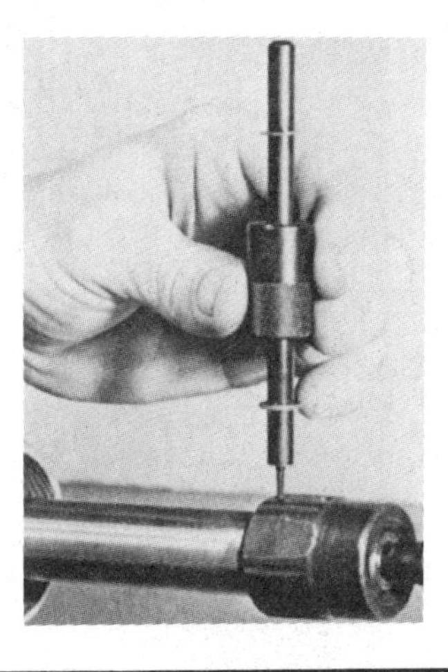

FIGURE 9-8B. Using a Tool to Remove the Hollow Roll Pin. (Courtesy of Moog-Canada Automotive, Ltd.)

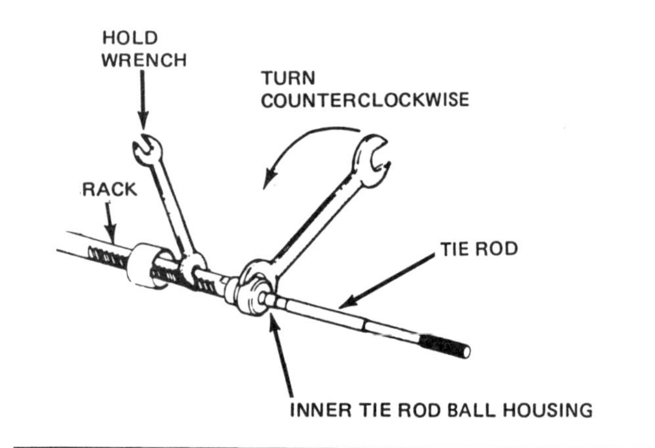

FIGURE 9-9. Using Tools to Prevent Damage. (Courtesy of General Motors of Canada, Limited)

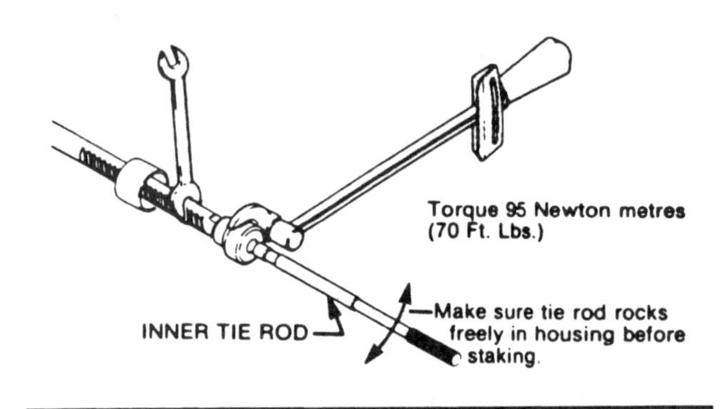

FIGURE 9-10. Using Torque Wrench to Tighten Ball Socket Housing. (Courtesy of General Motors of Canada, Limited)

Washing and Inspecting Internal Parts. Wash all internal parts in a nontoxic cleaning solvent. Thoroughly dry the parts by using compressed air. Place the parts on a clean surface and examine all bearings, bearing races, gear faces, and shaft surface areas for signs of metal fatigue, indentations, scoring, or damage. *Note:* If the rack, including the end threads, is damaged, worn, or has two drilled holes in it, a new rack or steering gear assembly will be required.

Assembling the Gear Mechanism. Assemble the parts of the gear mechanism by following the manufacturer's instructions (Figure 9-10). *Note:* Be sure to hold the rack as previously described and illustrated. Otherwise, damage and breakage of the rack or pinion gear teeth will result.

Caution: Do not overtorque the ball socket housing. Overtorquing will cause failure of the steering parts.

RECIRCULATING BALL MANUAL STEERING GEARBOX

Design and Operation

In the past, vehicle manufacturers used the *recirculating ball manual steering gearbox.* The wormshaft is connected to the steering wheel shaft by a flex coupling, which acts as a universal joint and insulator against noise and vibration. The coupling also permits removal of the steering gear assembly and steering column independently. The mechanical features of this steering gear design are the two recirculating ball bearing circuits in which steel ball bearings act as a rolling thread between the wormshaft and the ball nut. Figure 9-11 illustrates that the ball nut is geared (meshed) to the sector gear.

Every driver knows that the steering wheel, when turned to the left or right, directs the front wheels of a vehicle in that direction. To understand how this movement occurs, substitute a king-sized bolt for the steering wheel and shaft. As the bolt is turned to the left or right, the nut will move laterally along the threaded portion of the bolt (Figure 9-12).

Figure 9-13 illustrates steering gear operation. The head of the bolt (the steering wheel) is being turned to the left (counterclockwise). The ball nut is moving down toward the bottom of the threaded portion of the screw (toward the bottom of the wormshaft). The movement of the ball nut causes the sector gear and shaft to rotate. The sector shaft is attached to the **pitman arm**, which directs the lateral movement of the steering linkages and the front wheels.

We now understand the basic operation of the manual steering gearbox. Now we need to understand how the manufacturer reduces the friction between the various moving parts. We learned in Chapter 1 that early humans found it easier to move a loaded platform over round wooden logs instead of a sled. In other words, rolling friction is much less than sliding friction. Under these conditions, nothing rolls better than a round, steel ball bearing.

Modern manufacturers have designed threads (*helical spiral grooves*) to accommodate a number of ball bearings. These bearings are free to roll in a spiral circuit between the ball nut and wormshaft where steering forces are transferred with a minimum of friction. Possible binding of the bearings is prevented by the continuous recirculation of the

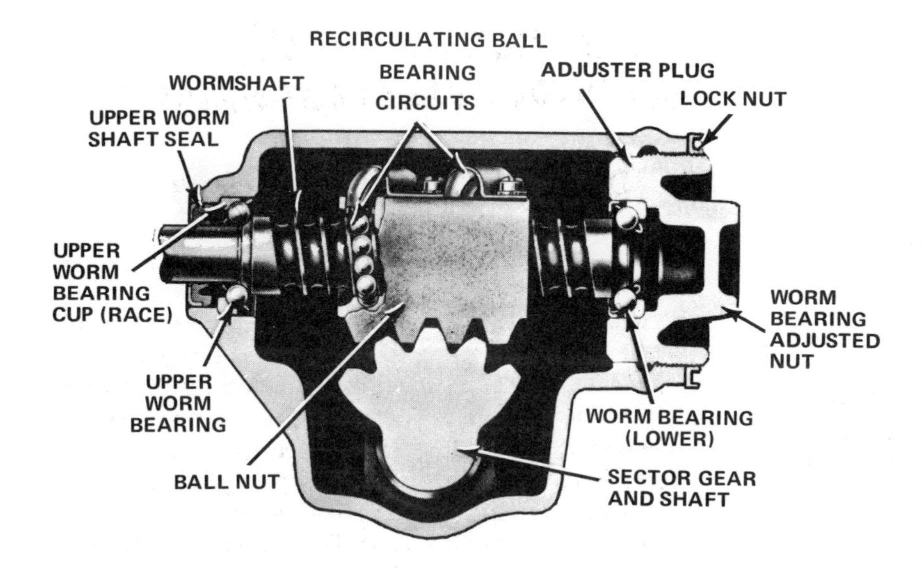

FIGURE 9-11. Recirculating Ball Manual Steering Gearbox. (Courtesy of General Motors of Canada, Limited)

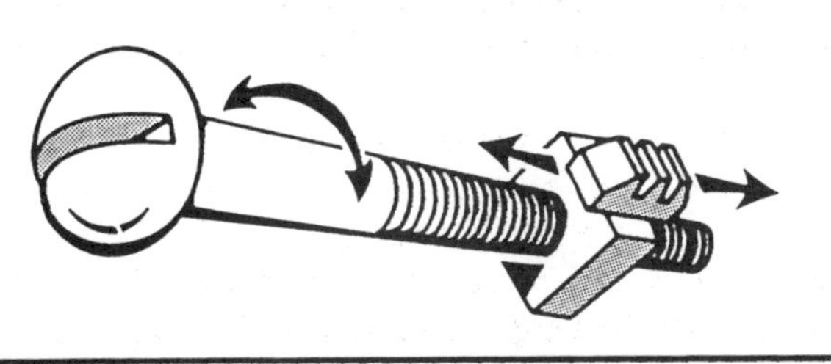

FIGURE 9-12. Basic Steering Wheel Operation. A nut and bolt are a good illustration of this operation.

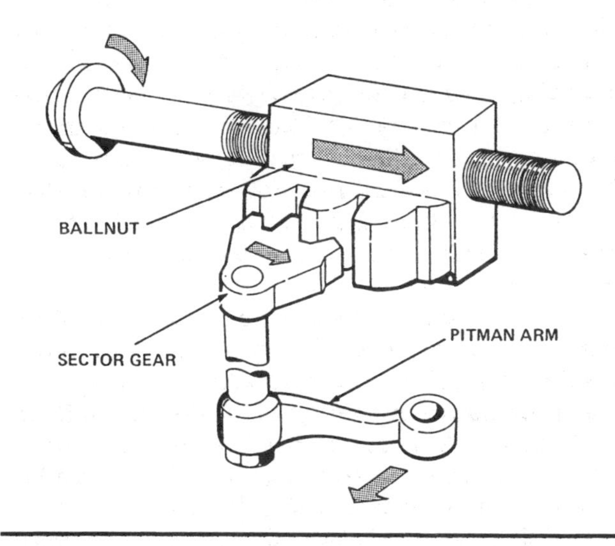

FIGURE 9-13. Steering Gear Operation.

bearings through ball circuit guides. To reduce rotary friction further, the manufacturer also uses caged upper and lower wormshaft bearings positioned between the wormshaft and the steering gear housing (Figure 9-14).

Figure 9-15 illustrates a front cross-sectional view of the steering gearbox. The gear faces of the ball and sector gear are compatibly tapered, thereby providing full gear face contact area. Vertical movement of the sector shaft is limited and controlled by the lash adjuster screw. Friction and lateral movement of the sector shaft is governed by the upper and lower sector shaft bushings. The upper wormshaft and lower sector shaft seals retain gear lubricant within the steering gear housing.

Diagnosing Problems

Problems generally associated with the malfunction of the manual steering gear mechanism can be discussed under the following headings. If any of these conditions exist, the steering wheel will not recenter itself after the vehicle has been steered around a corner.

Excessive Play Experienced at the Steering Wheel. This problem is often attributed to worn ball joints or loose tie-rod ends, when in fact it is caused by excessive clearance between the gear faces of the ball nut and sector gear. Insufficient torque of the wormshaft bearings or incorrect assembly of the ball bearings located in the ball nut will also produce a similar problem. In Figure 9-16, note the clearances between the gear faces of the ball nut and sector gear. Also, note the insufficient bearing preload of the wormshaft and the missing bearings in the ball nut circuit. All are factors that contribute to excessive play.

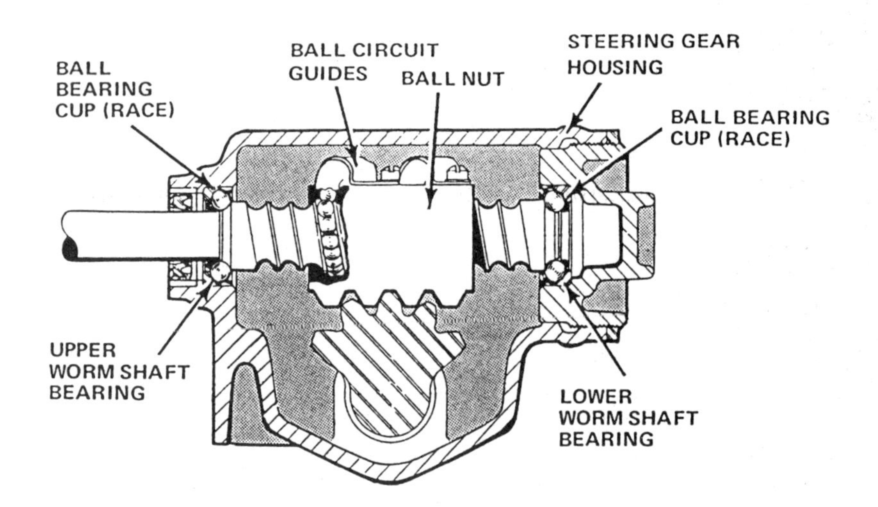

FIGURE 9-14. Cross-Sectional View of Gearbox Showing Recirculating Ball Bearings. (Courtesy of General Motors of Canada, Limited)

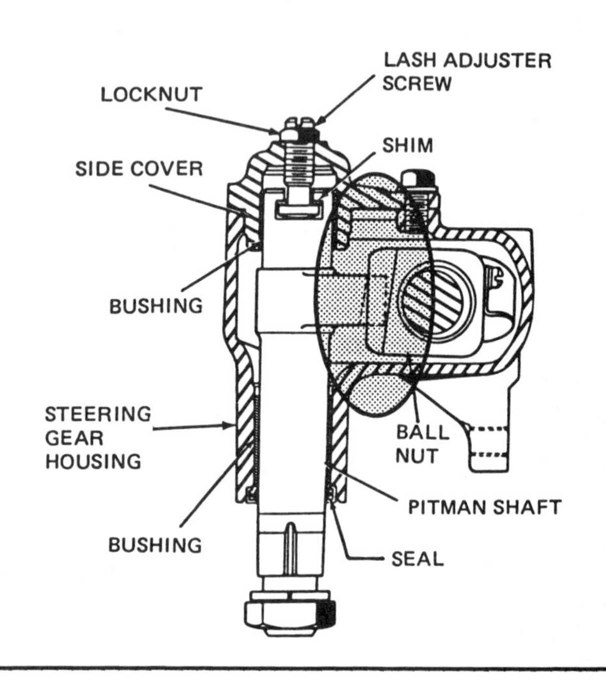

FIGURE 9-15. Front Cross-Section of Steering Gear Housing. (Courtesy of General Motors of Canada, Limited)

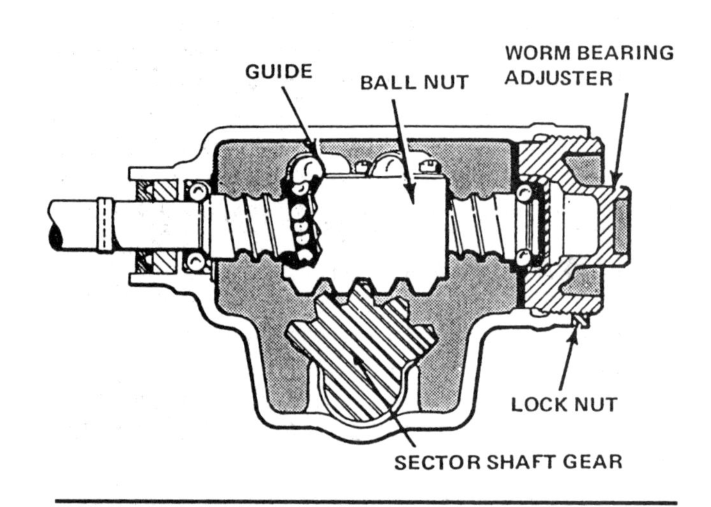

FIGURE 9-16. Steering Gearbox Defects. (Courtesy of General Motors of Canada, Limited)

Additional Steering Wheel Effort. Additional effort required to turn the steering wheel is not a common problem but will occur if there is a loss of lubricant or insufficient or improper lubricant in the steering gearbox mechanism. It is also possible for this problem to occur if an inexperienced person were to torque the wormshaft bearings excessively or cause excessive gear mesh between the ball nut and sector gear.

Steering Wheel Roughness. Roughness as the steering wheel is rotated is generally caused by metal fatigue at the wormshaft ball bearings or bearing races.

Loss of Steering Gearbox Lubricant. Although loss of gearbox lubricant is attributed to the sector shaft seal, it is normally caused by excessive clearance between the sector shaft and the upper and lower sector shaft bushings within the steering gearbox housing. The sector shaft (at the seal area) may also be subject to metal fatigue.. A brown, pitted rust area along the faces of the sector gear, ball nut, or sector shaft is caused by water entering the gear at the seal area (Figure 9-17).

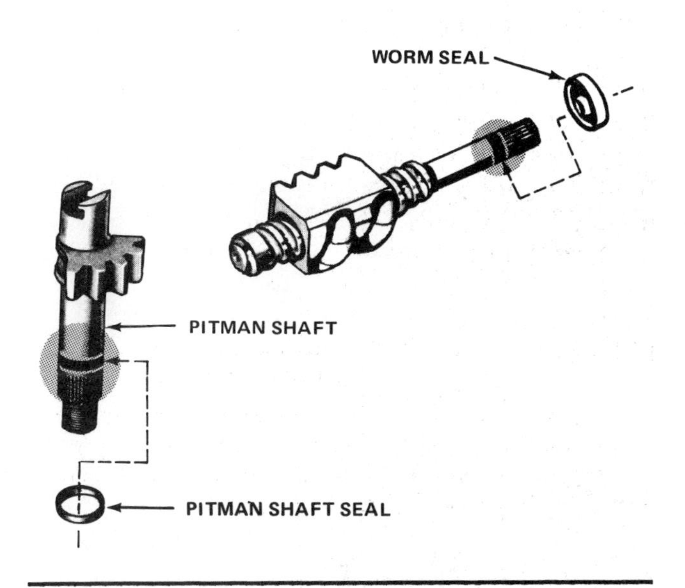

FIGURE 9-17. **Places for Possible Loss of Lubricant.** (Courtesy of General Motors of Canada, Limited)

Servicing the Manual Steering Gearbox

Removing the pitman Arm from the pitman Shaft. Endeavoring to remove the pitman arm from the pitman shaft can be difficult. Figure 9-18 illustrates how to separate these two parts. Since the eye of the pitman arm fits tightly over the tapered splines of the sector shaft, a puller is required to remove the pitman arm from the sector shaft. Attach a puller, as illustrated in Figure 9-18, and with an appropriate-sized wrench, turn the hex head of the puller's center bolt clockwise. This procedure applies tension against the bottom face of the sector shaft.

If you are unable to separate the pitman arm from the tapered splines of the sector shaft, you may carefully use an acetylene torch. First, obtain a class B fire extinguisher and know its operation. (Class B indicates that the extinguisher is applicable for use on flammable or combustible liquids, flammable gases, greases, or similar materials. The background of the symbol will be either metallic or red. The symbol will always be found on multipurpose dry chemical, carbon dioxide, and foam fire extinguishers.) Next, remove any excess grease from around the area where the torch is to be used. Wear dark safety glasses and use controlled heat to expand the eye of the pitman arm. Direct the flame away from the steering gearbox. Use the puller to assist in separating the arm from the sector shaft.

Disassembling the Gearbox. For disassembly instructions and descriptive procedures, follow the directions in the manufacturer's shop manual. As the gearbox and parts are disassembled, place all items in a parts tray.

Wash all parts in a nontoxic cleaning solvent, and thoroughly dry by using compressed air. Place the parts on a clean surface. Examine all bushings, bearings, bearing races, and shaft surface areas for signs of metal fatigue, indentation, or damage. Any part that shows signs of damage must be replaced.

Examine the gear faces of the sector gear and ball nut. The gear faces will have some indented lines, which are normal. These indented lines retain gear lubricant to reduce gear face contact wear. Inspect all seals. Any seal that is worn or has been removed from the steering gear housing must be replaced.

Examine the fit of the sector shaft bushing. First remove the sector shaft seal. Install the sector shaft in the gearbox housing and determine the lateral (side-to-side) clearance. Maximum clearance should not exceed 0.05 millimeter (0.002 inch). If these bushings are worn or tapered, a new sector shaft

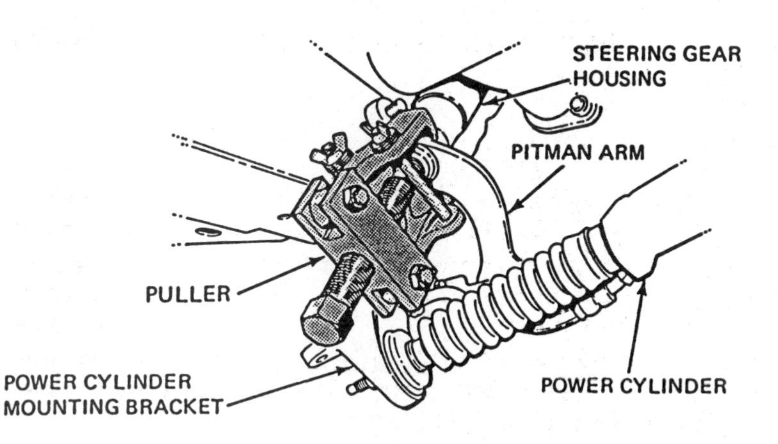

FIGURE 9-18. **Using a Puller to Remove Pitman Arm from Sector Shaft.** (Courtesy of Ford Motor Company)

cover and bushing assembly or housing bushing must be installed.

Assembling the Gearbox. Assemble the parts of the steering gearbox by carefully following the descriptive procedures outlined in the manufacturer's shop manual. Use the torque wrench to determine the bearing preload of the wormshaft after installation in the steering gear housing. Use a feeler gauge to determine the correct clearance between the slot of the sector shaft and the bottom of the lash adjuster screw. When the manual steering gearbox has been completely assembled, use the torque wrench to determine the gear mesh preload reading (between the ball nut and sector gear) while the wormshaft is being turned through the over-center straight-ahead position of its rotational travel. See the manufacturer's shop manual for correct adjustment and torque specifications.

Installing the Gearbox. Prior to installing the steering gearbox on the vehicle, fill the gear housing with the correct lubricant. Follow the manufacturer's shop manual installation procedures. When aligning the front wheels of the vehicle, you may be required to adjust the gearbox in the vehicle.

OTHER MANUAL STEERING GEARBOX MECHANISMS

Gemmer Worm and Roller

This type of steering gear is used by some industrial and commercial vehicle manufacturers (Figure 9-19). The steering shaft and worm assembly rotates within upper and lower bearings and meshes with the sector shaft roller, which is an integral part of the sector shaft. When the steering wormshaft is rotated, the sector shaft roller moves along the gear face of the worm gear. The worm and roller gear mesh adjustment is maintained by a lash adjusting screw which is threaded into the sector gear housing cover.

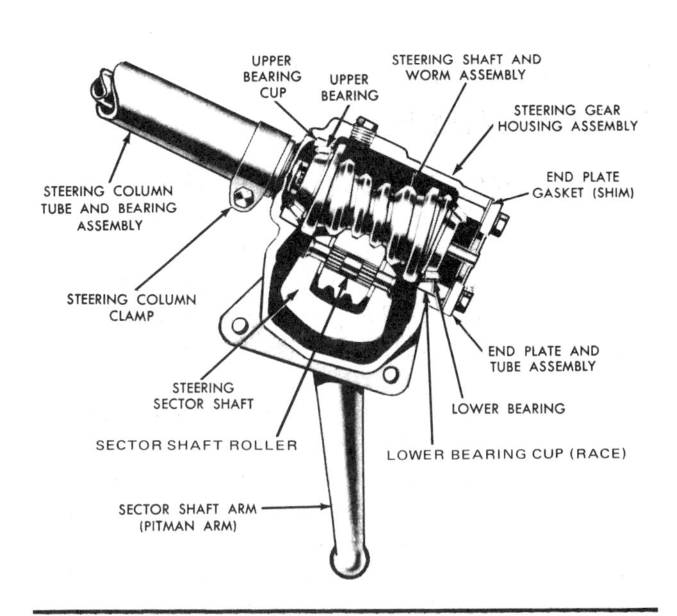

FIGURE 9-19. Gemmer Worm and Roller Steering Gear. (Courtesy of Ford Motor Company)

Ross Cam and Lever

This steering gear mechanism is used by some commercial truck manufacturers (Figure 9-20). The input wormshaft is designed similar to a rotary cam.

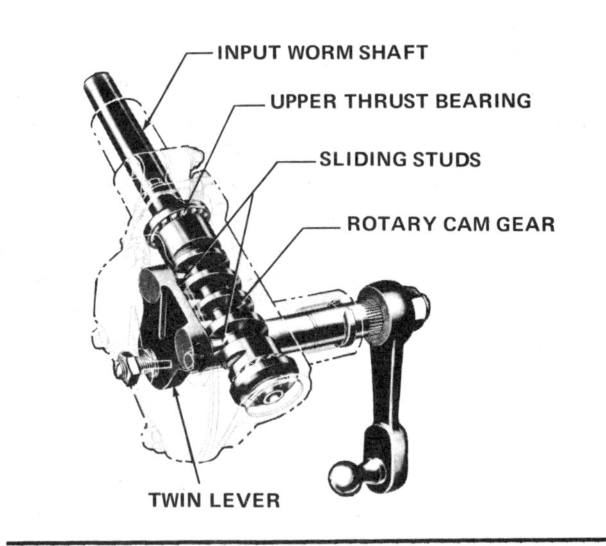

FIGURE 9-20. Ross Cam and Lever Steering Gear. (Courtesy of Ford Motor Company)

Both studs of the twin lever gear engage the cam during normal, straight-ahead driving. As the steering action moves away from the normal driving position into the parking range, one of the levers disengages the cam. The effective leverage of this single stud increases very rapidly. Therefore, the driver has a dual ratio gear. This gear automatically changes from one ratio to another to suit the steering requirements of highway speeds and parking maneuvers.

STEERING LINKAGE MECHANISMS

Design and Operation

The most common type of *steering linkage mechanism* used by North American vehicle manufacturers is the **parallelogram linkage** design. It is called this because the left and right tie-rods are parallel to the left and right lower control arms.

The linkage system consists of seven major parts, shown in Figure 9-21: two adjustable tie-rods, one horizontal relay rod, one pitman arm, one idler arm, and two steering arms. Most steering linkage parts are ball and socket design (similar to a ball joint). This design allows multiaxis movement, since the vehicle's suspension system moves vertically and horizontally. The adjustable tie-rods and tie-rod ends are designed with left- and right-hand threads, allowing for front wheel toe adjustment. (Toe adjustment is discussed in Chapters 17 and 20.) The pitman arm supports the left side of the relay rod since it is attached to the sector shaft. The **idler arm** supports the right side of the relay rod since it is attached to the right side rail of the vehicle's frame.

When the front wheels of a vehicle hit a bump, the wheels move up and down, resulting in a change of caster and camber. This movement causes the upper and lower control arms to move in their respective arcs. A short tie-rod from each steering arm is attached to the horizontal relay rod. The tie-rod (pivoted at a point on the relay rod) forms a radius to the arc followed by the tie-rod end at the steering arm. This arc differs only slightly from the arc followed by the lower control arm (Figure 9-22). As a result, the change in the toe position of the front wheels is minimal, thereby improving the vehicle's directional control and increasing the tread life of the front tires. This parallel relationship between the tie-rod and the lower control arm will not be maintained if the front springs of a vehicle sag, thus

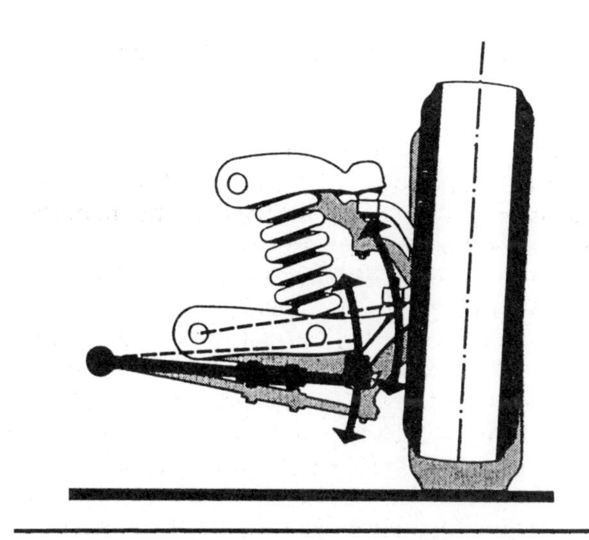

FIGURE 9-22. **Advantage of Parallelogram Design.** The design decreases the toe change in a vehicle's suspension. (Courtesy of Hunter Engineering Company)

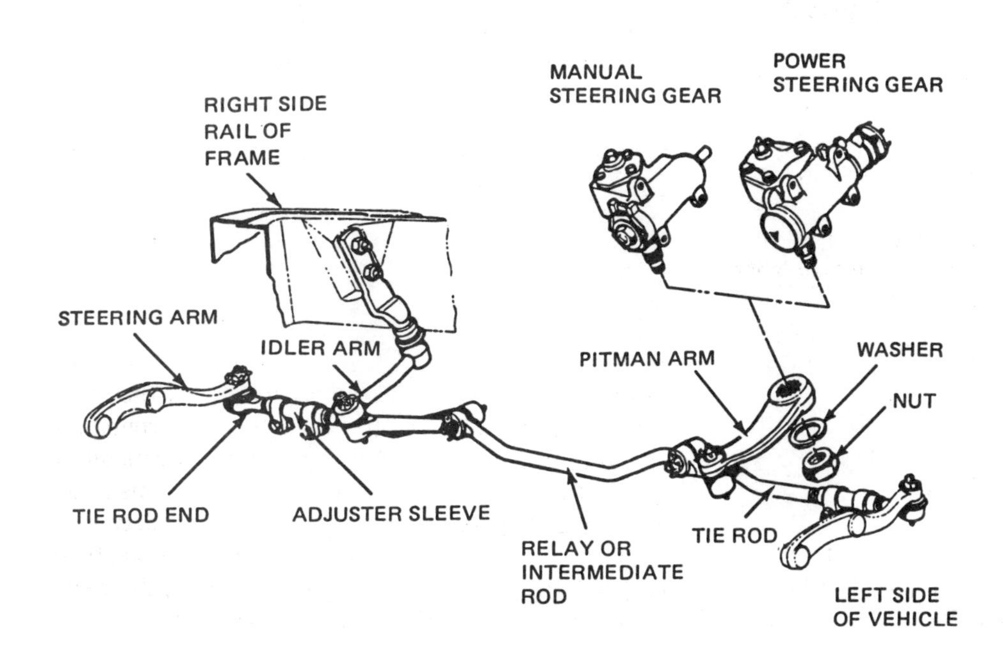

FIGURE 9-21. **Details of Parallelogram Steering Linkage Design.** (Courtesy of General Motors of Canada, Limited)

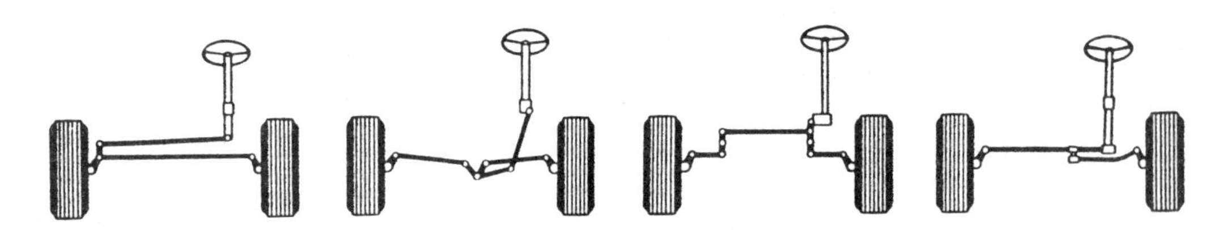

FIGURE 9-23. **Four Steering Linkage Designs.** (Courtesy of Hunter Engineering Company)

affecting the curb riding height. This condition is discussed more fully in Chapter 20.

Note that due to engineering design preferences, vehicle manufacturers may use other types of steering linkage mechanisms. Figure 9-23 illustrates four such steering linkage designs.

Diagnosing Problems

There are no established rules for determining the dependability of the steering linkage parts. However, the text will describe some valid procedures that can be used to inspect steering linkage parts.

Checking Relay Rod, Pitman Arm, and Tie-Rods. First, raise the vehicle and position it on safety stands under the front lower control arms. By using hand force, push on the steering linkage parts and visually check for excessive looseness caused by worn internal parts or weak or broken tension springs (Figure 9-24). At the same time, check for damaged or missing rubber dust shields or boots. If these protective parts are missing, the internal parts of the steering linkage have been subject to water and dirt. Replacement may be required.

The second part of the inspection is conducted with the aid of another service technician. If the vehicle is equipped with power steering, the vehicle's engine must be idling. Do not raise the vehicle. Position the front wheels of the vehicle in the straight-ahead position. Request the other service technician to cycle (rotate) the steering wheel from side to side through one-quarter revolution. While the steering wheel is being turned, once again visually observe and inspect the tie-rods and the other component parts for slack. This method of inspection allows the steering linkage to operate under maximum stress. If linkage parts indicate the slightest slack when they are moved, the parts must be replaced before the front wheels are aligned.

Checking the Idler Arm. A loose *idler arm* can be detected by grasping the relay rod near the idler arm and moving the relay rod vigorously up and down. Excessive looseness indicates that the idler arm is worn and should be replaced (Figure 9-25). Idler arms are often left unnoticed and considered relatively unimportant. However, the slightest

FIGURE 9-24. **Checking for Excessive Looseness.** (Courtesy of Moog-Canada Automotive, Ltd.)

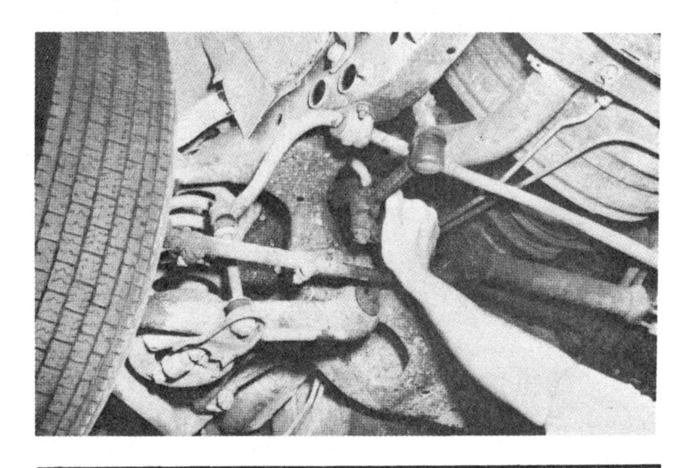

FIGURE 9-25. **Checking Idler Arms for Looseness.** (Courtesy of Moog-Canada Automotive, Ltd.)

looseness of an idler arm can cause an excessive amount of toe change while the vehicle is being driven. Any amount of toe change will scrub the tire treads and soon produce excessive tread wear.

When the front wheels of a vehicle have been lifted off the garage floor, shake the rear of the right front wheel in and out. While this is being done, examine the idler arm for looseness. If looseness in the idler arm allows the wheel to move, the idler arm must be serviced.

Checking the Steering Arms. *Steering arm* inspection is difficult. If the front wheel rim of a vehicle is excessively bent, carefully examine the arms. Measure carefully the distance from the tapered tie-rod stud to the wheel and compare left and right sides. Also visually inspect for disturbance of rust where the steering arm attaches to another part. If the toe-out-on-turns alignment reading does not conform to manufacturer's specifications, you may suspect that one or both of the steering arms is bent or has become distorted.

Servicing and Replacing Steering Linkage Parts

Most steering linkage parts cannot be repaired. If they are bent, worn, or defective, they must be replaced. Figure 9-26 illustrates two different methods of separating and removing a tie-rod end. When the tie-rod end has been removed, measure from the end of the sleeve to the center of the tapered stud. Record the linear measurement. When a new tie-rod end is threaded into the adjusting sleeve, be sure to duplicate the same dimensions to provide an approximate toe setting (Figure 9-27).

Next, loosen the bolt and nut that are part of the adjusting sleeve clamp. If possible, apply penetrating oil to the sleeve and unscrew the tie-rod end. Make sure that you inspect the tapered hole in the steering arm for possible distortion.

Because of their position, tie-rods and tie-rod ends often become seized in the adjustable sleeve. Loosen and remove the sleeve's bolt, nut, and clamp. Place a hammer at the back of the sleeve. Tap the front of the sleeve with another hammer; an air hammer may be used. This procedure will help break the rust between the inside of the sleeve and the part to be removed. Place a light film of lubricant on the threads of the tie-rod end and thread the tie-rod into the adjusting sleeve. Place the dust shield over the tapered stud and install the tapered stud into the steering arm. Thread the castellated nut onto the end of the tie-rod stud and torque the nut to the manufacturer's specifications. *Note:* The nut must never be backed off to insert the new cotter pin. If the castellated nut is backed off, the tie-rod end's stud will become loose in the steering arm, and further damage will result. When tightening the clamp bolt, the bottom of the housing on

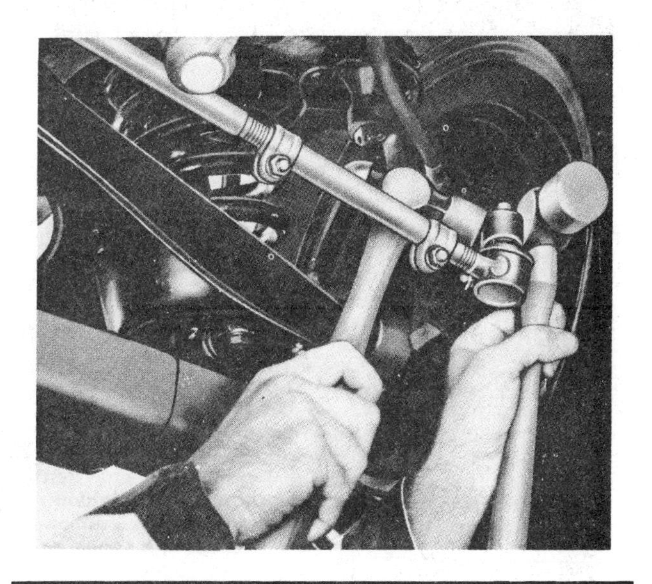

FIGURE 9-26A. Removing a Tie-Rod by Freeing the Ball Stud. (Courtesy of Ford Motor Company)

FIGURE 9-26B. Using a Puller to Remove a Tie-Rod. (Courtesy of Ford Motor Company)

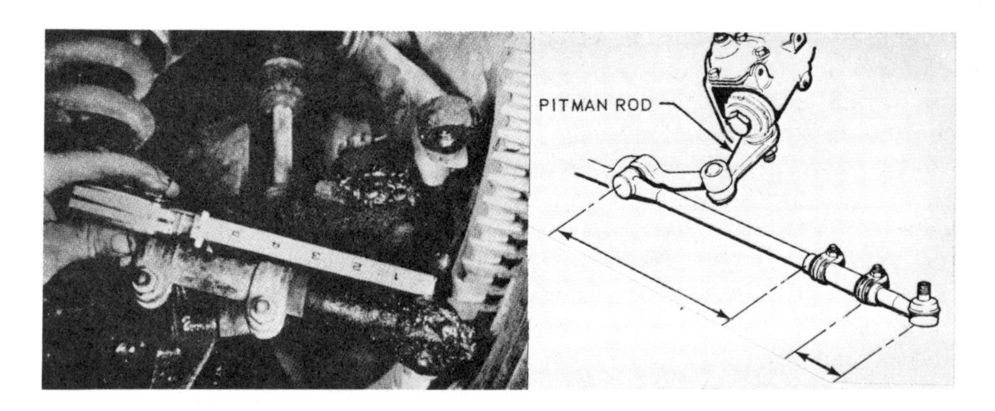

FIGURE 9-27. Measuring Tie-Rod Sleeve. (Courtesy of Moog-Canada Automotive, Ltd.)

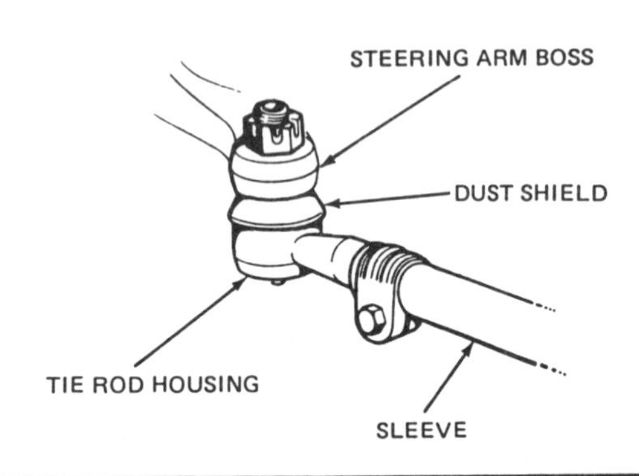

FIGURE 9-28. Positioning of Parts while Reassembling. (Courtesy of Moog-Canada Automotive, Ltd.)

the tie-rod end must be parallel with the steering-arm boss (Figure 9-28).

Some idler arms are similar to tie-rod ends and may be separated from the relay rod by using a tie-rod puller. Other types of idler arms are of rubber bushing design and will require a special tool to remove and install the rubber-mounted bushing. When installing a bushing-type idler arm, make sure that the front wheels are in the straight-ahead position before tightening the thrust locknut. This position will eliminate premature failure of the part and steering wheel pull.

Only some steering linkage parts are designed to be lubricated. On examination of the linkage parts, you may observe a hex head lube plug threaded into the housing of the part. The lube plug must be removed and a lube fitting installed into the housing. A manual lube gun designed to deliver lithium-base lubricant at low pressure 42 to 56 kPa (6 to 8 psi) is used to lubricate the parts, thereby avoiding damage to the special protective seal. Apply the lubricant slowly. There should be no visual evidence of lubricant escaping past the seal. When the lubrication is complete, remove the lube fitting and install the lube plug.

PRACTICAL SERVICE TIPS

1. All steering linkage parts are malleable and will bend, distort, or deflect rather than fracture under extreme shock.
2. Never attempt to heat, weld, or repair bent steering linkage parts.
3. If possible, let the customer see how you perform a steering suspension and linkage test. Explain the procedure in terms that nontechnicians can understand.
4. Although you may replace a vehicle's worn steering and suspension parts, you have not fully repaired the vehicle until you have restored the vehicle to the manufacturer's curb riding height specifications.
5. Always refer to the manufacturer's shop manual for correct repair procedures and specifications.
6. After replacing the steering or suspension parts, check and correct, if necessary, the alignment of the vehicle.

REVIEW TEST

1. Briefly explain what happens to the steering column in a severe frontal collision.
2. Define the terms *mechanical advantage* and *overall steering ratio.*
3. Why is rack and pinion steering used on small vehicles?

4. Describe the operation of a rack and pinion steering gear.
5. Summarize how to diagnose the operation of a rack and pinion steering gear and its linkages.
6. Describe how to separate a tie-rod from the rack gear.
7. List the major parts of a recirculating ball manual steering gearbox.
8. Describe the operation of a recirculating ball manual steering gearbox.
9. Briefly explain how maximum gear face contact area is maintained between the gear faces of the ball nut and sector gear.
10. Describe how to separate the pitman arm from the pitman shaft.
11. Describe the design of the Ross cam and lever gearbox.
12. List the parts of a parallelogram steering linkage design.
13. Describe how to diagnose wear in steering linkage parts.
14. Describe how to replace a defective tie-rod end.

FINAL TEST

This examination is multiple choice. Only one answer will be accepted. Carefully read every statement.

1. Mechanic A says that if the steering wheel turns six revolutions and the pitman shaft turns one-quarter of a revolution, the mechanical advantage is 6:1. Mechanic B says that the greater the weight supported by the front wheels of a vehicle, the greater the mechanical advantage of the steering gear mechanism. Who is right? (A) Mechanic A, (B) Mechanic B, (C) Both A and B, (D) Neither A nor B.
2. Mechanic A says that standard bolts and nuts may be used as fasteners when reassembling a steering column. Mechanic B says that if you are required to remove a steering wheel from a steering shaft, strike the shaft by using a ballpeen hammer. Who is right? (A) Mechanic A, (B) Mechanic B, (C) Both A and B, (D) Neither A nor B.

Note: Questions 3 to 8 refer to the rack and pinion steering gear.

3. Mechanic A says that rack and pinion steering has a lower mechanical advantage than a recirculating ball manual steering gearbox. Mechanic B says that because rack and pinion steering has fewer pivotable parts, it is more sensitive to rough road conditions. Who is right? (A) Mechanic A, (B) Mechanic B, (C) Both A and B, (D) Neither A nor B.
4. Mechanic A says that a machined input pinion gear meshes with the rack gear, which moves laterally inside the gear housing assembly. Mechanic B says that the movement of the rack gear is transmitted through ball housings that are threaded onto the ends of the rack. Who is right? (A) Mechanic A, (B) Mechanic B, (C) Both A and B, (D) Neither A nor B.
5. Mechanic A says that with the front wheels raised off the floor, free movement of the steering wheel must not exceed 9.5 millimeters (0.375 inch). Mechanic B says that with the front wheels raised off the floor, free movement of the steering wheel must not exceed 19 millimeters (0.750 inch). Who is right? (A) Mechanic A, (B) Mechanic B, (C) Both A and B, (D) Neither A nor B.
6. Mechanic A says that the only way to measure ball joint looseness accurately is to use a ruler. Mechanic B says that a slight amount of axial movement in a tie-rod end is acceptable. Who is right? (A) Mechanic A, (B) Mechanic B, (C) Both A and B, (D) Neither A nor B.
7. Mechanic A says that the acceptable way to determine looseness in the inner tie-rod socket is to feel the part. Mechanic B says that it is permissible to have a slight looseness in the rubber insulators that attach the gear housing to the frame. Who is right? (A) Mechanic A, (B) Mechanic B, (C) Both A and B, (D) Neither A nor B.
8. Mechanic A says that with the front wheels supporting weight and the steering wheel rotating while the vehicle is stationary, some knock is acceptable. Mechanic B says that a faint knock, produced by the steering gear, while driving on an extremely rough road, is not uncommon and will not affect the operation of the gear assembly. Who is right? (A) Mechanic A, (B) Mechanic B, (C) Both A and B, (D) Neither A nor B.

Note: Questions 9 to 12 refer to the recirculating ball manual steering gear.

9. Mechanic A says that inside the gearbox, the wormshaft meshes directly with the sector gear.

Mechanic B says that the gear faces of the ball nut and sector gear are tapered, thus providing full gear face contact. Who is right? (A) Mechanic A, (B) Mechanic B, (C) Both A and B, (D) Neither A nor B.

10. Mechanic A says that the vertical movement of the sector shaft is controlled by the lash adjuster screw. Mechanic B says that the lateral movement of the sector shaft is governed by upper and lower sector shaft bushings. Who is right? (A) Mechanic A, (B) Mechanic B, (C) Both A and B, (D) Neither A nor B.
11. Mechanic A says that excessive play at the steering wheel, which is caused by too much clearance between the ball nut and the sector shaft gears, is often attributed to worn ball joints. Mechanic B says that the steering wheel will not recenter itself if there is excessive gear mesh between the ball nut and sector gear. Who is right? (A) Mechanic A, (B) Mechanic B, (C) Both A and B, (D) Neither A nor B.
12. Mechanic A says that it is never necessary to check the clearance between the lash adjuster screw and the sector shaft. Mechanic B says that gear mesh preload is determined through the over-center straight-ahead position of the steer-wheel. Who is right? (A) Mechanic A, (B) Mechanic B, (C) Both A and B, (D) Neither A nor B.
13. Mechanic A says that the pitman arm is located on the left side, and the idler arm is positioned on the right side of the relay rod. Mechanic B says that if the caster, camber, and curb riding height are correct, the tie-rod will be almost parallel to the lower control arm. Who is right? (A) Mechanic A, (B) Mechanic B, (C) Both A and B, (D) Neither A nor B.
14. Mechanic A says that if you shake the right front wheel and the idler moves vertically less than 6 millimeters (0.236 inch), the part does not need to be replaced. Mechanic B says that if the toe-out-on-turns alignment reading does not conform to manufacturer's specifications, you may suspect a bent steering arm or tie-rod. Who is right? (A) Mechanic A, (B) Mechanic B, (C) Both A and B, (D) Neither A nor B.
15. Mechanic A says that if you cannot insert a new cotter pin through the nut and tie-rod stud, back off the nut. Mechanic B says that after installing a new rubber-bushed idler arm, turn the steering wheel to the extreme right and torque the thrust locknut. Who is right? (A) Mechanic A, (B) Mechanic B, (C) Both A and B, (D) Neither A nor B.
16. Mechanic A says that even though a vehicle's steering and suspension parts are new and the curb riding height is correct, the wheel alignment should be checked. Mechanic B says that since the steering parts are malleable and will bend, it is permissible to weld or straighten steering linkage parts. Who is right? (A) Mechanic A, (B) Mechanic B, (C) Both A and B, (D) Neither A nor B.

Power Steering

LEARNING OBJECTIVES

After studying this chapter, you should be able to:

- Explain the basic working principle of a hydrostatic hydraulic system,
- Explain the purposes of the various valves located in a power steering system and describe their operation,
- Describe how a power steering pump creates oil flow,
- Explain how to diagnose the performance capabilities of the Ford slipper-type power steering pump and steering gear assembly,
- Describe, with the aid of a shop manual, how to service or repair a Ford slipper-type power steering pump,
- Explain how to diagnose the performance of the Saginaw power steering pump.

INTRODUCTION

Power steering systems were originally developed by construction and industrial equipment manufacturers. After extensive use and some minor improvements in design, the modified systems were adapted to commercial highway vehicles. During the early 1950s, automotive manufacturers introduced optional power steering to the motoring public. Its benefits have been recognized to the point where most vehicles manufactured in North America are equipped with **hydraulic** power steering. There are two power steering systems used by vehicle manufacturers: nonintegral and integral. In a *nonintegral power steering system*, a belt, driven by the engine's crankshaft pulley operates the hydraulic power steering pump (Figure 10-1). Oil flow from the pump is transmitted through a flex line to a direc-

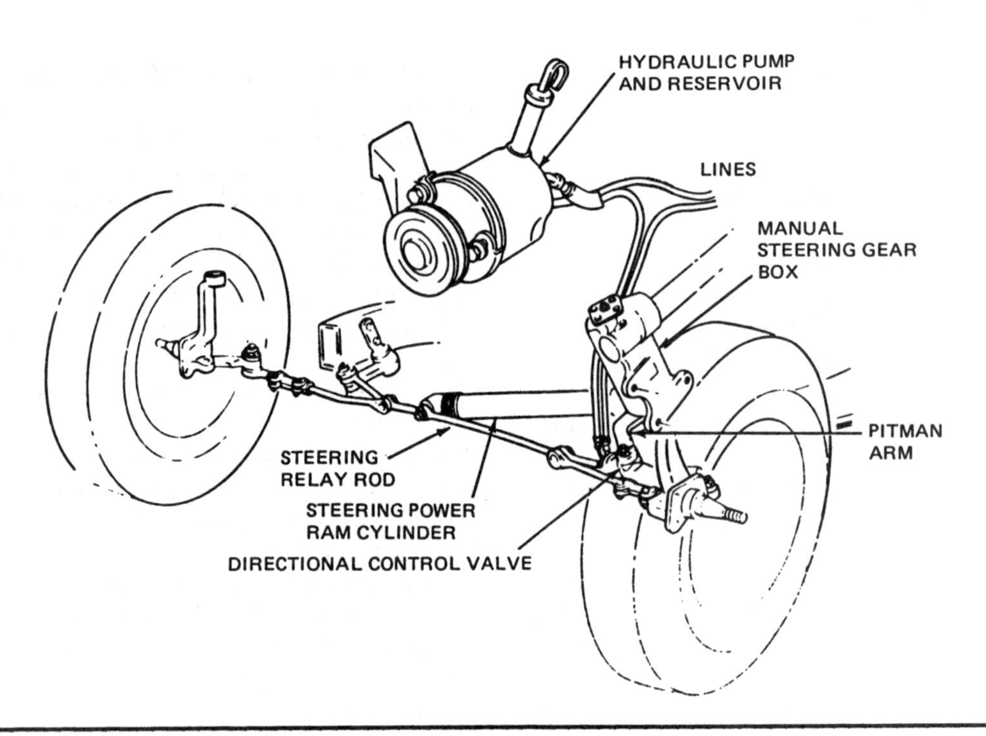

FIGURE 10-1. Nonintegral Power Steering System. (Courtesy of Ford Motor Company)

tional control valve attached to the steering relay rod. The valve, operated by the steering action of the driver, through the manual steering gear, pitman arm, and the turning resistance encountered by the front wheels, directs the flow of oil to the steering power ram cylinder. The power cylinder is connected to the steering linkage and to a bracket attached to the vehicle's frame. Once the oil has passed from the pump to the directional control valve and the power cylinder, it is returned to the pump's reservoir.

In an *integral power steering system*, a belt, driven by the engine's crankshaft pulley, operates the hydraulic power steering pump (Figure 10-2). Oil flow from the pump is transmitted through a flex line to a directional control valve, which is an integral part of the power steering gear. Inside the gearbox, oil is directed to the rack piston from the directional control valve based on the turning resistance encountered by the front wheels. Hydraulic oil from the power steering gear is returned to the pump's reservoir and is continuously recirculated through the hydraulic steering system.

Although rack and pinion power steering is different in design, it is also classified as integral power

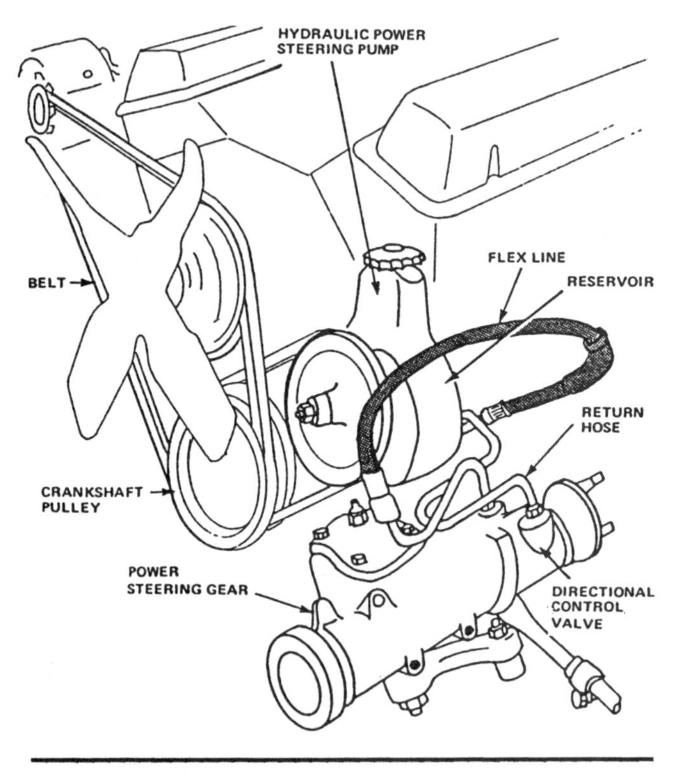

FIGURE 10-2. Integral Power Steering System. (Courtesy of General Motors of Canada, Limited)

steering. This type of steering mechanism is discussed further in Chapter 11.

PRINCIPLES OF HYDROSTATICS

Without question, power steering reduces the effort required to turn the steering wheel and is a definite asset in providing assistance when maneuvering a vehicle in a confined parking area. To understand how a power steering system actually functions will require a basic knowledge of a hydrostatic hydraulic system.

Hydrostatics is basically the study and the application of how the potential (static) energy of liquid under pressure is used to perform work. Examples are the hydraulic floor jack used to raise a vehicle and the power steering system used to assist the driver in steering a vehicle. To understand how flow and pressure are created in a hydraulic power steering system, let's first assume that liquids for practical purposes are almost incompressible. Although liquid is said to be incompressible, in a high-pressure hydraulic system, oil does compress slightly. For example, at 21,000 kPa (3000 psi), oil loses approximately 1.2 percent of its volume. This small volume depreciation is not significant in a vehicle's hydraulic power steering system.

The Basic Hydraulic System

To understand the design of a power steering system and the purpose of its parts, let's begin by examining a *basic hydraulic system.* In Figure 10-3, a force of 4.5 kilograms (10 pounds) is applied on the piston in cylinder A, having an area of 2.54 square centimeters (1 square inch). Then the pressure created on the oil just below the piston will be approximately 70 kPa (10 psi). Since a liquid is essentially incompressible, the displaced fluid will go to cylinder B. The pressure created in cylinder B and on the piston in this cylinder will be 70 kPa (10 psi). If the piston in cylinder B has an area of 64.5 square centimeters (10 square inches), the force that the piston will be able to support will be 45 kilograms (100 pounds). Thus, a small force applied to a small piston in cylinder A can be transformed to a much greater force acting on a larger piston. However, the movement of the larger piston through a given distance requires that the smaller piston be moved a greater distance.

For example, in Figure 10-3, if the piston in cylinder A were moved down 25.4 centimeters (10 inches), then the volume of oil displaced would be 25.4 centimeters (10 inches) times 2.54 square centimeters (1 square inch) (the area of the piston), or 163.8 cubic centimeters (10 cubic inches). To find the distance that the piston would move up in cylinder B, divide 163.8 cubic centimeters (10 cubic inches) by the area of the piston, 64.5 square centimeters (10 square inches). The distance will be 2.5 centimeters (1 inch).

Figure 10-4 illustrates a slightly more complex hydraulic system, where cylinder A is a simplified hand pump. Any force put on the piston in cylinder A will exert a greater force on the piston in cylinder B. Several strokes of the hand pump are required to move piston B through its complete stroke. Since the volume of cylinder A is smaller than that of cylinder B, it is necessary to have an additional quantity of oil, which can be obtained from a reservoir.

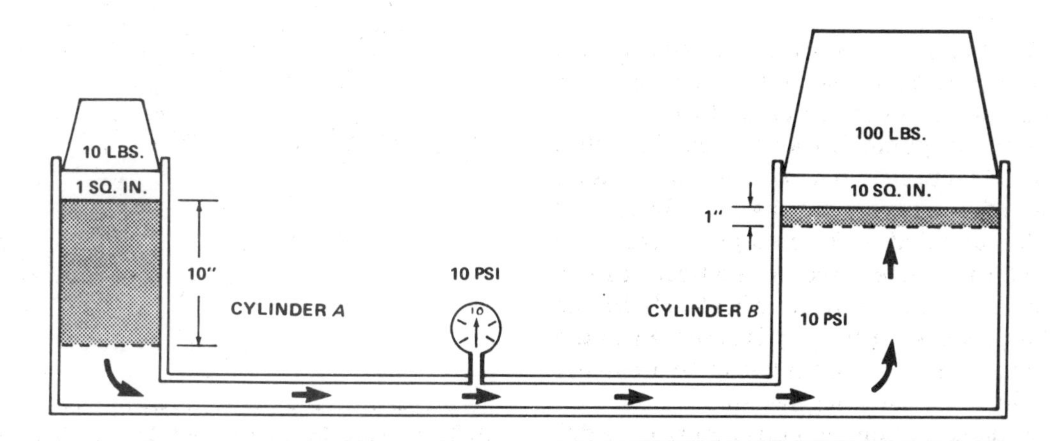

FIGURE 10-3. Simple Hydraulic Power Steering System.

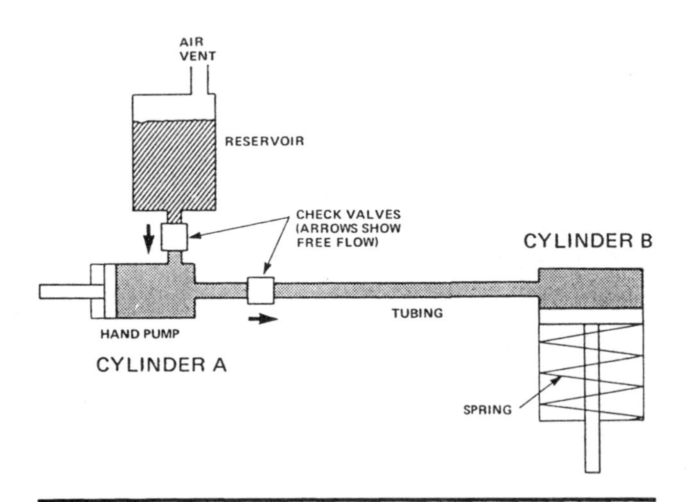

FIGURE 10-4. Hydraulic System Using Simple Hand Pump.

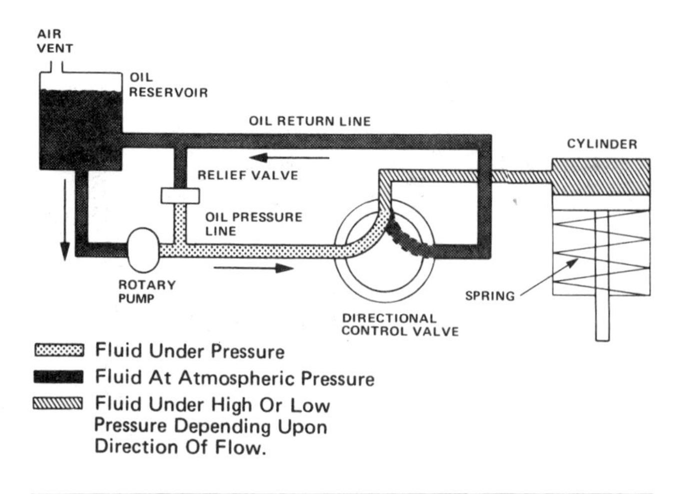

FIGURE 10-5. Schematic of Directional Control Valve. This valve will direct the hydraulic fluid where it is needed.

Check valves (one-way valves) are needed to permit the oil to be pumped from the reservoir to the cylinder at an increased pressure.

This system is impractical since once the piston is extended, the oil cannot return to the reservoir. For this reason, a directional control valve and a return line to the reservoir must be installed, as illustrated in Figure 10-5. A directional control valve is a device capable of directing the flow of oil in one of several directions. When the valve is in the position shown in the figure, the oil may be pumped into the cylinder, and the piston will be forced down. When the valve is in the position of passing oil indicated by the dots, the coil spring will force the piston up, and the oil will be delivered back to the reservoir.

Because it is desirable to operate such a cylinder at frequent intervals, a motor-driven pump is added. Because the *rotary pump* (power steering pump) supplies oil continuously, the check valves have been removed. However, a pressure relief valve has been added. The relief valve is a fundamental part of nearly all hydraulic systems. Without a relief valve, the pump would have to be stopped at the exact moment the piston reached the end of its stroke, or when the directional control valve is placed in a position that stops the flow of oil. If the pump were not stopped, the pressure in the system would quickly build up to a point where rupture of the parts or stalling of the driving motor would occur. The relief valve is set at a predetermined pressure according to the design of the system so that excessive pressure will cause it to open and allow oil to escape and return to the pump's reservoir.

The Hydraulic Power Steering System

Since we are now familiar with the component parts of a basic hydraulic system, let's discuss the purpose of the parts in a *hydraulic power steering system* (Figure 10-6). Every power steering system has three valves. The flow control valve and the pressure relief valve are located in the power steering pump. The directional control valve is separate from the pump.

When a vehicle proceeds along a road, the power steering pump is capable of producing a high rate of oil flow. For example, at 1500 revolutions per minute, the power steering pump is able to produce approximately 8 to 10 liters (1.8 to 2.2 imperial gallons) of oil flow per minute. This capacity of oil flow is more than is needed. To increase the service life of the pump, to reduce internal wear of the pump's component parts, and to prevent oil from overheating, the design engineer has placed a flow control valve into the hydraulic steering system. This valve allows oil that is not needed to return to the pump's reservoir.

When the driver is maneuvering a vehicle in a parking situation, the vehicle's front wheels are against the curb, and the steering wheel has been turned to its extreme rotational position. Due to the design and operation of the power steering gear mechanism, the flow of oil produced by the pump is restricted, and there is no return of oil to the reservoir. Since the pump is capable of producing excess flow and pressure, the pressure relief valve

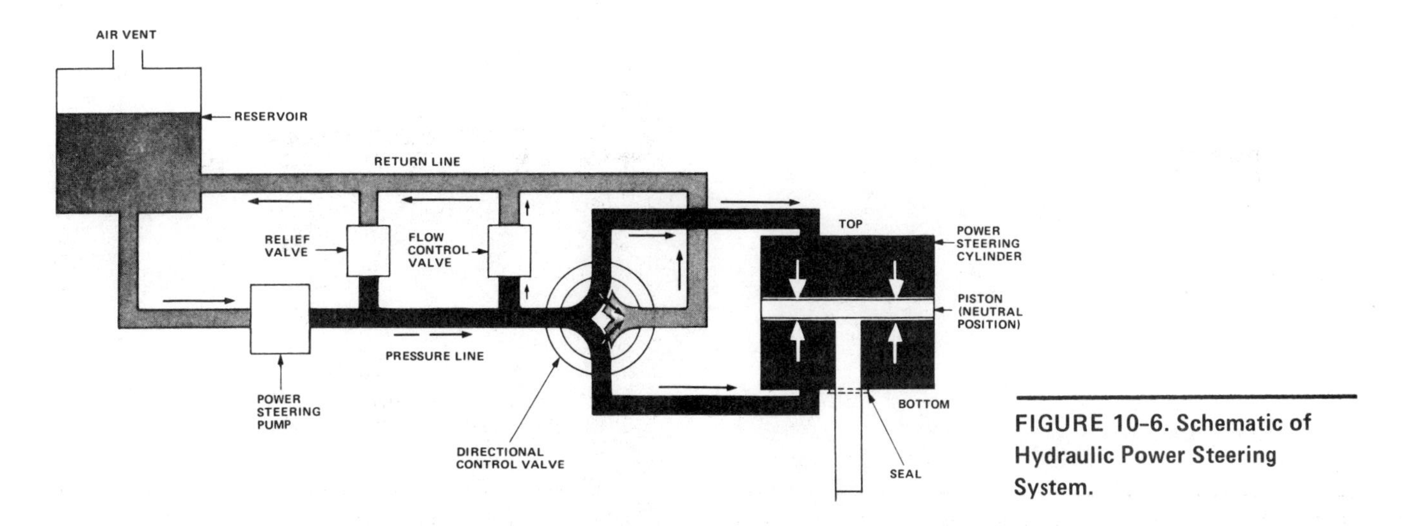

FIGURE 10-6. Schematic of Hydraulic Power Steering System.

opens, thereby controlling the excessive pressure and preventing a rupture of the flex lines and the seals.

As a vehicle proceeds along a road, the directional control valve directs the oil flow, under pressure, to either side of the piston in the power steering ram or the integral power steering gearbox. Based on steering requirements, this valve must also increase the oil pressure to one side of the piston and allow oil on the opposite side of the piston to be returned to the reservoir. The directional control valve is also a restricting device and allows pressure to be developed in the power steering system. The design and operation of the flow control and pressure relief valves is discussed later in this chapter. The design and operation of the directional control valve will be discussed in Chapter 11.

POWER STEERING PUMP OPERATION

Although power steering pumps may be different in design and appearance, they are similar in operation. The power steering pump illustrated in Figure 10-7 is a **constant displacement** slipper-type pump that is integral with the reservoir. The reservoir is attached to the rear of the pump housing. The pumping element is encased within the reservoir. The flow control valve and spring and the pressure relief valve can be externally removed from the pump after the outlet fitting has been unthreaded from the pump housing.

The parts of the pumping element are a rotor, slippers, and springs, which rotate inside a cam insert containing two lobes 180° from each other. These lobes form crescent-shaped chambers between the rotor and cam (see Figure 10-8); the rotor is contained by the inner and outer pressure plates, which mate with the end faces of the cam insert (see Figure 10-7). These ported pressure plates, in conjunction with the cam insert, provide a sealed chamber within which the slippers and rotor are free to move. The manufacturer's manual refers to these parts as the *campump pack.*

As the rotor turns, the slippers are forced outward against the inner surface of the cam insert by a combination of centrifugal force, slipper spring force, and oil pressure acting on the underside of the slippers. When the rotor rotates 90°, the slipper slides outward in its slot, riding on the cam. The volume of the sealed chamber increases, creating a vacuum and setting up a low-pressure area. With the inlet port placed in this area, the chamber will fill with oil. As the rotor rotates from 90° to 180°, the volume of the sealed chamber decreases, creating a pressure area. The pressure or outlet port is located in this area. While this pumping action is going on between 0° and 180°, the same conditions are occurring between 180° and 360°. This combination creates what is known as a *balanced rotor pump.* The two pressures and the two suction quadrants are diametrically opposite. Thus, the rotor is hydraulically balanced, and the drive shaft load is reduced to a minimum.

Idle

With the engine operating at idle speed, the pump output is sufficiently high to provide the desired power assist for turning the front wheels. The flow control valve operates within a chamber (Fig-

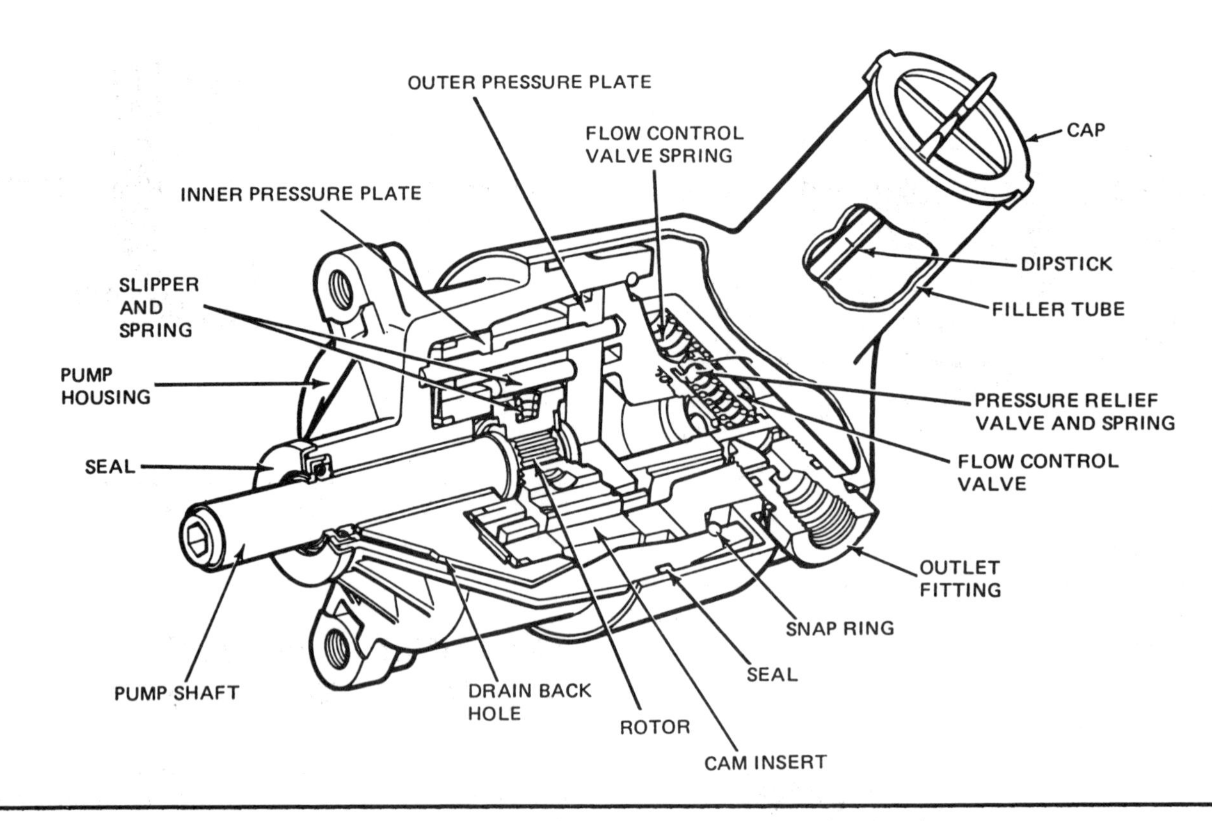

FIGURE 10-7. Ford Slipper-Type Power Steering Pump. (Courtesy of Ford Motor Company)

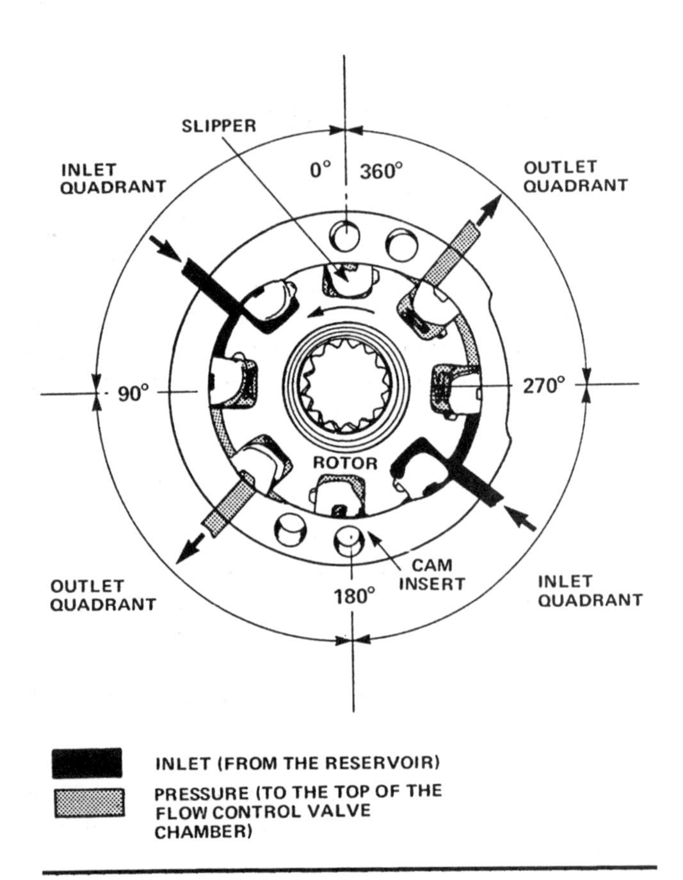

FIGURE 10-8. Rear View of Pumping Element. (Courtesy of Ford Motor Company)

ure 10-9). This chamber is open to pump pressure at one end and open to the pump outlet pressure line on the other end. A spring is enclosed within this chamber. Two pressures act on opposite sides of the control valve. Pump pressure is exerted on one side of the valve, while pressure from the pump outlet line and the enclosed spring exert pressure on the other side of the valve. At engine idle speed, the combined pressure of the outlet line and spring is greater than the internal pump pressure, resulting in the valve's being held in its closed position.

Flow Control

When the engine speed is increased, the pump speed also increases. At these higher speeds, the pump is capable of delivering more fluid than is needed to satisfy the power steering system demands. As this increased output tries to pass through an orifice ahead of the pump outlet line, a pressure differential is created at the two sides of the flow control valve. This difference in pressure causes the flow control valve to move to the position shown in Figure 10-10. When the valve moves as shown, a passage is opened that allows the excess fluid to return to the intake side of the pump.

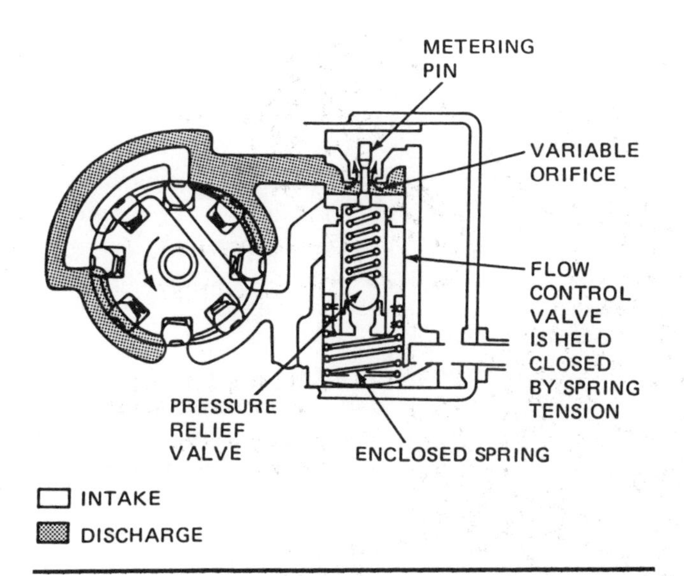

FIGURE 10-9. **Pump Operation: Idle.** (Courtesy of Ford Motor Company)

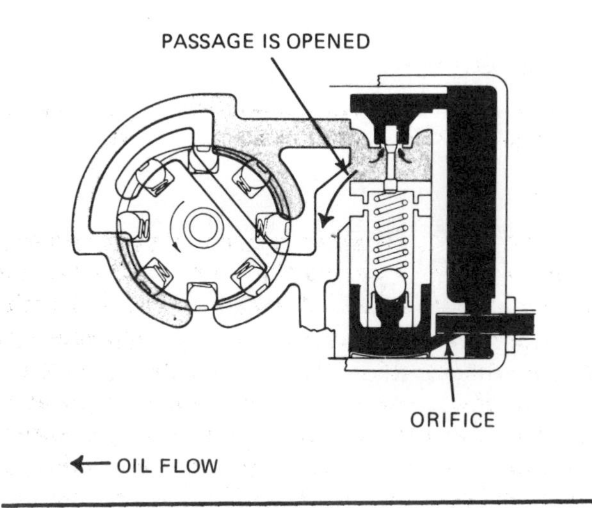

FIGURE 10-10. **Pump Operation: Flow Control.** (Courtesy of Ford Motor Company)

Pressure Relief

A pressure relief valve is built into the flow control valve to prevent the pump pressure from exceeding design limits. When the pump pressure reaches a predetermined limit, the relief spring compresses, and the relief valve is moved off its seat. With the relief valve off its seat, a pressure differential is created at the ends of the flow control valve. This pressure differential allows the flow control valve to open wider, allowing a greater volume of fluid to be recirculated and the pressure to be held at its maximum limit (Figure 10-11).

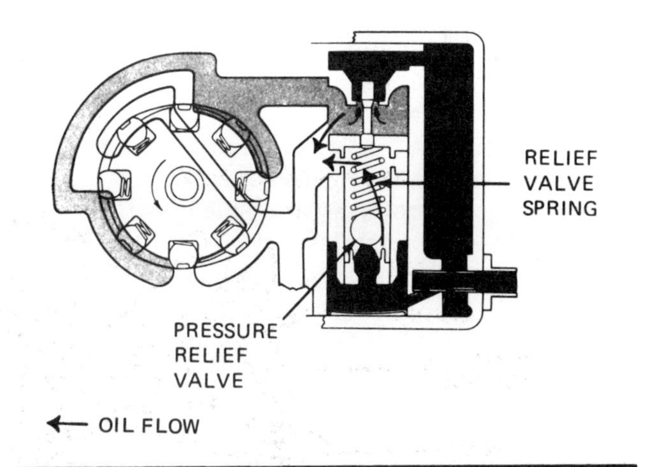

FIGURE 10-11. **Pump Operation: Pressure Relief.** (Courtesy of Ford Motor Company)

DIAGNOSING POWER STEERING PROBLEMS

Modern power steering systems seldom present problems. If the system is suspected of faulty operation, the source of the problem can be determined and corrected when the service technician follows proper diagnostic procedure.

Preliminary Check

Before dismantling the various units of the power steering system, try to obtain from the driver a verbal account of the performance of the steering system. The driver may provide a good description of the problem or may just state that the steering system doesn't operate. Whatever the case, it is recommended that you road test the vehicle, keeping the following questions in mind:

1. Is the steering effort equal on both sides of the steering wheel center position?
2. If the steering effort is not equal, in which direction does the vehicle steer?
3. Does the power steering system produce excessive mechanical or hydraulic noises?
4. Is there a loss of steering assist at low speeds similar to driving a vehicle equipped with manual steering?

If any of these conditions exists, return the vehicle to the service center and examine for obvious possible causes.

Unequal effort on either side of the steering wheel center position will not be caused by the faulty operation of the power steering pump. This complaint may be the fault of an off-centered directional control valve or other factors such as low tire pressure, front end wheel misalignment, a misalignment of a vehicle's frame, steering linkage, and/or suspension, or problems associated with the performance of tires (if radials have been installed).

Mechanical or hydraulic noises may be caused by external or internal problems directly related to the operation of the power steering pump. Mechanical noises may be the result of loose mountings, incorrect drive belt tension, or a glazed or cracked belt or caused by the internal parts of the pump like a broken slipper spring. If the noise is caused by the belt, the noise may be diagnosed by rubbing a bar of soap on the belt when the engine is not operating. Start the engine and determine if the sound has changed its tone or disappeared. If the belt is glazed, cracked, or worn, it must be replaced. Mechanical noises inside the pump may be diagnosed by using a technician's stethoscope (similar to the one used by medical doctors) to listen to the pump when it is operating. The only remedy for this complaint is to remove the pump from the engine, dismantle the pump, and service the unit. Hydraulic noises may be caused by **oil aeration,** low fluid level, oil not at operating temperature, or the pump's operation when the steering wheel has been turned to its extreme rotational position. A slight hydraulic hiss is normal in every power steering system and in no way affects the pump's operation. A noise caused by oil aeration is an uncommon complaint. If it exists, the sound will be a groan between 24 and 72 km/h (15-45 mph) on light acceleration. To correct the problem, the system must be bled or purged of air. This service procedure is outlined in the manufacturer's shop manual.

Loss of power steering assist may be intermittent. It may be caused by incorrect drive belt tension, momentary failure during certain stages of pump operation, or a bypass of oil in the power ram, directional control valve, or integral gear mechanism.

External Leakage Inspection

When the oil is low in the pump's reservoir, you should inspect the entire power steering system for probable external leakage. The following procedures may be used as a guide.

Step-by-Step Procedure

1. Clean the suspected area of any oily residue with a cleaning solvent, and dry by using compressed air.
2. Check all hose connections and cover plate bolts for possible looseness. Tighten if necessary to the manufacturer's specified torque.
3. If the vehicle is inside a building, attach an exhaust hose to the vehicle's tail pipe.
4. Start the engine and allow the oil in the power steering system to reach operating temperature—74°C (165°F) to 80°C (175°F).
5. To assist in the detection of a possible leak, turn the steering wheel to the extreme left and right rotational position.

Caution: Do not hold the steering wheel in the extreme rotational position for longer than 5 seconds. This position causes the pump to produce maximum pressure, and damage may result.

Pump Performance Testing

The primary purpose of a power steering pump is to create and regulate flow. Therefore, the first logical diagnostic step is to determine the capabilities of the steering pump.

Drive Belt. Since the steering pump is operated by a belt, determine the tension of the drive belt. If it is necessary to adjust the belt, loosen the adjusting and pivot bolts on the front face of the pump cover plate to allow the pump to pivot (on eight cylinders, also loosen the nut at the back of the pump) (see Figure 10-12). Then insert the special tool over the boss on the pump housing. Using a ½ inch drive ratchet that fits into the hole in the tool, pry upward to apply tension. When the tension is correct, torque bolts and nut to specification.

Do not pry on the reservoir housing since this action will induce external leakage. Adjust the belt tension to the manufacturer's specification. Tighten the necessary bolts. Manufacturers also recommend that a belt tension gauge be used to determine the tension of the drive belt. To measure belt tension, depress the gauge's plunger, affix the gauge to the power steering belt, release the plunger, and then read the gauge. A new belt or a belt that has been run for less than 15 minutes should have a tension of 534-667 newtons (120-150 pounds). A belt that has been run for longer than 15 minutes should have

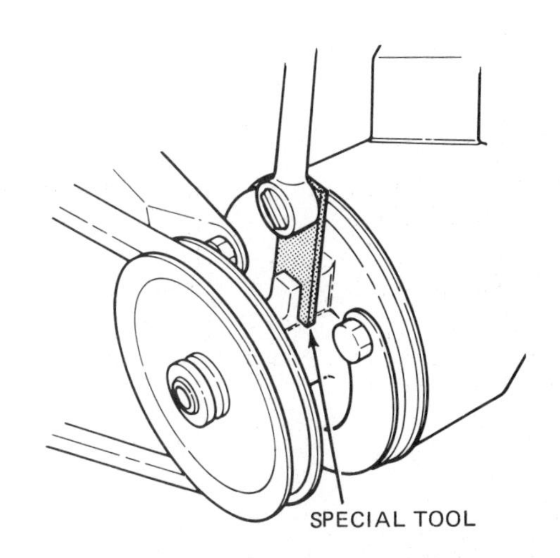

FIGURE 10-12. Adjusting Belt Tension. Insert the special tool over the boss on the pump's housing. Using a 1/2 inch drive ratchet, pry upward to apply tension. (Courtesy of Ford Motor Company)

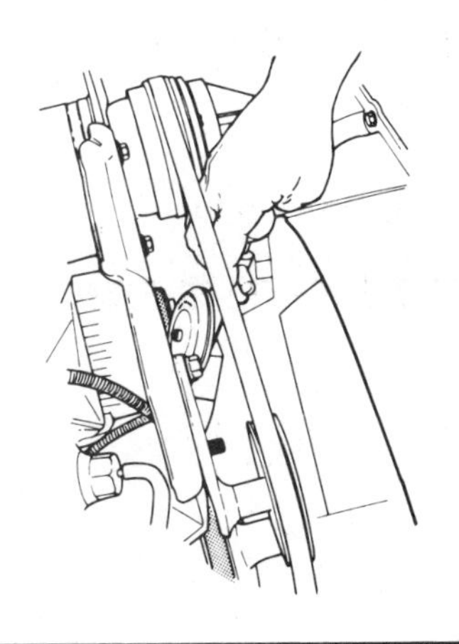

FIGURE 10-13. Measuring Belt Tension. (Courtesy of Ford Motor Company)

a tension of 401-533 newtons (90-120 pounds) (Figure 10-13).

Slipper-Type Pump Flow and Pressure. This test applies to the Ford slipper-type power steering pump. The following items are required:

- An engine (rpm) tachometer,
- Thermometer reading from 18 to 149°C (0 to 300°F),
- Flow pressure analyzer,
- Set of adapter fittings,
- Manufacturer's shop manual.

The test procedure used in conjunction with the Ford rotunda flow/pressure power steering analyzer provides a method for checking the complete power steering system. The analyzer can be used on integral or nonintegral power steering systems to determine the cause of hard steering and/or lack of assist problems.

The analyzer provides readouts for system back pressure, pump flow, and steering gear internal leakage. The interpretation of the readouts will determine which of the following conditions or components is the cause of the problem:

- Restriction in hoses or fittings,
- Sticking gear valve,
- Inefficient campump pack,
- Sticking flow control valve,
- Binding in the suspension.

Step-by-Step Procedure

1. To connect the analyzer to the steering system, remove the pressure fitting from the pump. Connect the fitting to the appropriate adapter on the analyzer.

2. Thread the other adapter of the analyzer into the pump.

3. Connect the analyzer to the adapters. Tighten both connections to 20 newton-meters (15 foot-pounds) maximum (Figure 10-14).

4. Add power steering fluid to the pump if required. Start the engine, and operate it for approximately 2 minutes with the idle set at manufacturer's specifications.

5. Record the following information: (a) flow, in liters per minute (gallons per minute) at 77° ± 2°C (170° ± 5°F) and (b) pressure, in kPa at 77° ± 2°C (psi at 170° ± 5°F) at idle with the gate valve fully open.

If flow is below 7.3 liters per minute (1.6 gallons per minute), the pump may require service. At this point, however, continue the diagnosis. Check the manufacturer's specifications for flow and relief pressure against the vehicle and engine being tested. If pressure is above 1034 kPa (150 psi), check hoses for restrictions.

6. Partially close the gate valve to build up 5100 kPa (740 psi). Observe and record flow. If flow drops to a level lower than the value in the specifications, disassemble the pump, and replace the cam pack (Figure 10-15). If the pressure plates are cracked or worn, replace them. Continue with the diagnosis.

7. Completely close and then partially open

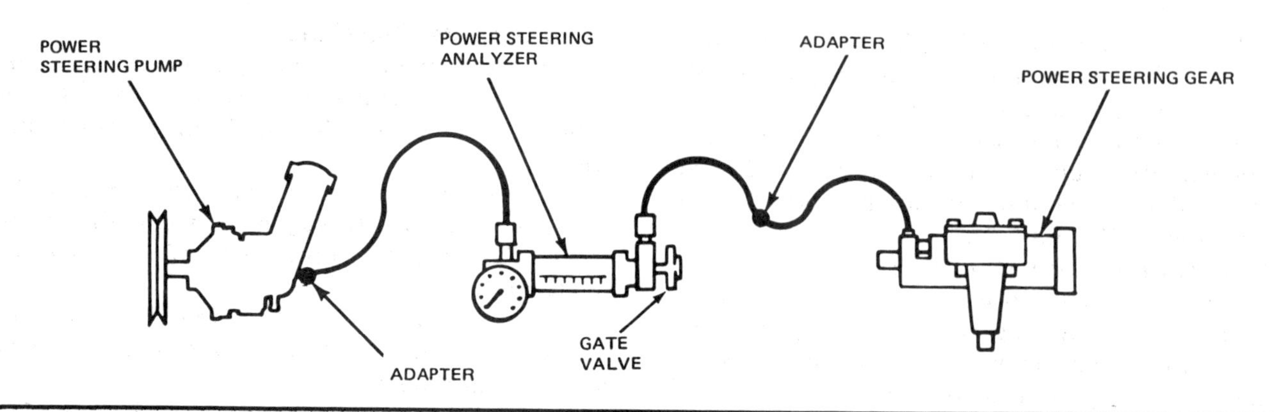

FIGURE 10-14. Connecting Ford Rotunda Flow/Pressure Power Steering Analyzer. (Courtesy of Ford Motor Company)

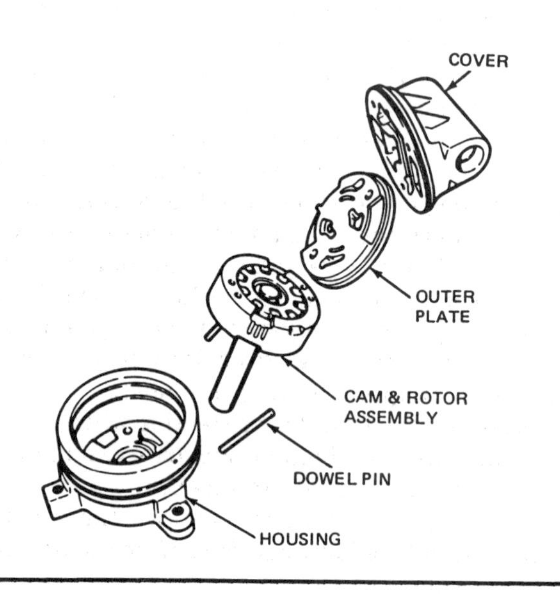

FIGURE 10-15. Disassembling Pump. (Courtesy of Ford Motor Company)

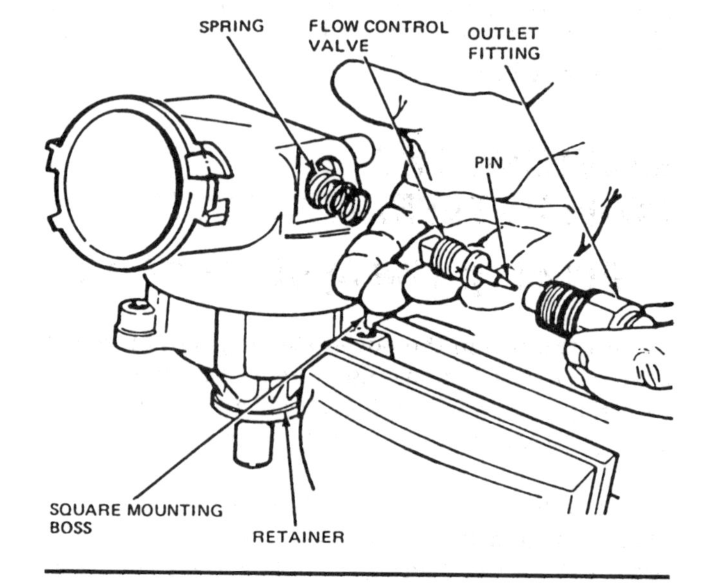

FIGURE 10-16. Removing Outlet Fitting, Flow Control Valve, and Spring. (Courtesy of Ford Motor Company)

the gate valve three times. Observe and record pressure in kPa (psi).

Caution: Do not allow the valve to remain closed for more than 5 seconds. Otherwise, the oil will overheat and reach a boil, resulting in damage to the pump.

8. Check the chart in the shop manual for the applicable pressure specification. If the pressure recorded is lower than the minimum specification listed, replace the flow control valve in the pump (Figure 10-16). If the pressure recorded is above the maximum specification listed, remove and clean or replace the flow control valve in the pump.

9. Increase the engine speed from idle to approximately 1500 revolutions per minute. Observe and record flow in liters per minute (gallons per minute). If the flow exceeds the maximum free flow stated in the specifications, the flow control valve in the pump should be removed and cleaned or replaced (see Figure 10-16).

10. Check the idle speed and set to specification if necessary. With the engine at idle, turn (or have someone else turn) the steering wheel to the left and right stops. Record the pressure and flow at stops. Pressure developed at both stops should be nearly the same as the maximum pump output pressure.

At the same time, the flow should drop below 1.89 liters per minute (0.5 gallon per minute). If the pressure does not reach maximum output or the flow does not drop below 1.89 liters per minute (0.5 gallon per minute), excessive internal leakage is occurring. Remove and disassemble the steering gear. Replace damaged or broken parts. Pay particular attention to the rack piston and valve seals for damage. (Rack and pinion power steering is discussed in Chapter 11.)

11. Turn (or have someone else turn) the steering wheel slightly in both directions, and release it quickly while watching the pressure gauge. The needle should move from the normal back pressure reading and snap back as the wheel is released. If the needle comes back slowly or sticks, the rotary valve in the steering gear is sticking.

12. Remove, disassemble, and clean the rotary valve. If the system is severely contaminated, the gear hoses, control valve, and pump must be completely disassembled and cleaned before reassembly. *Note:* If problems still exist, check the ball joints and steering linkage.

SERVICING THE POWER STEERING SYSTEM

Purging the System of Air

Air trapped in the power steering system causes a whining or moaning noise. The air can be removed by using a power steering pump air evacuator assembly (devac tool). Fabricate as shown in Figure 10-17.

Caution: Under no circumstances should the engine vacuum be utilized. The engine vacuum will draw the power steering oil out of the pump into the engine.

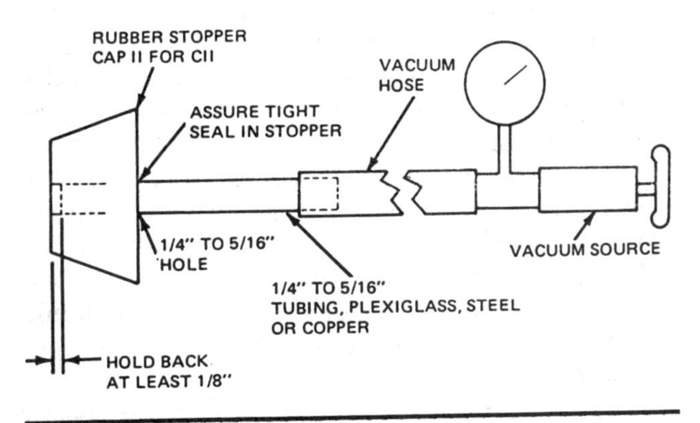

FIGURE 10-17. Power Steering System Purging Tool. (Courtesy of Ford Motor Company)

Step-by-Step Procedure

1. Check and fill the pump reservoir with type F automatic transmission fluid ESW-M2C33-F or equivalent to the cold full mark on the pump dipstick.

2. Remove the engine coil wire. Ground the end of the wire by using a jumper wire to the engine block. Then raise the front wheels of the vehicle off the floor.

3. Crank the engine with the starter motor. Recheck the fluid level. Do not turn the steering wheel at this time.

4. Refill the pump reservoir to the cold full mark on the dipstick. Recrank the engine with the starter motor while cycling the steering wheel from lock to lock. Recheck the fluid level.

5. Insert the rubber stopper of the air evacuator assembly tightly into the pump reservoir fill neck. Reinstall the coil wire.

6. Apply 51 kPa (15 inches mercury) maximum vacuum on the pump reservoir for a minimum of 3 minutes with the engine idling. As air purges from the system, the vacuum will fall off. Maintain adequate vacuum with the vacuum source.

7. Release the vacuum and remove the vacuum source. Refill the reservoir to the cold full mark.

8. Reapply 51 kPa (15 inches mercury) vacuum to the pump reservoir with the engine idling. Cycle the steering wheel from lock to lock every 30 seconds for approximately 5 minutes.

Caution: Do not hold the steering wheel on the stops while cycling. Maintain adequate vacuum with the vacuum source as the air purges.

9. Release the vacuum. Remove the vacuum equipment. Add additional fluid if necessary, and reinstall the dipstick.

10. Restart the engine. Cycle the steering wheel and check for oil leaks at all connections. In severe cases of fluid aeration, it may be necessary to repeat steps 5 through 9.

Flushing the System

Always flush the power steering gear when replacing the pump because of fluid contamination.

Step-by-Step Procedure

1. Remove the pump and pulley. Then service the pump as described in the shop manual.

2. Install the pulley on the serviced pump. Install the pump. Flush the pressure line before installing. *Note:* Connect only the pressure hose.

3. Place the fluid return line from the gear in a container, and plug the reservoir return line.

4. Fill the reservoir with type F automatic transmission fluid ESW-M2C22-F or equivalent.

5. Disconnect the coil wire and ground the wire. Then raise the front wheels of the vehicle off the floor.

6. While adding approximately 2 liters (2 quarts) of fluid, turn on the ignition (using the ignition key). Crank the engine with the starter while turning the steering wheel from left to right.

7. When all the fluid has been added, turn off the ignition. Attach the line to the reservoir.

8. Remove the plug from the reservoir return line, and attach the line to the reservoir.

9. Check the fluid level and, if necessary, add more fluid. *Note:* Do not overfill the reservoir.

10. Lower the vehicle.

11. To expel any trapped air, turn the steering wheel from side to side without hitting the stops.

Replacing Slipper-Type Pump Components

Wash all parts except the seals in a chlorinated solvent. Dry the parts with compressed air. Inspect all parts carefully to determine whether they can be reused or need to be replaced. Note that some components must be replaced regardless of their condition.

Step-by-Step Procedure

1. Reuse the outlet fitting if the corners are not rounded and the threads are intact.

2. Replace all seals except the rotor shaft seal. Reuse the rotor shaft seal if it does not leak.

3. Reuse the housing or housing assembly if the O-ring and the snap ring surfaces are not damaged.

4. Reuse the upper and lower pressure plates if there is no scoring on the wear surface. Polish the phosphate coating, if necessary, but do not remove it.

5. Reuse the rotor and cam assembly if wear is limited to removal of the phosphate coating on the cam contour. Do not disassemble the unit. Push rotor partway through the cam insert, being careful not to dislodge the slippers and springs. Check the cam inside diameter for scoring or burring. Check the rotor faces and outside diameter for scoring and chipping. *Note:* Do not service or refinish the upper and lower pressure plates, cam, or rotor assembly. If wear or burring is evident, replace the parts with new components.

6. Install a new rotor and cam assembly if the slippers are worn. Replace the springs if they are bent or broken.

7. Reuse the rotor shaft if the thrust faces, bushing diameter, and shaft seal diameter are not excessively worn or scored.

8. Reuse the housing and bushing assembly if all the threaded holes are not damaged beyond service and the bushing diameter is not scored or worn 0.01 millimeter (0.0005 inch) over the 18 millimeter (0.6897 inch) maximum. Service threaded holes by drilling out the damaged threads and installing helicoil inserts. If the bushing is scored or excessively worn, install a new housing and bushing assembly.

9. Reuse the valve body if the valve bore is free of nicks and scoring. The valve must fall freely in the valve bore. Replace the valve housing and/or the valve if it sticks in the bore.

Vane-Type Pump Operation and Testing

A *Saginaw power steering pump* is a vane-type constant displacement power steering pump (Figure 10-18). The internal operation is similar to the slipper pump.

The housing and internal parts of the vane-type pump are inside the reservoir so that the pump parts are submerged in oil. The reservoir is sealed against the pump housing, leaving the front pump housing face and the shaft hub exposed. A large hole in the rear of the pump housing body contains the pumping element parts: rotor, vanes, pump ring, and pressure plates. A smaller hole contains the flow control and the pressure relief valve assembly. Oil flow is created in the power steering system when the pump's drive shaft turns the rotor. The vanes contact and follow the inner surface of the pump ring moving outward and inward twice during each revolution of the rotor, thereby producing flow. If after careful preliminary inspection of the hydraulic steering system, it is considered necessary to test the operation of the pump, use the following procedure that applies to the Saginaw power steering pump.

Step-by-Step Procedure

1. Disconnect the pressure hose at the pump line union. Use a container to catch any fluid that escapes.

2. Connect a spare pressure hose to the pump's union.

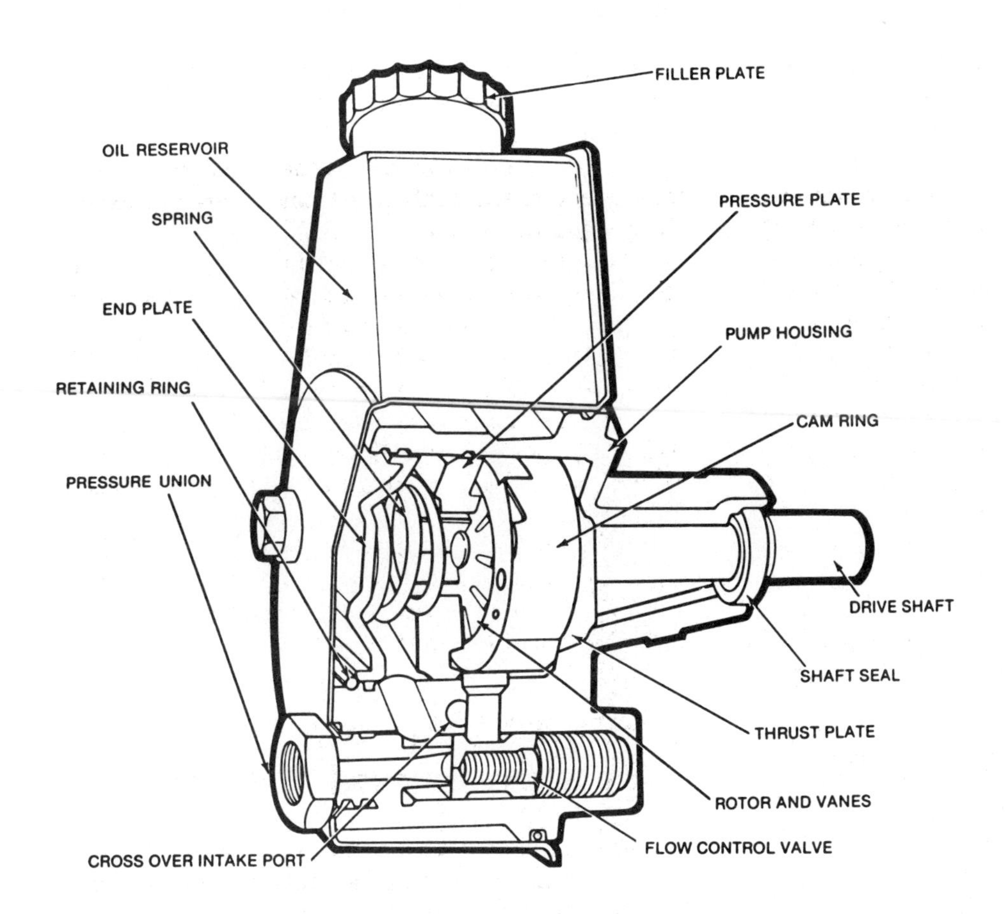

FIGURE 10-18. Sectional View of Vane-Type Power Steering Pump. (Courtesy of Ford Motor Company)

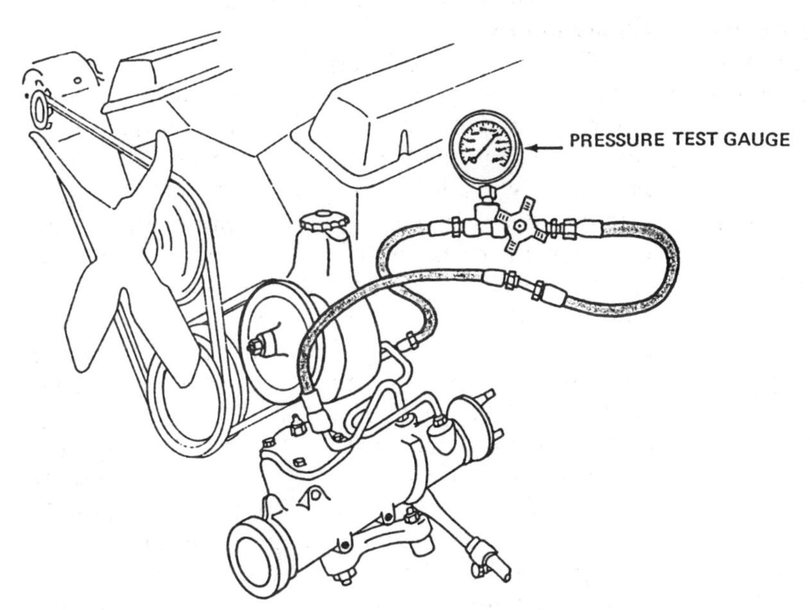

FIGURE 10-19. Proper Installation of Hydraulic Pressure Gauge. (Courtesy of General Motors of Canada, Limited)

3. Using a hydraulic pressure gauge and a special line adapter fitting, connect the gauge to both hoses (Figure 10-19). The gauge must be capable of reading 14,000 kPa (2000 psi).

4. Open the valve at the gauge. Fill the pump's reservoir to the minimum acceptable level.

5. Start the engine. Allow the system to reach operating temperature. Then check the fluid level and add fluid if necessary.

6. When the engine is at normal operating temperature, the pressure reading on the gauge (valve open) should be in the 550-860 kPa (80-125 psi) range. If the pressure is above 1380 kPa (200 psi), check the hoses for restrictions and/or the poppet

valve (located in the directional control valve) for proper installation and assembly.

7. Fully close the valve at the gauge three times, and record the highest pressure attained each time.

Caution: Do not leave the valve fully closed for more than 5 seconds, or pump damage may result.

8. Check the manufacturer's specifications for pressure readings. If the pump's recorded pressure readings are within 344 kPa (50 psi) of the manufacturer's specifications, the pump is considered fully operational.

If the pump's recorded pressure readings are high but do not repeat within 344 kPa (50 psi) of each other, the flow control valve in the pump is sticking. Thoroughly wash and clean the area around the pressure union with a suitable solvent. Dry by using compressed air. Remove the union and withdraw the valve by using a magnet. Clean the valve and remove any burrs by using **crocus cloth** or a fine hone. If the system contains some grit, flush the system. The valve must slide freely in its housing. When installing the union, install a new O-ring seal on the union.

If the pump's recorded pressures are constant but more than 689 kPa (100 psi) below the manufacturer's specifications, replace the flow control valve assembly and recheck. *Note:* This part is serviced as a unit and is factory calibrated. If the pressure is still low, the pump must be removed from the vehicle and repaired.

If the power steering pump conforms to the manufacturer's specifications and there is still an intermittent loss of power steering assist, conduct the following test.

Step-by-Step Procedure

1. Leave the valve at the pump gauge open and turn (or have turned) the steering wheel to both left and right rotational stop positions.

2. Record the highest pressure readings. Compare the readings. If the recorded pressure readings are not the same, the steering gear mechanism is internally leaking and must be disassembled and repaired.

3. Shut off the engine, remove the pressure test gauge and hoses, check the fluid level, and make the necessary repairs to the steering system.

PRACTICAL SERVICE TIPS

1. When performing repair service to the power steering pump, you can remove and install the pump's pulley providing you have the necessary removing and installing tools. **Caution:** Never attempt to remove the pulley by using a hammer and punch. You may eventually remove the pulley and in so doing cause serious internal damage to the pump.
2. Service and assembly of the component parts of the hydraulic steering system must be made on a clean workbench. *Note:* Cleanliness is of utmost importance. The workbench, tools, and parts must be kept clean at all times. Handle all parts very carefully to avoid nicks, burrs, scratches, and dirt that could make the parts unfit for use. Never use old seals a second time; always replace. Do not wash or soak new seals in a cleaning solvent prior to installation. Only power steering oil may be used to prelubricate a seal.
3. A suspected faulty power steering system can be analyzed providing you follow proper diagnostic procedure.
4. When repairing a power steering system, follow the procedures described in the manufacturer's shop manual.
5. When refilling the pump's reservoir, use only the oil specified by the manufacturer to prevent deterioration of flex lines and seals.

REVIEW TEST

1. Define the term *hydraulic pressure.*
2. Briefly explain the purpose of a power steering pump.
3. Describe how it is possible to transmit force and motion through the use of a hydraulic system.
4. Describe how it is possible to multiply force through the use of a hydraulic system.
5. Name three valves that are part of a power steering system.
6. Explain the purpose of each valve in a power steering system.
7. Name the component parts of a power steering system.
8. Describe how a power steering pump produces oil flow.
9. Describe the operation of the flow control and pressure relief valves shown in Figure 10-20.
10. Prior to road testing for a power steering complaint, what questions should you keep in mind?
11. Prepare a list of things that may cause an intermittent loss of power steering assist.

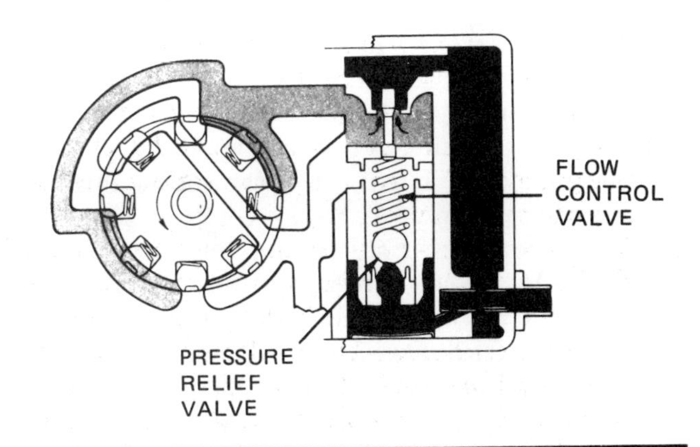

FIGURE 10-20. Illustration for review question 9.

12. Briefly describe how to test the capabilities of a power steering pump.
13. How is it possible to determine if a flow control valve is the cause of low pressure inside a power steering pump?
14. When is it necessary to flush a power steering system?
15. When testing the operation of a power steering system, why should the steering wheel not be held against its stop for more than 5 seconds?

FINAL TEST

This examination is multiple choice. Only one answer will be accepted. Carefully read every statement.

1. Mechanic A says that the primary purpose of a pump is to produce oil flow. Mechanic B says that the primary purpose of a pump is to create hydraulic pressure. Who is right? (A) Mechanic A, (B) Mechanic B, (C) Both A and B, (D) Neither A nor B.
2. Mechanic A says that in a nonintegral power steering system, oil flow is transmitted directly to a steering power ram cylinder. Mechanic B says that in an integral power steering system, oil flow is first transmitted to a directional control valve. Who is right? (A) Mechanic A, (B) Mechanic B, (C) Both A and B, (D) Neither A nor B.
3. Mechanic A says that because of hydraulics, it is possible to transmit motion and force. Mechanic B says that because of hydraulics, it is possible to transmit motion as well as increase motion. Who is right? (A) Mechanic A, (B) Mechanic B, (C) Both A and B, (D) Neither A nor B.
4. Mechanic A says that the purpose of a pressure relief valve is to prevent possible rupture in the power steering pressure hose. Mechanic B says that the purpose of a pressure relief valve is to allow fluid to escape and return to the reservoir. Who is right? (A) Mechanic A, (B) Mechanic B, (C) Both A and B, (D) Neither A nor B.
5. Mechanic A says that the purpose of a flow control valve is to prevent oil from overheating and to regulate flow. Mechanic B says that the purpose of a flow control valve is to limit the maximum pressure in a power steering system. Who is right? (A) Mechanic A, (B) Mechanic B, (C) Both A and B, (D) Neither A nor B.
6. Mechanic A says that a power steering pump is capable of producing more flow than the system requires. Mechanic B says that the pressure relief valve and the flow control valve never operate at the same time. Who is right? (A) Mechanic A, (B) Mechanic B, (C) Both A and B, (D) Neither A nor B.
7. Mechanic A says that the directional control valve is located inside the power steering pump. Mechanic B says that the pressure relief valve is located inside the power steering gear to prevent damage to the seals. Who is right? (A) Mechanic A, (B) Mechanic B, (C) Both A and B, (D) Neither A nor B.
8. Mechanic A says that the first thing to check when diagnosing power steering problems is belt tension. Mechanic B says that the tension on a power steering belt that has been operating for longer than 15 minutes should be 401–533 newtons (90–120 pounds). Who is right? (A) Mechanic A, (B) Mechanic B, (C) Both A and B, (D) Neither A nor B.
9. Mechanic A says that when a power steering pump is tested correctly, the technician will be able to determine system back pressure, pump flow, and steering gear internal leakage. Mechanic B says that based on the interpretation of the test gauge readouts, the technician will be able to determine only an inefficient cam-pump pack and a defective flow control valve. Who is right? (A) Mechanic A, (B) Mechanic B, (C) Both A and B, (D) Neither A nor B.
10. Mechanic A says that when you are testing the operation of a Ford power steering pump, al-

ways perform a pressure test before a flow test. Mechanic B says that when you are conducting a maximum pressure test, do not have the steering wheel held against the stop for more than 5 seconds. Who is right? (A) Mechanic A, (B) Mechanic B, (C) Both A and B, (D) Neither A nor B.

11. Mechanic A says that when the steering wheel is rotated to its two maximum stop positions, the pressure readings must be within 1400 kPa (200 psi) of each other. Mechanic B says that when the steering wheel is held against its stop position and if the pressure is high and the flow is in excess of 4.5 liters per minute (1 gallon per minute), the steering gear does not require service. Who is right? (A) Mechanic A, (B) Mechanic B, (C) Both A and B, (D) Neither A nor B.
12. Mechanic A says that after a power steering pump has been repaired and before the pump is allowed to operate, remove the engine's coil wire. Mechanic B says that to assist in removing trapped air from a power steering system, it is permissible to use engine vacuum. Who is right? (A) Mechanic A, (B) Mechanic B, (C) Both A and B, (D) Neither A nor B.
13. Mechanic A says that when the engine is idling and the steering wheel is not rotating, the average pressure in a steering system is 550–860 kPa (80–125 psi). Mechanic B says that when testing a power steering pump and if the recorded pressure readings are within 344 kPa (50 psi) of the specifications, the pump is fully operational. Who is right? (A) Mechanic A, (B) Mechanic B, (C) Both A and B, (D) Neither A nor B.
14. Mechanic A says that before new seals are installed in a power steering pump, always soak the seals in a cleaning solvent. Mechanic B says that the best way to remove burrs from a power steering part is to use a fine metal file. Who is right? (A) Mechanic A, (B) Mechanic B, (C) Both A and B, (D) Neither A nor B.

Gear Mechanisms for Power Steering Systems

LEARNING OBJECTIVES

After studying this chapter, you should be able to:

- Illustrate and describe the design and operation of the integral power steering gear,
- Explain how the driver senses road feel through the steering gear mechanism,
- Describe the design features of the Saginaw, Chrysler, and Ford rack and pinion power steering systems,
- Define the terms *constant ratio* and *variable ratio gear designs*,
- Describe the design and operation of the nonintegral power steering systems.

INTRODUCTION

Modern power steering systems are designed to reduce steering effort; however, a slight resistance when rotating the steering wheel must be retained to provide the driver with some road feel. The feeling is essential to the driver in sensing and determining the vehicle's steering responses. An experienced driver can detect the point at which the wheels of a vehicle start to drift sideways and whether more or less steering effort is required to stabilize the vehicle's directional control and stability.

A power steering system that required no steering effort would not possess road feel and would result in a hard-to-control, unsafe vehicle. To understand how the power steering system provides road feel, let's become acquainted with various designs of power steering gear mechanisms.

Modern power steering gear mechanisms can be divided into two general design classifications: (1) integral (internal directional control valve and rack piston and (2) nonintegral (external directional control valve and power ram). The basic difference between the integral and nonintegral steering systems is the application of applied force to the sector shaft or rack gear (rack and pinion power steering) in integral systems and the applied force to the steering linkage relay rod or drag link in nonintegral systems. See Figures 11-1 and 11-2. The integral and nonintegral types of steering systems are used by automotive, light truck, and commercial truck manufacturers.

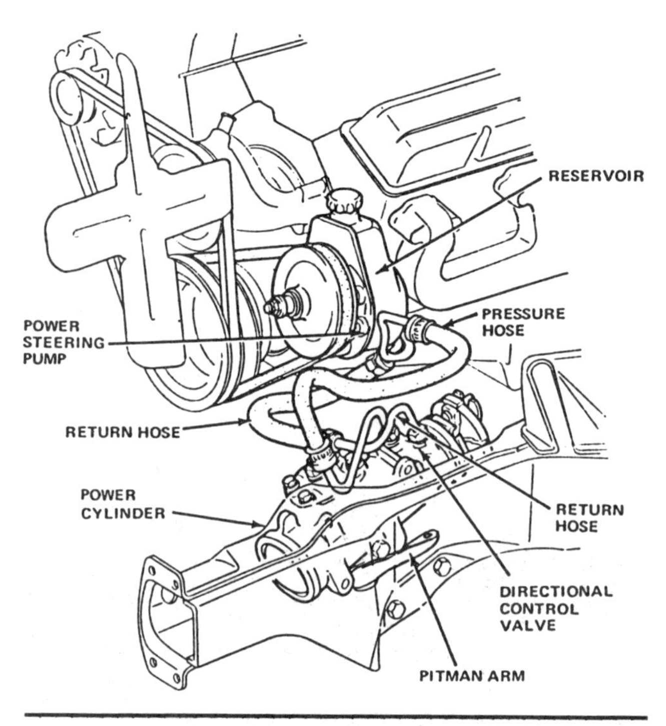

FIGURE 11-1. Integral Power Steering System. (Courtesy of Ford Motor Company)

SAGINAW INTEGRAL POWER STEERING

Design

Figure 11-3 provides a side view of the Saginaw *rotary valve* integral power steering gear. The gear assembly is always filled with fluid, and all internal components of the gear are immersed in fluid. Therefore, periodic lubrication is unnecessary. Although the gear housing is one complete part, the housing is actually divided into two compartments. The lower part of the housing (the left side of Figure

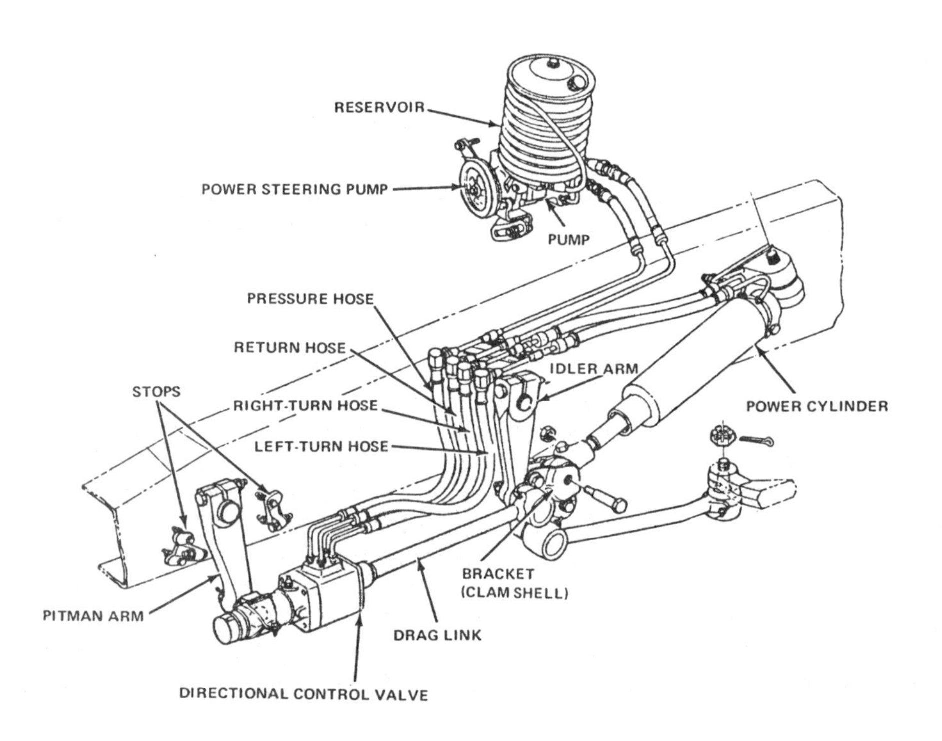

FIGURE 11-2. Nonintegral Power Steering System (Truck Installation). (Courtesy of Ford Motor Company)

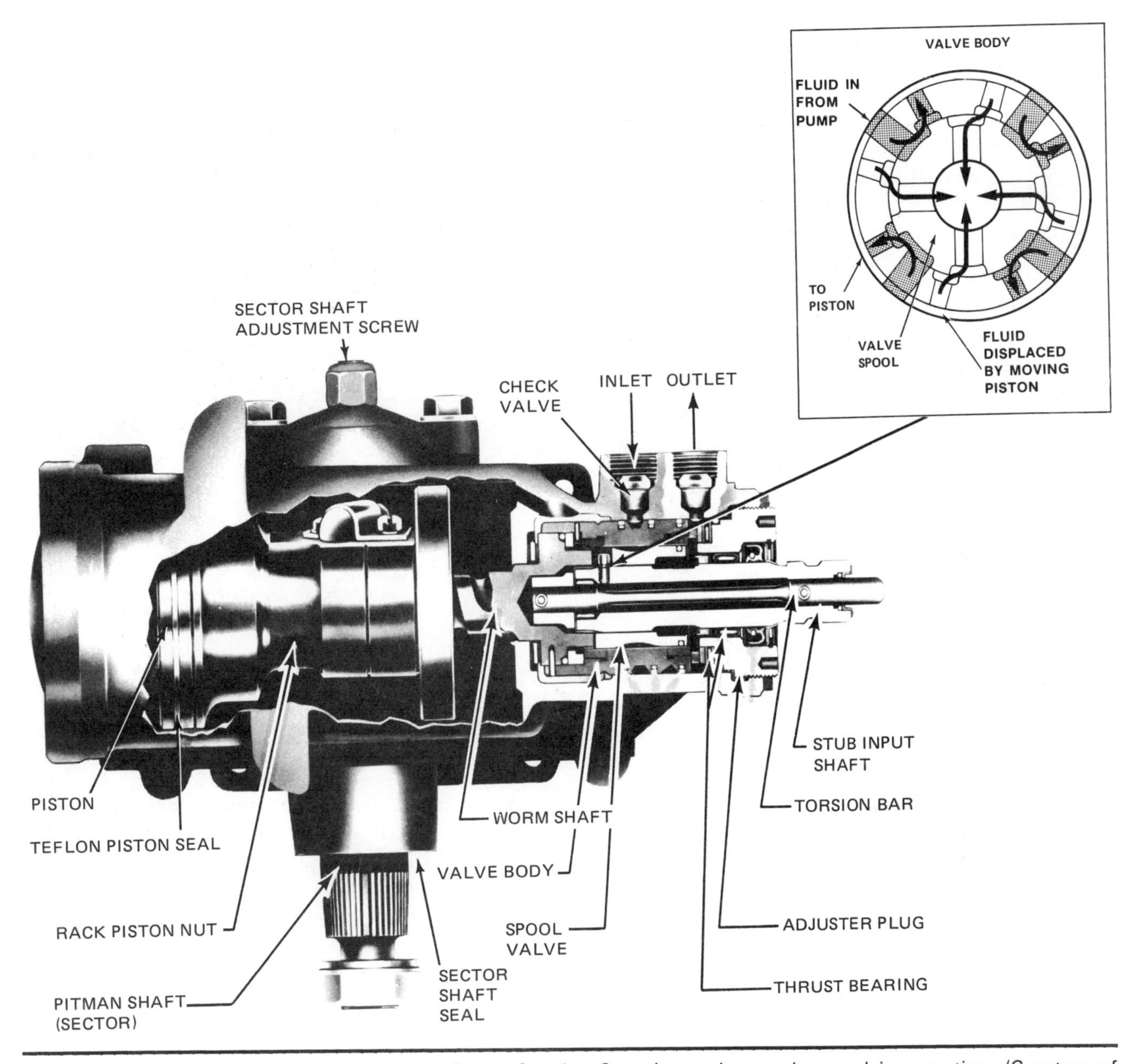

FIGURE 11-3. Saginaw Rotary Valve Integral Power Steering Gear. Insert shows valve spool in operation. (Courtesy of General Motors of Canada, Limited)

11-3) contains the pitman (sector) shaft and gear, rack piston nut, and wormshaft. The sector shaft and rack piston gear mesh is maintained by the sector shaft adjustment screw. The upper part of the housing (the right side of Figure 11-3) contains the stub input shaft, torsion bar, spool valve, and valve body, all component parts of the directional rotary control valve.

In this type of power steering gear, the torsion bar plays an important role in the operation of the gear mechanism. The twisting of the torsion bar allows the valve spool to displace or move from its neutral position in relation to the valve body, thereby directing oil flow to either side of the rack piston. This causes the rack piston to move, thus applying a rotational turning effort to the sector shaft gear.

The Saginaw power steering gear system utilizes either a constant ratio or variable ratio design. In a *constant ratio* system, approximately 17° of steering wheel movement is required to turn the front wheels 1°. In a *variable ratio* system, the gear ratio varies continuously from a moderate 16:1 for straight-ahead driving to a low 13.1:1 in full turns (Figure 11-4). From the straight-ahead driving position, the steering ratio stays constant for the first 40° of steering wheel movement and then decreases very gradually. This type of gear design provides precise steering control for highway driving. When the vehicle is being parked, however, the steering wheel may be turned to near its extreme rotational position. The ratio then decreases to approximately 13.1:1. As the ratio decreases, the maneuverability of the vehicle increases. This is seldom noticed in a power steering system since the system is designed to decrease steering effort through hydraulic assistance.

In a constant ratio conventional sector and rack gear design, the gear teeth are all of equal length (Figure 11-5). Since the sector, like any other gear, is basically a series of levers, the movement of the rack piston will always cause the sector gear to swing the pitman arm in the same ratio—that is, to turn the pitman arm the same number of degrees with each tooth in the sector.

In a variable ratio sector gear design, the sector gear uses a short tooth on either side of a long center tooth rather than three teeth of equal length as in a constant ratio gear (Figure 11-5). Companion changes are also made in the design of the rack piston gear. Since the center gear teeth are larger, when the rack piston has moved enough to engage one of the other teeth of the sector gear, the leverage is changed. As a result, the steering ratio is reduced, causing the pitman arm to move noticeably further for a given steering wheel movement.

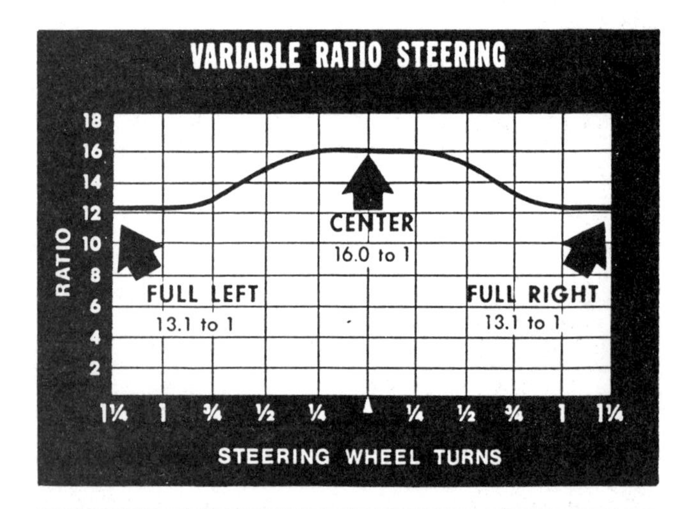

FIGURE 11-4. Graph of Variable Ratio Steering. (Courtesy of General Motors of Canada, Limited)

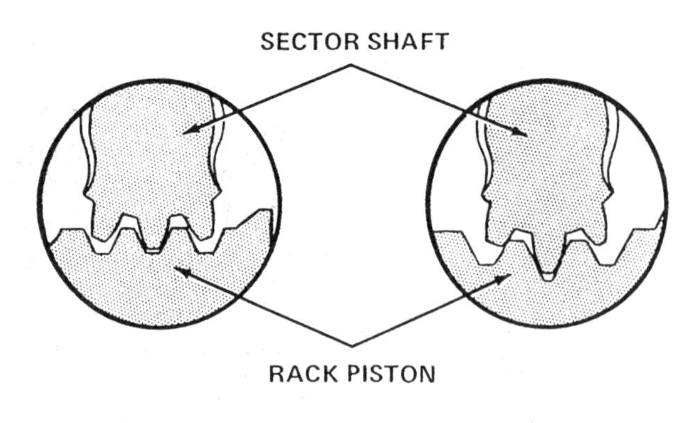

FIGURE 11-5. Sector Shaft Gears for Constant and Variable Ratio Systems. (Courtesy of General Motors of Canada, Limited)

With the steering wheel turned one-half turn, the steering ratio is reduced to approximately 14.2:1. With a three-quarter turn of the steering wheel, the leverage has been further reduced to approximately 13.3:1. From three-quarters to one full turn of the steering wheel, the ratio continues to diminish to 13.1:1. This design feature provides the driver with more precise steering wheel control and a faster response when maneuvering a vehicle in sharp turning situations.

Operation

Neutral (Straight-Ahead) Position. When turning effort is not being applied at the steering wheel, the slots in the spool valve are positioned so that oil entering the valve body from the housing high-pressure inlet port passes through the slots in the spool valve to the oil return port in the housing. (See Figure 11-6.) The chambers at both ends of the rack piston and around the pitman shaft are always full of oil. The oil acts as a cushion to absorb road shock.

In addition, this oil lubricates all the internal components of the gear. A **check valve** located under the high-pressure inlet connector seat hydraulically dampens the shock transmitted to the steering gear when driving on washboard (wavy) roads. During this stage of operation, equal pressure exists on both sides of the piston.

Right Turn Position. When the steering wheel is turned to the right, the wormshaft resists being turned because of the resistance encountered at the

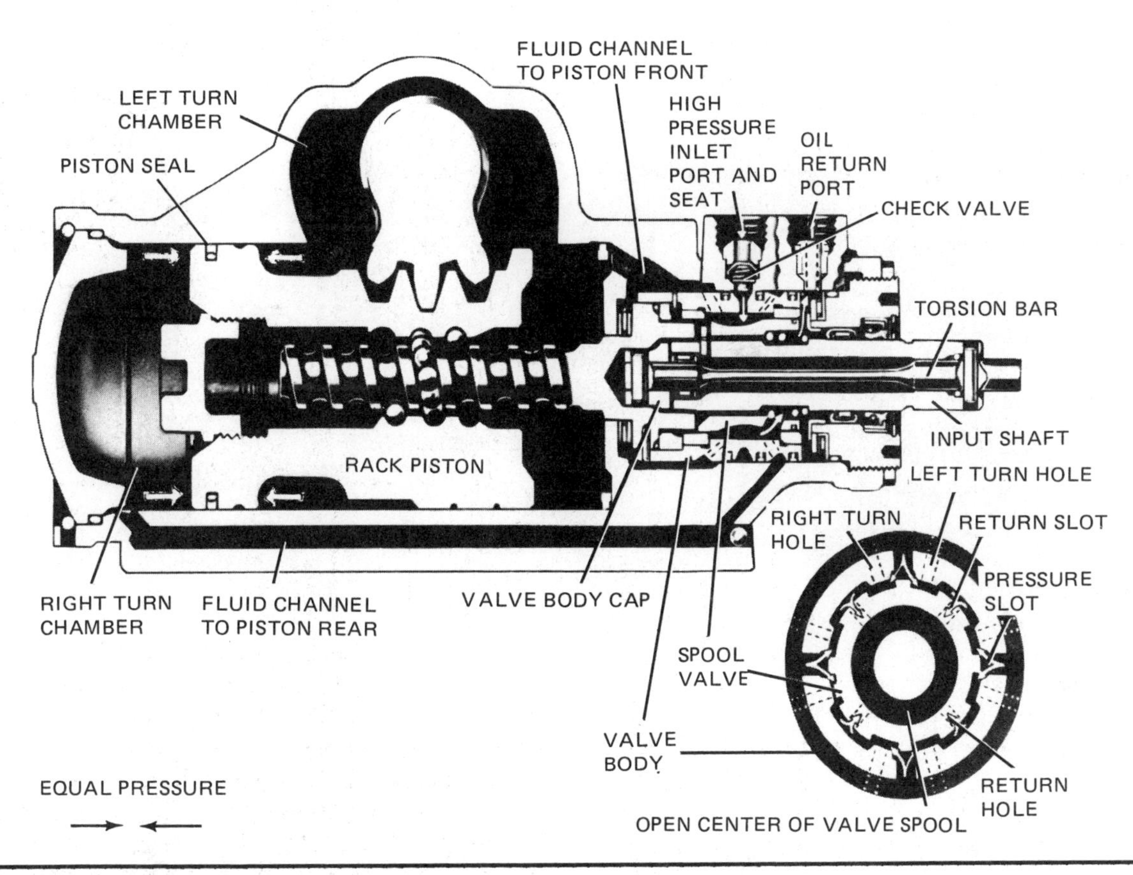

FIGURE 11-6. Steering Gearbox in Neutral (Straight-Ahead) Position. (Courtesy of General Motors of Canada, Limited)

front wheels. (See Figure 11-7.) The valve body also resists turning because it is pinned or brazed to the worm. Driver force exerted at the steering wheel turns the lower shaft and spool valve a slight amount in relation to the valve body because of the twisting action of the torsion bar. This slight amount of turning of the spool valve is sufficient to position the slots in the valve body and spool valve for power assist.

The right turn slots in the spool valve are closed off from the return (wide) slots in the valve body and opened more to the pressure (narrow) slots in the valve body. The left turn slots in the spool valve are closed off from the pressure slots in the valve body and opened more to the return slots in the valve body.

Pressure immediately begins to build up against the lower end of the rack piston, forcing it upward to apply turning effort to the pitman shaft. The oil in the chamber at the upper end of the rack piston is then forced out through the valve body and spool valve through the oil return port to the pump reservoir.

The instant the driver stops applying turning effort to the steering wheel, the spool valve is forced back into its neutral position by the torsion bar. Oil pressure on the lower end of the rack piston then decreases so that pressure is again equal on both sides of the rack piston.

Under normal driving conditions, oil pressure does not exceed 1400 kPa (200 psi) except when turning corners, where it does not ordinarily exceed 3150 kPa (450 psi). Oil pressure, when parking, ranges from 6300 to 9100 kPa (900 to 1300 psi) depending on road conditions and car weight. The steering effort during normal driving ranges from 4.5 to 9 newton (1 to 2 pounds) and, during parking, from 9 to 15.75 newton (2 to 3½ pounds), again depending on road conditions.

Left Turn Position. When the steering wheel is turned to the left, the relationship between the spool valve slots and valve body slots is again changed through the twisting of the torsion bar (Figure 11-8). Pressure immediately builds up against the upper end of the rack piston, forcing it downward

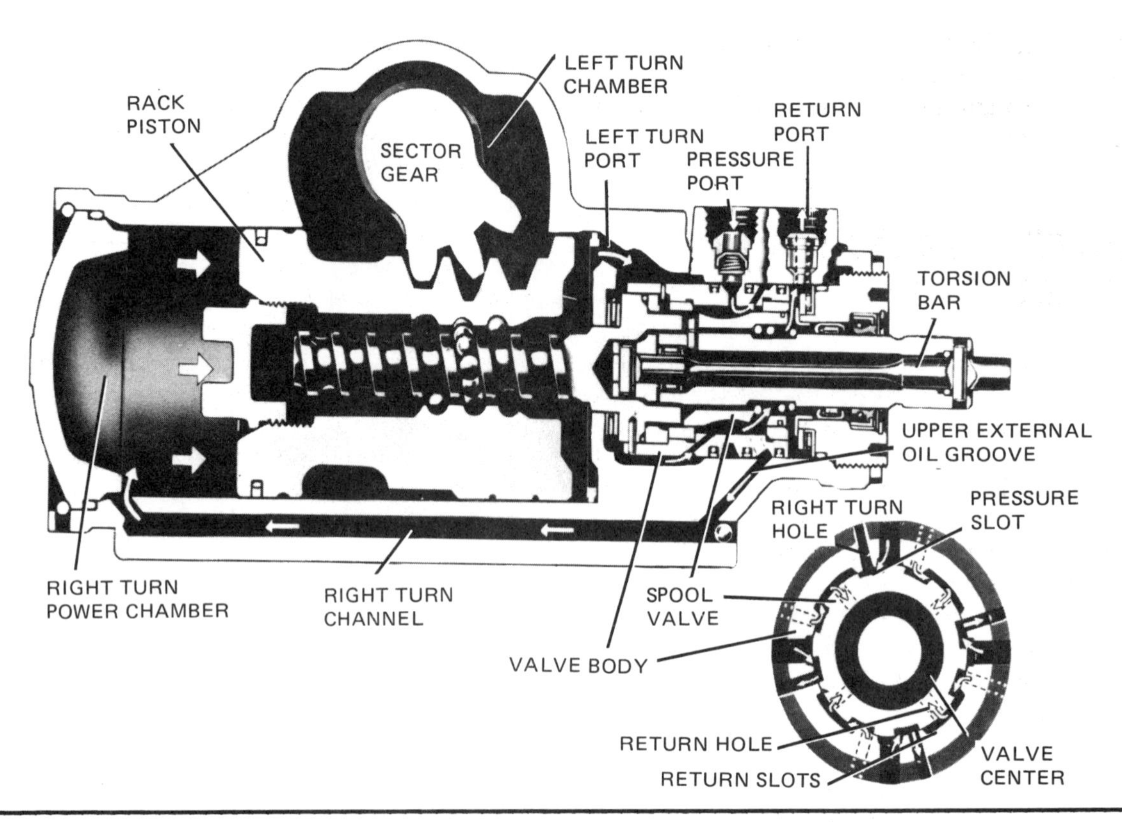

FIGURE 11-7. Steering Gearbox in Right Turn Position. (Courtesy of General Motors of Canada, Limited)

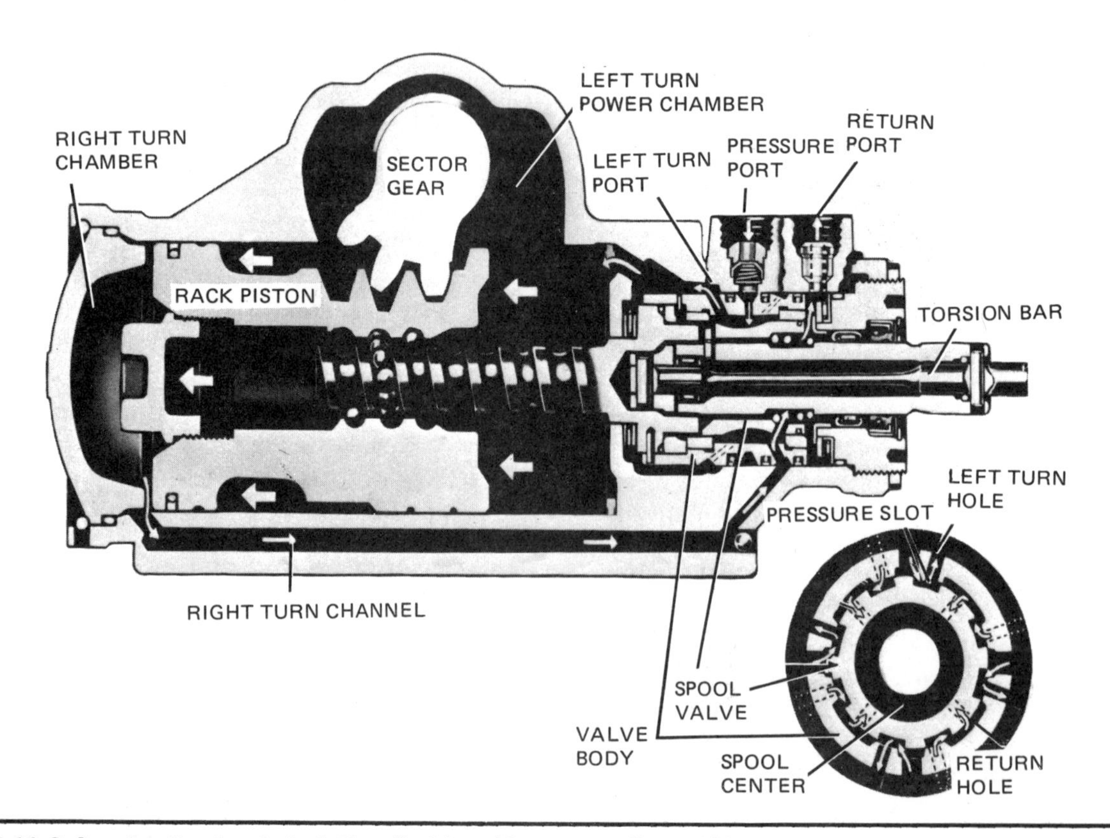

FIGURE 11-8. Steering Gearbox in Left Turn Position. (Courtesy of General Motors of Canada, Limited)

to apply turning effort to the pitman shaft. The oil in the chamber at the lower end of the rack piston is forced out through the valve body and spool valve to the pump reservoir.

CHRYSLER INTEGRAL POWER STEERING

The Chrysler *constant control* integral power steering gear uses a different design of the directional control valve, as compared to the Saginaw integral system. The directional control valve assembly is mounted above the gearbox housing, through a design preference of the manufacturer. The benefit of this design feature provides easy access to adjustments (Figure 11-9).

The power steering gear housing contains a sector shaft and gear. A power piston with gear teeth milled into the side of the piston is in constant mesh with the sector shaft gear. A wormshaft connects the movement of the steering wheel to the power piston gear assembly.

Mechanically, the power piston moves toward the splined portion of the wormshaft when the steering wheel is turned to the left and moves toward the left (see left side of Figure 11-9) when the steering wheel is turned to the right. In the steering gear, there is a natural resistance to any movement of the power piston because the front wheels have a resistance to deviation from the straight-ahead position. When the wormshaft is turned, the shaft must move up or down a slight amount before the rack piston starts to move. This movement is transferred to the center thrust bearing race, causing the pivot lever to move laterally in line with the wormshaft. As the pivot lever moves, the spool valve shifts from its neutral position, thereby directing fluid flow to either side of the power piston. When the turning force on the steering wheel is relieved (after the turn has been made), the directional control spool valve recenters automatically, and the power steering assist stops. With the directional spool valve cen-

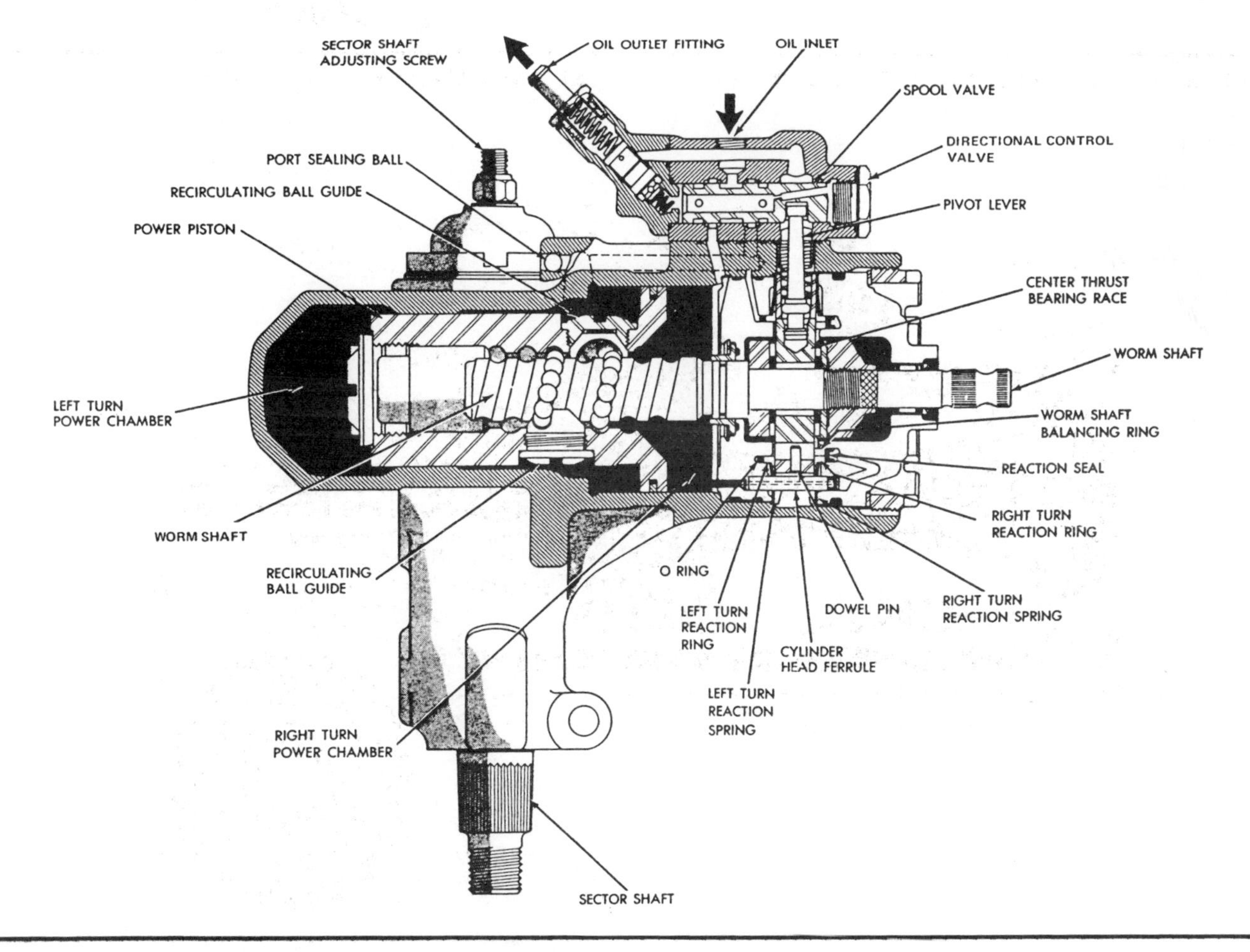

FIGURE 11-9. Chrysler's Constant Control Integral Power Steering Gear. (Courtesy of Chrysler Corporation)

tered, the flow and pressure from the power steering pump is reduced. Pressure is then equal on both sides of the rack piston.

FORD INTEGRAL POWER STEERING

The Ford *rack and pinion* integral power steering gear has become widely adopted by both domestic and import vehicle manufacturers (Figure 11-10). This type of steering gear is available in two ratios: 20:1 variable ratio and a 15:1 constant ratio. Normally, the heavier the car, the higher the ratio needed to manipulate and control the steering. The different-sized gear assemblies are basically the same except for minor dimensional differences in the rack teeth and valve assembly. Service procedures for both types of rack and pinion mechanisms are the same.

The gear housing and valve housing are combined into a one-piece aluminum die casting. The gear design incorporates quick connect fittings for the pressure and return lines that allow the lines to swivel. This play is normal and does not indicate loose fittings.

The gear, a hydraulic mechanical unit, uses an integral piston and rack design to provide power-assisted vehicle steering control. Internal valving directs pump flow and controls pressure, as required, to reduce steering effort during operation. The unit contains a rotary control valve assembly integrated to the input shaft and a power cylinder integrated with the rack gear (Figure 11-11).

The rotary design control valve uses relative rotational motion of the input shaft and valve sleeve to direct fluid flow. When the steering wheel is turned, wheel resistance and the weight of the vehicle cause the torsion bar to deflect (twist). This deflection changes the position of the valve spool and sleeve ports, directing fluid under pressure to the appropriate end of the power cylinder. The difference in pressure forces on the piston helps move the rack to assist turning effort. The piston is attached directly to the rack, and the housing functions as the power cylinder. The oil in the opposite ends of the power cylinder is forced to the control valve and back to the pump reservoir.

When the driver stops applying steering effort, the valve is forced back to a centered position by the torsion bar. When this occurs, pressure is equalized on both sides of the piston, and the front wheels tend to return to a straight-ahead position through the influence of steering axis inclination. This factor and how it assists in steering control is discussed in Chapter 18 of this text.

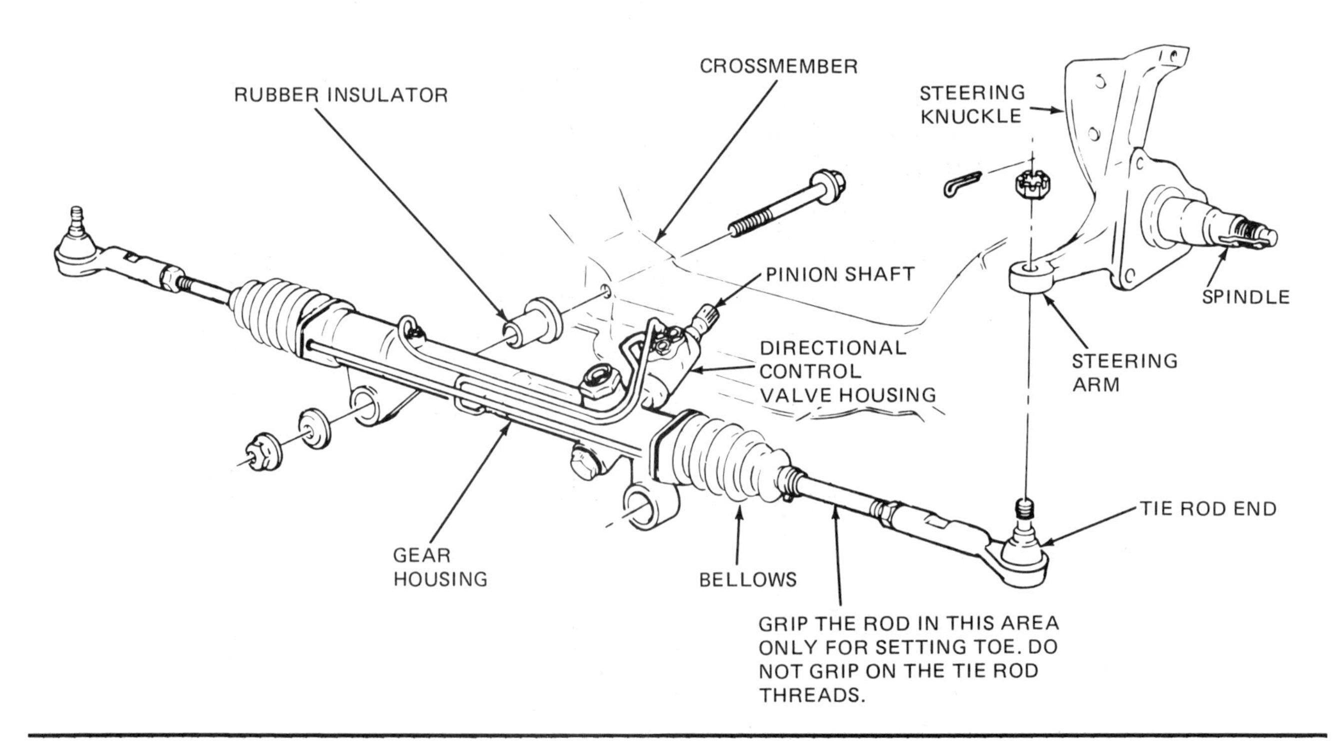

FIGURE 11-10. External View of Ford's Rack and Pinion Integral Power Steering Gear. (Courtesy of Ford Motor Company)

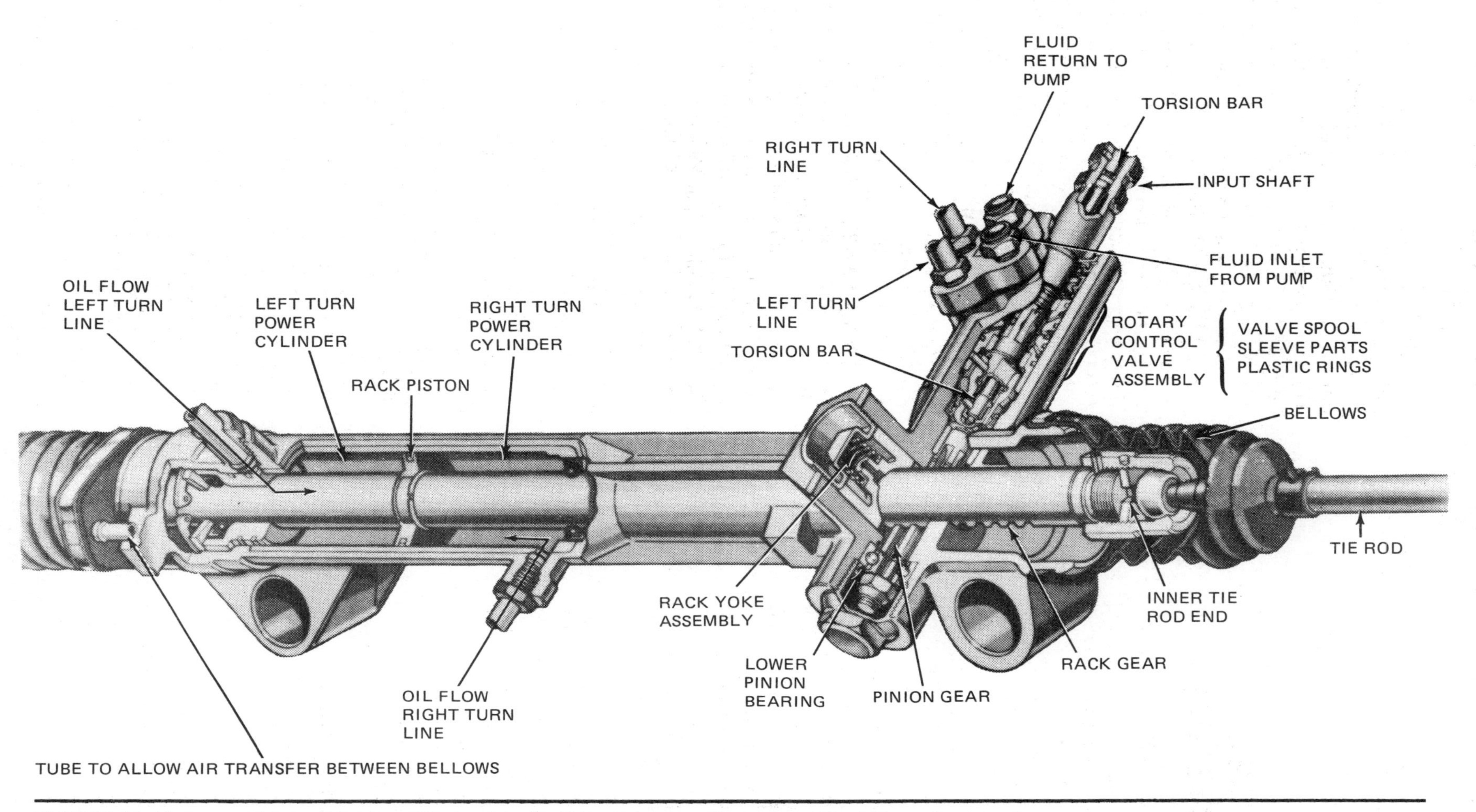

FIGURE 11-11. Internal View of Rack and Pinion Steering Gear Assembly. (Courtesy of Ford Motor Company)

NONINTEGRAL POWER STEERING

Design

The nonintegral power steering system (Figure 11-12) incorporates a directional control valve and a hydraulic power steering cylinder. Depending on the design of the steering system, the directional control valve may be attached to the steering relay rod or the drag link assembly (in trucks). A spool valve, built into the directional control valve housing, is controlled by two coil springs that hold the valve in its centered position for straight-ahead driving. The power steering cylinder is attached to the steering linkage. The end of the piston rod is anchored to a bracket attached to the vehicle's frame.

Any movement of the pitman arm will cause the directional control spool valve to move laterally. The movement of the spool valve then directs fluid flow, under pressure, to either side of the power cylinder piston. Two hydraulic steel lines connect the cylinder to the control valve housing.

Operation

Although the nonintegral power steering system is different in design from the integral power steering system, the operation of the system is similar. For straight-ahead driving, when no steering effort is applied to the steering wheel, the control valve main spool and reaction limiting plunger are held in their center or neutral positions by their respective centering springs (Figure 11-13). In this position, the three grooves in the valve housing are interconnected by the main valve spool lands and grooves. The large areas of the main spool and the reaction limiting valve are interconnected by passages and grooves in the small plunger, which are connected to the main valve cylinder passages. Fluid from the pump flows through the control valve and returns to the pump. Balanced pressure is maintained on both sides of the power piston and both sides of the control valve main spool. Fluid in excess of the volume required to keep the power cylinder and reaction chamber filled is returned from the control valve to the pump's reservoir.

When the steering wheel is turned left and the turning effort at the pitman arm becomes great enough to overcome the resistance of the centering spring, the control valve main spool is moved to the right (Figure 11-14). With the valve spool in this position, fluid flow is directed to the right side of the power cylinder and to the right side of the control valve reaction chamber. The fluid in the left side of the power cylinder is free to return to the fluid reservoir. Fluid pressure in the right side of the cylinder forces the cylinder and the attached steering linkage toward the right, providing the desired power assist for a left turn. Fluid in the left side of

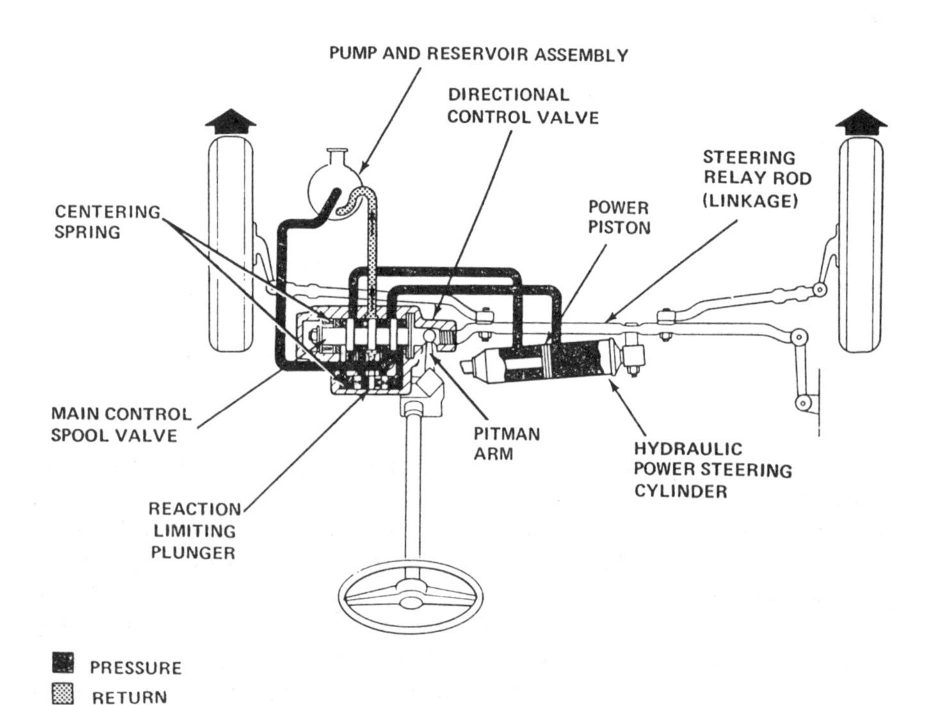

FIGURE 11-12. Nonintegral Power Steering System. (Courtesy of Ford Motor Company)

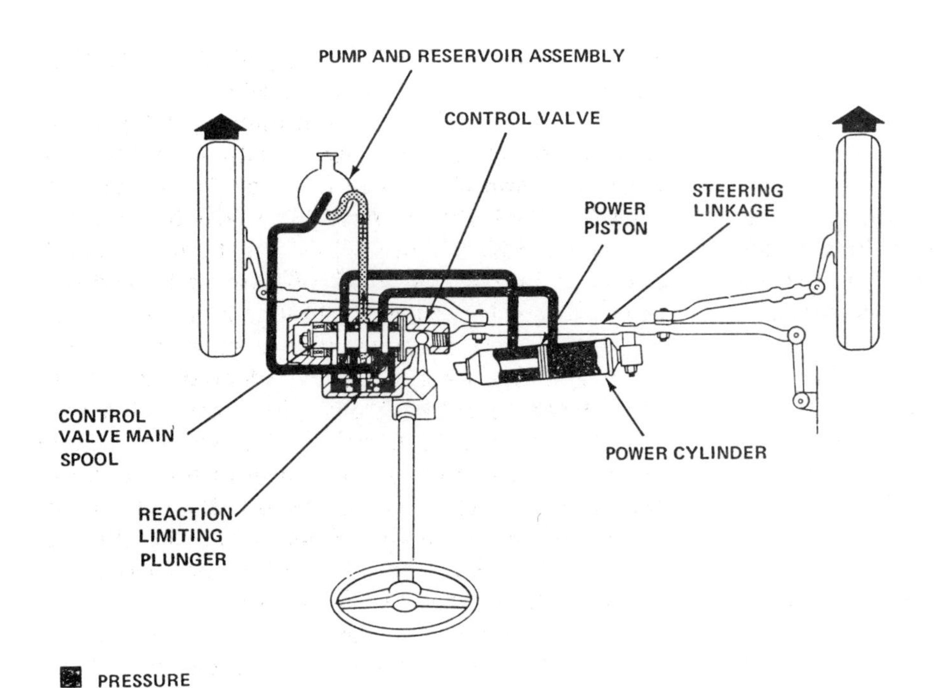

FIGURE 11-13. Nonintegral Power Steering Operation, Straight-Ahead Driving. (Courtesy of Ford Motor Company)

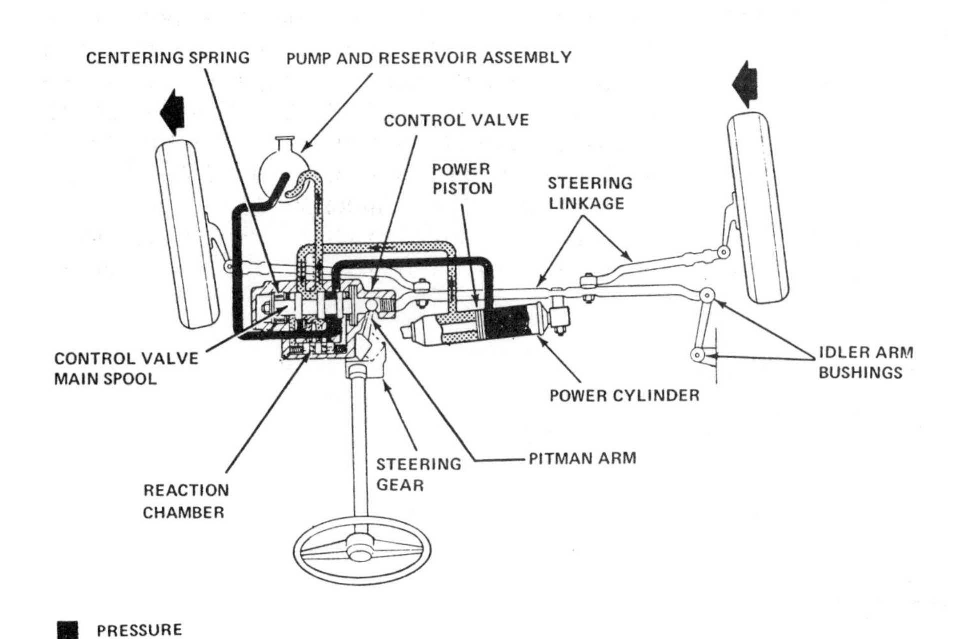

FIGURE 11-14. Nonintegral Power Steering Operation, Left Turn. (Courtesy of Ford Motor Company)

the cylinder is displaced by the piston and cylinder movement and is forced to return to the reservoir. A combination of forces from the main spool centering spring and the fluid pressure in the reaction chamber at the right end of the main spool tend to return the spool to its center position.

Pressure in the reaction chamber is proportional to the pressure being applied to the power cylinder until it reaches a predetermined value. The effort required to turn the wheel against this reaction pressure provides the driver with the feel of the road. When the force exerted by the pitman arm drops below the centering spring force and reaction force, the valve spool returns to its center position, ending the power assist. No more power assist will be provided until the driver exerts enough force through the steering gear to pitman arm to overcome the centering spring force.

When the steering linkage is moved away from the straight-ahead position in either direction, the torsion-type rubber idler arm bushings are twisted. After the turn is completed, the twist in the bushings and the normal effect of the front end geometry return the wheels to the straight-ahead position. (Front end geometry is the interrelationship of camber, caster, toe, kingpin or ball joint inclination, and toe-out on turns. Kingpin or ball joint inclination is the major factor that influences the front wheels to return to the straight-ahead position. Steering geometry and the individual factors are discussed and illustrated in Chapters 14 to 20.) Operation for right turn is similar. Pressure is directed by control valve to opposite side of piston in power cylinder.

Modern power steering systems seldom produce problems. If the system is suspected of faulty operation, the source of the problem can be determined and corrected when the service technicians follow proper diagnostic procedures. These procedures were described in the previous chapter and are outlined in all manufacturer's shop manuals.

PRACTICAL SERVICE TIPS

1. Assembly of the steering gear or power steering pump must be made on a clean workbench. This rule applies to all hydraulically operated units.
2. Handle all parts very carefully. Avoid nicks, burrs, scratches, and dirt, which could make the parts unfit for use.
3. When repairing a power steering unit, always follow the service procedures described in the manufacturer's shop manual.

REVIEW TEST

1. Name the parts that are common to both the integral and nonintegral power steering systems.
2. Briefly explain why road feel is important to the driver of a vehicle equipped with power steering.
3. Briefly describe the operation of a Saginaw power steering gear during a right turn.
4. Explain the term *variable steering gear ratio.*
5. Describe the design features of a Chrysler integral power steering gear and a Ford rack and pinion power steering gear.
6. Briefly describe the operation of a nonintegral power steering mechanism during a left turn.

FINAL TEST

This examination is multiple choice. Only one answer will be accepted. Carefully read every statement.

1. Mechanic A says that when a vehicle is equipped with nonintegral power steering, one end of the power ram is attached to the frame, and the opposite end is connected to the steering linkage. Mechanic B says that when a vehicle is equipped with integral power steering, the directional control valve is located on or in the steering box. Who is right? (A) Mechanic A, (B) Mechanic B, (C) Both A and B, (D) Neither A nor B.
2. Mechanic A says that road feel is obtained in a power steering system by the turning resistance of the front wheels. Mechanic B says that the directional control valve cannot restrict fluid flow until the front wheels produce turning resistance. Who is right? (A) Mechanic A, (B) Mechanic B, (C) Both A and B, (D) Neither A nor B.

Note: Questions 3 to 6 pertain to the Saginaw power steering gear.

3. Mechanic A says that when the steering wheel is rotated to the right, the spool valve must shift laterally toward the stub input shaft. Mechanic B says that the twisting of the torsion bar causes the spool valve to move from its position within the valve body. Who is right? (A) Mechanic A, (B) Mechanic B, (C) Both A and B, (D) Neither A nor B.
4. Mechanic A says that when a steering gearbox is variable ratio, the mechanical advantage is less in the straight-ahead position than in a full

left turn. Mechanic B says that when a steering gearbox is variable ratio, the gear teeth are all of equal length and taper. Who is right? (A) Mechanic A, (B) Mechanic B, (C) Both A and B, (D) Neither A nor B.

5. Mechanic A says that the prime purpose of the directional control valve is to restrict and direct oil flow when the steering wheel is rotated. Mechanic B says that when the steering wheel is rotated, oil pressure must be high on one side of the piston and low on the opposite. Who is right? (A) Mechanic A, (B) Mechanic B, (C) Both A and B, (D) Neither A nor B.
6. Mechanic A says that in a normal straight-ahead driving condition, pressure should not exceed 1400 kPa (200 psi) in a power steering system. Mechanic B says that in an extreme left-hand turn, oil pressure will not exceed 3500 kPa (500 psi) in the power steering system. Who is right? (A) Mechanic A, (B) Mechanic B, (C) Both A and B, (D) Neither A nor B.
7. This question pertains to the Chrysler integral power steering system. Mechanic A says that when the wormshaft is rotated, the shaft must move up or down a slight amount before the rack piston starts to move. Mechanic B says that with the spool valve centered, the flow and pressure from the pump is then equal on both sides of the rack piston. Who is right? (A) Mechanic A, (B) Mechanic B, (C) Both A and B, (D) Neither A nor B.
8. This question pertains to the Ford power rack and pinion power steering system. Mechanic A says that when the rack gear moves to the right within the gear housing, the piston must move toward the left. Mechanic B says that after the front wheels have made their turn, the front wheels are moved to the straight-ahead position through the influence of the steering axis inclination. Who is right? (A) Mechanic A, (B) Mechanic B, (C) Both A and B, (D) Neither A nor B.
9. This question pertains to the nonintegral power steering system. Mechanic A says that any movement of the pitman arm will cause the directional control valve spool to move laterally. Mechanic B says that when the steering wheel is rotated left and overcomes the resistance of the centering spring, the valve spool is moved to the right. Who is right? (A) Mechanic A, (B) Mechanic B, (C) Both A and B, (D) Neither A nor B.

Frames and Frame Damage

LEARNING OBJECTIVES

After studying this chapter, and with the aid of frame gauges and the manufacturer's shop manual, you should be able to:

- State the main purposes of a vehicle's frame,
- Name three common types of frames and describe their design,
- Explain the meaning of the terms *tracking*, *wheelbase*, *centerline*, *datum line*, and *comparable measurement*,
- Describe five different types of frame damage and analyze the extent of the damage by using frame gauges and measuring equipment,
- Describe how to determine the tracking of a vehicle by using a track gauge.

INTRODUCTION

The first motor vehicles of the early 1900s were simply built in comparison to the vehicles of today. The frame assembly was a heavy, rigid, almost flat metal structure shaped like a large ladder. The frame provided the vehicle with strength and served as a platform on which the engine and passenger compartment, or body, rested. By the late 1930s, the frame members had become massive. When a heavy frame member sustained damage in a collision, the accepted repair procedure was to force the damaged area back into shape and then reinforce it with a heavy steel plate welded to the original metal.

In recent years, manufacturers have incorporated major changes into vehicle frames. Significant changes include the use of lighter and more durable alloys in frame construction, and, in some models, the separate frame has been replaced by partial or complete **unitized body** design.

Although there has been a constant change in style and design, the original purposes of a vehicle's frame or unitized body have not changed. The frame or unitized body can be compared to the skeleton of the human body; without bones, the body could not stand erect. The same applies to the vehicle; without the vehicle's frame, the automobile or truck could not support its weight or the weight of the driver, passenger, or cargo load. When you have the opportunity to examine a vehicle's frame, it may appear to be a rigid, unyielding part of an automobile or truck. This is not the case. The modern frame must not only be able to support the vehicle; it must also be able to flex by absorbing torsional stress and torque produced by the wheels and the vehicle's motor. Based on our preliminary discussion of the vehicle's frame, let's list its main purposes:

- To enable the vehicle to support its total weight,
- To enable the vehicle to absorb torsional road stress and torque produced by the motor and the wheels,
- To provide a main member for the attachment of other component parts.

FRAME CONSTRUCTION AND DESIGN

In modern vehicles, most manufacturers incorporate a combination of the channel (partial box), complete box, and tubular type of frame construction. Combining these types of construction drastically reduces the sprung weight of a vehicle yet still allows for strength in the overall frame design (Figure 12-1).

While inspecting and repairing suspension systems, you have no doubt noticed the different types of frame design. The following discussion will familiarize you with their various shapes and trade names.

Ladder Design

The *ladder frame design* (Figure 12-2) is well suited to its name, is best adapted to commercial vehicles. The side rails have little or no offset and are built on a more direct line between the front and rear wheels. This design has generally more crossmembers and is reasonably rigid since it must support considerable weight.

Perimeter Design

Perimeter frame design (Figure 12-3) is separate from the body and forms a border surrounding the passenger compartment. At the cowl area, the front frame rails are stepped inward to provide a sturdy foundation for the engine mounts and front end assembly and to provide clearance for the movement

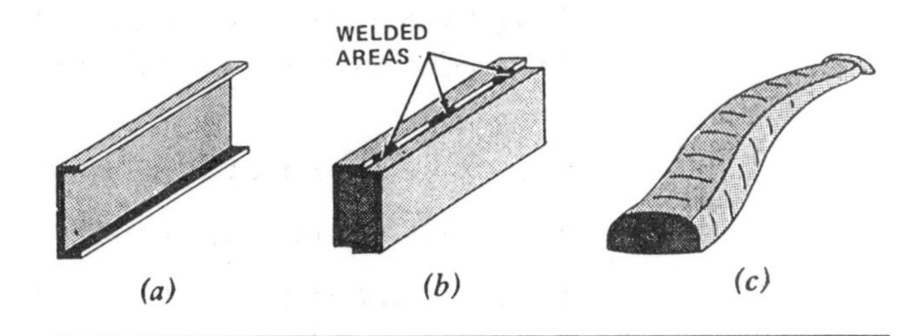

FIGURE 12-1. Three Common Types of Frame Construction. (a) Channel or partial box; (b) complete box; (c) tubular crossmember.

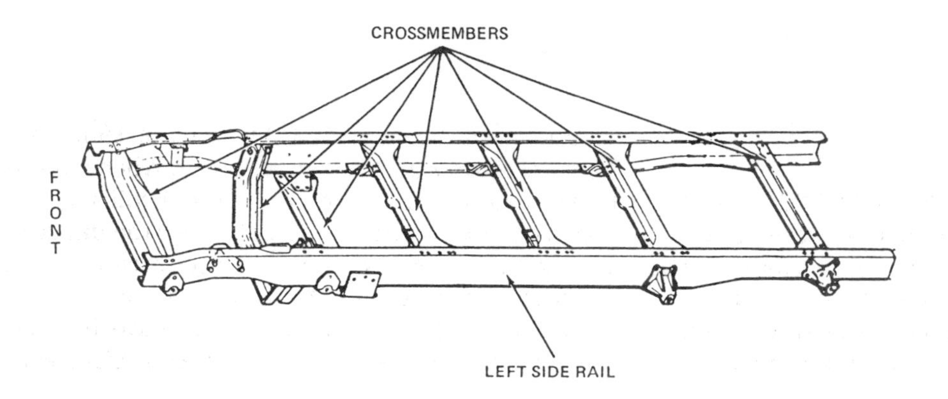

FIGURE 12-2. Ladder Frame Design. (Courtesy of Chrysler Corporation)

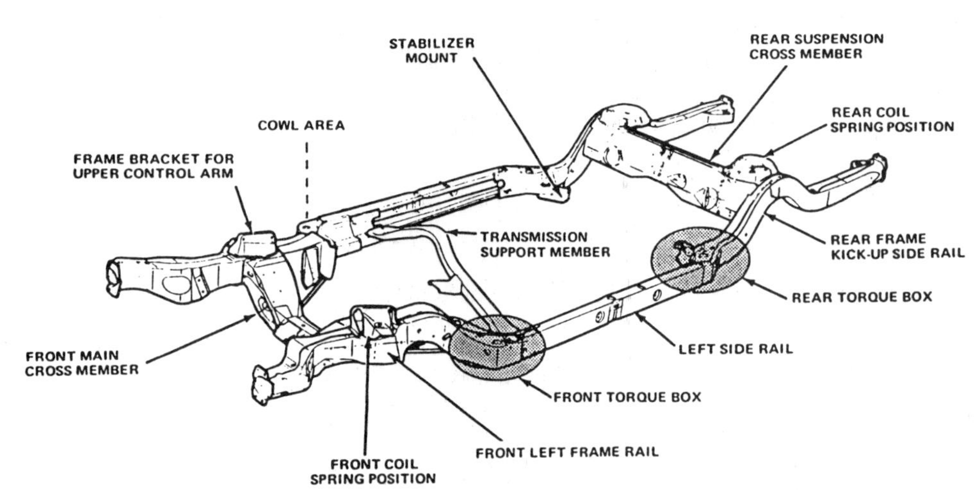

FIGURE 12-3. Perimeter Frame Design. (Courtesy of General Motors of Canada, Limited)

of the front wheels. The torque boxes of the frame are in an S configuration.

As you study Figure 12-3, you will see how the front and rear frame sections are joined to the left and right side rails. When a vehicle is involved in a collision, the torque box is designed to absorb a major portion of the impact, thereby reducing the damage to other sections of the frame and body. The rear frame kick-up side rails provide support for the body and rear suspension system. Lateral support is provided by welded-in crossmembers.

Unitized Body Design

In the late 1940s, American Motors Corporation introduced the first real departure from conventional frame and body construction. The accepted term for this new type of construction quickly became "unitized body," a definitive term because the body and frame were an indistinguishable unit. The frame assembly, as such, was eliminated completely. The overall strength and rigidity of the car was built into the body. This design was achieved by the extensive use of body sheet metal fabricated into a box configuration. Strength was achieved through shape and design rather than just mass and weight of metal.

In the *unitized body frame design*, each member is joined to another so that all sections of the body become load-bearing members (Figure 12-4). The floor pan, inner aprons, rocker panels, and even the roof and other sheet metal parts of the body are integrally joined so that all parts form a single unibody frame structure.

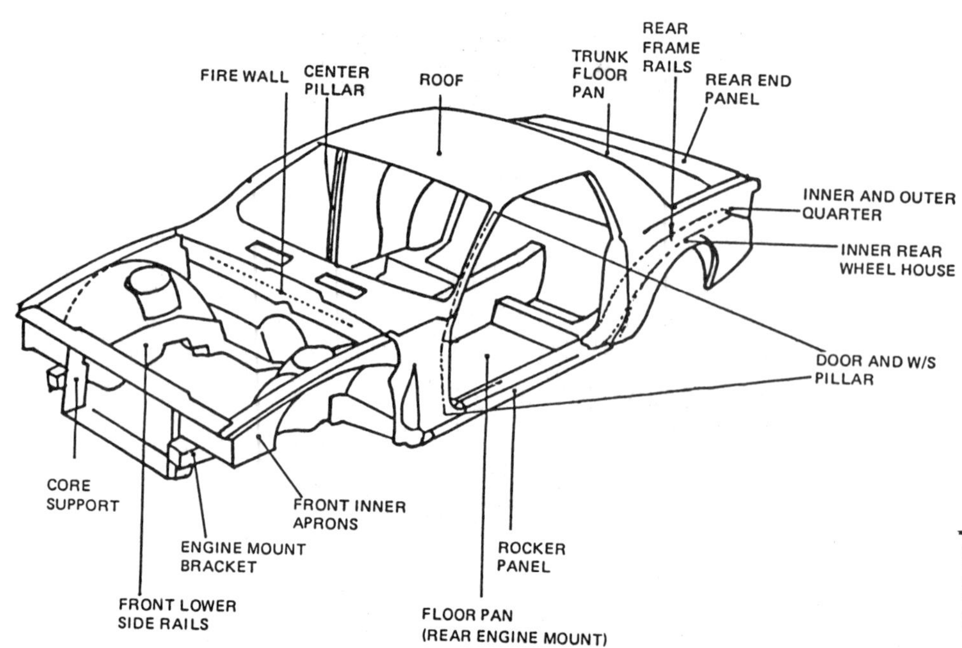

FIGURE 12-4. Unitized Body Frame Design. (Courtesy of Chart Industries, Ltd.)

Within a very few years after the development of unitized body design, virtually every domestic manufacturer of automobiles had converted to some degree to this method of construction. Vehicles of this era are considered first-generation unitized bodies (the AMC Rambler, for instance).

Problems, however, were encountered in full unitized construction. The car designers then combined the perimeter and unitized construction types and developed unitized construction with bolt-on stub-frame sections. This type of construction was referred to as second-generation unitized bodies (the GMC Nova is an example).

Since 1980, unitized bodies have progressed through another period of evolution. They include the X cars and the K cars, among others and can be referred to as third-generation unitized bodies.

Many of the metals used in these modern cars are generally referred to as high-strength steel (HSS). The thickness of this HSS ranges from 0.61 to 2.6 millimeters (0.024 to 0.102 inch). Depending on where this type of metal is used, special handling and tools are required while restoring a damaged vehicle body. Because some parts are galvanized on both sides (zinc coated), additional precautions are required when repairing these sections. Some parts made of HSS can withstand only a modest amount of heat—371°C (700°F) maximum—to facilitate strengthening and repair.

Some door beams and bumper reinforcements are often martinistic or ultra HSS. These parts are designed so that their metal is exceptionally hard and cannot be straightened cold with typical shop equipment. Any reheating or repair of these parts destroys their metal characteristics and reduces the strength to a very mild steel. Therefore, any damaged body part made of ultra HSS cannot be repaired; it must be replaced.

A unibody car is a complex structure. When such a vehicle is involved in a collision, the collision forces are spread throughout the vehicle's body. When the vehicle is being repaired, it demands a number of different pulls. The method of repairing unitized bodies is discussed in Chapter 13.

VEHICLE CONTROL AND STABILITY

Regardless of the type of body or frame, a vehicle's suspension and steering systems are attached to the frame. Thus, the vehicle's directional control and stability depend on the condition of the frame.

To understand how the frame affects directional control and stability, you must become familiar with two important terms. First is the **tracking** of a vehicle—that is, the parallel relationship of the front and rear wheels when the vehicle is moving forward.

The definition is very simple; however, there is more to it than meets the eye. If the wheels of a vehicle are to track, each front wheel must be the same distance from the vehicle's centerline, just as each rear wheel must also be an equal distance from the same centerline. The front wheels need not necessarily have the same track width as the rear wheels, but all four wheels must be in a parallel relationship (Figure 12-5).

The second term you need to understand is **wheelbase**—that is, the distance between the front and rear wheel centers when the front wheels are pointing straight ahead.

To measure a vehicle's wheelbase correctly, the axis (center) of the front spindles and the axis of the rear axles must be square with the centerline of the vehicle (Figure 12-6). For this condition to exist, the vehicle's frame must be square, and the axis centerlines through the front and rear wheel axles must be perpendicular (90° to the vehicle's

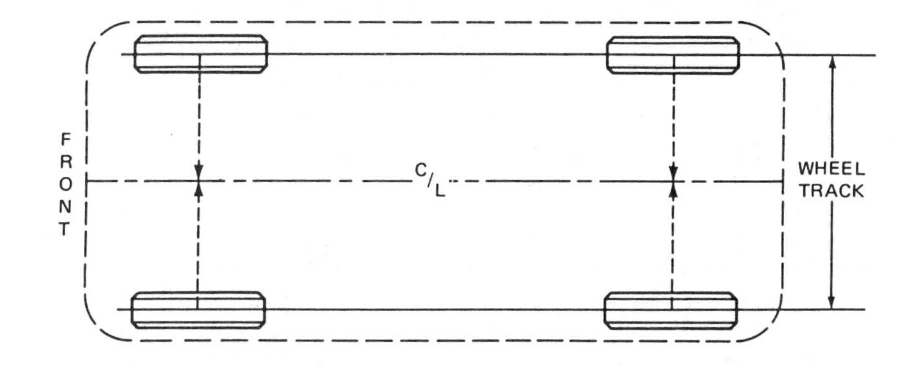

FIGURE 12-5. How Correct Tracking Is Achieved. To track properly, the front and rear wheels of a vehicle must be parallel and equidistant from the vehicle's centerline.

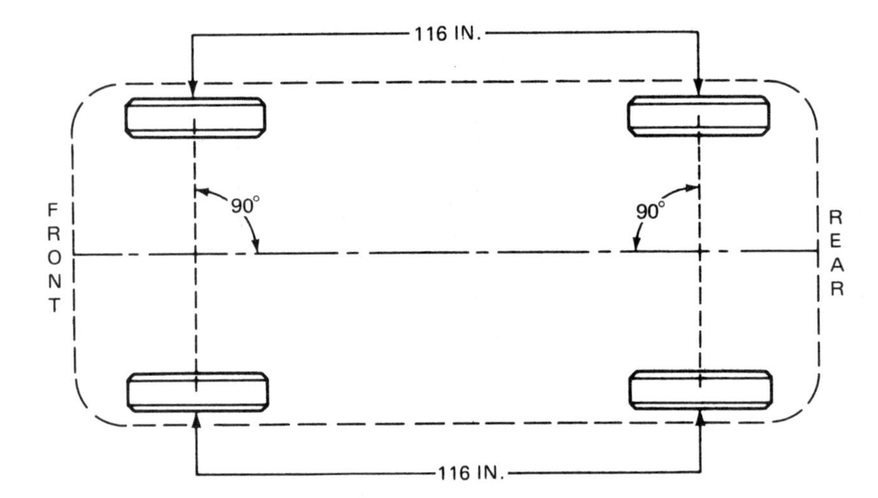

FIGURE 12-6. **Measuring Vehicle Wheelbase.** Proper measurement of a vehicle's wheelbase requires that the vehicle's centerline and the axis centerlines of the front and rear wheels be perpendicular.

centerline. Proper determination of the wheelbase is necessary to ensure that the automobile or truck will have correct tracking and thus aid the vehicle's directional control and stability.

Effects of Incorrect Tracking

Tracking is adversely affected by frame and/or suspension conditions. These conditions are referred to as wheel setback, axle offset, and axle side-set.

Wheel Setback. *Wheel setback* is a condition where one spindle is rearward of the opposite spindle. This condition can happen on the front wheels of a solid axle suspension or on any of the four wheels of an independent suspension system. Although the axis line in the center of the two front wheels is perpendicular to the center of the vehicle (see Figure 12-7), the two front wheels are not sharing a common centerline, and the wheelbases for the left and right sides are not the same. As the vehicle proceeds down the road, it will have a natural tendency to steer to the left. The driver constantly will have to turn the steering wheel to the right in order to maintain a straight vehicle path.

Axle Offset. *Axle offset* is a condition where the rear axle has been rotated so the drive direction (thrust line) is not parallel with the geometric centerline. This centerline is established as the line that connects the midpoints of both axles. For the example shown in Figure 12-8, the steering wheel must be turned left to maintain a straight vehicle path, and the vehicle will "dog track" due to the axle offset. (The term *dog tracking* describes a vehicle that is subject to incorrect tracking. Can you recall having seen a dog walk along the street with its front end out of line with its rear end? The body of the animal was diagonal, or off-center; hence, the phrase.)

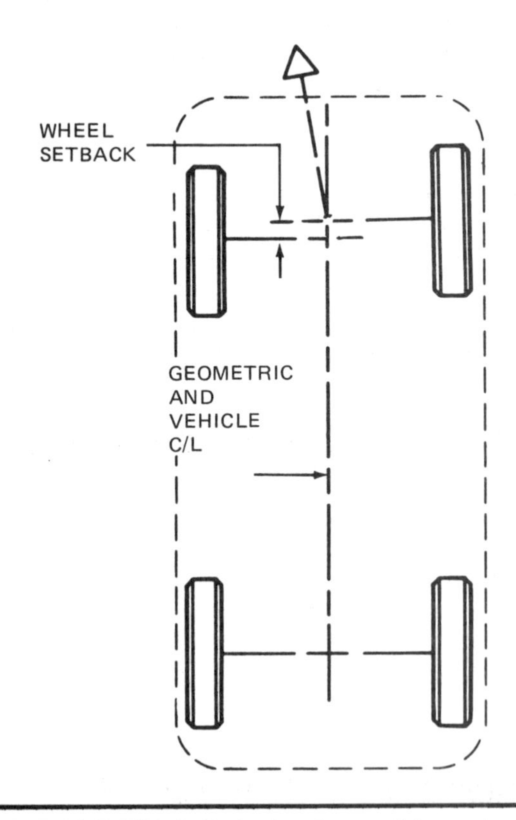

FIGURE 12-7. Wheel Setback: A condition where one spindle is rearward of the opposite spindle. (Courtesy of Hunter Engineering Company)

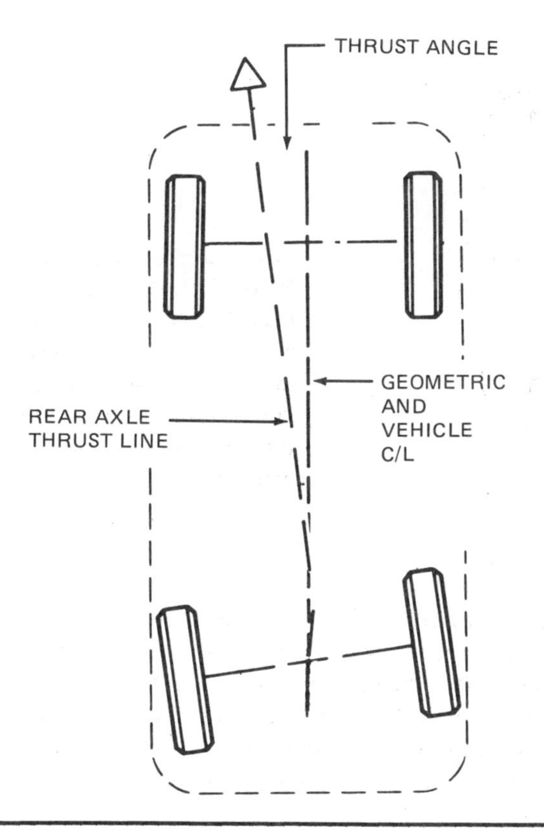

FIGURE 12-8. **Axle Offset: A condition where the rear axle thrust line is not perpendicular with the geometric and vehicle centerline.** (Courtesy of Hunter Engineering Company)

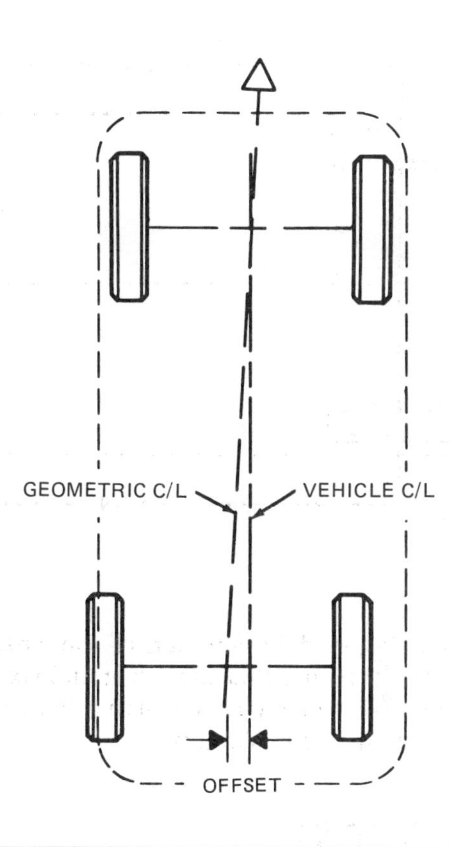

FIGURE 12-9. **Axle Side-Set: A condition where axle centers are not aligned to the vehicle's centerline.** (Courtesy of Hunter Engineering Company)

Axle Side-Set. *Axle side-set* occurs when all four wheels are parallel to each other, but the axle centers do not lie on the same track (Figure 12-9). The centerlines of the front and rear axles do not follow each other when the vehicle is moving. This condition will also cause the vehicle to dog track.

FIVE TYPES OF FRAME DAMAGE

Now that you are familiar with important related conditions and realize how they aid in the vehicle's directional control, you need to become familiar with five types of frame damage. As you examine the figures in this section, try to determine how each type of damage would affect a vehicle's directional control and stability and tracking.

Side Sway Condition

The first type of frame damage is called *side sway* (Figure 12-10). A front or rear side sway condition is usually present when the vehicle has been pushed sideways at either the front or rear frame position. Sway is a lateral misalignment that affects the frame or body centerline. Overall side sway is present when the vehicle has sustained damage at the middle of the left or right side rail.

It is obvious that the left and right wheelbase measurements in Figure 12-10 are not the same. How would this affect a vehicle's directional control and stability? Since the left wheelbase is greater than the right, this condition will influence the vehicle to steer (pull) to the right.

A FACT TO REMEMBER

A vehicle will always steer (pull) to the side that has the shorter wheelbase.

Sag Condition

A *sag,* or bending, condition usually results when the vehicle has sustained a direct collision

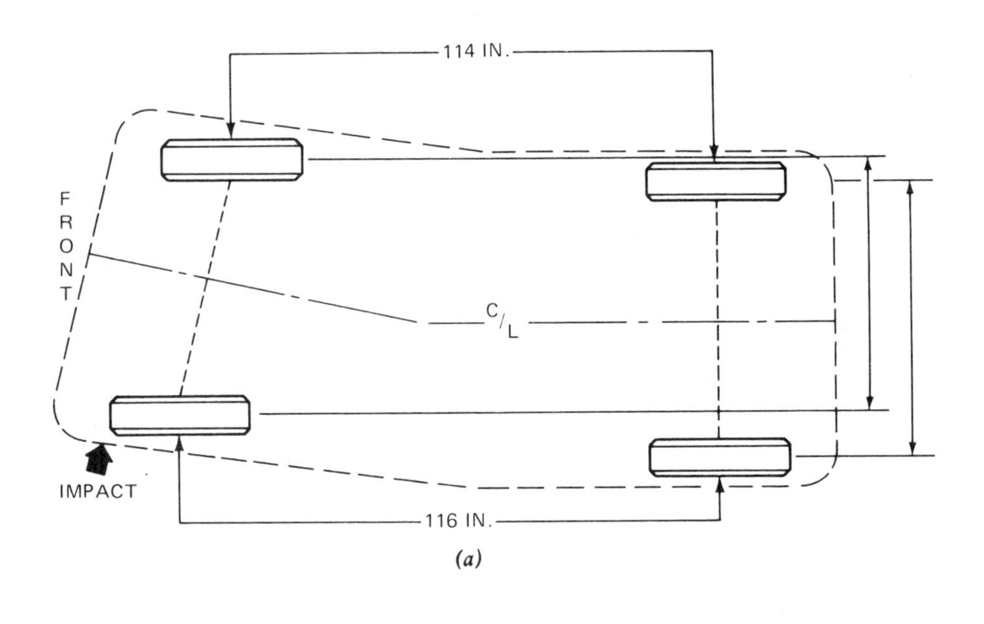

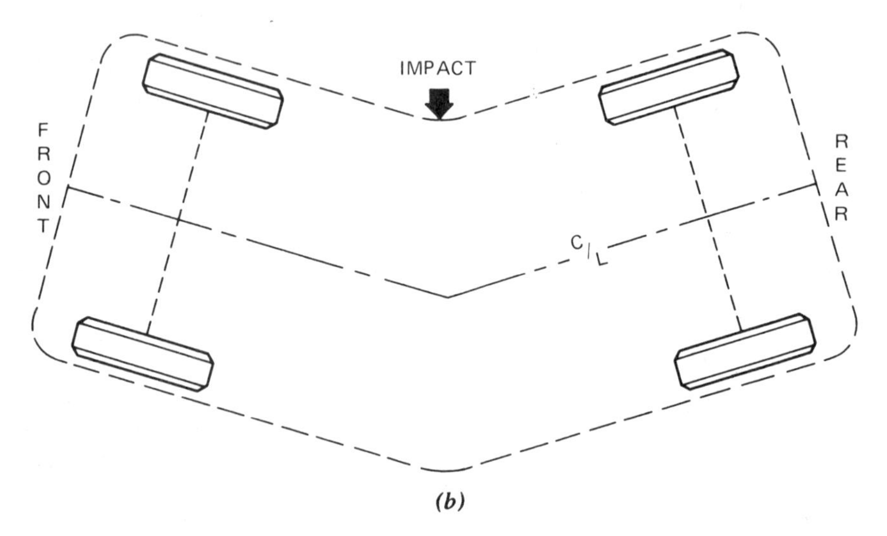

FIGURE 12-10. Side Sway Conditions. (a) Front side sway; (b) overall side sway.

from the front or the rear (Figure 12-11). At first glance, it appears that the left, right, or both side rails of the vehicle have literally sagged; however, sag is normally caused by the upward movement of front or rear frame rails. *Note:* The word *sag* refers to a vertical (up or down) frame misalignment. If only one side of the vehicle sustains damage, the left and right wheelbase measurements will be different.

Another part of the frame susceptible to sag is the front main crossmember. This type of sag occurs when the collision narrows the distance between the left and right upper control arms or the MacPherson strut towers (unibody construction), creating a negative (-) camber condition in the front end (Figure 12-12). The repair of this type of frame damage is discussed in Chapter 13.

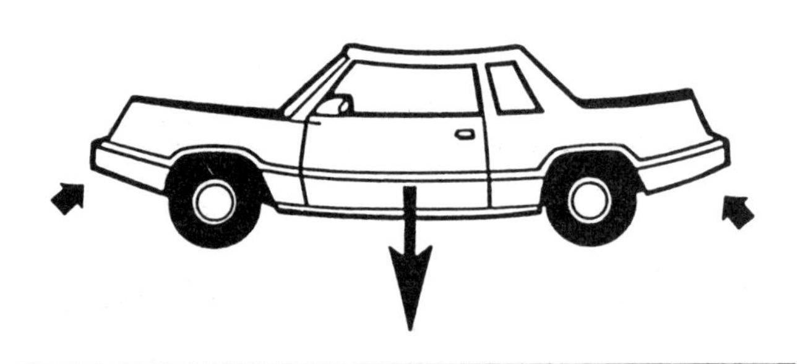

FIGURE 12-11. Sag Condition of Overall Frame: A result of a direct front or rear collision. (Courtesy of Chrysler Corporation)

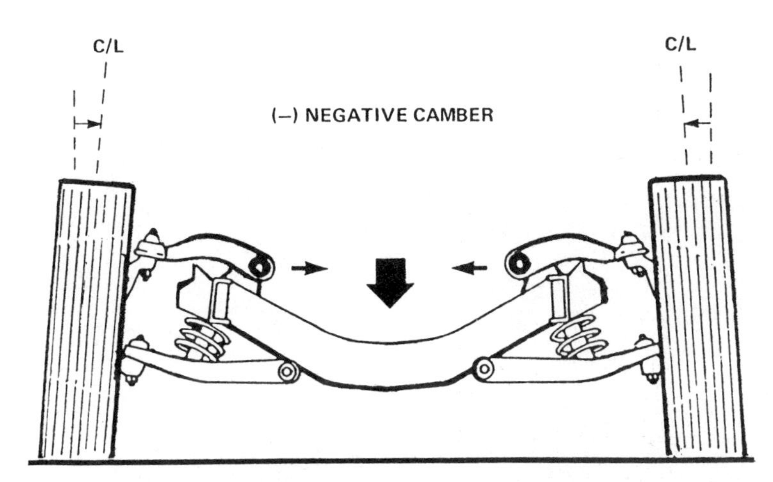

FIGURE 12-12. Sag Condition in the Front Main Crossmember.

Mash Condition

Figure 12-13 illustrates a type of frame damage called *mash*, or crushing. A mash condition is present when the length of the frame is shorter from the cowl to the front bumper or from the rear wheels to the rear bumper. Mash is usually caused by direct damage (head-on or rear-end collisions). In most cases, the left or right side of the frame is shortened, thus changing the wheelbase. When a unibody-type vehicle is hit from the rear, the forces of impact may cause the side of the structure to bow out away from the passengers (never in), distorting the side rails and door openings.

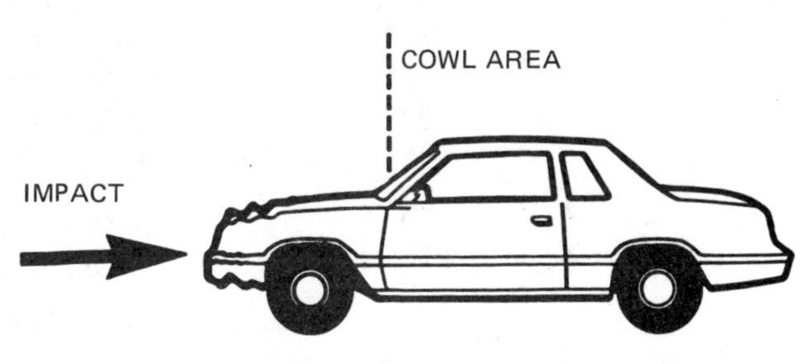

FIGURE 12-13. Mash Condition in Front Portion of Automobile Frame. (Courtesy of Chrysler Corporation)

Diamond Condition

The fourth type of frame damage to be discussed is called *diamond*. A diamond condition is present when one side rail of the vehicle has been moved in such a way to cause the frame and/or body to be out of square (that is, to be shaped like a parallelogram). Figure 12-14 illustrates that the right side rail of the frame or unitized body has been driven back, causing the shape of the frame to be distorted. As you study the illustration, try to determine how this type of frame or body damage would affect a vehicle's tracking and directional control.

Diamond damage places the frame and rear axle at an angle to the direction of travel, causing a severe abrasive scuffing on all four tire treads. In addition, the rear wheels have a tendency to steer the rear of the vehicle toward the right, thus forcing the front end of the vehicle to the left. Naturally, the front wheels have to be continually held to the right to enable the vehicle to proceed down the road. A unitized frame seldom sustains diamond damage because of its construction and design.

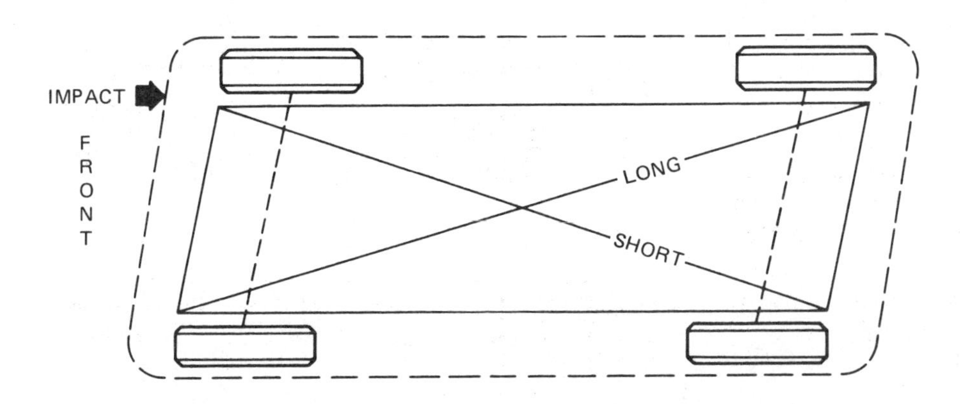

FIGURE 12-14. Diamond Condition.

Twist Condition

The fifth type of frame damage that you need to be concerned with is called *twist*. A twist condition is present when one corner of the vehicle is higher than the other corners (Figure 12-15). This type of frame damage normally results when a vehicle carrying an excessive load turns over or tends to turn over on its side. Twist damage affects the vehicle by not allowing it to sit in a proper relationship to the surface of the road, thus affecting the vehicle's alignment. If after a major collision or rollover, one corner of the vehicle appears to be sagging while it is positioned on a level surface (as though the vehicle had a defective spring), the vehicle's frame should be checked for twist damage.

FIGURE 12-15. Twist Condition. (Courtesy of Chrysler Corporation)

ANALYZING FRAME DAMAGE

Just as a medical doctor examines a patient, the vehicle repair technician must also be able to examine, locate, and diagnose frame damage. Since the sheet metal and frame are so closely interrelated, certain forms of sheet metal damage can indicate frame misalignment. The following symptoms suggest possible frame damage:

- Misaligned door fit (irregular door fit),
- Fender-to-door gap (uneven vertical space from top to bottom),
- Hood misalignment (excess gap along the fender on one side while overlapping on the other),
- Irregular gap at trunk opening,
- Irregular bumper fit (with no impact damage),
- Buckles in fenders that show no sign of direct impact,
- Dents and buckles in aprons and rails,
- Buckles in quarter panels (with no evidence of impact),
- Buckles in top of quarter panel at upper deck panel,
- Cracked or broken windshield (with no evidence of impact),
- Damaged floor pans and rack and pinion mounts (unitized bodies),
- Misaligned suspension and motor mounts,
- Cracked paint and undercoating,
- Split seams and seam sealer,
- Pulled or broken spot welds,
- Wheels off-center in wheel openings.

The presence of any of these conditions could indicate that the vehicle has sustained frame damage. However, before beginning any type of frame repair, a thorough inspection and analysis of the results must be made.

Inspection Procedure

When inspecting frame damage follow this recommended procedure:

1. Relate all visual damage to a point of impact.
2. Visually trace the force from impact through the vehicle to a point where there is no further evidence of damage.
3. Make a rough sketch of the frame.
4. On the sketch, mark the damages found by your visual inspection.
5. Verify your findings with a gauge analysis.

Gauge Analysis

Gauge analysis of frame damage requires your familiarity with two important terms, "datum line" and "comparable measurement." The *datum line* is an imaginary reference line or plane established by vehicle manufacturers located a fixed distance below the vehicle from which vertical measurements can be made. Figure 12-16 illustrates sample height measurements made from the datum line. The datum line is similar to the centerline in that it cannot be seen and exists only on frame specification charts. Its purpose is to provide a baseline for determining whether the vehicle's frame has sustained damage.

Comparable measurement is the reference measurement established by the manufacturer for determining the comparable distances from point to

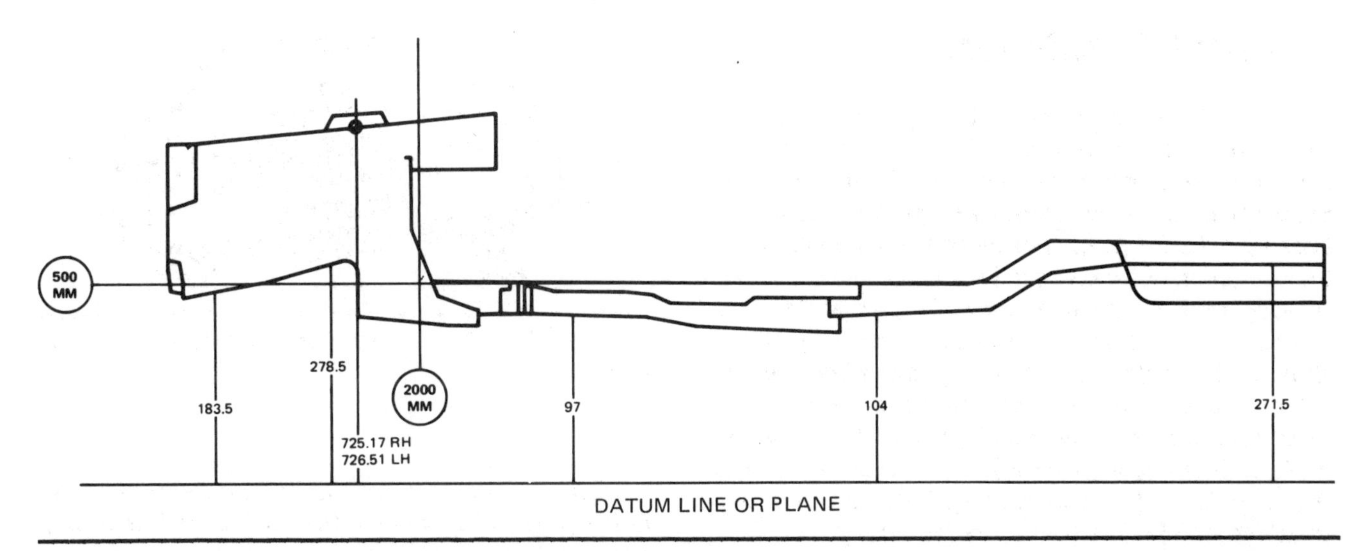

FIGURE 12-16. Datum Line: A line established by the manufacturer as a reference point from which all vertical frame measurements are made. (Courtesy of Ford Motor Company)

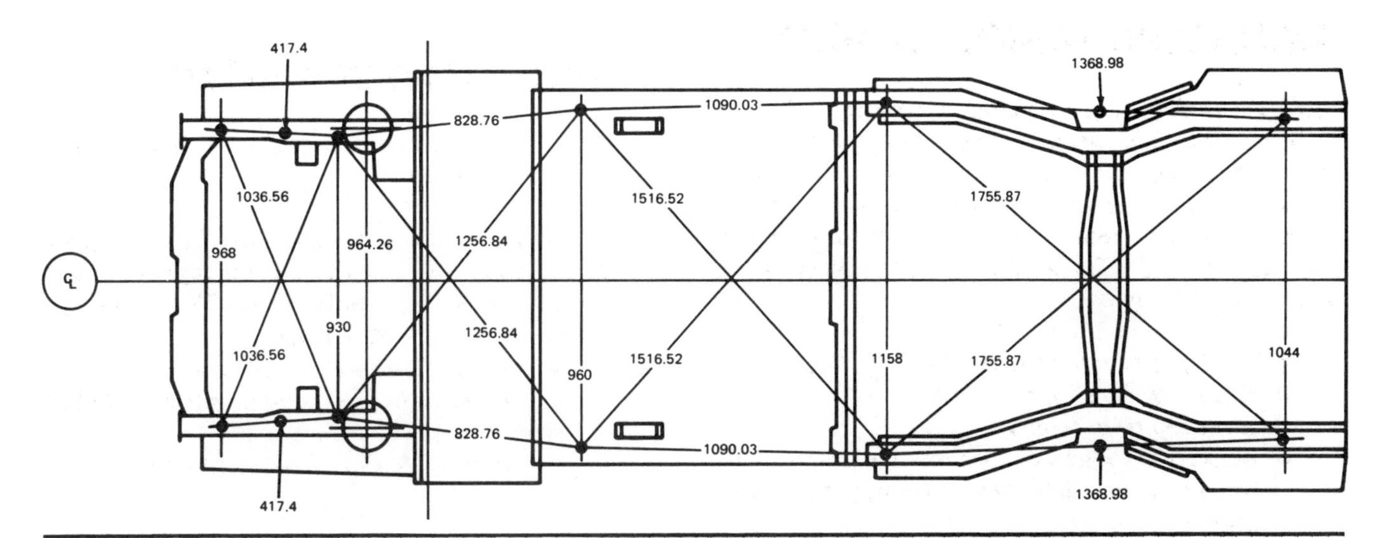

FIGURE 12-17. Comparable Measurement: A reference measurement established by the manufacturer for determining comparable distances from point to point on either side of the vehicle frame. (Courtesy of Ford Motor Company)

point on either side of the vehicle's frame. Figure 12-17 illustrates the bottom view of a vehicle's frame and the comparable measurements.

As you study Figure 12-17, observe that if only one side of a vehicle's frame were damaged, you could compare the measurements from the opposite side, thereby assisting yourself in analyzing the extent of frame damage. For example, in Figure 12-17, you will see the midpoint in the frame, a distance of 1516.52 millimeters (59.7 inches). If the frame has not sustained damage, the opposite diagonal distance should be the same. At the extreme right of Figure 12-17 is a crosswidth distance of 1044 millimeters (41.1 inches). If you are required to measure from the centerline to either the left or right rear frame rails, the distances of the centerline would be the same. These measurements are referred to as being symmetrical. Some vehicles with unique structural design have asymmetrical measurements. This means that the distances from the centerline, when measuring across the frame, will be different.

Traditional tools for frame damage measurements are tram track gauges and self-centering frame gauges. When applied to specific frame reference points, these tools will assist you in visually locating and verifying various types of frame damage. The tram track gauge is used to measure straight, diagonal, and wheelbase distances. Self-centering gauges

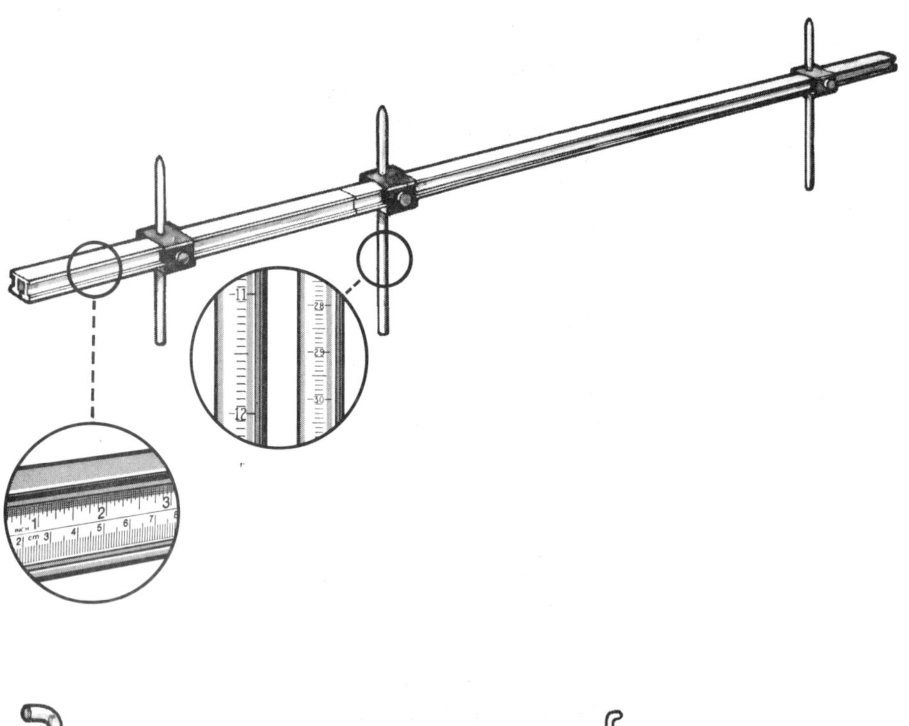

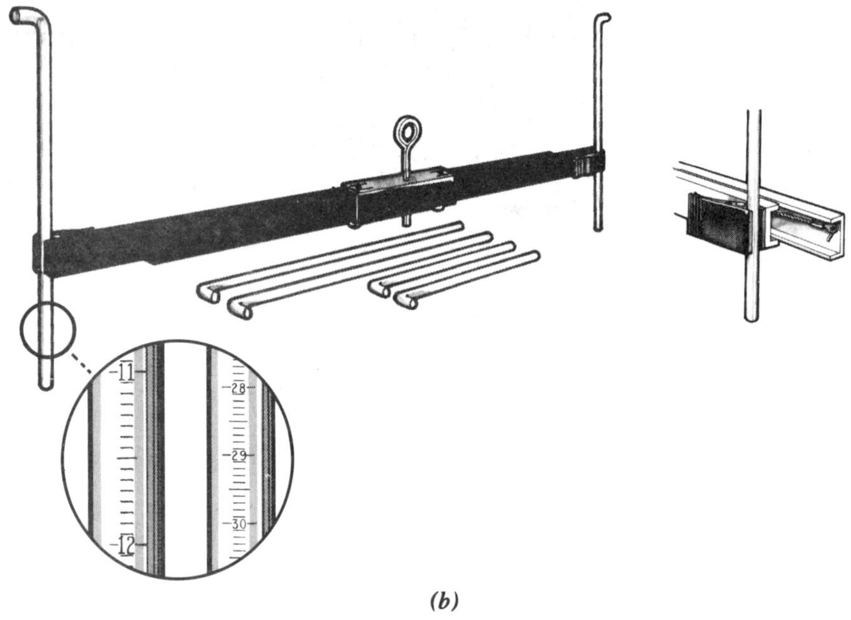

(b)

FIGURE 12-18. Frame Damage Gauges. (a) Adjustable track gauge; (b) self-centering frame gauge. (Courtesy of Blackhawk Automotive Division, Applied Power Canada, Ltd.)

are used to establish a vehicle's centerline, as well as the lateral and vertical positions of a frame's side rails and crossmember. These gauges are shown in Figure 12-18. More recent developments in collision damage measurements are further discussed in Chapter 13.

When you suspect frame damage, use tram gauges or self-centering gauges as follows:

- Mash (frame length): Use a tram gauge.
- Mash (frame width): Use a tram gauge.
- Side sway: Use self-centering gauges.
- Sag: Use self-centering gauges.
- Kick-up, kick-down: Use self-centering gauges.
- Twist: Use self-centering gauges.
- Diamond: Use diamond attachment on self-centering gauges.

In all cases, a tram gauge should be used to check frame length and for diagonal distances that cannot be measured with self-centering gauges.

Using a Tram Track Gauge. A *tram track gauge* (often referred to as a tram gauge) performs the same function as a tape measure. However, when measuring long distances, a tape measure will not remain rigid and will encounter the obstacles of other vehicle parts. A tram gauge is an adjustable linear bar with two adjustable pointers and is used to measure around the obstacles presented by other vehicle parts.

Before measuring any distance, preadjust the linear bar and the vertical pointers to the approximate distance to be measured. When using the gauge, always place the bar in a parallel position to the crossmember or frame rail to be measured. Then adjust the gauge to obtain the exact distance. Figure 12-19 illustrates the practical use of the tram gauge.

Using a Self-Centering Frame Gauge. Although *self-centering frame gauges* may appear complicated, they are in fact simple to understand and relatively easy to use. With some well-invested time and instruction, any student or apprentice can learn how to use one.

Self-centering frame gauges are designed to be used when the vehicle has been placed on a frame alignment rack or when the vehicle has been raised and placed on elevated safety support stands. If you are a student and have been requested by your shop instructor to inspect and analyze a vehicle's frame condition, use the following procedure.

Step-by-Step Procedure

1. To prepare the damaged vehicle for measurement, jack it up and place safety stands under the front and rear torque boxes. Make sure the stands are placed on a level floor and are all exactly the same height (Figure 12-20A).

FIGURE 12-19. Using Tram Track Gauge to Check Frame Measurements around Obstacles. (Courtesy of Chart Industries, Ltd.)

FIGURE 12-20A. Safety Stands Placed under Front and Rear Torque Boxes. (Courtesy of Blackhawk Automotive Division, Applied Power Canada, Ltd.)

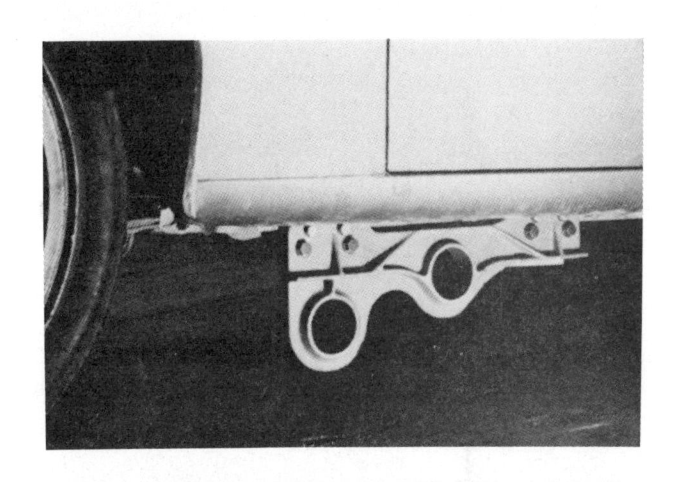

FIGURE 12-20B. Clamps Attached to Pinch Weld. (Courtesy of Blackhawk Automotive Division, Applied Power Canada, Ltd.)

FIGURE 12-20C. Crosstube Placed through Safety Stand and Underbody Clamp. (Courtesy of Blackhawk Automotive Division, Applied Power Canada, Ltd.)

Caution: Modern automobile construction is extremely flexible. Therefore, it is important when body frame misalignment is being diagnosed that the car is not supported by its suspension. Otherwise, the effect of sagged springs, bent suspension parts, flat tires, and other conditions may show up in the gauging procedure as twist, sag, or other conditions that do not really exist in the body frame.

Keep in mind that when stands are used to support the central section of the vehicle and because the frame is designed to flex, the front and rear sections of the vehicle may sag from 6 to 19 millimeters (¼ to ¾ inch). Front wheel drive cars may sag even more in the front than in the rear. This sagging will have a bearing in the datum line height measurement when gauging the frame.

If the vehicle is a unitized body type, attach the underbody clamps to the vehicle pinch weld at both ends of the rocker sill (Figure 12-20B). Then place the crosstube through the safety stand and underbody clamp, as shown in Figure 12-20C.

2. Obtain the manufacturer's shop manual and refer to the datum line height specifications found in the manual's frame section.

3. Adjust and lock the support rod for the self-centering gauges to the datum line height specification.

4. To adjust the gauges to the frame width, always grip the gauges by the two vertical uprights. Then push or pull the horizontal bar to the desired width.

5. The frame gauges have a self-centering 41.2 millimeter (1 5/8 inch) diameter alignment target (Figure 12-21). The height from the inside of the hook to the top of the frame gauge bar must be set at the heights specified in the frame dimension book. *Note:* Gauges must be placed at identical points on both sides of the frame.

6. Attach the first self-centering gauge, following the datum line specifications.

7. Place the second gauge at the vehicle's own cowl area.

8. Place the third gauge in front of the rear suspension.

9. If necessary, place the fourth gauge at the rear of the vehicle. When all the gauges have been hung from the frame, double-check the height by using a tape measure (Figure 12-22).

Note: All frames or unitized body designs are not exactly alike. The placement of the gauges and the method of dividing the frame into three sections is a standard procedure used by all experienced

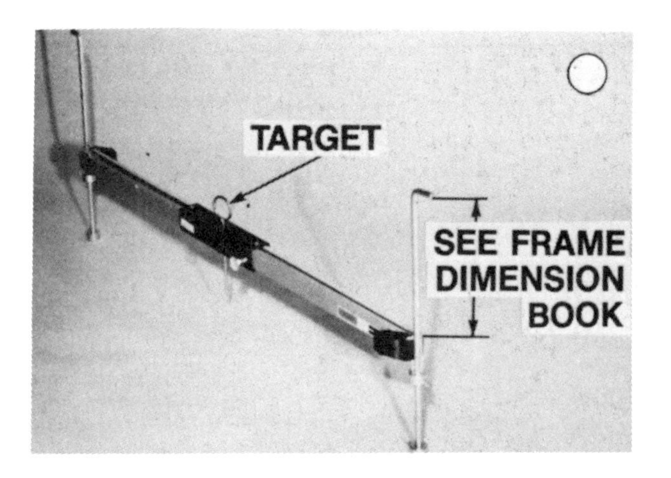

FIGURE 12-21. Alignment Target on Self-Centering Frame Gauge. (Courtesy of Blackhawk Automotive Division, Applied Power Canada, Ltd.)

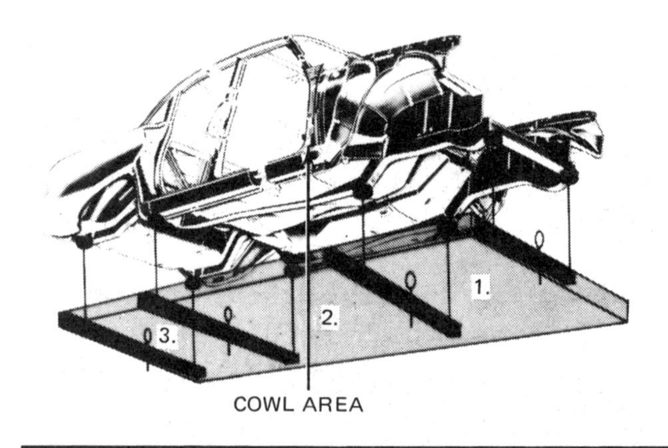

FIGURE 12-22. Placement of Gauges for Diagnostic Purposes. (Courtesy of Blackhawk Automotive Division, Applied Power Canada, Ltd.)

frame technicians, however. Figure 12-23 illustrates three self-centering gauges attached to the vehicle's frame. Since the bars of the gauges are horizontally level, the gauges indicate that the frame is in perfect alignment at all points where the gauges are positioned. The round centerline pins form concentric circles similar to a bull's-eye, indicating that the frame's centerline is also in perfect alignment.

Diagnosis

When a vehicle has been in a collision, it seldom sustains only one type of frame damage. It will probably receive a combination of damages, such as mash and sag, or side sway, mash, sag, and twist. When you suspect, or know, that a vehicle's

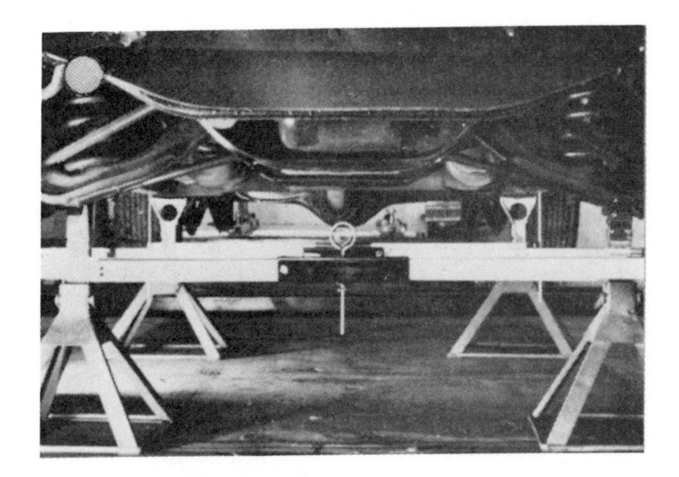

FIGURE 12-24. Side Sway Damage. The front gauge is offset indicating front side sway. (Courtesy of Blackhawk Automotive Division, Applied Power Canada, Ltd.)

frame has been damaged, the type of damage must be diagnosed before any repairs are attempted.

Diagnosing Side Sway. Side sway usually can be located by sighting along the side of the vehicle. It is recommended that two gauges be attached to the frame in an undamaged area and a third or fourth gauge in the damaged region. The position of the gauges will determine exactly where the frame is bent and out of alignment. The gauge bars should be parallel, but the centerline pin sights will be out of line sideways (Figure 12-24).

Diagnosing Sag. The location and extent of a sagged frame can be determined through the use of frame gauges. Position the gauges as described for side sway. When the gauges have been set and attached to the frame, check to see if the parallel bars are in their correct alignment. If sag damage is present, the center gauge will appear to be lower than the other two gauges, as shown in Figure 12-25. Move the gauges to various locations to determine the location and extent of the damage.

Diagnosing Mash. Mash damage is usually limited to the area in front of the cowl or in back of the rear window. Wrinkles and severe distortion will be found in the front or rear fenders, and the frame will generally rise upward at the top of the wheel arch. With mash damage, there is very little vertical displacement of the bumper since the damage is usually the result of a direct collision.

When the gauges are attached to the frame, they may indicate no damage near the front bumper or rear bumper but will show a definite upward rise when hung at the front or rear wheel arches. Close visual inspection may also show wrinkles in the bottom of the frame over the wheel arches. On unitized vehicles, mash damage may extend into the floor pan, showing wrinkles in the floor pan around the rear of the frame rails. The extent of this type of damage can also be detected by using a tram gauge (Figure 12-26A). Simply adjust the gauge and measure from a midframe point to the front or rear of the vehicle's frame. Compare the distance found to the distance shown in the manufacturer's frame specifications manual (Figure 12-26B). The amount of mash is determined by subtracting the actual measurement from the specified measurement.

FIGURE 12-25. Sag Damage. The center gauge is hanging below the other two, indicating sag damage. (Courtesy of Blackhawk Automotive Division, Applied Power Canada, Ltd.)

FIGURE 12-26A. Mash Damage. The tram gauge measures linear distance from a midpoint on the frame to the front of the vehicle.

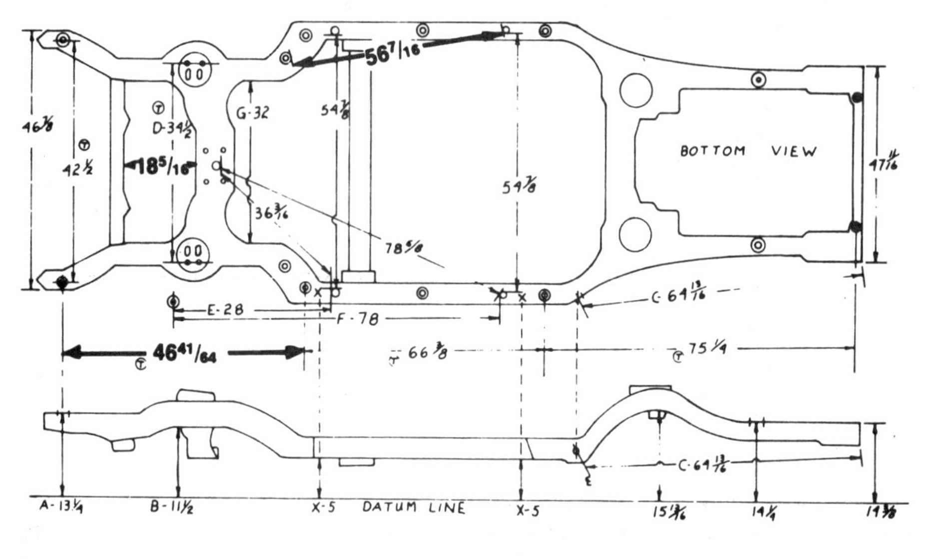

Note: Solid dot always indicates a center measurement; an outline, unless marked otherwise, indicates an edge measurement.

FIGURE 12-26B. Mash Damage Determinants. Actual measurements are compared with the specified measurements in the frame dimension book.

Diagnosing Diamond. Diamond damage is caused by a hard impact on a corner of a vehicle. Visual indications are hood and trunk lid misalignments and buckles in the quarter panel near the rear wheel housing or at the roof to quarter panel joints. Wrinkles and buckles will also appear in the passenger and trunk compartments. Mash and sag damages are normally combined with a diamond condition. Since diamond damage is severe, it is usually sustained in the entire length of a vehicle's frame. This type of damage may be detected by using a diamond sight gauge and by cross-checking the diagonal measurements of a frame with a tram gauge. The diamond sight gauge is an attachment to the self-centering gauges and sits on a supporting bar. If the two vertical aligning pins are not in alignment with the centering pin of another gauge, the frame has sustained a diamond condition (Figure 12-27).

The oldest method of determining diamond damage is similar to using a tram gauge to measure diagonal distance. Place the vehicle on a clean, level section of the garage floor with the parking brake applied. Select five or more corresponding points on each side rail of the frame. With a **plumb bob,** project and carefully chalk mark each position of the plumb bob onto the garage floor. Move the vehicle away from the chalk marks. Then measure and compare the diagonal distances. These measurements should be within 3 millimeters (1/8 inch) of each other (Figure 12-28).

Diagnosing Twist. A twist condition can be

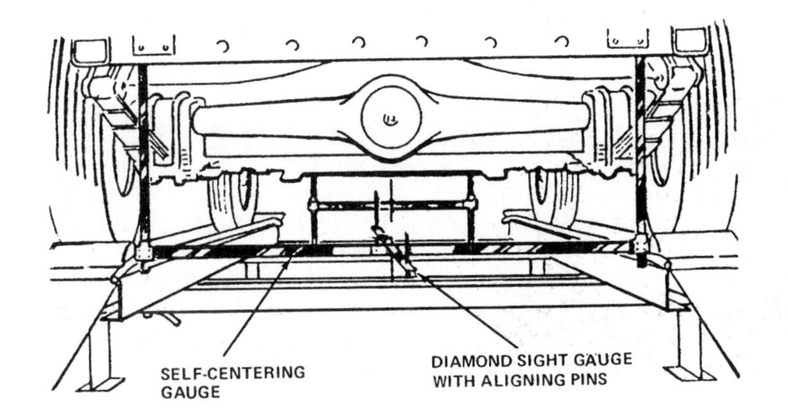

FIGURE 12-27. Using a Diamond Sight Gauge to Locate Diamond Frame Damage. (Courtesy of FMC Corporation, Automotive Service Equipment Division)

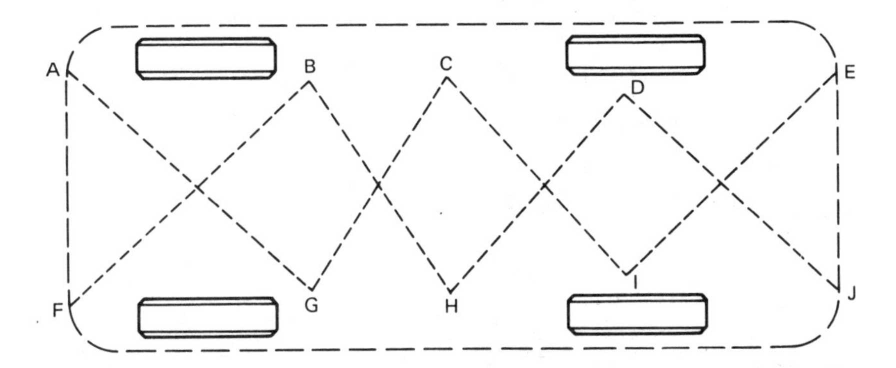

FIGURE 12-28. Measuring Diagonals to Determine Diamond Damage. To determine if diamond damage is present, measurements are made along the corresponding diagonals—for example, E to I with D to J.

determined by attaching four or more gauges to the vehicle's frame. As you study Figures 12-29A and 12-29B, you will observe that the horizontal bars of the gauges are tilted when compared to each other, even though the centering pins are in perfect alignment. The amount that the gauges are unparallel to each other shows the amount of twist (Figure 12-29B). By determining the amount of twist, the body-frame mechanic can determine the extent of distortion and the force required to correct the problem.

Remember that a twist condition usually extends the full length of a vehicle if it is present in a ladder frame. If the condition appears to be localized, the damage is likely to be a combination of twist and sag or kick-up or kick-down (damage at the kick-up section of the frame) rather than a pure twist condition.

Diagnosing Combination Damage. Nearly all collision damage involves two or more types of frame damage. The majority of collisions occur at the vehicle's front end from a side angle, resulting in combinations such as front end side sway with sag and mash or front end side sway with a kick-up and mash. When combination damages occur, frame gauges will indicate the various types of frame damages.

Diagnosing Swung Rear Axle. The loss of directional control and stability is not always caused by the misalignment of a vehicle's frame. In some cases, it may be caused by incorrect wheel alignment or a fault in the vehicle's front or rear suspension systems. Figure 12-30 illustrates a condition where the rear axle has shifted back on the left side. This condition is described as a *swung rear axle* and has been caused by the shearing of the spring tie bolt or a damaged rear suspension control arm. As you study the illustration, you will notice that the vehicle's frame is in correct alignment. The shift of the rear axle has changed the wheelbase measurements of the vehicle, thereby influencing its tracking and directional control. If you suspect that a vehicle's tracking is incorrect, it may be measured by using a tracking gauge. A tracking gauge is basically a tram gauge with an added pointer.

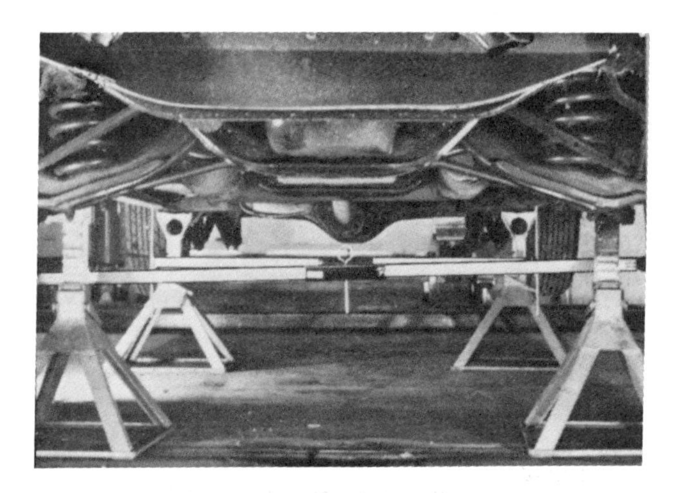

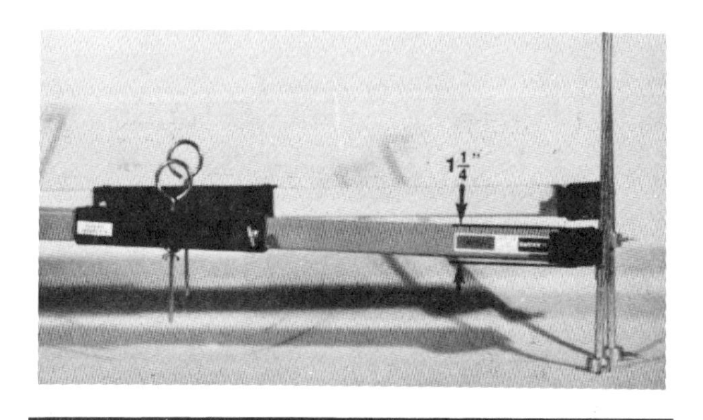

FIGURE 12-29B. **Establishing the Amount of Twist by Comparing Gauges.** (Courtesy of Blackhawk Automotive Division, Applied Power Canada, Ltd.)

FACTS TO REMEMBER

1. When reading self-centering frame gauges, standing too close to the vehicle changes your line of sight. If possible, stand back and observe the overall view of the gauges in line with the vehicle's centerline.
2. When sighting the horizontal position of the frame gauges, remember that your height will distort your view of the gauges. When reading gauges, your eyes must be on a level plane with the gauges. In some cases, reading the gauges from the end opposite the damaged frame section will assist you in your analysis of frame damage.
3. When viewing and analyzing the position of the frame gauges, inspect the vertical position and displacement of the support rods in relation to the horizontal bars of the gauges.
4. When viewing the gauges, always consider the lateral deflection of the centerline pins and the circle line-up of the bull's-eyes. Determine if one circle is out of line or pushed to the side.
5. Check to see if one or more of the horizontal bars is up or down from the horizontal position.

MEASURING A VEHICLE'S TRACKING

The track of a vehicle can be measured by using the following procedure.

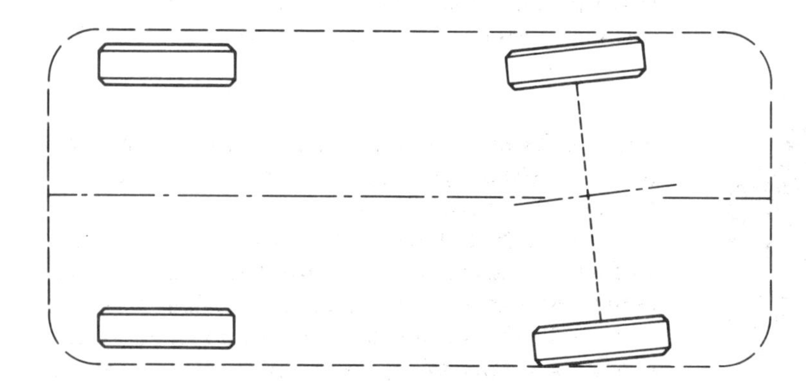

FIGURE 12-30. **Swung Rear Axle.** A swung rear axle alters a vehicle's wheelbase and, thus, its tracking and directional control.

Step-by-Step Procedure

1. Obtain a floor jack and a lateral runout gauge.
2. Lift one corner of the car so that the wheel is free and can rotate.
3. Place the lateral runout gauge against the wheel. Then, by rotating the wheel, determine if the wheel has sustained any runout. If the wheel has runout, indicate that position by placing a chalk mark on the sidewall of the tire.
4. The maximum position of lateral runout (indicated by the mark) is then positioned at the very top or the very bottom of the wheel.
5. Lower the wheel back to the garage floor. Follow the same procedure for the other three wheels. *Note:* If the vehicle is equipped with power steering, center the front wheels with the steering in operation. If any wheel is bent due to a collision, it must be replaced before using the tracking gauge.
6. Adjust the three pointers of the tracking gauge so they contact the edge of the wheel rim, as illustrated in Figure 12-31. *Note:* The long bar of the gauge must be kept parallel to the side of the vehicle.
7. Move the gauge to the other side of the vehicle to obtain a comparison measurement. Make sure the pointers and the tubing of the gauge are secure. Set the pointers against the wheels to determine any length variations or unparallel misalignment positions of the front or rear wheels. Figure 12-32 illustrates how a tracking gauge may be used to determine a swung rear axle, a side sway, or diamond damage.

All conditions that pertain to a vehicle's directional control and stability should be verified before any repairs are attempted. If you suspect the tracking of a vehicle, you may confirm your suspicions by conducting two checks.

FIGURE 12-31. Positioning Track Gauge and Three Pointers against the Wheels. (Note position of chalkmarks.)

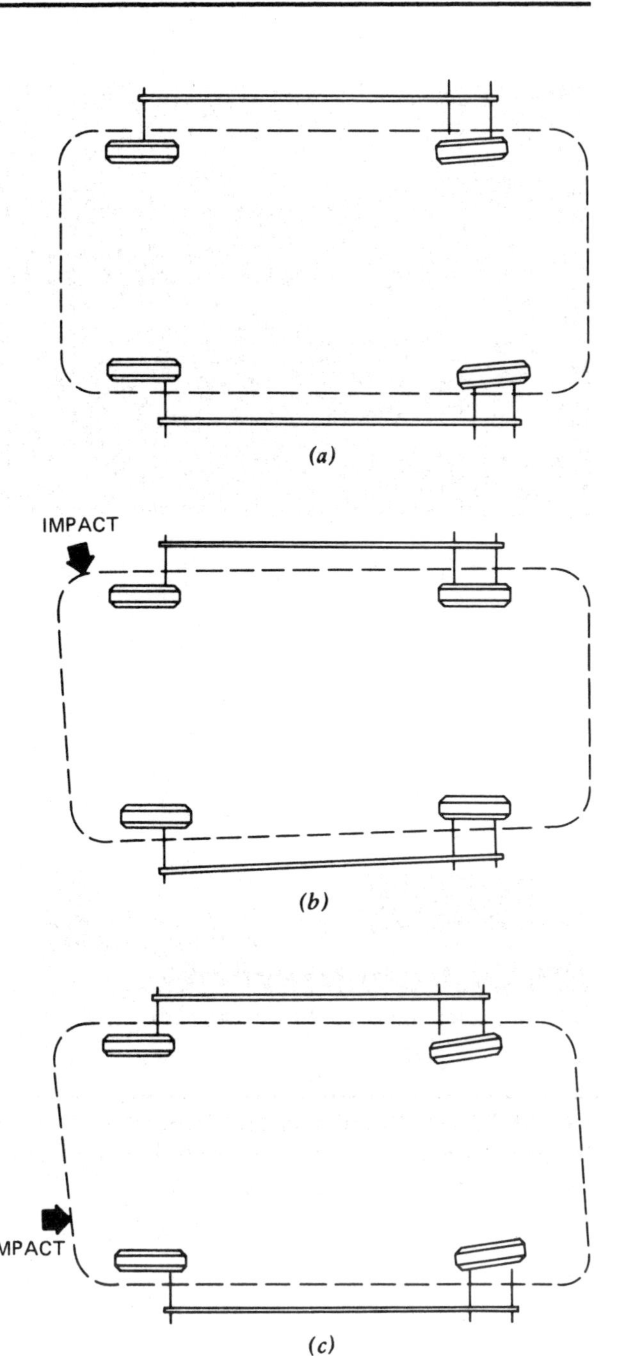

FIGURE 12-32. Using Tracking Gauge to Identify Three Types of Damage. (a) Swung rear axle; (b) side sway; (c) diamond.

For the first check, wet down an area on a road surface or garage floor. Drive the vehicle slowly across the wet area as straight as possible. Stop the vehicle. Walk to the rear of the vehicle and visually check the marks left by the tracks of the tire treads.

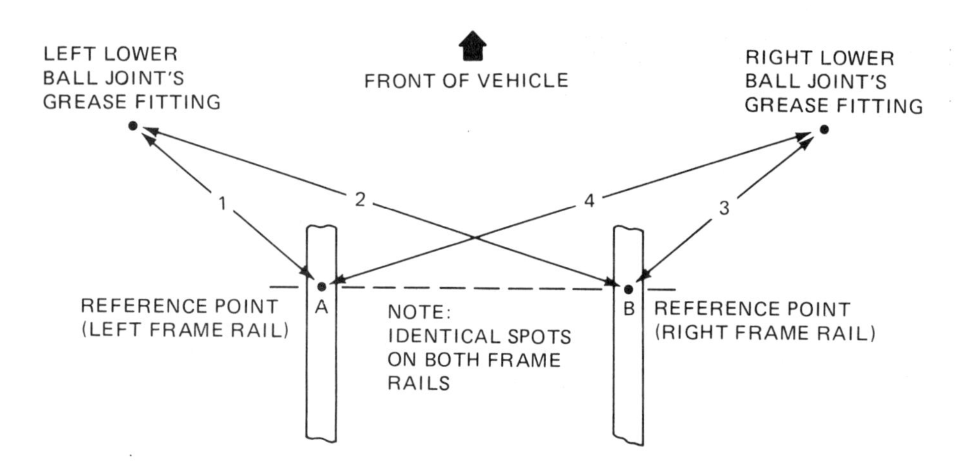

FIGURE 12-33. Using Tram Gauge to Check Comparable Measurements to Verify (Diagnose) Tracking Problems.

If the tracks are not parallel, the wheels of the vehicle are not tracking.

Check 2 is best conducted when the vehicle has been positioned on an alignment rack. Obtain a tram gauge and measure the diagonal distance (point 1 in Figure 12-33) from the left lower ball joint's grease fitting to reference point A on the vehicle's left frame rail. With the tram gauge, measure the diagonal distance (point 2) from the left lower ball joint's grease fitting to reference point B on the right frame rail. Record these entries on a piece of paper.

Again, using a tram gauge, measure the diagonal distance (point 3) from the right lower ball joint's grease fitting to reference point B on the right frame rail. With the gauge, measure the diagonal distance (point 4) from the right lower ball joint's grease fitting to reference point A on the left frame rail.

If the control arms are in their correct relationship to the vehicle's frame, distances 1 and 3 will be the same, and distances 2 and 4 will be the same. If they are not the same, this information will assist you in diagnosing the cause for incorrect vehicle tracking. A similar procedure can be used for determining tracking problems in the vehicle's rear suspension.

REVIEW TEST

1. State the three main purposes of a vehicle's frame.
2. Define the terms *tracking* and *wheelbase.*
3. List ten reasons why a vehicle may not possess correct tracking.
4. Describe how to diagnose a front main cross-member sag condition.
5. Describe how to diagnose a twist condition in a vehicle's frame.
6. Explain the meaning of the terms *datum line* and *comparable measurement.*
7. Describe how self-centering gauges are used to detect body-frame damage.
8. Describe how a diamond sight gauge is used to detect body-frame damage.
9. Describe how you would detect a swung rear axle. List three possible causes of this condition.
10. You have placed a vehicle on a frame alignment rack or on safety stands for frame and/or body analysis. You have attached self-centering gauges to the frame and are now viewing the vehicle from the front. Explain, in point form, the types of frame damages as they appear in Figure 12-34.

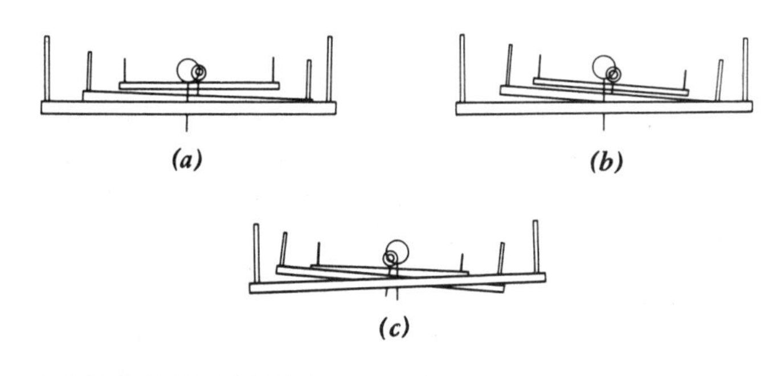

FIGURE 12-34. Illustration for review question 9. (Courtesy of Blackhawk Automotive Division, Applied Power Canada, Ltd.)

FINAL TEST

This examination is multiple choice. Only one answer will be accepted. Carefully read every statement.

1. Mechanic A says that the frame of a modern vehicle must be able to absorb torsional stress. Mechanic B says that the frame of a modern vehicle is made of light, durable alloy materials. Who is right? (A) Mechanic A, (B) Mechanic B, (C) Both A and B, (D) Neither A nor B.
2. Mechanic A says that the primary purpose of a vehicle's frame is to support the weight of the engine. Mechanic B says that the frame, regardless of its design, is the main (base) sustaining member of a vehicle. Who is right? (A) Mechanic A, (B) Mechanic B, (C) Both A and B, (D) Neither A nor B.
3. Mechanic A says that the torque box area of a frame is designed to absorb a major portion of the impact when the vehicle is involved in a head-on collision. Mechanic B says that to aid a vehicle's tracking, the axis centerline through front wheels must be perpendicular with the vehicle's centerline. Who is right? (A) Mechanic A, (B) Mechanic B, (C) Both A and B, (D) Neither A nor B.
4. Mechanic A says that all frames, regardless of their design, must be able to flex. Mechanic B says that perimeter frame design is best adapted to commercial vehicles. Who is right? (A) Mechanic A, (B) Mechanic B, (C) Both A and B, (D) Neither A nor B.
5. Mechanic A says that a sagged front main crossmember will always produce positive (+) camber. Mechanic B says that a sagged frame condition may be determined only by diagonal measurements. Who is right? (A) Mechanic A, (B) Mechanic B, (C) Both A and B, (D) Neither A nor B.
6. Mechanic A says that a diamond frame condition can always be detected by the use of self-centering frame gauges. Mechanic B says that dog tracking and diamond frame damage will affect a vehicle's directional control and stability. Who is right? (A) Mechanic A, (B) Mechanic B, (C) Both A and B, (D) Neither A nor B.
7. Mechanic A says that if the right front corner of a vehicle is not level with the adjacent corners, the frame has probably sustained a diamond frame condition. Mechanic B says that a wheel off-center in its wheel opening may indicate a mashed frame condition. Who is right? (A) Mechanic A, (B) Mechanic B, (C) Both A and B, (D) Neither A nor B.
8. Mechanic A says that a bulge or buckle along the side of a fender that shows no sign of direct impact is normally an indication of frame misalignment. Mechanic B says that hood misalignment (excess gap along the fender on one side while overlapping on the other) is an indication of side sway frame misalignment. Who is right? (A) Mechanic A, (B) Mechanic B, (C) Both A and B, (D) Neither A nor B.
9. Mechanic A says that the ladder frame has little or no offset and is reasonably rigid in design. Mechanic B says that in unitized frame construction, every member and part is a load-bearing member. Who is right? (A) Mechanic A, (B) Mechanic B, (C) Both A and B, (D) Neither A nor B.
10. Mechanic A says that the front and rear wheel track of a vehicle must be parallel in order to assist the vehicle's directional control and stability. Mechanic B says that for a vehicle to possess correct tracking, the front and rear wheels of a vehicle need not be the same distance from the vehicle's centerline. Who is right? (A) Mechanic A, (B) Mechanic B, (C) Both A and B, (D) Neither A nor B.
11. Mechanic A says that a side sway frame or body condition affects the lateral misalignment of the vehicle's centerline. Mechanic B says that overall side sway is present when the vehicle has sustained damage at either the left or right middle side rail frame location. Who is right? (A) Mechanic A, (B) Mechanic B, (C) Both A and B, (D) Neither A nor B.
12. Mechanic A says that a sagged frame condition may always be determined by using self-centering frame gauges or using a tram gauge. Mechanic B says that a sag condition may be the result of a mashed frame and may be diagnosed by using a tram gauge. Who is right? (A) Mechanic A, (B) Mechanic B, (C) Both A and B, (D) Neither A nor B.
13. Mechanic A says that a side sway condition in the rear section of a vehicle's frame may be detected by the alignment of the self-centering gauges' centering pins and comparable diagonal

distances. Mechanic B says that when you are analyzing the position of the self-centering gauges, the vertical position and displacement of the support rods and the position of the gauges' horizontal bars indicate frame misalignment Who is right? (A) Mechanic A, (B) Mechanic B, (C) Both A and B, (D) Neither A nor B.

14. Mechanic A says that a vehicle will always steer (pull) to the side that has the longer wheelbase. Mechanic B says that a swung rear axle may be caused by the shearing of a spring tie bolt or by a damaged rear suspension lower control arm. Who is right? (A) Mechanic A, (B) Mechanic B, (C) Both A and B, (D) Neither A nor B.

Body-Frame Repairs

LEARNING OBJECTIVE

After studying this chapter, you should be able to:

- Describe how body-frame equipment is used to restore a damaged vehicle to manufacturer's specifications.

INTRODUCTION

This chapter provides an overview of the modern body-frame equipment that is used to restore damaged vehicles to the manufacturer's specifications. Body-frame design and construction have been changing constantly. These innovations have presented a whole new set of collision damage repair problems. The long-accepted stationary frame rack push principle could not be applied to unitized structures. There simply was not enough material in any one place on unitized structures to push against to do the basic body alignment job. Ever so slowly, a new approach evolved: pulling out a damaged area instead of pushing it out. A portable frame machine called a *Damage Dozer®* (Figure 13-1), was developed to meet the new and different concept of repair.

The Damage Dozer works by raising the vehicle and fastening the vehicle with pinch weld clamps to the two horizontal pipes shown in Figure 13-1. The car may then be fastened to the floor by chains. Then the horizontal beam of the dozer is fastened to the frame of the vehicle by using the two fixtures. A clamp (bracket) is attached to the damaged area of the body of the vehicle, and a chain is attached to the body bracket and wrapped around the vertical beam. The hydraulic ram pushes the vertical beam away from the vehicle, pulling the damaged part of the vehicle outward. Therefore, the damaged area of the vehicle is now being pulled out rather than pushed out by a pushing device.

The manufacturers of stationary-type frame equipment converted from the push concept to the pull concept by adding pull towers to their equipment that could be located in different places around the unit. By this conversion, frame units were kept functional, but they also became more massive (Figure 13-2).

For several years, these two types of equip-

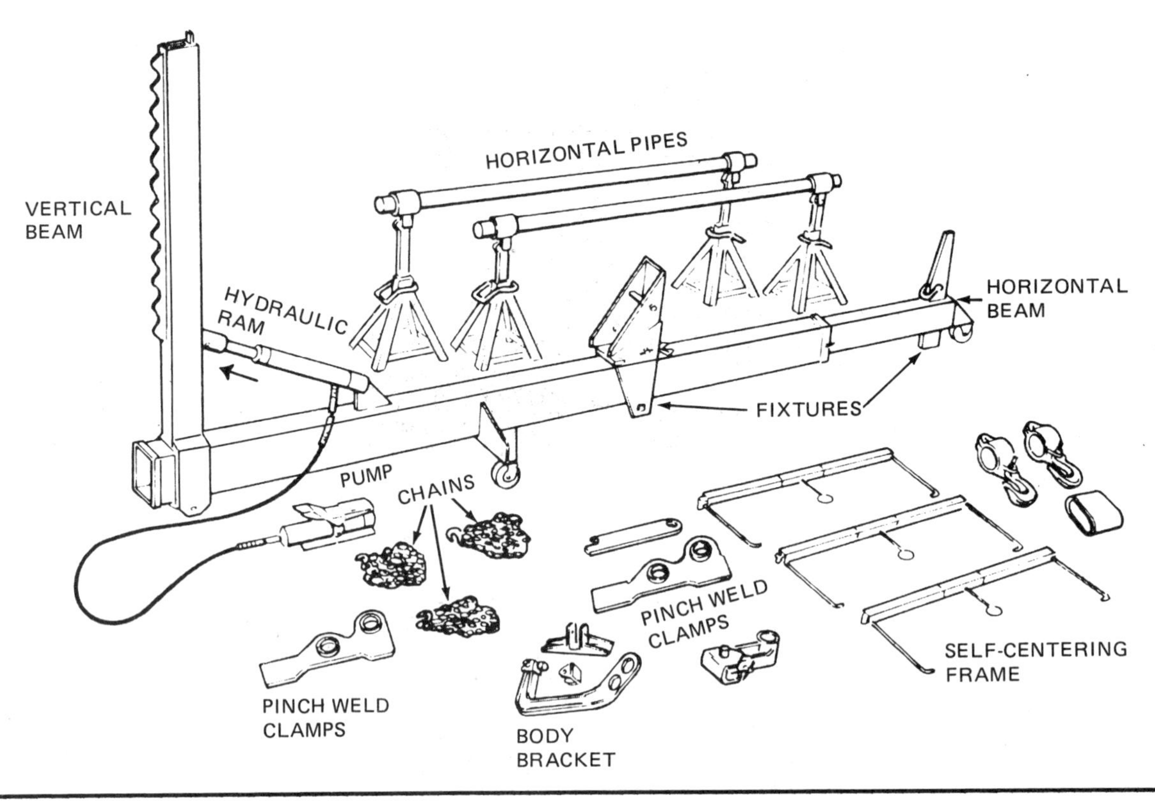

FIGURE 13-1. Damage Dozer® Portable Frame Machine. (Courtesy of Blackhawk Automotive Division, Applied Power Canada, Ltd.)

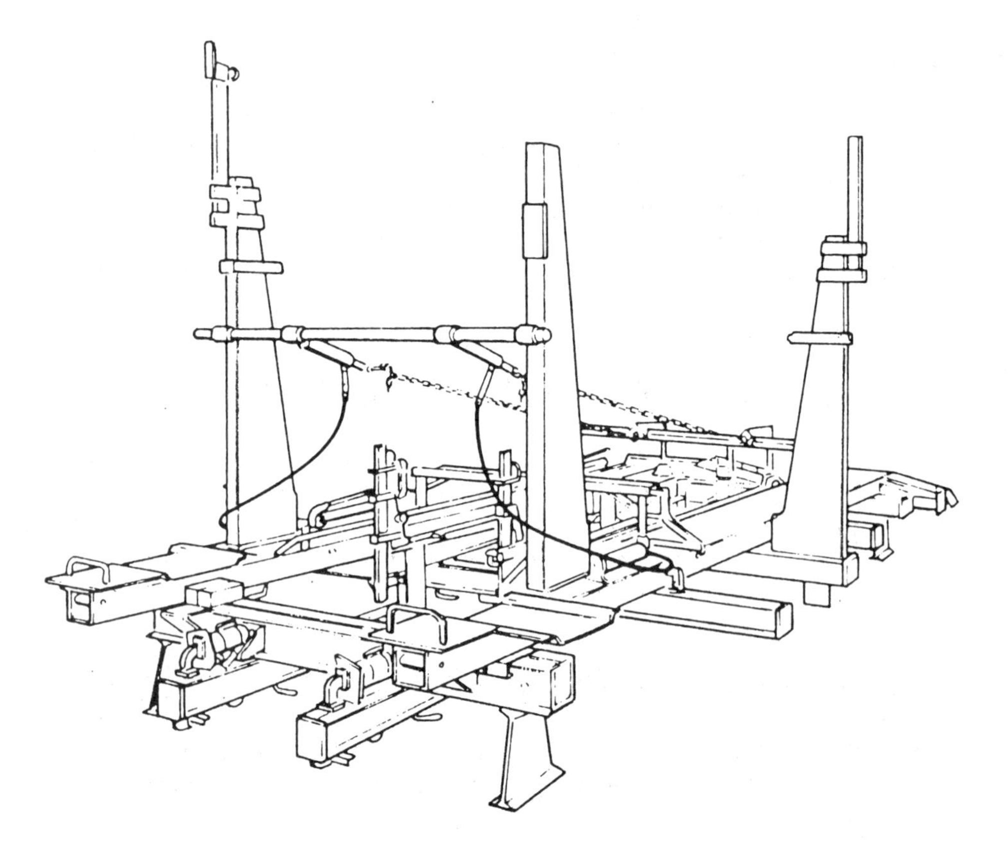

FIGURE 13-2. Updated Stationary Frame Machine. (Courtesy of Blackhawk Automotive Division, Applied Power Canada, Ltd.)

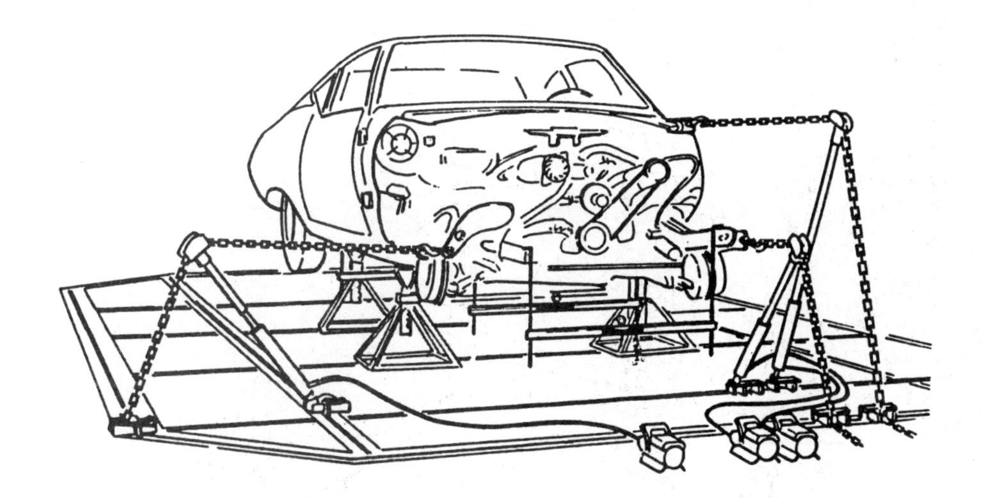

FIGURE 13-3. Korek Body-Frame Repair System. (Courtesy of Blackhawk Automotive Division, Applied Power Canada, Ltd.)

ment—stationary and portable—were used to achieve basic body-frame alignment. However, car designers and engineers again changed their basic construction approach.

The overall strength and rigidity of car structure was achieved through a more sophisticated unitized body construction than ever before. The frames and frame sections were very complicated in shape and presented a number of problems in controlling the metal motion during the repair procedure. In short, there were two sizable collision damage repair problems rolled into one.

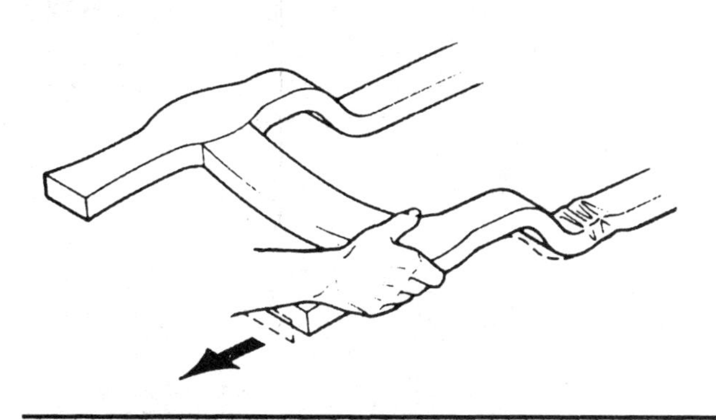

FIGURE 13-4. Pulling Damaged Frame Back into Place. (Courtesy of Blackhawk Automotive Division, Applied Power Canada, Ltd.)

THE KOREK BODY-FRAME REPAIR SYSTEM

The *Korek body-frame repair system* was designed to solve two repair problems: those associated with unitized frames and bodies and those associated with light but extremely complex frame design. The system consists of steel beams welded into the design, as shown in Figure 13-3. Reinforced concrete is used to strengthen the system further.

The advantages of Korek system equipment are its light weight and its use of hydraulic rams, support posts, and chains. These features allow technicians to pull the frame in any number of lateral and vertical angles to restore the damaged frame to its original dimensions. Holding chains are attached to various points around the vehicle, allowing for a full working area. Indeed, accessibility is the unique feature of this equipment.

A good technician can look at a major collision repair job and quickly determine what needs to be done to follow the basic rule of reverse direction and sequence, which can be explained in this way: A car that is hit in the front will be pushed in. The repair technician must analyze the type of damage that the vehicle has sustained and pull the damaged areas out. The Korek equipment functions in just the same way the technician would, if the technician were physically able to pull the bent frame member or cowl section into proper position (Figure 13-4).

ANCHORING VEHICLES

For any major pulling with the Korek system we must first have the capability to anchor the car securely. Every car made today has four key anchor points, the location of which depends on the frame design.

Perimeter Frame Design

In the case of perimeter frame design, the four anchor points are the front and rear torque boxes,

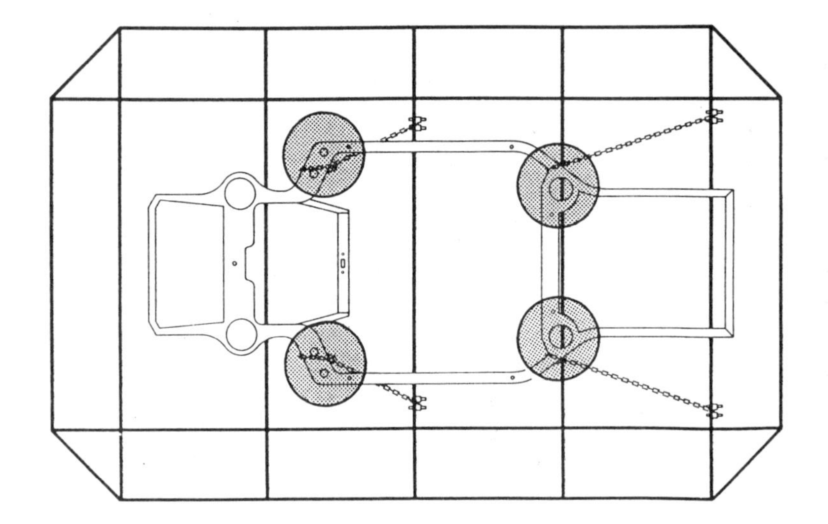

FIGURE 13-5. Four Key Frame Anchoring Points. (Courtesy of Blackhawk Automotive Division, Applied Power Canada, Ltd.)

the circled areas in Figure 13-5. For cars with a bolt-on stub frame (the front section of a perimeter-type frame that is bolted to the rest of the vehicle's unitized body), the Korek system uses a combination of both chain wrap and underbody clamps, always at these four key areas. Depending on the amount of pulling force required and its direction, any or all four anchor points are used effectively. Thus the anchor load is split, and the possibility of causing damage at the attaching points is reduced.

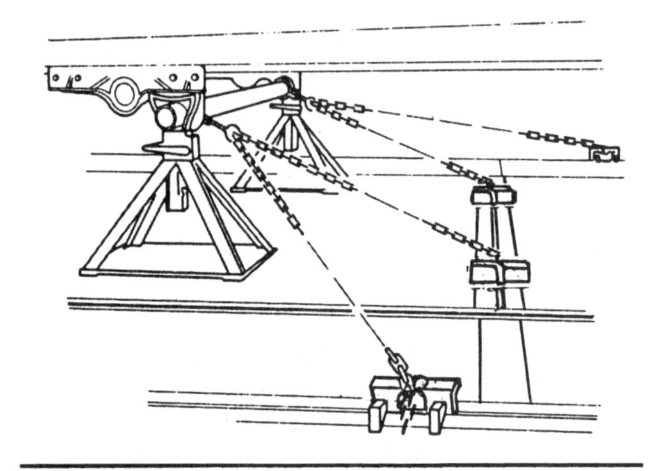

FIGURE 13-6. Anchoring a Unitized Body with Bolt-On Crossmember. (Courtesy of Blackhawk Automotive Division, Applied Power Canada, Ltd.)

Unitized Frame Design

Because unitized bodies do not have a massive frame to anchor onto, the chance of causing additional damage to the frame once pulling begins is much higher. In the case of unitized design, the four strongest anchor points are at both ends of the rocker sills. Two methods are used for anchoring unitized cars. First, they may be anchored by attaching four individual underbody clamps and two crosstubes, forming bolt-on crossmembers (Figure 13-6). The second method, which is much quicker, uses the Blackhawk Quadri-Clamp. This device, like the individual clamps used in the other method, also lessens the chance of causing additional damage when pulling on the body. It uses a four-sided box frame attached to the pinch welds. It is fully portable, rolls on wheels, and is adjustable to accommodate most unitized cars without the use of additional equipment (Figure 13-7).

With the Korek system, the anchor tie-down position is automatic in any direction from the required hold position on the clamping device. The anchor chains also can be pretightened to distribute

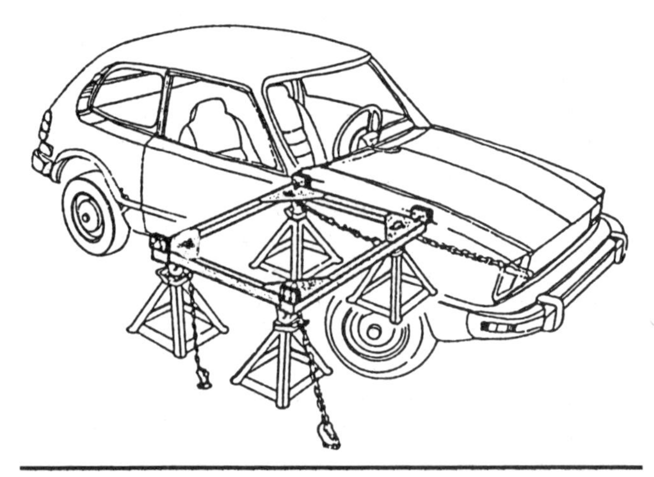

FIGURE 13-7. Anchoring a Unitized Body with a Blackhawk Quadri-Clamp®. (Courtesy of Blackhawk Automotive Division, Applied Power Canada, Ltd.)

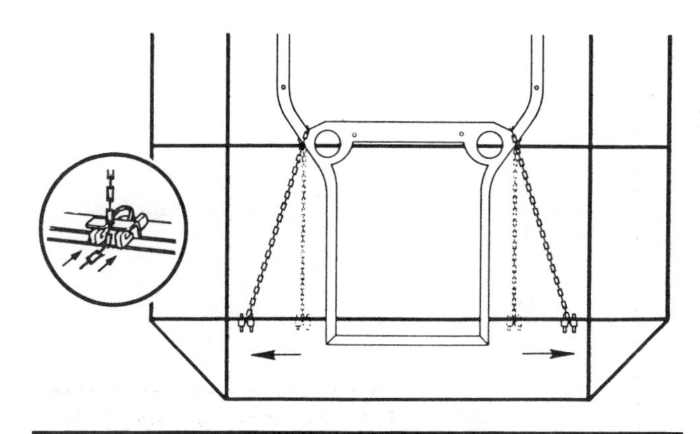

FIGURE 13-8. **Pretightening Anchor Chains.** (Courtesy of Blackhawk Automotive Division, Applied Power Canada, Ltd.)

the load evenly on all anchor positions by simply sliding the chain anchors along the open rail slots in the Korek equipment. When the anchor chains are tight, simply lock the anchors in their positions with wedges (Figure 13-8).

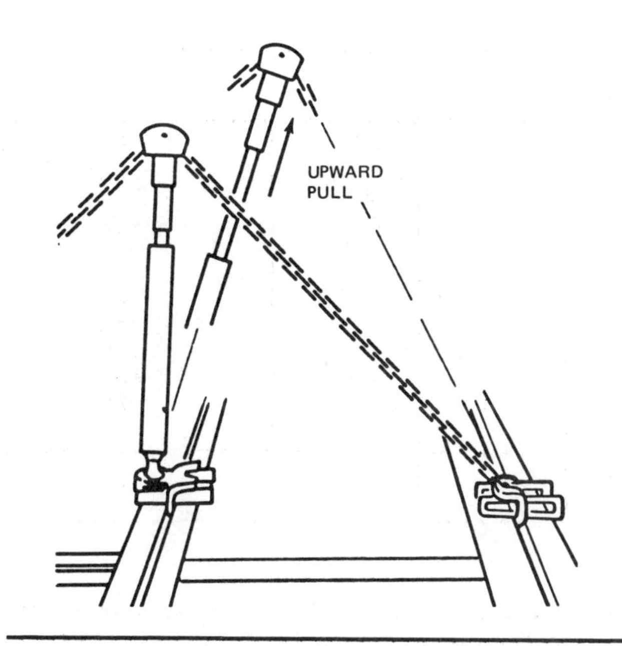

FIGURE 13-9. **Pulling Arrangement: Upward.** (Courtesy of Blackhawk Automotive Division, Applied Power Canada, Ltd.)

PULLING ARRANGEMENTS

The pulling arrangement for the Korek system is determined by a simple triangle. Illustrated in Figure 13-9 is a frequently used setup for pulling upward and outward. The ram, the base unit, and the chain form the triangle. As the ram is extended, one side of the triangle becomes longer than it was before. Therefore, it must swing over to the right because the chain is locked at the top of the ram. As the ram swings over to its new position, it brings the damaged vehicle upward and over with it. This simple procedure is based on the principle of vectors. Tremendous forces can be exerted in a carefully planned direction using this principle.

Figure 13-10 shows the triangular arrangement that will provide more of a straight-out pull with a slight upward movement. Note that the frame is placed in the ram foot at an angle to the right from true vertical. (Placing the ram farther down from the true vertical line will produce more of a straight-out pull.) As force is applied, the ram will swing to the right, pulling with it the damaged sections of the car. The metal will be pulled outward and slightly upward. It is important that the ram part of the triangle be at proper height. Height is controlled by adding the proper lengths of tubing to arrive at the correct height before the hook-up is complete. Adjustments are made easily.

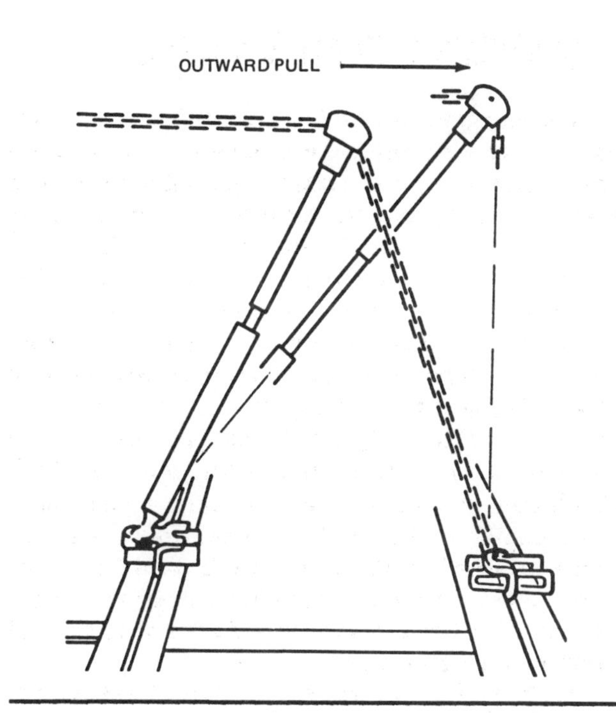

FIGURE 13-10. **Pulling Arrangement: Outward.** (Courtesy of Blackhawk Automotive Division, Applied Power Canada, Ltd.)

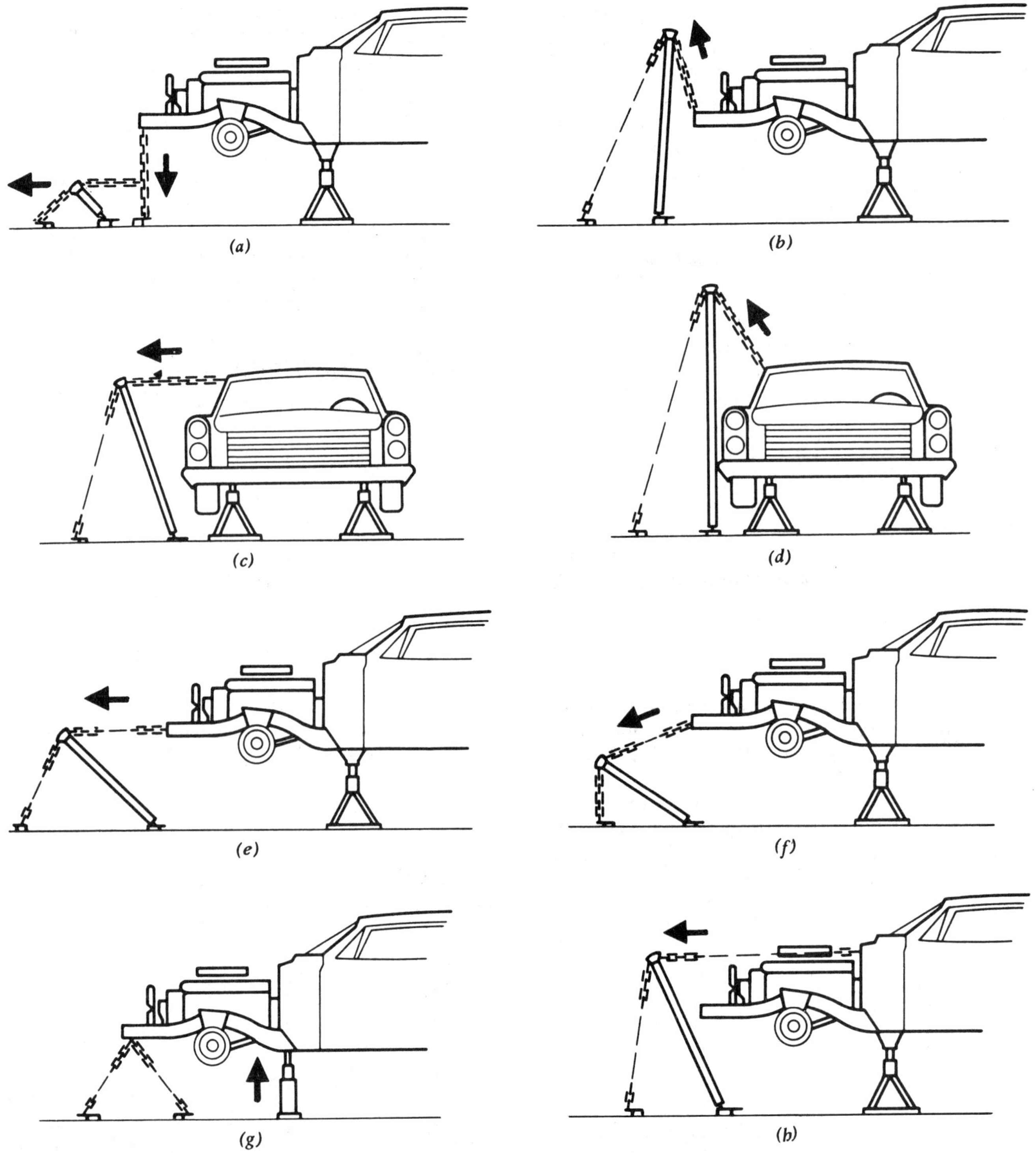

(a)

(b)

(c)

(d)

(e)

(f)

(g)

(h)

FIGURE 13-11. Typical Pull Setups. Note the various uses of the chains and rams in each example. (a) Down pull with chain bridge; (b) upward and outward pull with ram and extension tubing; (c) horizontal roof line pull with ram and extension tubes; (d) upward and outward roof pull with ram strut positioned higher than roof area; (e) straight-out pull with ram at a 45° angle; (f) down-and-out pull pattern with ram at a 25° angle; (g) upward cowl pull with front end tied down; (h) horizontal cowl pull with extension tubing and ram at 65° angle. (Courtesy of Blackhawk Automotive Division, Applied Power Canada, Ltd.)

One can easily pull up, out, or down with this system by controlling the angle of the ram and by adjusting the length of the tubing used with the ram. For a high pull, more tubing is required. For a downward pull, the ram must be placed on an angle low to the floor. Figure 13-11 illustrates how the Korek equipment is used to facilitate various types of body-frame pulls.

Multiple Pulling Points

The Korek system provides a number of air-hydraulic power units, each of which can be located anywhere around or under the car. Because of the design of the base unit, a pull point is available automatically at any angle or direction desired in the repair operation. A high and low pull in the same direction can be accomplished by setting up two power units side by side (Figure 13-12). If a multiple pull approach is required, a third and fourth power unit can be set up at the desired angles and controlled in conjunction with the first two (Figure 13-13).

The multiple pull point approach accomplishes these objectives:

1. The exact desired direction of pull can easily be achieved from three or four points at one time, providing the control needed in the repair of modern vehicles.
2. The use of multiple pull points reduces considerably the amount of force required at any single point, thus reducing the risk of tearing the new lightweight metals. The design of many cars does not provide enough material in any one place to attach to in order to transmit sufficient force to complete a repair. Again, as in the anchoring system, the pull load must be distributed through several attaching points.

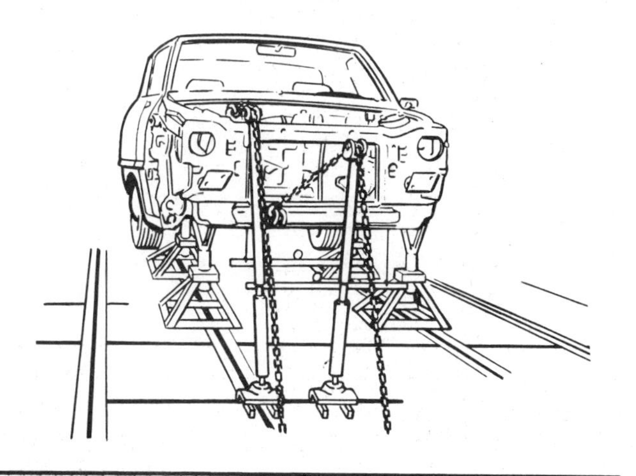

FIGURE 13-12. Combination High and Low Pull. (Courtesy of Blackhawk Automotive Division, Applied Power Canada, Ltd.)

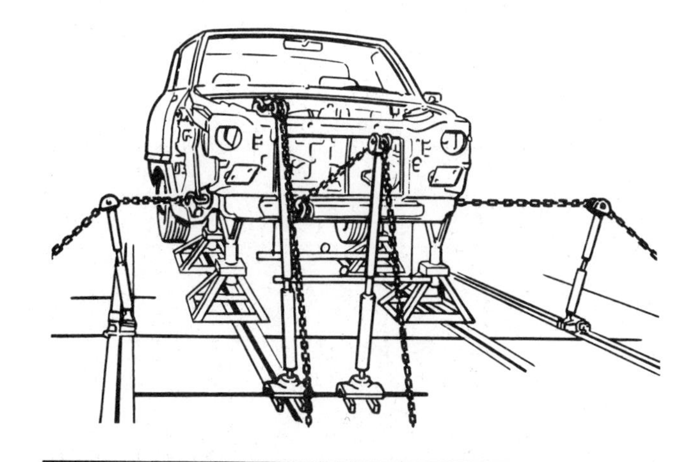

FIGURE 13-13. Multiple Pull Point Setup. (Courtesy of Blackhawk Automotive Division, Applied Power Canada, Ltd.)

You will note in Figure 13-13 that the area under the vehicle is clear. Therefore, gauges can be hung from under the vehicle for measuring and diagnosis, eliminating a great deal of wasted time in replacing gauges.

Correcting Sagged Front Main Crossmember

Figure 13-14 illustrates how the Korek equipment may be used to correct a sagged front main crossmember. This damage occurs when an impact narrows the width distance between the left and right upper control arms. Since the alignment of the front wheels and the directional control of the vehicle are dependent on this area of the body-frame, accurate repair and restoration of the damaged area is critical.

As you view Figure 13-14, you will see that a hydraulic jack has been placed under the center of the front main crossmember. The frame at the upper inner control arm position is being pulled outward, thus increasing the distance between the upper inner control arms, which are attached to the frame. This, in turn, forces the top of the wheels to a more

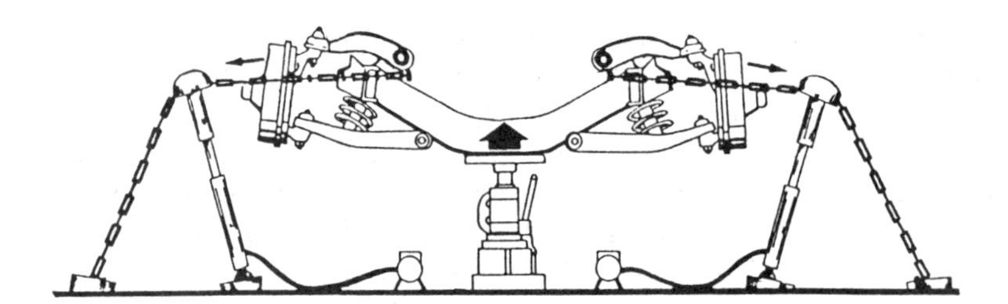

FIGURE 13-14. Repairing a Sagged Front Main Cross-member. (Courtesy of Blackhawk Automotive Division, Applied Power Canada, Ltd.)

positive camber position. A good safety practice when anchoring a vehicle in preparation for heavy pulling is to lean toward overanchoring and place a heavy blanket around the chain to dampen the flying force of the chain if it should break.

A tram gauge must be used to verify all width and linear distances relative to the manufacturer's specifications. Care must be taken not to damage any hydraulic brake or fuel lines.

MODERN UNIBODY VEHICLE REPAIR

The modern unibody vehicle is a complex structure. In a typical body-on-frame car, the major fore-and-aft strength and alignment is a simple, thick-gauge box section or C channel frame rail (a length of metal or a beam in the shape of a C). In the unibody, this same function may require a dozen or so lightweight members welded together. Thus, in a collision, there are numerous unibody parts that can bend, twist, or pull loose, causing misalignment. The reason is that in a unibody, without a stub frame, even the front sheet metal is precision fitted to establish and maintain vital locating points (holes on the unibody that the fixtures fit into). Also, steering alignment depends on that complex positioning structure. All locating points must be precisely restored during collision repair.

Structure

Since modern unibody vehicles are essentially a shell made of different high-strength steel metals, they react differently from second-generation vehicles when involved in a collision. Since the structure of a modern vehicle is stiffer, more impact is distributed throughout the vehicle when it is involved in a collision (Figure 13-15).

This design characteristic, which helps protect occupants during a collision, causes damage patterns that differ from those of frame-type vehicles. The heavy gauge of the mild steel frame tends to dampen and localize damage. The stiffer sections used with unibody design tend to transmit and distribute impact energy throughout more of the vehicle, causing misalignment in areas remote from the impact point. Even sections that are buckled or torn loose may have passed along heavy forces before deforming. Worse, much of this remote damage can easily be overlooked in casual inspection but still be sufficient to cause problems later.

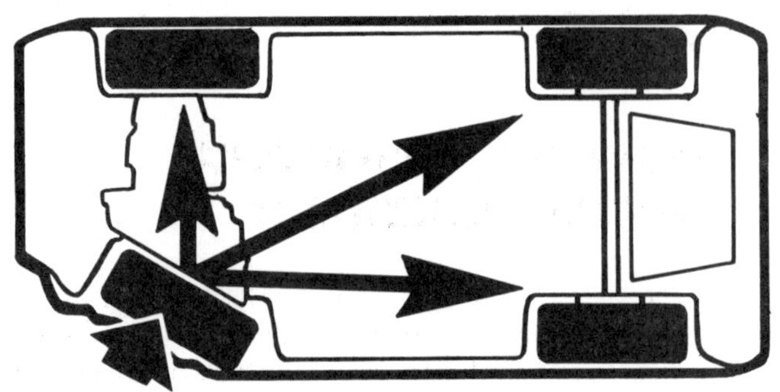

FIGURE 13-15. Distribution of Impact. More impact is spread throughout the vehicle when a unibody vehicle is involved in a collision. (Courtesy of Blackhawk Automotive Division, Applied Power Canada, Ltd.)

The extra complexity and stiffness of the structure is especially critical in the front end, which houses not only the front suspension and steering linkage but also the entire drive train, engine, transaxle, drive shafts, and constant velocity U joints. To keep all these parts in proper alignment requires support, including that supplied by the front end sheet metal.

An example of the importance of complete restoration of the unibody vehicle dimensions is the condition known as *bump steer*. Although a vehicle's frame rails, track width, wheelbase, and wheel alignment may have been restored, the geometrical relationship between the tie-rod ends and the lower control arms may still not be correct due to a misalignment of the rack and pinion assembly, where it is mounted to the vehicle's firewall (the vertical sheet metal located between the engine and the front of the passenger compartment).

When the vehicle is driven over a series of bumps, one front wheel deflects differently from the opposite wheel when the suspension is compressed or extended. This wheel problem may cause the car to steer violently to one side when the wheel goes over a bump, causing the driver to lose vehicle control.

New Diagnostic and Repair Equipment

Because of the complex structure of unitized bodies, the autobody repair industry has developed new gauging procedures for determining damage and has developed new concepts in the repair of these vehicles.

Korek Bench System. One of the modern types of body-frame repair equipment developed to meet the needs of repairing modern vehicles is the *Korek bench system* (Figure 13-16). The bench consists of a rigid steel framework with a precision-machined top surface. The top of the bench is drilled with a series of holes 100 millimeters (3.94 inches) apart along both sides. These holes are numbered from 1 to 40. The bench is mounted on four heavy-duty swivel casters, two of which are adjustable for leveling on an imperfect floor. The wheels are urethane covered for easy maneuverability on most floor surfaces. The wheels can be locked in place to prevent the bench from moving on slight inclines and also while pulling.

Six transverse beams are included with the bench system. The body anchoring system consists of two specially designed body clamp assemblies. The assemblies bolt directly to the surface of the bench and clamp to the pinch welds on the car. The anchoring loads are transmitted through the clamps. Because the fixtures are not required to carry the anchoring load, the fixtures can function as measuring devices. The clamps are self-centering to provide a degree of control in positioning the car on the bench.

The fixtures are the heart of the bench system (Figure 13-17). They are designed from the car manufacturer's drawings to duplicate locations of the body control points that are used in manufacturing the suspension mountings or other key locations of the underbody. The fixtures must fit into these points (holes) that are designed into the unitized body.

Fixtures bolt to the transverse beams, creating a true datum plane and mating with the underbody of the car. If the fixtures fit the car properly, the car is in perfect alignment to the car manufacturer's specifications. All that is necessary is to straighten the car until the factory control points match the fixtures. The fixtures can also be used when replacing the welding structural underbody members (Figure 13-18).

In replacing vehicle structural parts, a welding method should be used that produces the strength of a factory weld. Welding methods such as MIG (wire feed), TIG, stick electrode, and compression

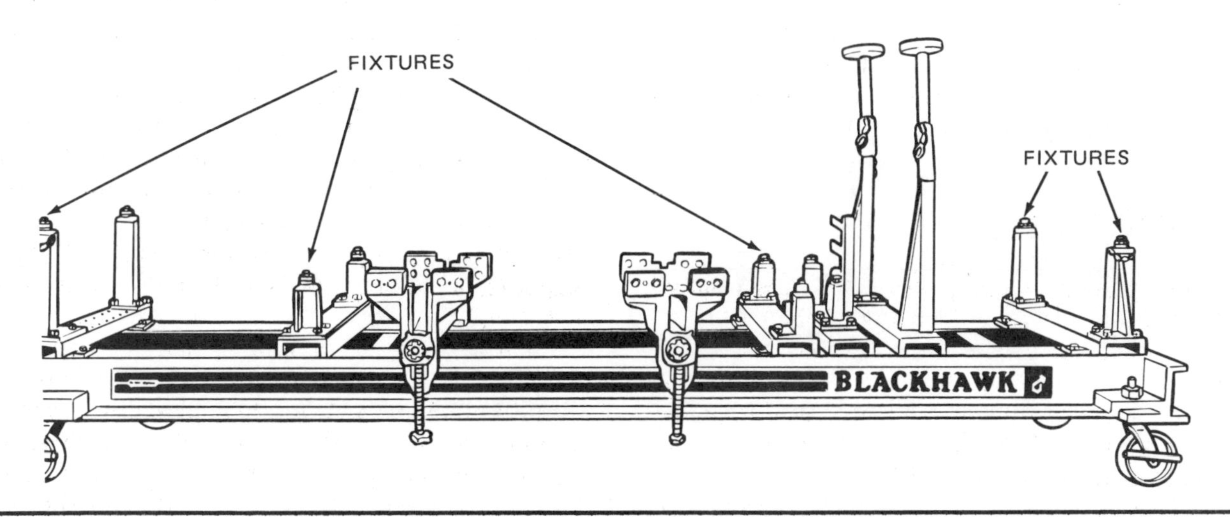

FIGURE 13-16. Korek Bench System. (Courtesy of Blackhawk Automotive Division, Applied Power Canada, Ltd.)

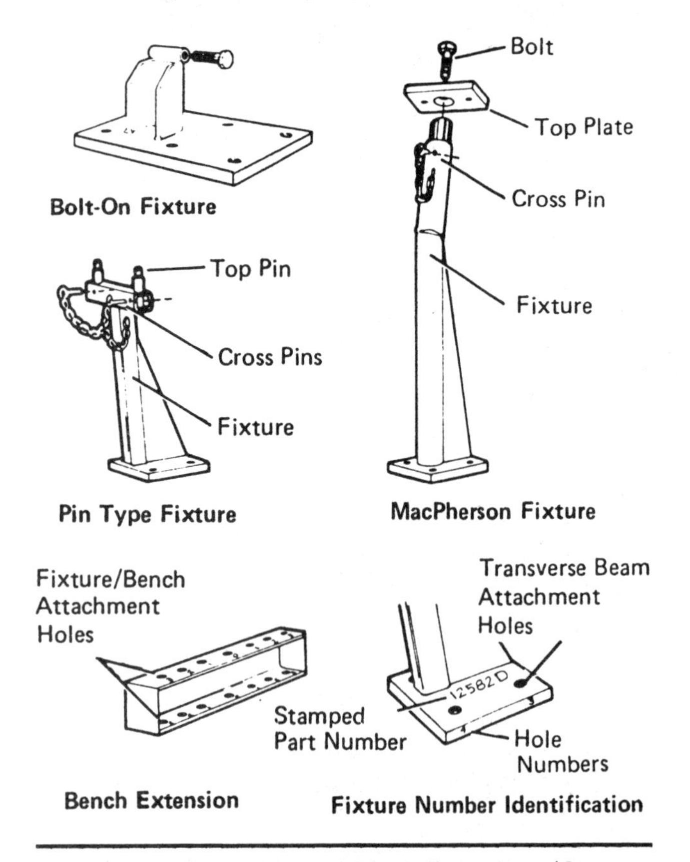

FIGURE 13-17. **Fixtures and Bench Extensions.** (Courtesy of Blackhawk Automotive Division, Applied Power Canada, Ltd.)

FIGURE 13-18. **Welding Components into Place.** (Courtesy of Blackhawk Automotive Division, Applied Power Canada, Ltd.)

spot welding are acceptable welding methods. *Note:* Brazing, gas welding, and noncompression resistance spot welding are not recommended.

Before the car is put on the bench, you should determine the degree of damage and begin making a plan for the way you will repair it. As a general rule, the bench system should be used whenever the damage affects the suspension, steering, or power train mounting points. This, of course, would include situations such as side collision where the suspension components and their mountings are not damaged directly but because of deformation in the center section of the car's structure, the whole body is out of alignment.

You can determine whether a particular collision meets this rule either by eye where there is obvious damage or by making some general measurements with a tape measure or tram gauge. These would include diagonal measurements to check for diamond and length measurements and to check for mash. Try to get as good an idea as possible of where the damage begins and ends. You can use whatever dimensions you have available, including body-frame dimension books or car manufacturer's data or by checking against an undamaged car. *Note:* Carefully analyze the vehicle and the damage to determine what parts must be removed. Disassemble only those components that must be removed to get to the area of the car being repaired.

Car-O-Liner Bench Measuring System. Although unibody design is somewhat new to North America, it was adopted earlier by most foreign vehicle manufacturers. The autobody repair business in those countries found that it was necessary to develop a different type of measuring and alignment system in order to be able to restore their vehicles to specifications. This situation led to the development of the Car-O-Liner system, which requires no jigs or fixtures and has been introduced to North America (Figure 13-19). With this system, you need only dismantle the parts necessary to allow you access to the damaged sections of the chassis. The chassis is secured to the equipment by means of four sturdy clamps at the sill members.

When the chassis is secured to the bench, the measuring system can be fit. This system consists of a measuring bridge with movable measuring tape for longitudinal measurements, measuring slides for lateral measurements, and measuring scales for height measurements. You can establish the centerline of the chassis from the undamaged points on

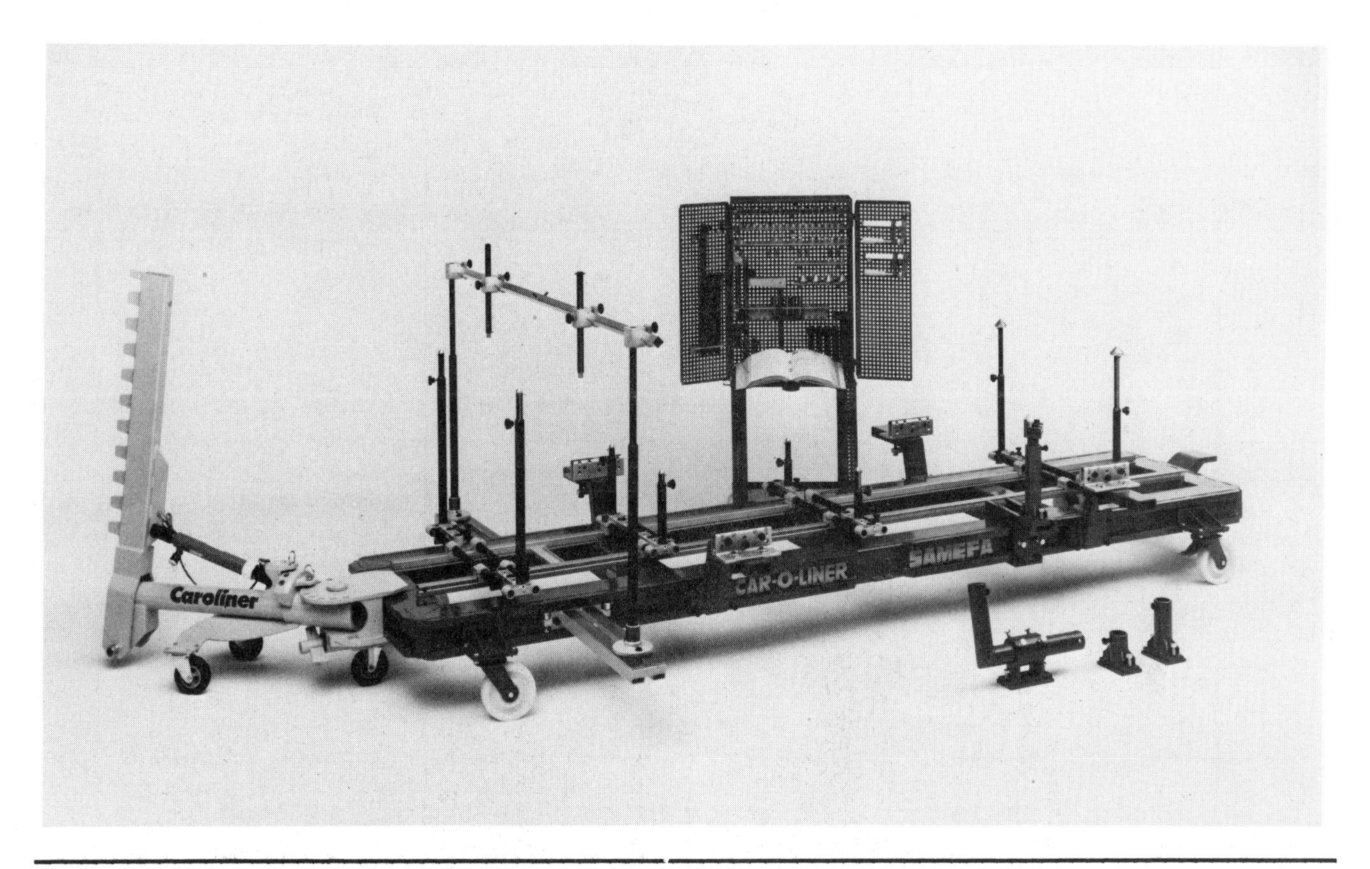

FIGURE 13-19. Car-O-Liner Bench Measuring System. (Courtesy of Car-O-Liner Company)

the chassis. You are then ready to start measuring symmetrical and asymmetrical points from the centerline using the movable measuring slides. You then set the correct longitudinal, lateral, and height dimensions for the chassis from the particulars in the appropriate data sheet.

Data sheets (Figure 13-20) are compiled from measurements of a number of cars of the same model and from the manufacturer's specifications. Height, width, and length measurements are specified on each data sheet for a number of points on the car body.

This measuring system is sturdy, lightweight, and easy to read. It is left on the workbench throughout the alignment work. Therefore, when you align damaged parts, you can immediately check the results of the work. The measuring system never needs to be recalibrated during the measuring and alignment work.

The draw aligner is an extension of the technician's arm. It is the vertical beam and puller that is operated by a hydraulic ram. This piece of equipment can be attached at any place to the Car-O-Liner bench. The technician must be able to gain access to pull a damaged part, however difficult its location. The draw aligner helps with this work. It can quickly be locked to the workbench at any point around the chassis in order to align any type of damage. It can be inclined laterally and can be angled in relation to the bench. Once the chassis has been measured, aligned, and checked and the heavily damaged parts have been removed, new parts are secured into position and welded.

Laser Beam Aligner. Because unitized vehicles are frameless, precise restoration is critical to the proper and safe operation of the vehicle. The *laser beam aligner* provides a means to accurately measure the vehicle to exacting specifications and can be used on all types of vehicles. The laser rays consist of electromagnetic waves. These waves differ from ordinary light in that the rays are transmitted at vertical and horizontal planes in fairly narrow beams. The rays are invisible until they meet resistance. They then become obvious in the form of a round red dot.

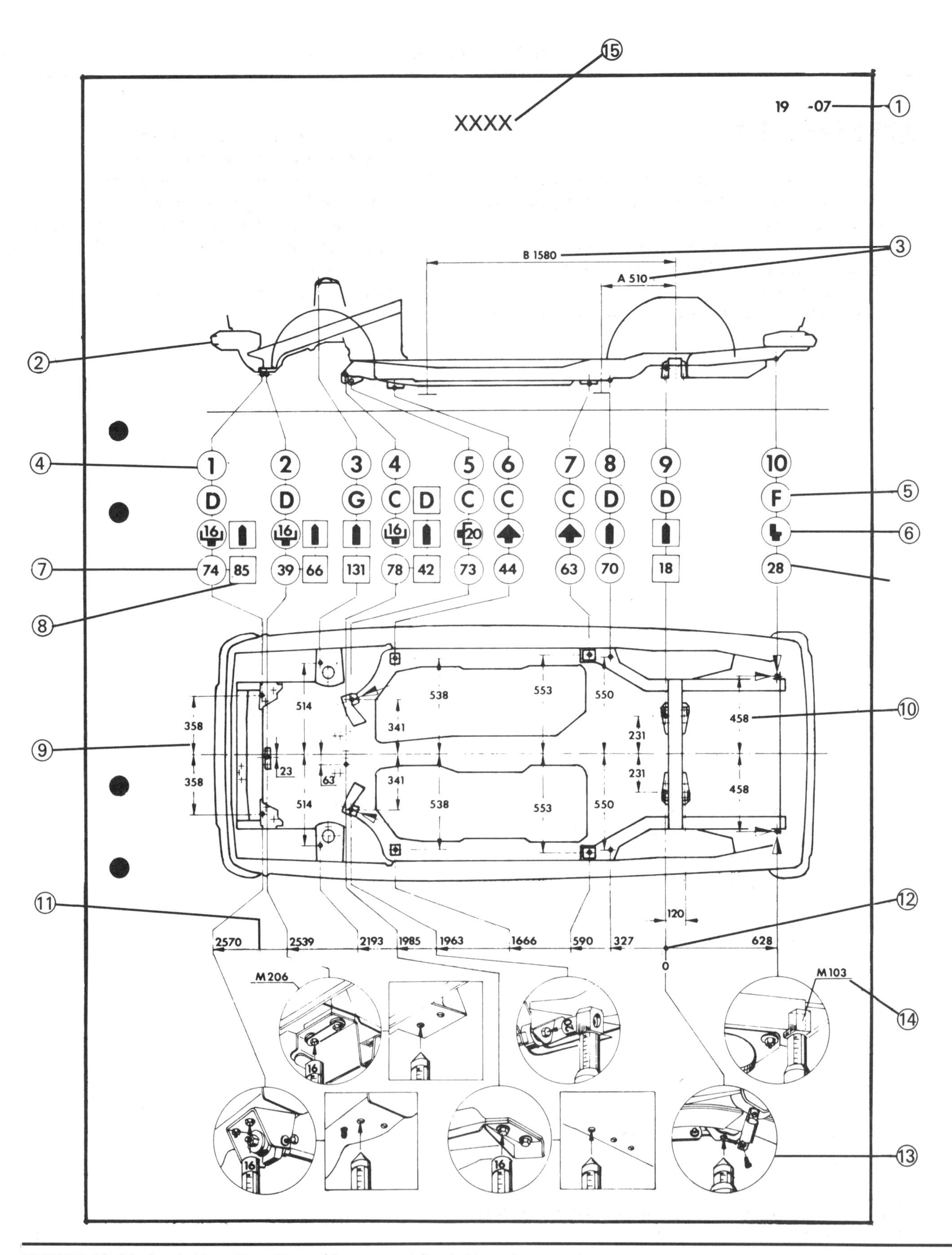

FIGURE 13-20. Car-O-Liner Data Sheet. (Courtesy of Car-O-Liner Company)

REVIEW TEST

1. Describe, using a point form answer, how to repair a vehicle that has sustained a sagged front main crossmember.
2. The left front frame rail of a vehicle has sustained a mash condition. Describe how to restore the frame.
3. The front of the left front frame rail has been raised due to an accident. Describe how to pull the rail down and lengthen the rail.
4. Describe how a unibody vehicle reacts when it is involved in a major collision.
5. Describe how a data sheet is used to determine if the frame of a vehicle is within the required specifications.
6. Explain what causes bump steer.

(1) Date of issue
(2) Side view
(3) Dimensions A and B for positioning clamps (no dimensions are specified for special clamps)
(4) Number of measuring point
(5) Type of measuring socket
(6) Symbol for type of measuring stud
(7) Height for unstripped access point
(8) Height for stripped out access
(9) Underneath view
(10) Width
(11) Length
(12) 0 point
(13) Picture of measuring point*
(14) Footnote if special tools required*
(15) Car model

*Figures in circles are with parts in place. Figures in squares are for parts removed.

FIGURE 13-20. Continued

FINAL TEST

This examination is multiple choice. Only one answer will be accepted. Carefully read every statement.

1. Mechanic A says that with the Korek body-frame repair system, the frame technician can pull, push, and anchor 360° around the vehicle. Mechanic B says that repairing a vehicle's damaged body-frame is literally reversing the sequence and direction of the collision damage. Who is right? (A) Mechanic A, (B) Mechanic B, (C) Both A and B, (D) Neither A nor B.
2. Mechanic A says that every vehicle made today has four key frame anchoring positions. Mechanic B says that to anchor a unitized body-frame, underbody clamps are positioned at the ends of the rocker sills. Who is right? (A) Mechanic A, (B) Mechanic B, (C) Both A and B, (D) Neither A nor B.
3. Mechanic A says that to obtain a straight pull on a vehicle's frame, the two angles between the chain and the hydraulic ram should be equal. Mechanic B says that a chain bridge and the vertical extended position of the hydraulic ram may be used to raise a vehicle's frame. Who is right? (A) Mechanic A, (B) Mechanic B, (C) Both A and B, (D) Neither A nor B.
4. Mechanic A says that a track gauge may be used to determine the width distance between the left and right upper control arms. Mechanic B says that using the multiple pull approach while repairing a vehicle's frame reduces the amount of force required at any single pull position. Who is right? (A) Mechanic A, (B) Mechanic B, (C) Both A and B, (D) Neither A nor B.
5. Mechanic A says that raising the center of a crossmember and anchoring the frame at the upper control arms moves camber toward the positive position. Mechanic B says that when anchoring a vehicle in preparation for heavy pulling, lean toward overanchoring. Who is right? (A) Mechanic A, (B) Mechanic B, (C) Both A and B, (D) Neither A nor B.
6. Mechanic A says that if a vehicle's firewall is not straightened and if the rack and pinion is mounted to the firewall, a bump steer condition will probably exist. Mechanic B says that if a vehicle's firewall must be replaced due to a severe accident, the part can be attached by brazing. Who is right? (A) Mechanic A, (B) Mechanic B, (C) Both A and B, (D) Neither A nor B.
7. Mechanic A says that modern cars made of high-strength steel metals react differently from older vehicles when involved in a collision. Mechanic B says that a modern vehicle is flexible and that when the body sustains damage, the impact is not distributed to other parts of the body. Who is right? (A) Mechanic A, (B) Mechanic B, (C) Both A and B, (D) Neither A nor B.
8. Mechanic A says that when a modern car is involved in a collision, remote damage can easily be overlooked during a casual inspection. Mechanic B says that the front end sheet metal of a modern vehicle has no bearing on the vehicle's front suspension and steering linkage. Who is right? (A) Mechanic A, (B) Mechanic B, (C) Both A and B, (D) Neither A nor B.
9. Mechanic A says that when a vehicle has been restored to its correct shape, the bench fixtures will fit the car properly. Mechanic B says that gas welding may be used to attach a new structural part to a modern vehicle. Who is right? (A) Mechanic A, (B) Mechanic B, (C) Both A and B, (D) Neither A nor B.

Wheel Alignment Factors

- Explain the importance of wheel alignment to vehicle handling,
- Define these five wheel alignment factors: camber, caster, toe, kingpin or ball joint inclination, and toe-out on turns or turning radius,
- Explain the unit of measurements for the five wheel alignment factors,
- State which front end wheel alignment factors are considered adjustable and nonadjustable.

INTRODUCTION

The excellent roadability and handling characteristics typical of modern automobiles and trucks are primarily due to improved steering and suspension design. Although other things, such as radial tires and riding devices (examples are shocks and springs), affect the vehicle's steering control, manufacturers depend to a large extent on the benefits derived from the front end wheel alignment factors.

Wheel alignment is more than the simple tracking of the front and rear wheels to ensure that they roll freely on a straight-ahead course. Wheel alignment must also be maintained in turns and other maneuvers even though bumps, holes, and other surface irregularities cause the wheels to move up and down almost constantly. In addition to steering control, wheel alignment provides the directional stability that helps the driver hold a straight course without making continuous steering corrections. Ideally, very light force at the steering wheel should be enough to keep the vehicle headed on a straight course.

Figure 14-1 illustrates a vehicle being driven on a rural road. The road surface is high (crowned) in the center to aid the runoff of water when the surface is wet. Since the curb side of the road is lower than the crown, the vehicle, affected by gravity, is influenced to steer to the right side of the road. To correct this unfavorable influence, the driver must countersteer by applying a constant steering wheel effort to the left. In this situation, the driving of an automobile is not a pleasure. Correct wheel alignment compensates for these unfavorable steering influences.

To aid the vehicle's steering control and stability, the design engineer also utilizes the friction between the tires and the road surface. Wheel alignment and the operating angles (alignment factors) of the steering and suspension systems determine the amount of friction between the tires and road. The force generated by a rolling tire as a result of tire friction and wheel alignment is called *rolling friction*. When a vehicle's wheel alignment is correct, the turning load placed on the tire tread is evenly distributed, thereby reducing wear. In short, the wheel alignment story boils down to understanding how the various operating angles—for example, camber and caster—are designed into a vehicle's steering and suspension system to enable it to have excellent directional stability, easy steering, and normal tire tread life expectancy. Some of the mystery that surrounds wheel alignment factors is due to a lack of understanding of what they are, what they do, and how they work together. The five wheel alignment factors that aid the steering and control of a vehicle are as follows:

- Camber,
- Caster,
- Toe,
- Kingpin or ball joint inclination,
- Toe-out on turns or turning radius.

On earlier cars, camber, caster, and toe were adjustable. In recent years, however, depending on the make of the vehicle, only camber and toe are adjustable. In some cases, only toe is adjustable. How these factors are adjusted will be discussed later in the text. Factors 4 and 5 are nonadjustable.

The remainder of this chapter introduces and defines each of the wheel alignment factors. The following chapters will discuss each factor separately.

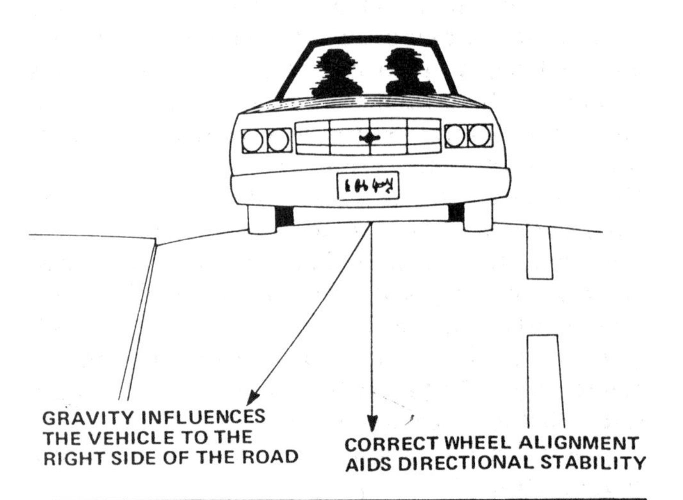

FIGURE 14-1. How Correct Wheel Alignment Offsets Undesirable Effects of Irregular Road Surfaces. (Courtesy of Chrysler Corporation)

CAMBER

Camber is the inward or outward tilt of the top of a wheel from the true vertical position as viewed from the front of a vehicle. Since a wheel can tilt in or out, the definition of camber has two parts. Positive (+) camber is the outward tilt of the top of the wheel from the true vertical position. Negative (-) camber is the inward tilt of the top of the wheel from the true vertical position.

The amount of tilt that a wheel may lean is measured in degrees from the true vertical position; deviation from that vertical position is called the *camber angle* (∢) (Figure 14-2). The zero degree (0°) camber position of a wheel exists when the centerline (C/L) of a tire and wheel is at its true vertical position (in other words, straight up and down). Actually, while driving a vehicle, the zero degree camber position is seldom present. The reasons for this are the design of a vehicle's suspension system, road surface deviation, shifting body weight, centrifugal force, and acceleration and braking forces. Since excessive position and/or negative camber produces tire tread wear and is affected by defective suspension parts, camber is never used to compensate for wear in a defective suspension part.

At this point, you should know the following about camber:

- The definitions of the terms *positive* and *negative camber*,
- That camber is measured in degrees or fractions of a degree,

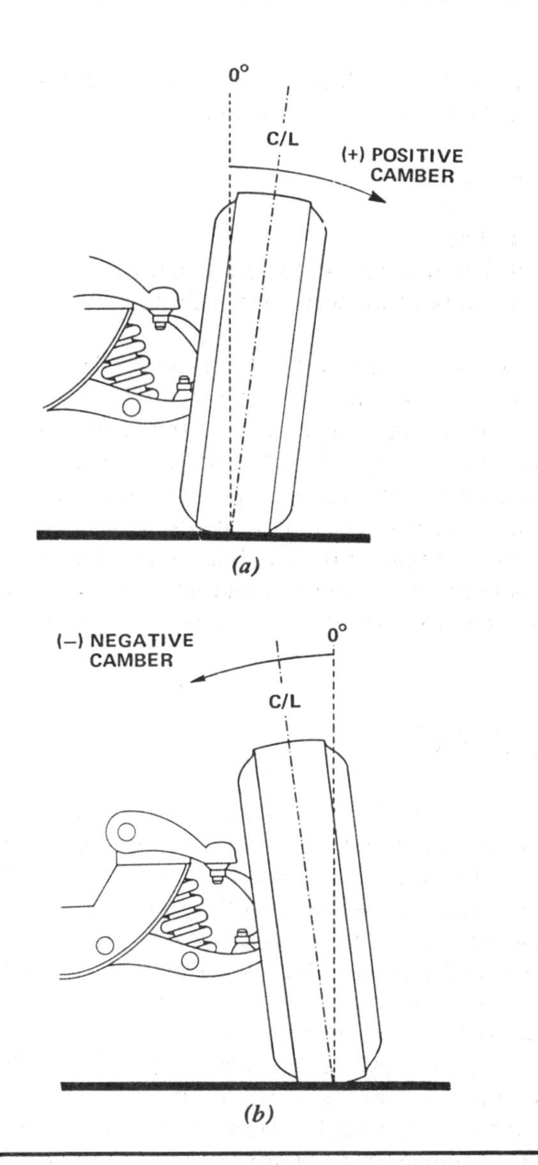

FIGURE 14-2. Two Types of Camber. (a) Positive camber; (b) negative camber. (Courtesy of Hunter Engineering Company)

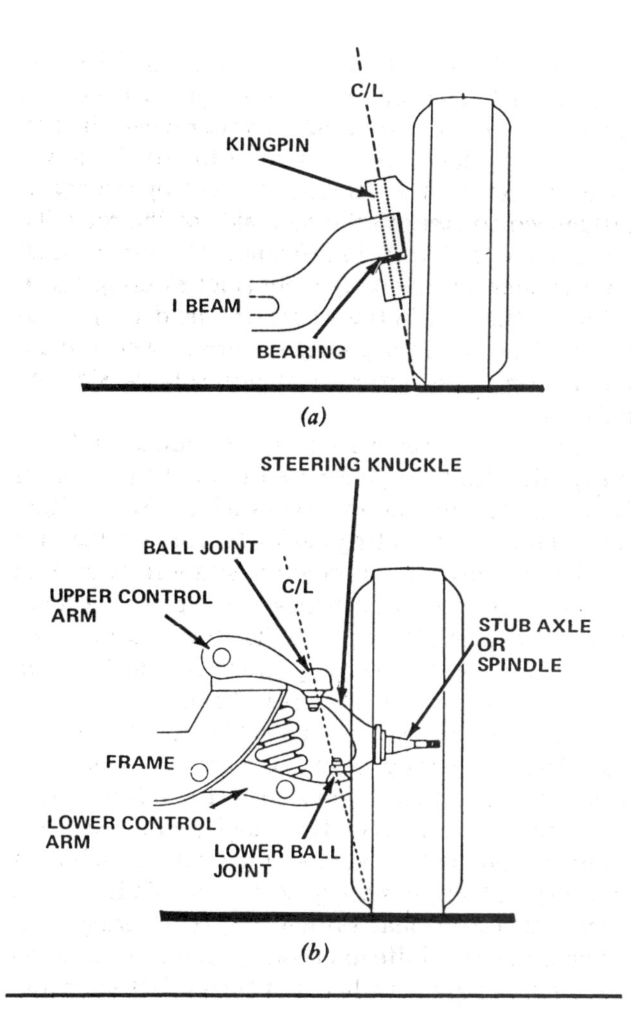

FIGURE 14-3. Establishing Steering Axis Centerline. (a) In kingpin suspension, (b) in ball joint suspension. (Courtesy of Hunter Engineering Company)

- That the camber angle of a wheel seldom remains at the 0° position when the vehicle is being driven.

CASTER

Caster is the rearward or forward tilt of the top of the kingpin or ball joint steering axis centerline from the true vertical position as viewed from the side of a vehicle. Figure 14-3 will assist you in understanding the position of the centerline through a kingpin and ball joint type suspension system. The establishment and measurement of the position of the steering axis centerline is what caster is all about. Since a kingpin's or ball joint's centerline may tilt forward or backward, the definition of caster has two parts. Positive (+) caster is the rearward tilt of the top of the kingpin or ball joint centerline. Negative (-) caster is the forward tilt of the top of the kingpin or ball joint centerline. The amount the centerline tilts is measured in degrees from the true vertical position. Deviation from the vertical position is called the *caster angle* (∢) (Figure 14-4). When a vehicle is being driven, caster seldom remains at a definite position because of the same forces that cause camber to change.

Let's briefly review what you should know about caster:

- The definitions of the terms *positive* and *negative caster*,

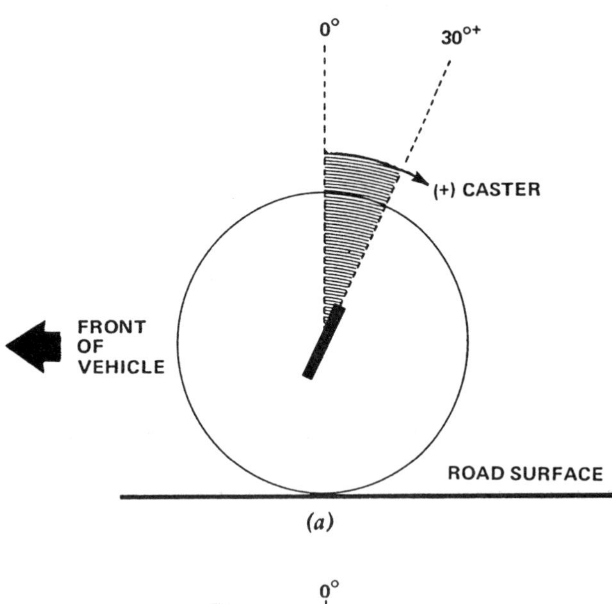

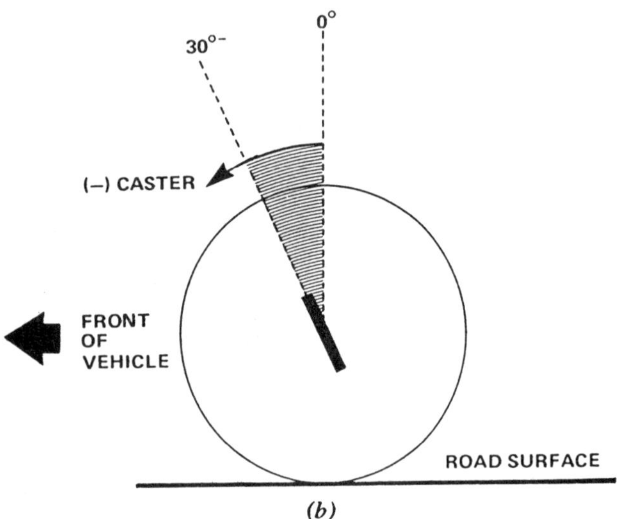

FIGURE 14-4. Two Types of Caster. (a) Positive caster; (b) negative caster.

- That caster is measured in degrees or fractions of a degree,
- That the caster angle of a kingpin or ball joint steering axis centerline seldom remains at a definite position when the vehicle is being driven.

When learning technical information, it is often a help to use some connecting idea in assisting yourself to learn a definition. To help yourself apply the definitions of negative camber and caster, use your hands to represent the left and right upper control arms of an automobile's suspension system. Place the palms of your hands face down; extend the thumbs toward each other. The directions that the thumbs are pointing will indicate negative camber. Extend the first fingers on both hands so that they are pointing away from you; curl the remaining fingers under so that they are touching the palms of your hands; the direction that the index fingers are pointing will indicate negative caster (Figure 14-5). When you know the negative positions for camber and caster, the positive positions, of course, will be in the opposite directions.

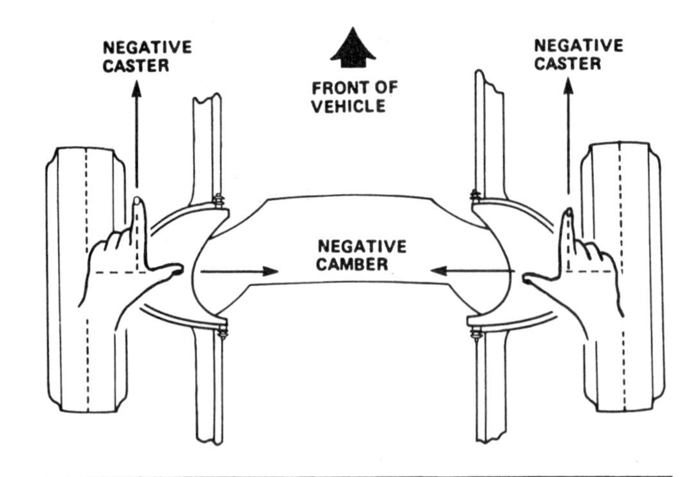

FIGURE 14-5. How to Remember Negative Positions for Camber and Caster. The extended thumbs and first fingers indicate the directions of negative camber and caster respectively.

TOE

The word "toe" should provide a clue to the meaning of this term. When we stand with our feet together, we normally stand with our toes outward. *Toe* is the difference in the linear distance between the front of the front wheels of a vehicle at axle height and the distance at the rear of the front wheels at axle height. (This definition can also apply to the rear wheels of a modern automobile.)

Figure 14-6 illustrates that the distance between the front of the front wheels is 1500 millimeters (59 inches). The distance between the rear of the front wheels is 1508 millimeters (59 5/16 inches). Since the linear distance is 8 millimeters less at the front than at the rear, the toe measurement is stated as 8 millimeters (5/16 inch) **toe-in**. (All distances are measured from the centerline of the tire's tread at axle height.) If the linear distance were wider at the front of the front wheels as compared to the rear of the front wheels, the toe measurement would be stated as **toe-out**.

Since incorrect toe is a major influence in tire

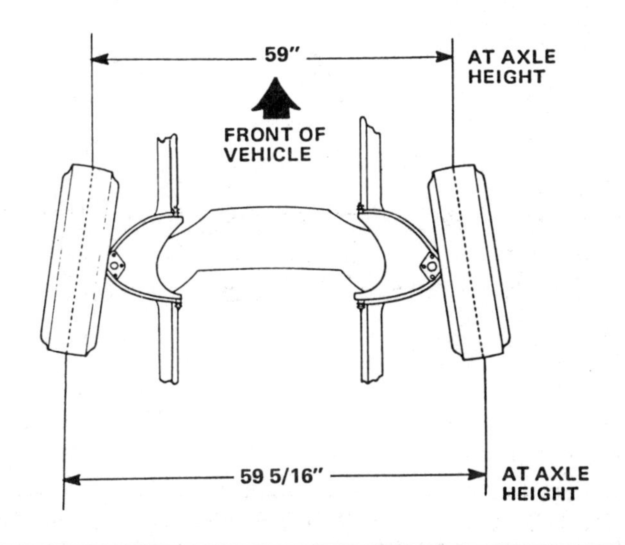

FIGURE 14-6. Top View of Front End Showing Proper Points for Taking Toe Measurements.

tread wear, it must be set to the manufacturer's specification for that particular vehicle. Because defective steering linkage parts can alter toe setting, the toe setting is not used to compensate for worn steering linkage parts.

Let's briefly review what you should know about toe:

- The definition of the term *toe*,
- That it is measured in millimeters or a fraction of an inch, not degrees,
- That toe is calculated and measured from a centerline position of the tire tread surface at axle height.

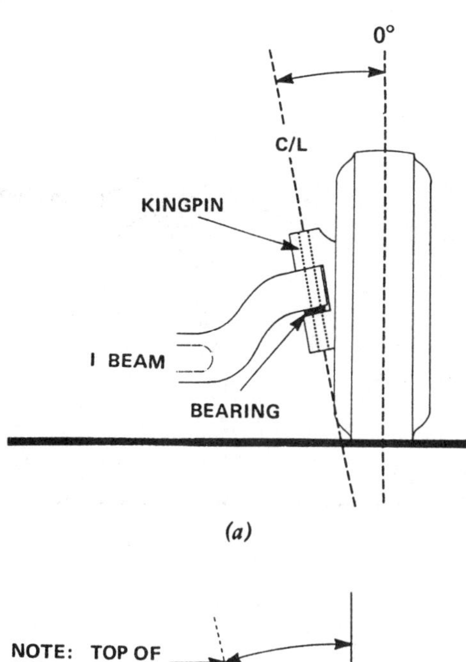

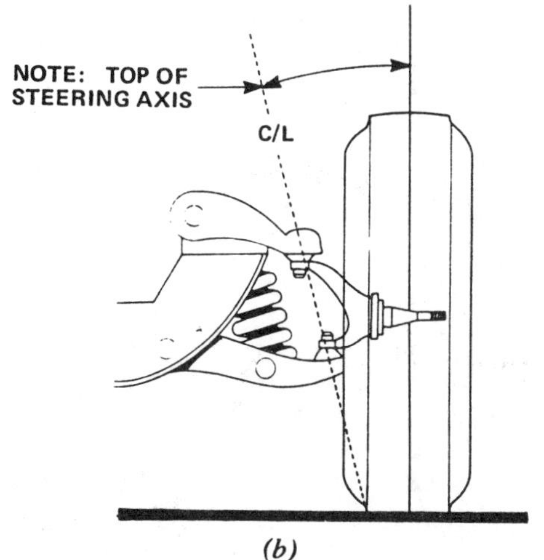

FIGURE 14-7. Measuring Angle of Inclination. (a) Kingpin; (b) ball joint. (Courtesy of Hunter Engineering Company)

KINGPIN OR BALL JOINT INCLINATION

During the pioneering years of automotive design and manufacturing, vehicles were designed with kingpins. Since the early 1960s, however, automobiles have been designed with ball joint suspensions. As you progress through the following chapters, you will learn that there are similarities in the purposes of the caster and the kingpin or **ball joint inclination factors.** Both are concerned with the tilt of the steering axis centerline of the kingpin or the upper and lower ball joints. You will recall that caster is the tilt of the top of the kingpin or ball joint centerline as viewed from the side of a vehicle. Do not confuse the definition of kingpin or ball joint inclination with any other definition.

Kingpin or ball joint inclination is the inward tilt of the top of the steering axis centerline through the kingpin or ball joints as viewed from the front of the vehicle. Figure 14-7 illustrates that the kingpin or ball joint inclination angle is determined from a true vertical position of the front wheels to a centerline position through the kingpin or ball joint. Since this is an angle, it is calculated in degrees.

Let's briefly review what you should know about kingpin or ball joint inclination:

- The definition of the terms *kingpin inclination* and *ball joint inclination,*
- That it is measured in degrees,
- That the terms *kingpin inclination* (kpi), *ball*

joint inclination (bji), and *steering axis inclination* (sai) are similar terms with a common definition.

TOE-OUT ON TURNS OR TURNING RADIUS

The final wheel alignment factor is called *toe-out on turns.* Toe-out on turns is the number of degrees the front inside wheel (on a corner) turns out as compared to the number of degrees the opposite wheel turns in. This factor is necessary to assist in the directional control of a vehicle when turning a corner. Figure 14-8 provides an illustration of a vehicle's front wheels positioned in a turning situation. As you view the diagram, notice that the left front wheel is required to turn out at a greater angle than the right front wheel turns in. This alignment factor aids the tracking of a vehicle and reduces tire tread slippage when cornering. Because the front wheels turn at different angles, the unit of measurement is in degrees.

Do not confuse this factor with toe. Although the names are similar, toe is a different factor and is measured in millimeters or a fraction of an inch. *Toe-out on turns* and *turning radius* are similar terms, and their definitions are held in common.

Let's briefly review what you should know about toe-out on turns:

- The definition of the terms *toe-out on turns* and *turning radius*,
- That it is measured in degrees,
- That when a vehicle turns to the right, the right front wheel turns out a greater number of degrees than the opposite wheel turns in.

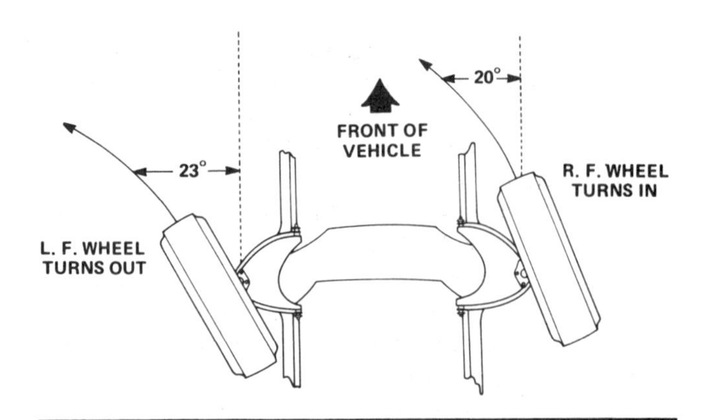

FIGURE 14-8. Measuring Toe-Out on Turns or Turning Radius. The dotted line is the centerline from which the measurements are taken.

REVIEW TEST

1. Define the terms *positive camber* and *negative camber.*
2. Define the terms *positive caster* and *negative caster.*
3. Sketch a front view of a wheel that is subject to negative camber.
4. Sketch a side view of a kingpin that is subject to positive caster.
5. State the definition of toe.
6. Sketch a view of a vehicle's front suspension that is subject to steering axis inclination.
7. Define the term *toe-out on turns.*
8. Sketch an overhead view of the two front wheels of a vehicle as the vehicle turns left.

FINAL TEST

This examination is multiple choice. Only one answer will be accepted. Carefully read every statement.

1. Mechanic A says that positive camber is the outward tilt of the top of the kingpin. Mechanic B says that positive camber is the outward tilt of the top of the wheel. Who is right? (A) Mechanic A, (B) Mechanic B, (C) Both A and B, (D) Neither A nor B.
2. Mechanic A says that positive caster is the rearward tilt of the top of the wheel. Mechanic B says that positive camber is the backward tilt of the top of the steering axis centerline. Who is right? (A) Mechanic A, (B) Mechanic B, (C) Both A and B, (D) Neither A nor B.
3. Mechanic A says that toe is an adjustable front-end alignment factor. Mechanic B says that toe is measured and calculated in degrees. Who is right? (A) Mechanic A, (B) Mechanic B, (C) Both A and B, (D) Neither A nor B.
4. Mechanic A says that toe is used to compensate

for worn steering linkage parts. Mechanic B says that camber is used to compensate for loose front wheel bearings. Who is right? (A) Mechanic A, (B) Mechanic B, (C) Both A and B, (D) Neither A nor B.

5. Mechanic A says that ball joint inclination is an adjustable wheel alignment factor. Mechanic B says that ball joint inclination is measured and calculated in degrees. Who is right? (A) Mechanic A, (B) Mechanic B, (C) Both A and B, (D) Neither A nor B.
6. Mechanic A says that correct wheel alignment prolongs tire tread life. Mechanic B says that correct wheel alignment compensates for unfavorable steering influences. Who is right? (A) Mechanic A, (B) Mechanic B, (C) Both A and B, (D) Neither A nor B.
7. Mechanic A says that when a vehicle turns left, the left front wheel toes out more than the opposite wheel toes in. Mechanic B says that when a vehicle turns right, the right front wheel toes in more than the opposite wheel toes in. Who is right? (A) Mechanic A, (B) Mechanic B, (C) Both A and B, (D) Neither A nor B.
8. Mechanic A says that toe is always determined from the tire's centerline position at axle height. Mechanic B says that toe-out on turns is always determined from the straight-ahead position of the front wheels. Who is right? (A) Mechanic A, (B) Mechanic B, (C) Both A and B, (D) Neither A nor B.
9. Mechanic A says that when a wheel is positioned at 1° positive camber, the wheel is tilted out. Mechanic B says that when a wheel is subject to 1° negative caster, the top of the steering axis centerline is tilted rearward. Who is right? (A) Mechanic A, (B) Mechanic B, (C) Both A and B, (D) Neither A nor B.
10. Mechanic A says that the camber angle is always determined from the true vertical position. Mechanic B says that the caster angle is always determined from the true horizontal position. Who is right? (A) Mechanic A, (B) Mechanic B, (C) Both A and B, (D) Neither A nor B.

Camber

LEARNING OBJECTIVES

After studying this chapter, you should be able to:

- Explain why and how the camber of a wheel constantly changes on a vehicle designed with unequal length control arm independent suspension,
- Explain how incorrect camber causes tire tread wear,
- Explain why a wheel rolls in the direction of its tilt,
- Explain how vehicle weight, passenger weight, cargo load, and gravity affect the directional control and stability of a vehicle,
- Explain how correct camber may be used to counteract the influence of a crowned road,
- Describe how to measure the camber of a wheel,
- Analyze the camber readings for the left and right front wheels of a vehicle and state its steering direction.

INTRODUCTION

Camber is the first wheel alignment factor that we will study in depth. Since camber influences the directional control and stability of a vehicle, wheel alignment technicians must possess a comprehensive understanding of the camber factor. Be sure you know the definitions of positive and negative camber. To understand camber fully, we will first discuss some related facts about the wheel.

A revolving wheel that is positioned perpendicular on a horizontal road surface will not roll to the right or to the left. We can assume that the wheels of a vehicle are always at a true vertical position, but in reality we know that this is not so. In average driving, a wheel encounters such things as bumps, holes, curves, and **road crown.** On curves, centrifugal force causes load shift, which varies with speed. In addition, body tilt caused by passenger load affects the vehicle's suspension system.

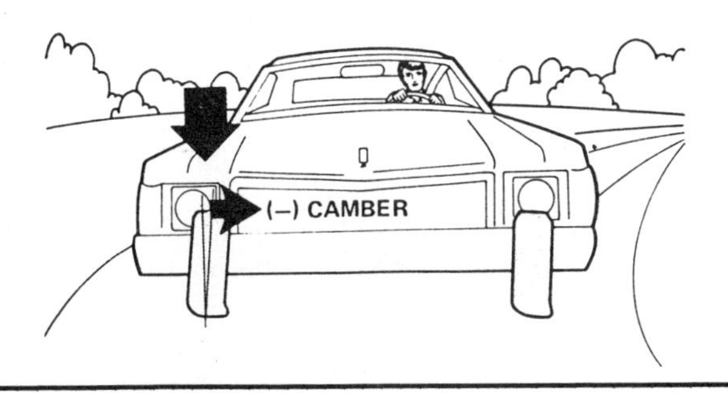

FIGURE 15-2. Negative Camber Produced by Downward Movement of Body. (Courtesy of Ford Motor Company)

WORKING PRINCIPLES OF CAMBER

Let's review what happens to (1) a revolving front wheel of an automobile with unequal length upper and lower control arms and (2) to a front wheel that is attached to a MacPherson strut.

Unequal Length Control Arms

When a front wheel hits a bump, the wheel moves upward in what is called a *jounce movement.* Because the upper control arm is shorter than the lower control arm, the paths or arcs traveled by the upper and lower ball joints are not the same. As the wheel rises, the swing arc of the shorter upper control arm causes the top of the wheel to tilt inward, producing a negative camber condition (Figure 15-1).

As the vehicle is driven into a fast, smooth turn, the body-chassis rolls toward the outside wheel. This movement places a force on the outside wheel that acts like jounce. In other words, with the upward movement of the front wheel over a bump or with the downward movement of the body due to load shift on a curve, the camber angle of the wheel becomes negative (Figure 15-2). At the same time, the roll effect of the body-chassis lifts the weight from the opposite wheel, causing it to lean slightly outward at the top.

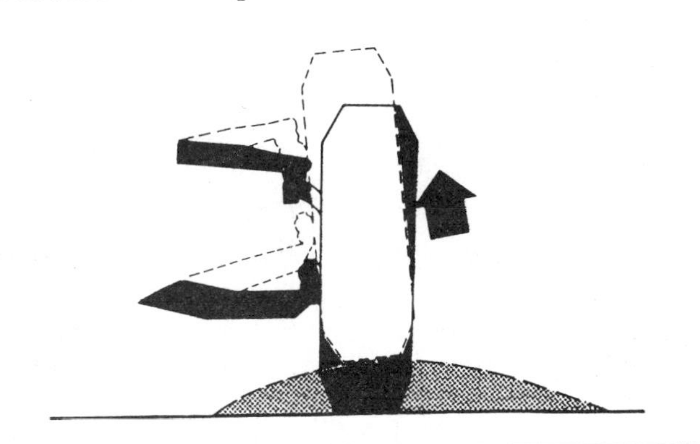

FIGURE 15-1. Negative Camber Produced by Jounce Movement. (Courtesy of Ford Motor Company)

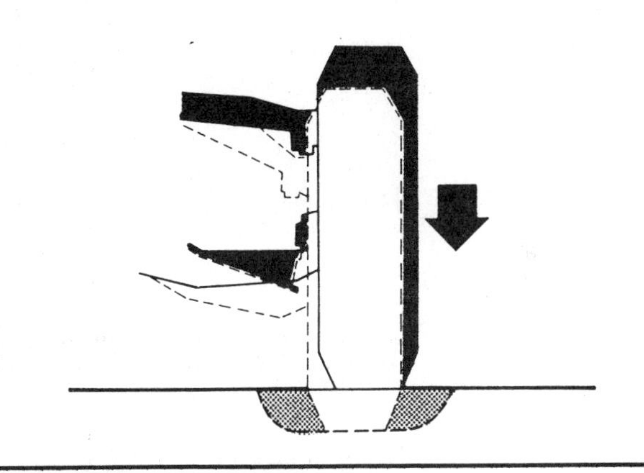

FIGURE 15-3. Zero or Slightly Positive Camber Produced by Rebound Movement. (Courtesy of Ford Motor Company)

When a front wheel drops into a hole, the downward movement is called *rebound.* This rebound action produces zero camber or a slight increase in the positive camber angle because the geometry of the upper and lower control arms causes the wheel to move inward at the top and bottom (Figure 15-3).

As you might expect from the wheel jounce, body lean relationship, there is a comparable action on rebound. Upward body movement produces the same relative effect as when the wheel drops on rebound. This means that we get zero or a slight positive camber at the inside wheel as the body lifts away from the wheel on a fast turn (Figure 15-4). Camber changes that occur as a result of wheel jounce or rebound have some directional stability; however, the way in which the camber factor influences the directional characteristics of a vehicle is more obvious in turns or on crowned road surfaces because a sideways weight shift is added to the up-and-down movement of the wheel.

FIGURE 15-4. Zero or Slightly Positive Camber Produced by Upward Movement of Body. (Courtesy of Ford Motor Company)

FIGURE 15-5. How Camber Produces Skid Resistance. (Courtesy of Ford Motor Company)

MacPherson Strut

The front wheels of a vehicle that are attached to a MacPherson strut are also subject to camber change. When a vehicle is carrying a passenger load, the two front wheels move from a slight positive camber position to an almost true vertical position. This is because of the design of the suspension, which has only lower control arms and struts that are attached to the vehicle's body, allowing the top of the suspension to rotate.

Resistance to Side Forces

Regardless of the design of the suspension system, as the vehicle's body leans throughout a turn, centrifugal force tries to move the vehicle sideways. This tendency to move the vehicle toward the outside of the turn is resisted by the traction of the tires. On a turn, especially at higher speeds, body tilt or lean produces negative camber at the outside wheel. The resultant action is that the inner section of the tire tread digs in and braces itself against sideways skidding. This bracing counteracts centrifugal force trying to move the vehicle sideways. The downward weight shift on the outside wheel also increases traction to resist skidding. While the outside wheel tilts in a negative camber position, the zero or slight positive camber at the inside wheel also adds a bracing force against skidding. In fact, it does not resist as much as the outside wheel since the load and traction on the inside wheel are not as great (Figure 15-5).

TIRE TREAD WEAR

When a vehicle is being maneuvered through a road curve, it is a definite advantage for the wheels to tilt, thereby aiding the vehicle's directional control and stability. However, tilting can also be a disadvantage to a tire's tread life when it is subject to an incorrect camber angle. A tire and wheel assembly is subject to its various rolling circumferences.

The best way to illustrate this statement is to perform an experiment. Find a deserted road, and measure and mark the distance of exactly 1.6 kilometers (1 mile). Obtain two wheels of different diameters (one large and one small). Roll both wheels over the marked course and record the revolutions of each wheel. Which wheel assembly revolved the greatest number of revolutions? The answer: the small one.

That simple experiment may be applied to a revolving wheel that is subject to excessive camber. Figure 15-6A illustrates that the outside of the wheel, being smaller in circumference, rotates faster than the inside, since it must cover the same distance in a greater number of revolutions. Since the outside and the inside portions of the tread are on the same tire, the outside portion of the tread must slip (scuff) while the wheel is revolving. This condition induces tread wear at the outside of the tire's tread surface, as seen in Figure 15-6B.

Manufacturers, when designing a vehicle, know that a tire's tread surface will provide normal life expectancy when the camber angle of the wheel is kept within alignment specifications. Tread life may also be extended when the vehicle is driven at mod-

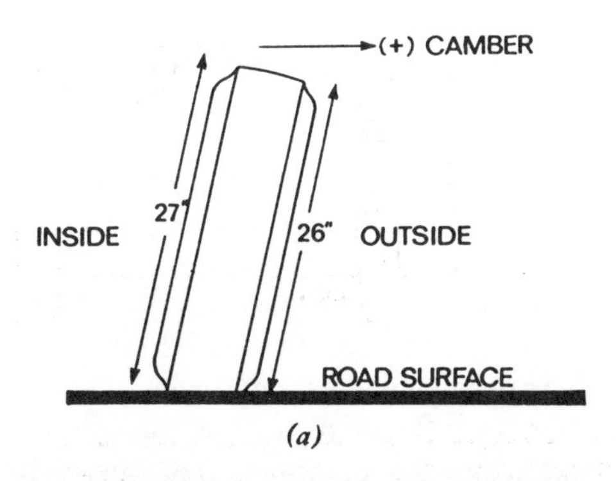

FIGURE 15-6A. Differences in Inner and Outer Tire Diameters. (Courtesy of Snap-on Tools of Canada Ltd.)

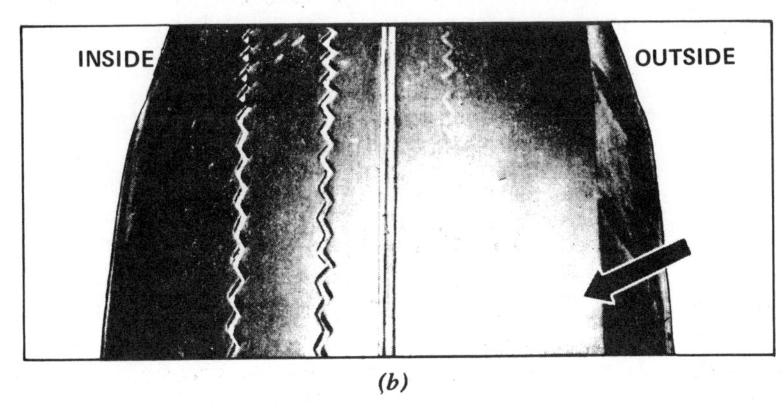

FIGURE 15-6B. Tread Wear Caused by Incorrect Camber. (Courtesy of Snap-on Tools of Canada Ltd.)

erate highway speeds, tires are inflated at the correct pressure, steering and suspension systems are serviced at the recommended intervals, and wheel balance and alignment are checked at least once a year.

VEHICLE CONTROL AND STABILITY

Coning Influence of the Wheel

To appreciate fully how camber may be used to aid the directional control and stability of a vehicle, we need to understand the *coning influence* (tilt in or out) of a wheel. A change in the camber of a wheel influences a tire and wheel assembly to roll in the direction of its tilt. This can be visualized by laying a cone-shaped object (for example, an ice cream cone) on a flat surface and attempting to roll it in a straight-ahead direction. You will notice that the movement of the cone describes a circle. To understand the principle further, recall having ridden a bicycle with no hands and making a turn simply by tilting the bike in the direction of travel. In effect, you made the bicycle turn by increasing the camber or tilt of the front wheel. Let's apply this principle of tilt to a vehicle's front wheel. Figure 15-7 illustrates a front wheel that is subject to positive camber. Since the wheel is tilted out, its direction of rolling travel is also out (around the apex of the cone).

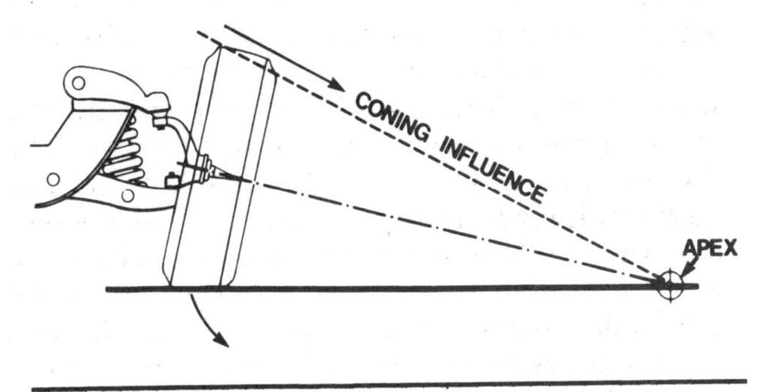

FIGURE 15-7. Coning Influence of Wheel with Incorrect Positive Camber. (Courtesy of Hunter Engineering Company)

FACTS TO REMEMBER

1. When a front wheel is subject to positive camber, it will be influenced to roll outward.
2. When subject to negative camber, a front wheel will be influenced to roll inward.
3. The principle and the benefits of the coning influence of the wheel are used by a vehicle manufacturer to aid the directional control and stability of a vehicle when driven on a crowned road.

Effect of a Crowned Road

A vehicle driven on the right side of a crowned road is pulled to the right (curb side) by the shift of the vehicle's mass (weight) from the force of gravity. If the front wheels have been aligned correctly, weight transfer to the right front wheel has resulted in negative camber at that wheel and slight positive camber to the left front wheel. Since the front wheels of a vehicle tend to roll and steer in the direction of their lean, the directional influence of the wheels counteracts the influence of weight and gravity. This partially explains how a vehicle maintains a straight course on a crowned road surface (Figure 15-8).

Effect of Vehicle Load

In addition to the effects of shifting vehicle weight, the camber angle of a vehicle's front wheels is also changed by variations in passenger load and cargo weight. In most automobiles, the driver alone is the passenger load, so we will consider only the effects of this condition. When an automobile is being driven on a crowned road, the driver's weight causes the left side of the automobile to drop slightly lower than the right. Since the vehicle is on a crowned road, some of the driver's weight is also transferred to the right side of the vehicle's suspension by the influence of gravity. To compensate for road crown and a slight shift of body-chassis and driver weight, the camber on the left front wheel is set positive to assist the vehicle in maintaining a straight course on a crowned road surface. If the driver's weight is excessive or the right and left wheel camber angles not correctly aligned, the camber can cause some steering pull, resulting in a loss of directional control and stability.

CAMBER ALIGNMENT METHOD

Some vehicle manufacturers use the working principles of camber outlined in this chapter exclusively

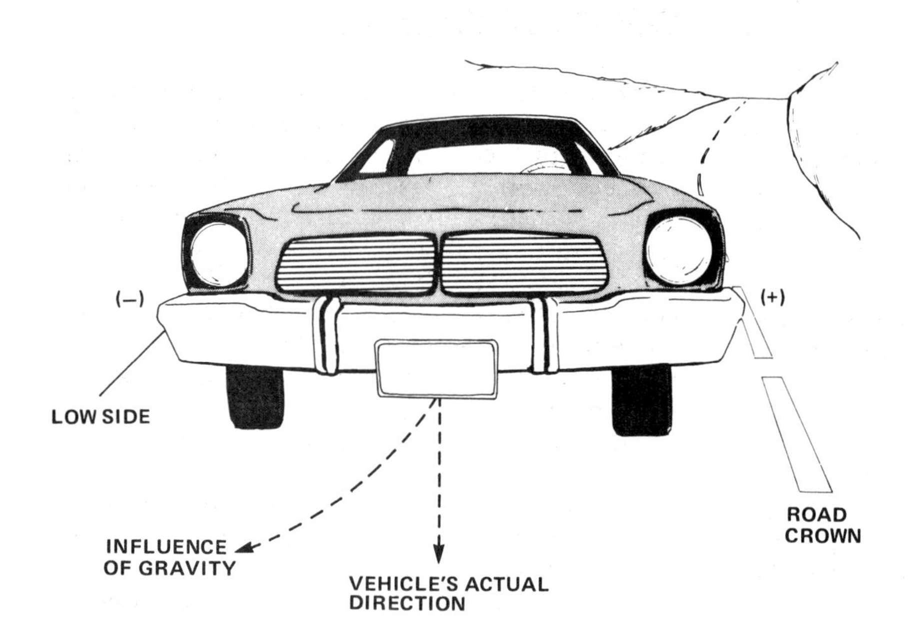

FIGURE 15-8. Using Camber to Counteract Influence of Weight and Body. A crowned road surface causes gravity to affect a vehicle's direction.

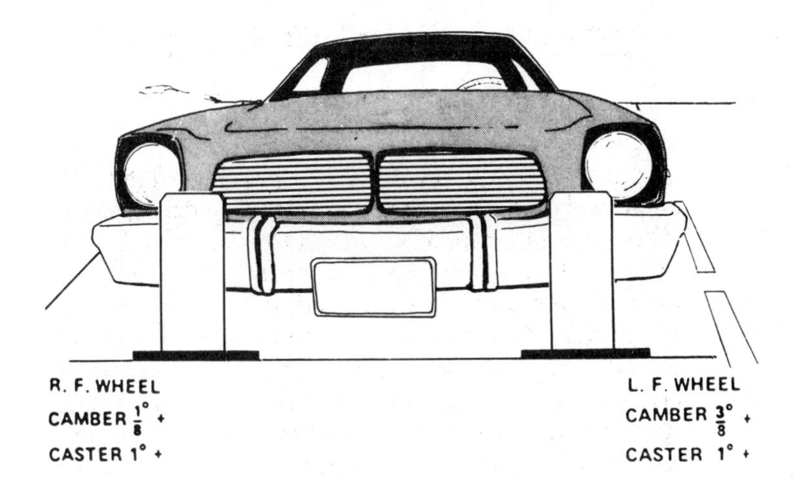

FIGURE 15–9. Caster and Camber Settings for Camber Alignment Method.

to offset the effects of a crowned road. When the camber method is used, the wheel alignment technician must also set the caster angles the same for both the left and right sides. Figure 15–9 illustrates the alignment settings for an automobile when the camber method has been specified by the manufacturer. As you study the illustration, pay particular attention to the actual figures (settings) of the camber and caster.

Camber Measurement

It is not the intent of this chapter to explain how the camber angle of a wheel is adjusted. This subject will be discussed in a later chapter. However, it is important to know that the camber angle of a wheel is measured only when the wheel has been placed in the straight-ahead position. Also, before a vehicle can be aligned, the curb riding height must be set to the manufacturer's specification.

In some rare instances, a rear wheel will indicate an incorrect camber condition. The camber for a rear wheel is measured in a manner similar to front wheel camber. The curb riding height and the curb weight of a vehicle are also relative to the camber angles of the front wheels. If the front end of a vehicle is raised, the camber angles of the front wheels change. Recall the descriptions of jounce and rebound.

FACTS TO REMEMBER

1. Do not set the front wheels of a vehicle with negative camber (unless specified by manufacturer's specifications).
2. When using the camber method, set the right front wheel slightly positive (preferably 1/8° positive).
3. Do not set the left front wheel camber excessively positive (always refer to the manufacturer's specifications).
4. The closer the wheel camber setting is to the 0° position, the greater the tire tread life expectancy.
5. Set the caster angles the same for both sides.
6. If the driver's weight is excessive, it may be necessary to request the driver to sit in the vehicle when the wheels are being aligned.

ADVANTAGES OF CORRECT CAMBER

When a customer brings a vehicle to a repair shop for alignment service, it would be to his or her advantage if the vehicle's loss of direction or tread wear problem could be discussed with the alignment technician. However, this is not always possible. In some instances, it will be necessary for the technician to drive the vehicle in order to diagnose the complaint. Problems such as unstable steering, steering wander, and excessive tire tread wear normally are indicators of incorrect wheel alignment caused by worn, defective parts.

Let us review the advantages of correct camber:

1. Correct camber is used to aid a vehicle's directional control and stability when cornering.
2. Correct camber may be used to counteract

the influence of a crowned road while driving a vehicle.
3. Correct camber reduces tire tread wear.
4. Correct camber reduces steering effort, thereby reducing the physical fatigue of constant steering control.

REVIEW TEST

1. Briefly explain how the camber of a front wheel constantly changes on a vehicle designed with unequal length control arm and MacPherson strut suspension.
2. Explain why a tire's tread wears when it is subject to incorrect camber.
3. Briefly list how tire tread life may be extended.
4. Describe how correct camber is used to offset for road crown.
5. List the facts that the alignment technician must keep in mind when using camber to offset for road crown.
6. List the four advantages of correct camber.
7. Briefly describe how camber is measured.
8. Figure 15-10 illustrates a vehicle being driven on a rural road. Based on what you have learned, study the settings for the left and right camber readings. State the direction the vehicle will steer. Support your answer with a written explanation.
9. Figure 15-11 illustrates another vehicle being driven on a rural road. Study the settings for the left and right camber readings. State the direction the vehicle will steer. Support your answer with a written explanation.

FIGURE 15-10. Illustration for review question 8.

FIGURE 15-11. Illustration for review question 9.

FINAL TEST

This examination is multiple choice. Only one answer will be accepted. Carefully read every statement.

1. Mechanic A says that positive camber is the outward tilt of the top of the wheel. Mechanic B says that positive camber is the outward tilt of the top of the steering axis centerline. Who is right? (A) Mechanic A, (B) Mechanic B, (C) Both A and B, (D) Neither A nor B.
2. Mechanic A says that the excessive positive camber causes inside tire tread wear. Mechanic B says that excessive negative camber causes outside tire tread wear. Who is right? (A) Mechanic A, (B) Mechanic B, (C) Both A and B, (D) Neither A nor B.
3. Mechanic A says that the camber angle is measured in millimeters or a fraction of an inch. Mechanic B says that the camber angle of a wheel seldom changes even when the vehicle is turning a corner. Who is right? (A) Mechanic A, (B) Mechanic B, (C) Both A and B, (D) Neither A nor B.
4. Mechanic A says that regardless of suspension design, on a fast, smooth turn to the left, the camber of the right front wheel moves negative. Mechanic B says that regardless of suspension design, if the right front wheel is subject to positive camber, the wheel has a tendency to roll out. Who is right? (A) Mechanic A, (B) Mechanic B, (C) Both A and B, (D) Neither A nor B.
5. Mechanic A says that when camber is used to offset for road crown, the left front wheel's camber is set less positive than the right. Mechanic B says that when camber is used to offset for road crown, the caster angle settings must be the same. Who is right? (A) Mechanic

A, (B) Mechanic B, (C) Both A and B, (D) Neither A nor B.

6. Mechanic A says that before a vehicle is aligned, the curb riding height must be within the manufacturer's specifications. Mechanic B says that correct camber reduces steering effort and aids a vehicle's directional control and stability. Who is right? (A) Mechanic A, (B) Mechanic B, (C) Both A and B, (D) Neither A nor B.
7. Mechanic A says that in order to prolong tire tread life, the right front wheel's camber should be set 1° negative. Mechanic B says that it should be set 2° positive. Who is right? (A) Mechanic A, (B) Mechanic B, (C) Both A and B, (D) Neither A nor B.

Note: The following camber settings pertain to Question 8: right front wheel camber is -1° and left front wheel camber is +1°.

8. Mechanic A says that the camber settings will not influence a vehicle's directional control. Mechanic B says that as the vehicle is driven, it will steer to the left. Who is right? (A) Mechanic A, (B) Mechanic B, (C) Both A and B, (D) Neither A nor B.
9. Mechanic A says that the camber of a front wheel is measured when the wheel is located in the straight-ahead direction. Mechanic B says that if the outside of the rear tire is subject to wear, excessive positive camber probably is the reason. Who is right? (A) Mechanic A, (B) Mechanic B, (C) Both A and B, (D) Neither A nor B.

Caster

LEARNING OBJECTIVES

After studying this chapter, you should be able to:

- Explain how positive caster assists the front wheels of a vehicle to return to the straight-ahead position,
- Explain why negative caster does not aid the front wheels of a vehicle to return to the straight-ahead position,
- Explain how positive caster aids the directional control and stability of a vehicle,
- Explain why negative caster does not aid the directional control and stability of a vehicle,
- Explain why excessive positive caster influences the front wheels of a vehicle to shake or shimmy at low speeds,
- Explain how correct negative caster reduces the tendency for the front wheels of a vehicle to shake or shimmy at low speeds,
- Explain why excessive positive caster increases the effect of road shock on a vehicle,
- Explain how correct negative caster reduces the influence of road shock on a vehicle,
- Explain how a vehicle's riding height affects the caster angle,
- Explain how correct positive caster may be used to counteract the influence of a crowned road,
- Explain how correct negative caster may be used to counteract the influence of a crowned road.

INTRODUCTION

As you study wheel alignment factors, you will realize that all the factors influence a wheel when it is in motion. Caster is no exception. Do you recall the definitions of positive and negative caster?

To understand how caster affects the front wheels of a vehicle, you must realize that positive caster influences a front wheel to toe-in, and negative caster influences a front wheel to toe-out. By varying the setting of the caster angle, it is possible to influence a vehicle's steering direction to the left (high side of a crowned road) or to the right (low side of the road), thereby adversely affecting its directional control and stability. To understand how this factor actually affects a vehicle, let's first study the basic operating principles of caster.

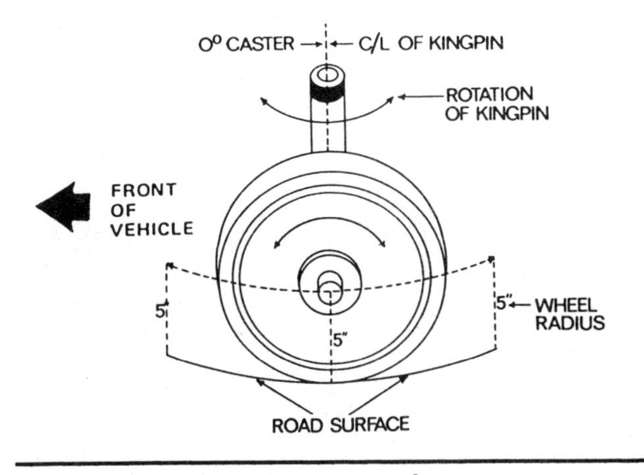

FIGURE 16-1. 0° Caster Angle. A 0° caster angle does not affect the movement of the wheel or kingpin.

PRINCIPLES OF CASTER

The principles of caster are as follows:

1. Positive caster assists the front wheels of a vehicle to return to the straight-ahead position.
2. Negative caster causes the front wheels of a vehicle to fail to return to the straight-ahead position.
3. Correct positive caster produces a trailing effect that aids the directional control and stability of a vehicle.
4. Negative caster produces a pushing effect that does not aid the directional control and stability of a vehicle.
5. Excessive positive caster influences the front wheels of a vehicle to shake or shimmy at low speeds.
6. Correct negative caster reduces the tendency for the front wheels of a vehicle to shake or shimmy at low speeds.
7. Excessive positive caster increases the influence of road shock on a vehicle.
8. Correct negative caster reduces the influence of road shock to a vehicle.
9. Correct positive caster may be used to counteract the influence of a crowned road.
10. Correct negative caster may be used to counteract the influence of a crowned road.

Principle 1: Positive caster assists the front wheels of a vehicle to return to the straight-ahead position. Figure 16-1 illustrates the left front wheel of a vehicle designed with a kingpin suspension. The caster angle is 0° (the kingpin pivot axis centerline is vertical). The radius of the wheel is 12.5 centimeters (5 inches). If the wheel were turned in or out at the front, movement, to the right or left, would not cause the kingpin to lift or lower because the caster angle is 0°.

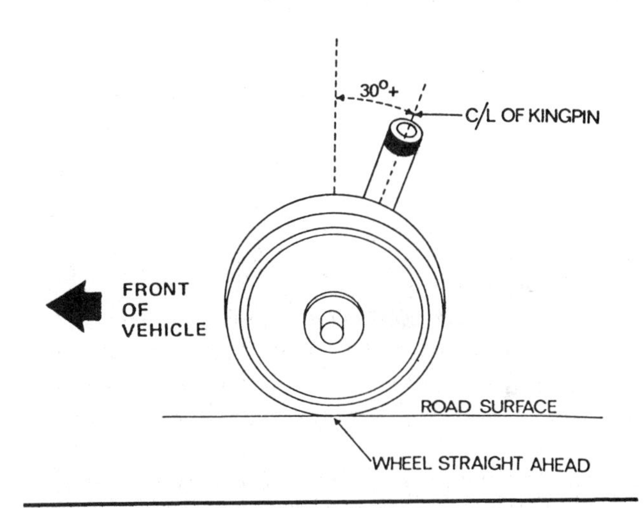

FIGURE 16-2. 30° Positive Caster Angle with Wheel in Straight-Ahead Position.

Figure 16-2 illustrates an exaggerated caster angle. The caster angle is purposely excessive to exaggerate the effect of this important factor. As you view the diagram, notice that the caster angle is 30° positive. The top of the kingpin is tilted toward the rear of the vehicle, and the wheel is positioned straight ahead.

The caster angle in Figure 16-3 is still 30° positive, but the left front wheel has been turned out.

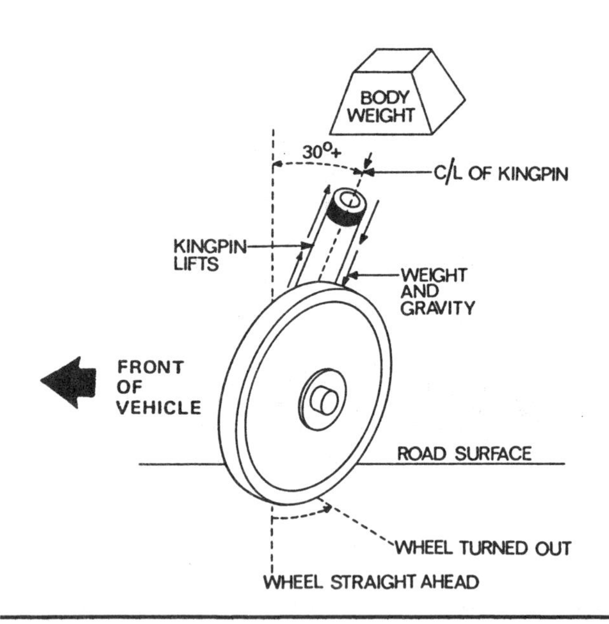

FIGURE 16-3. 30° Positive Caster Angle with Wheel Turned out. The positive caster angle causes the kingpin to lift when the wheel is turned out.

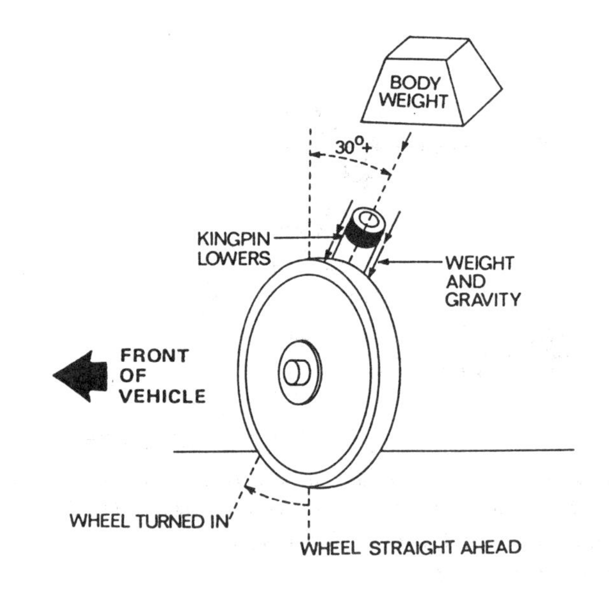

FIGURE 16-4. 30° Positive Caster Angle with Wheel Turned in. Positive caster forces the kingpin to seek its lowest possible position and toe-in.

As you compare Figure 16-2 and Figure 16-3, you will notice that Figure 16-3 shows the lift of the kingpin. You are probably wondering why the kingpin has lifted. The reason is as follows. In Figure 16-2, the kingpin is subject to 30° positive, and the wheel is located in the straight-ahead position. The spindle (axle), the part that the wheel is attached to, is parallel to the road surface. As the front of the wheel is turned out, as illustrated in Figure 16-3, the end of the spindle, because of the backward tilt of the centerline of the kingpin, will endeavor to move closer to the road surface. But this cannot happen because of the radius of the wheel. Instead, as the front of the wheel is turned out, the kingpin is forced to lift. Since the wheel and the kingpin support a portion of the vehicle's weight, gravity pulls the kingpin down, causing the left front wheel to return to the straight-ahead position.

As a driver turns the steering wheel and directs the vehicle around a corner, it is necessary for the front wheels to return to the straight-ahead steering position. Positive caster provides this assistance.

In Figure 16-4, the caster angle is still 30° positive, and the left front wheel has turned in. What has influenced the wheel to move from the straight-ahead position to its present position? Recall that the wheel and the kingpin support a portion of the vehicle's weight. Since the weight is equal to approximately one-quarter of the total vehicle weight—180 to 225 kilograms (400 to 500 pounds)—the kingpin rotates on its axis to its lowest possible position. Hence, the left front wheel is forced to toe-in.

Principle 2: Negative caster causes the front wheels of a vehicle to fail to return to the straight-ahead position. The caster angle in Figure 16-5 is 30° negative. The top of the kingpin's centerline is tilted toward the front of the vehicle, and the wheel is straight ahead.

Compare Figure 16-5 and Figure 16-6. Notice that Figure 16-6 shows the kingpin in a lower position. The vehicle's weight has forced the kingpin to rotate to its lowest possible position. Hence, the front of the left front wheel has been forced to toe-out. Figure 16-7 illustrates that the caster angle is still 30° negative, and the wheel has moved from the turned-out position (Figure 16-6) to the turned-in position. Notice the lift of the kingpin.

At this point, it would appear that both positive and negative caster aid the directional control and stability of a vehicle equally; however, this is not true. Negative caster does not aid the front wheels of a vehicle to return to the straight-ahead steering position; therefore, negative caster does not aid a vehicle's directional control and stability. To explain why, assume that a vehicle is turning left; the caster angles for the left and right wheels are 30° negative. The left front wheel is turned out, and the right front wheel turned in.

In Figure 16-8, the left front wheel scribes

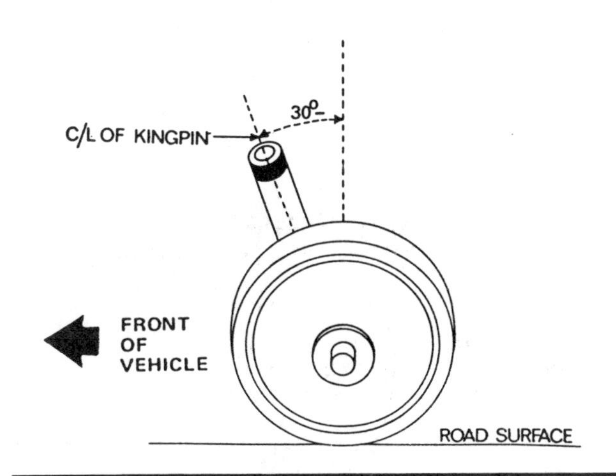

FIGURE 16-5. 30° Negative Caster Angle with Wheel in Straight-Ahead Position. The kingpin is tilted toward the front of the vehicle.

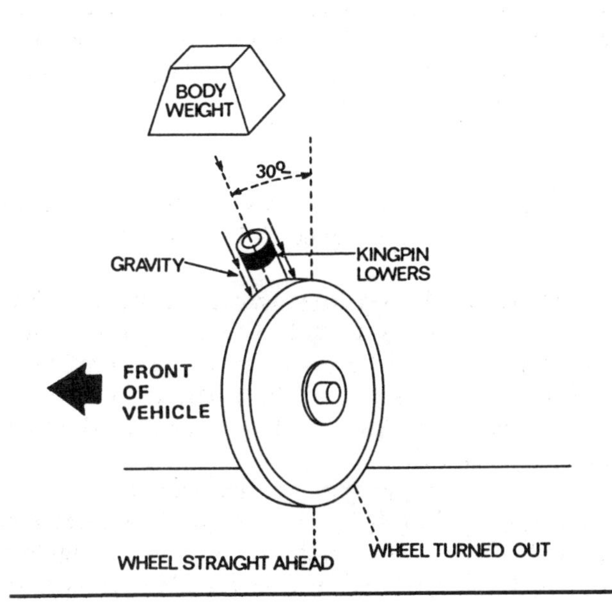

FIGURE 16-6. 30° Negative Caster Angle with Wheel Turned out. The vehicle's weight causes the kingpin to drop.

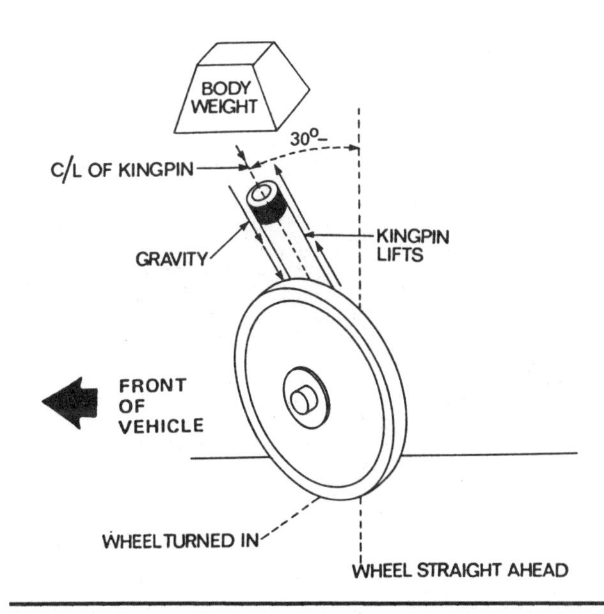

FIGURE 16-7. 30° Negative Caster Angle with Wheel Turned in. By turning the wheel in, negative caster causes the kingpin to lift.

an arc of 23°, and the right front wheel scribes an arc of 20° (recall the definition of toe-out on turns from Chapter 14, and keep in mind that both front wheels are subject to negative caster). The left front portion of the vehicle's body will drop, and the right front portion of the vehicle's body will lift. Because the left front wheel has turned out a greater number of degrees than the right front wheel has turned in, the vehicle's body weight will not allow the front wheels of the vehicle to return to the straight-ahead position.

At this point, you are no doubt wondering how manufacturers obtain directional control and stability of a vehicle when it is designed with negative caster. Directional control, and the ability of the vehicle's steering wheel to return to the straight-ahead position after cornering, is achieved through the kingpin or ball joint inclination. This important factor will be discussed later in the book.

It is true that caster angles tend to cause the vehicle's body-chassis to lift. However, modern vehicle caster angles are small, and the lift they create is negligible. When the front wheels of a vehicle have been turned, their influence has become a minor force in the steering stability of a vehicle. Although caster angles are small, they nevertheless have an influence on the front wheels.

Caster is not an easy alignment factor to understand. Part of the reason for the difficulty is that we cannot actually see the angle; we can only experience its influence and effects. This is very similar to the problems students encounter when studying electricity for the first time. No one would deny that electrical energy is not real, but one does not see it flowing through the wires. When you have studied caster, you will realize that the effects created by the caster angle are also real.

FACTS TO REMEMBER

1. Positive caster tends to influence a vehicle's front wheels to toe-in.
2. Negative caster tends to influence a vehicle's front wheels to toe-out.

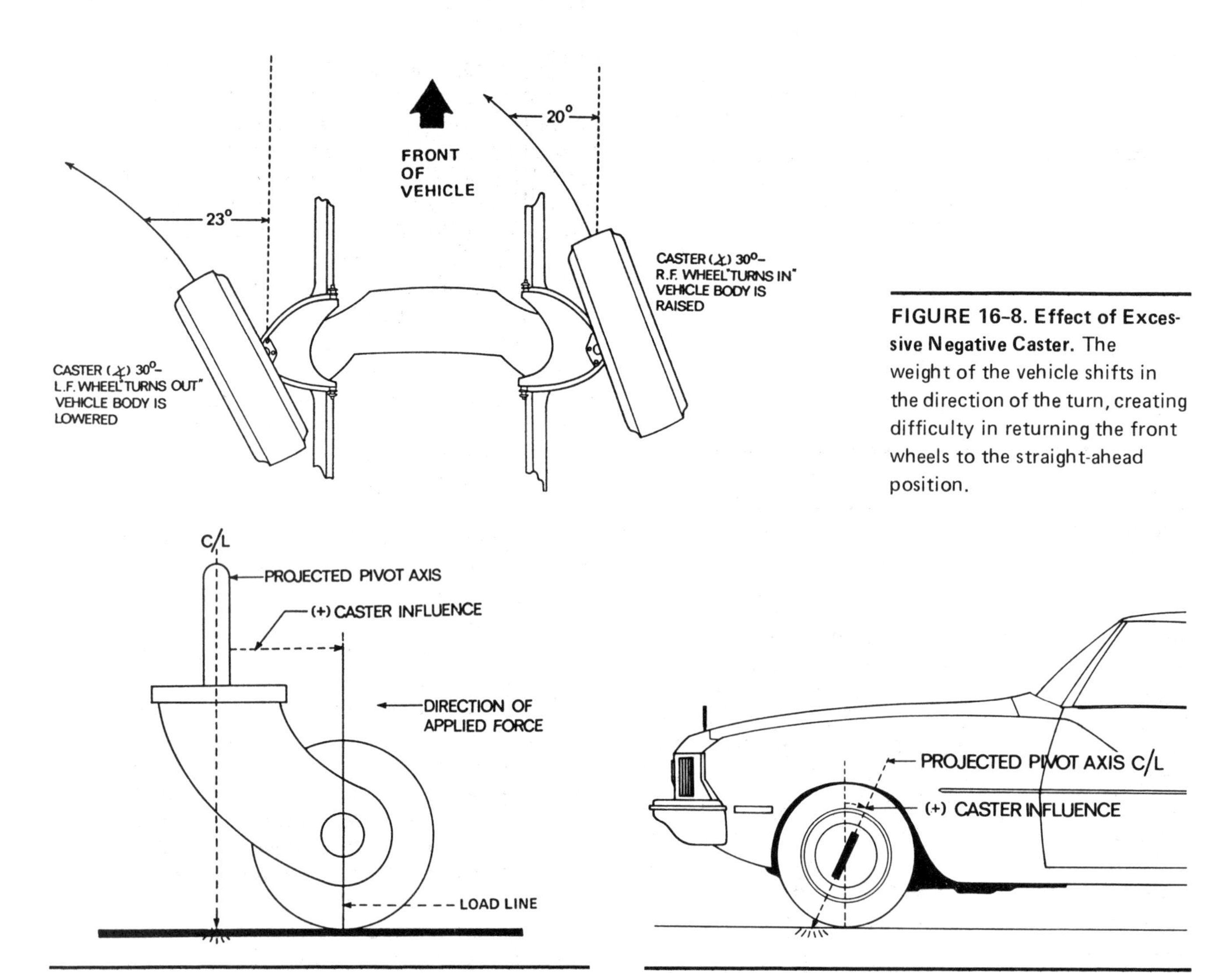

FIGURE 16-8. Effect of Excessive Negative Caster. The weight of the vehicle shifts in the direction of the turn, creating difficulty in returning the front wheels to the straight-ahead position.

FIGURE 16-9. Trailing Effect on a Shopping Cart Wheel Produced by Positive Caster.

FIGURE 16-10. Trailing Effect on a Vehicle's Wheels Produced by Positive Caster.

Principle 3: Correct positive caster produces a trailing effect that aids the directional control and stability of a vehicle. A shopping cart used at a food market illustrates another useful purpose of caster. When a shopping cart is pushed, the front caster wheels (Figure 16-9) turn on their pivots until the wheels are in line with the applied force. This produces a trailing effect on the wheel because the wheel is behind the projected pivot axis centerline. In other words, the pivot axis is leading so it pulls the caster wheel behind it. This influence is illustrated by extending the centerline of the caster pivot and noting that it strikes the floor in front of the contact point of the wheel. The wheel is subject to positive caster.

The same principle applies to the front wheel of a vehicle, as shown in Figure 16-10. When the steering pivot axis is tilted backward at the top, the projected pivot axis centerline strikes the road ahead of the point of tread contact, producing the same effect as the shopping cart caster. The pivot axis centerline creates a trailing effect on the wheel, caused by positive caster, assisting the vehicle's directional control and stability. As the caster angle is increased in the positive direction, the effort required to turn the vehicle from the straight-ahead course and hold it there is greatly increased. The tendency of the front wheels to straighten out rapidly when leaving a turn is also increased by increasing the positive caster angles.

Principle 4: Negative caster produces a pushing effect that does not aid the directional control and stability of a vehicle. When the pivot axis center-

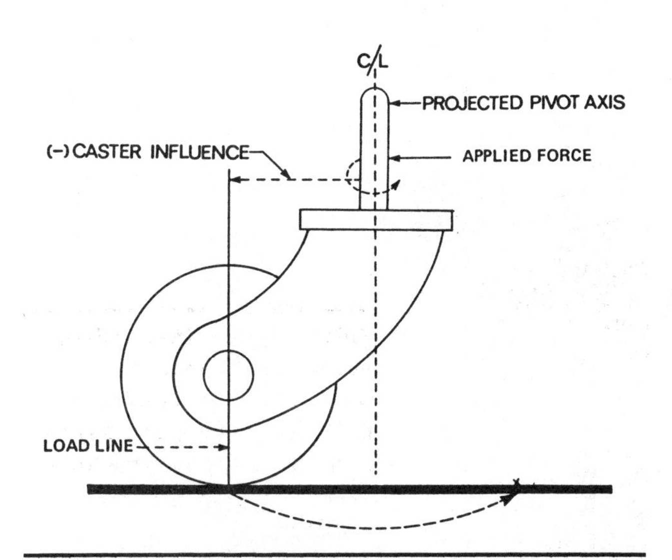

FIGURE 16-11. Rotational Effect on a Shopping Cart Wheel Produced by Negative Caster.

line is tilted forward at the top, the wheel is subject to the influence of negative caster. As you study Figure 16-11, notice that the caster wheel contacts the floor ahead of the projected pivot axis centerline, producing a pushing influence on a wheel. The caster will not remain in its present straight-ahead position; the pivot axis will endeavor to turn 180°, allowing the caster wheel to trail the pivot axis centerline and producing the same influence on a front wheel of a vehicle. Therefore, negative caster does not aid directional control and stability of a vehicle (Figure 16-12).

Principle 5: Excessive positive caster influences the front wheels of a vehicle to shake or shimmy at low speeds. Recall the last time you shopped and

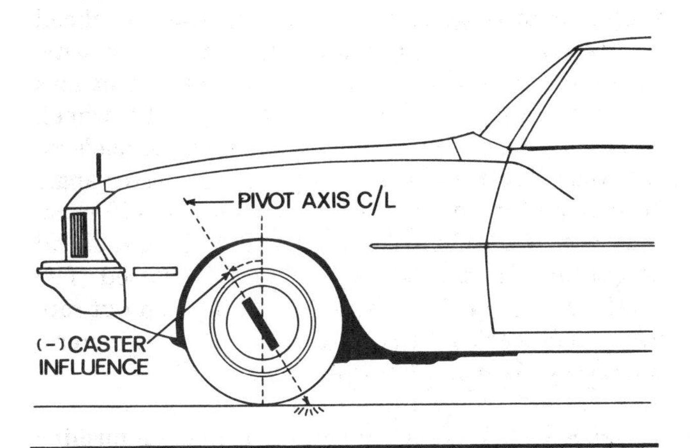

FIGURE 16-12. Pushing Force on a Vehicle's Wheels Produced by Negative Caster.

pushed a shopping cart. Remember that as the front wheels began to rotate, the speed of the cart increased the lateral shaking of the front wheels. The caster wheels were trailing the pivot axis centerline, and the wheels were influenced by the effect of positive caster. Because of suspension design and the influence of excessive positive caster, the front wheels of a vehicle will also shake or shimmy at low speeds. Many vehicles designed with rigid I-beam suspension or front wheel drive are susceptible to this problem. To correct the shake condition of a vehicle's front wheels, manufacturers have installed a device called a *steering damper.* A steering damper is similar to a shock absorber. It connects to the steering linkage to reduce the lateral movement of the steering parts.

Principle 6: Correct negative caster reduces the tendency for the front wheels of a vehicle to shake or shimmy at low speeds. Mr. and Mrs. Adams enjoy the outdoor life and spend considerable time traveling to various recreational areas. They also appreciate the good life and will not go camping unless all the modern conveniences are packed in the trunk of the vehicle. The additional weight in the trunk as well as the weight of the camping trailer influences the directional control and stability of the vehicle. Mr. and Mrs. Adams have noticed the tendency for the steering wheel to oscillate while driving (Figure 16-13). The reason for the steering wheel shimmy is an increase of the positive caster angles, caused by the excess weight's lowering the vehicle's rear end and thus raising the front end. The tendency for the steering wheel to shake has reduced the pleasure of driving the vehicle.

When Mr. and Mrs. Adams returned from the camping trip, they mentioned the problem to the wheel alignment technician. The technician has corrected the problem by aligning the vehicle's front end and has changed the caster angles for both front wheels to slight negative settings (within the manufacturer's specifications). The technician has also installed load leveler shocks in the rear suspension, correcting the oscillation of the steering wheel.

A FACT TO REMEMBER

Negative caster is a pushing influence on a front wheel since the tire's road contact area is in front of the projected pivot axis centerline. This influence reduces the tendency for a wheel to shake or shimmy at low speeds.

FIGURE 16-13. Front End Shimmy. Excess weight in the rear changes the caster angles, producing front end shimmy.

Principle 7: Excessive positive caster increases the influence of road shock on a vehicle. The caster settings on a vehicle are very important. If they are incorrect due to incorrect alignment and excessive loading, steering stability will decrease, increasing driver fatigue.

Figure 16-14 illustrates a wheel with excessive positive caster about to fall into a road depression. Since the pivot axis centerline is ahead of the tread contact area, the results of the tire's contacting the depression will be transmitted through the tire, wheel assembly, suspension parts, frame, and body. A wheel subject to excessive positive caster increases the influence of road shock to a vehicle by providing a direct line through which the force can act.

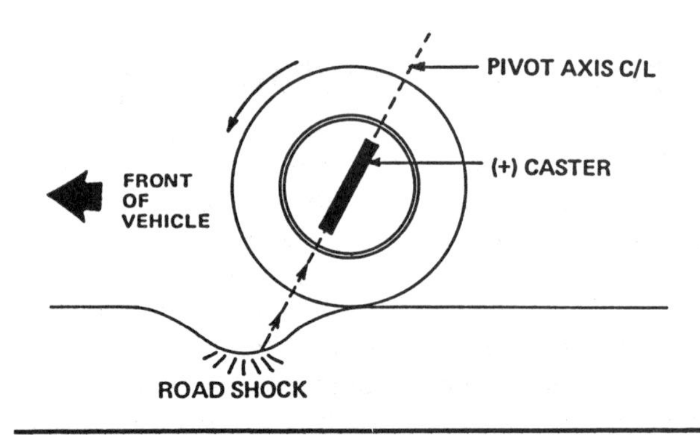

FIGURE 16-14. Effect of Excessive Positive Caster. Road shock is transmitted directly through the suspension.

Principle 8: Correct negative caster reduces the influences of road shock to a vehicle. Although negative caster does not assist the front wheels of a vehicle to return to the straight-ahead position, it does benefit the ride control of the vehicle. Figure 16-15 illustrates a wheel with negative caster about to fall into a road depression. The pivot axis centerline is behind the tread contact area. With negative caster, the transmitted road shock is forced through the pneumatic tire. Hence, the negative caster angle, plus the ability of the tire to flex, reduces the influence of road shock on the body of a vehicle.

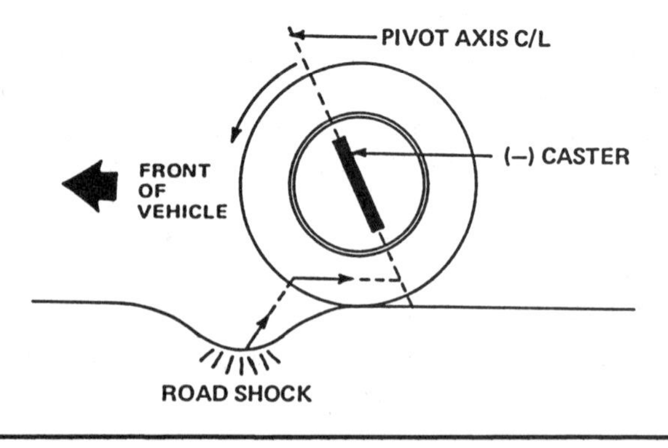

FIGURE 16-15. Effect of Correct Negative Caster. Road shock is deflected.

CASTER ANGLE CHANGE

Rear Vehicle Loading and Spring Sag

When a heavy load is placed in the trunk of a vehicle or when the rear suspension springs sag, the rear suspension height becomes lower. This results in a change of the front wheel caster angle to an increased positive position and will contribute to steering and road handling problems.

Rear Suspension Height

New or stiff rear springs will raise the rear riding height of a vehicle, changing the caster angle in a negative direction. The caster change will be even more pronounced when extended high-lift spring shackles or load leveler shocks are installed at the rear of the vehicle. The experienced wheel alignment technician will measure the rear as well as the front riding height of a vehicle before measuring and setting the various alignment factors. *Note:* All wheel alignment specifications are stated at a definite front and rear manufacturer's suspension height and are found in the shop manual.

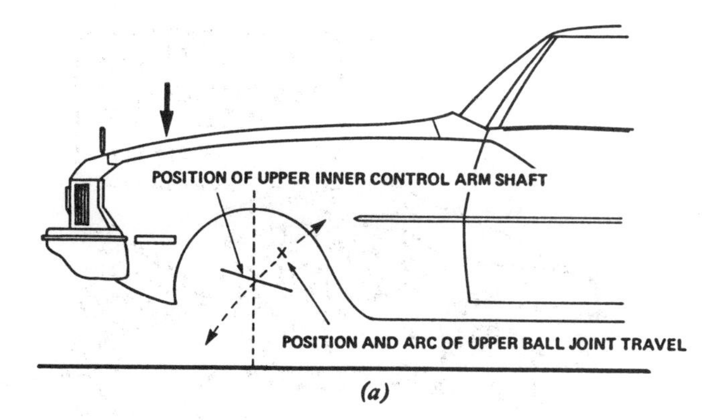

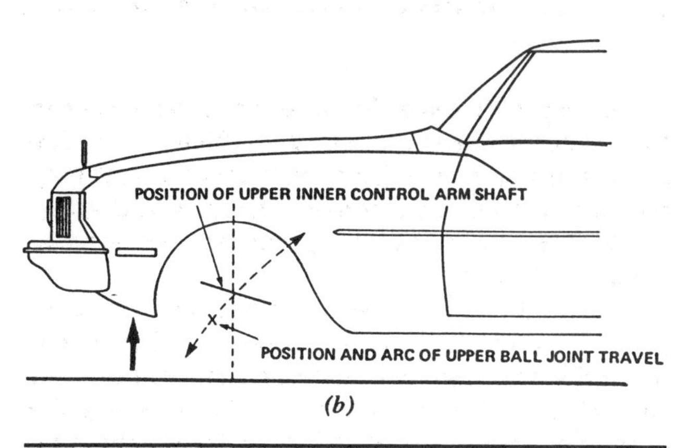

FIGURE 16-16. How Front Suspension Height Affects Caster. (a) Lower vehicle front riding height moves caster positive; (b) raised vehicle front riding height moves caster negative.

Front Suspension Height

Unequal Length Control Arms. Since you are familiar with the unequal length control arm independent suspension, you have no doubt noticed how the manufacturer attaches the upper inner control shafts to the vehicle's frame. The control arm shaft is not mounted parallel to the frame; it is higher at the front and lower at the rear. This design feature causes the caster angle to change constantly under acceleration, braking, and centrifugal forces and road surface deviation. In other words, the top of the steering axis centerline (which determines the caster angle) is constantly moving in a prescribed arc as the vehicle is driven along the road.

MacPherson Strut Suspension. Just as incorrect curb riding height affects a vehicle with unequal length control arms, so does incorrect curb riding height affect a vehicle with MacPherson strut suspension. Some MacPherson strut suspensions are tilted back at the top. When the vehicle is equipped with a modified MacPherson strut design and if the springs sag, the front of the vehicle's body is lowered, and caster becomes less positive. If the front of the vehicle is raised, the caster becomes more positive. You should now be able to understand why a vehicle must have its curb riding height restored prior to aligning the wheels.

FACTS TO REMEMBER

1. Positive caster increases with additional rear loading of a vehicle or rear suspension spring sag.
2. Caster moves negative with increased rear vehicle riding height.
3. Caster moves to a positive position when the front of the vehicle's body height is lowered (see Figure 16-16).
4. Caster moves to a negative position when the front of the vehicle's body height is raised (see Figure 16-16).

CROWNED ROAD SURFACES

Recall that when you began your study of the caster factor, the text illustrated how positive caster influenced a wheel to toe-in and how negative caster influenced a wheel to toe-out. By using correct caster, it is possible to counteract the effects of a crowned road. Let's first examine how this is accomplished by returning to our list of principles of caster.

Principle 9: Correct positive caster may be used to counteract the influence of a crowned road. By varying the amount of the positive caster angle that

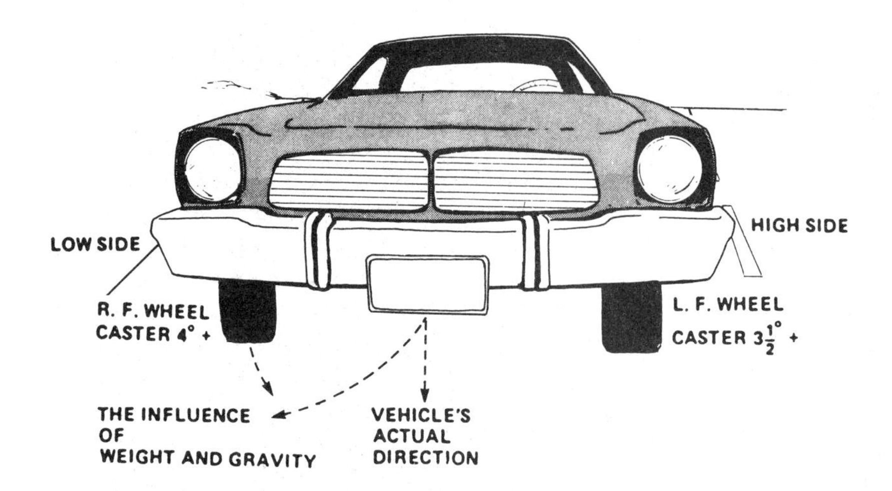

FIGURE 16-17. Using Positive Caster to Offset for Effects of Road Crown. The difference in caster angles influences the right front wheel toward the crowned side of the road.

a front wheel is subject to, it is possible to increase or decrease the influence of the caster angle in a particular wheel. Let's be realistic. Up to now we have been using large caster angles, but no North American vehicle manufacturer subjects a front wheel to 20° or 30° caster angles. Those large figures were used only to illustrate a point. What would be a realistic angle for caster? Figure 16-17 shows the answer. The caster angle for the right front wheel has been set at 4° positive, and the left front wheel has been set at 3 ½° positive. Why the difference? The right front wheel (since it is subject to increased positive caster) is influenced to roll toward the crowned (high) side of the road. This influence on the right front wheel counteracts the influence of gravity that tends to influence the vehicle to steer toward the curb (low) side of the road.

FACTS TO REMEMBER

1. Correct positive caster may be used to counteract the influence of a crowned road.
2. When positive caster is used to counteract the influence of a crowned road:
   - The caster setting must be more positive on the right front wheel and less positive on the left front wheel,
   - The caster setting differential must not exceed 1/2°,
   - The camber angles for both front wheels are set the same and close to 0°.

Principle 10: Correct negative caster may be used to counteract the influence of a crowned road. By varying the amount of the negative caster angle that a front wheel is subject to, it is also possible to increase or decrease the influence of the caster angle on a particular wheel. Figure 16-18 illustrates that the caster angle setting for the left front wheel is 1° negative and for the right front wheel is 1/2° negative. Why the difference? The left front wheel (since it is subject to increased negative caster) is influenced to roll toward the crowned (high) side

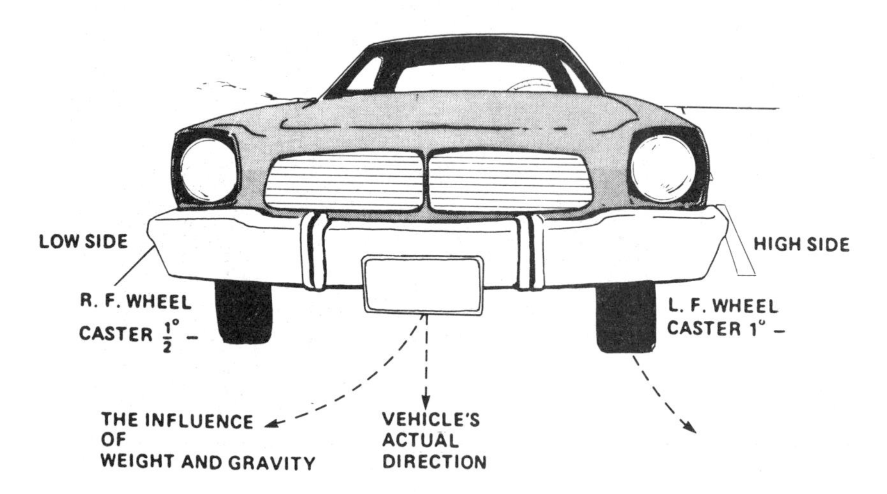

FIGURE 16-18. Using Negative Caster to Offset for Effects of Road Crown. The difference in negative caster angles can be used to counteract the effects of a crowned road.

of the road. This influence on the left front wheel will counteract the influence of gravity that causes the vehicle to steer toward the curb (low) side of the road.

FACTS TO REMEMBER

1. Correct negative caster may be used to counteract the influence of a crowned road.
2. When negative caster is used to counteract the influence of a crowned road:
   - The caster setting must be more negative on the left front wheel and less negative on the right front wheel,
   - The caster setting differential must not exceed 1/2°,
   - The camber angles for both front wheels are set the same and close to 0°.

An Alignment Situation To Be Avoided

The alignment of a vehicle's wheels demands precision workmanship by the technician. *Note:* There is no room for sloppy work. Excessive unequal caster can create a tendency for the vehicle to steer toward the wheel with the least positive caster angle or the most negative caster angle. If there is positive caster at one front wheel and negative caster at the other, watch out. When the brakes are applied, the wheel with negative caster tries to turn outward, pulling the vehicle strongly in that direction. That can be a terrifying experience in tight traffic. When a vehicle's directional control and stability are affected and the vehicle is sent to the alignment shop for correction, it would be an advantage if both the driver and the technician could discuss the problem. Unfortunately, this is not always possible.

CASTER MEASUREMENT

All wheel alignment equipment obtains the caster angle reading by actually measuring the camber angle of a front wheel at two rotational positions. For example, the front wheel is turned 20° from the straight-ahead position; the caster scale is at zero, and the wheel is turned 40° in the opposite direction. The caster angle is actually the difference between these two rotational positions calibrated and shown on a caster scale (Figure 16-19). How the caster angle is changed will be discussed in Chapter 20.

FACTS TO REMEMBER

1. Never attempt to measure the caster angle without first knowing the operating instructions of the equipment.
2. Normally, the best directional control and stabil-

FIGURE 16-19. Determining Caster Angle Using a Caster Scale. (Courtesy of Snap-on Tools of Canada Ltd.)

ity will be obtained by adjusting the caster angle to the preferred manufacturer's specifications.

3. The main purpose of correct positive caster is to aid the directional control and stability of a vehicle.
4. The main purpose of correct negative caster is to assist the ride and improve the handling of a vehicle.

REVIEW TEST

1. Define the terms *negative caster* and *positive caster.*
2. Describe how positive caster assists the front wheels of a vehicle to return to the straight-ahead position.
3. Describe why negative caster does not assist the front wheels of a vehicle to return to the straight-ahead position.
4. Describe how positive caster influences a front wheel to toe-in.
5. Describe how negative caster influences a front wheel to toe-out.
6. List three advantages of correct positive caster.
7. List three disadvantages of incorrect positive caster.
8. List three advantages of correct negative caster.
9. List three disadvantages of incorrect negative caster.
10. Explain how the rear riding height of a vehicle influences the caster angle.
11. Explain how the front riding height of a vehicle influences the caster angle.
12. Explain how correct positive caster is used to counteract the influence of a crowned road.
13. Explain how correct negative caster is used to counteract the influence of a crowned road.
14. Describe how to measure the caster angle.
15. Figure 16-20 illustrates a vehicle being driven on a rural road. Study the settings for the left and right camber and caster angles. Based on what you have learned, state the direction the vehicle will steer. Explain your answer.
16. Figure 16-21 illustrates a vehicle being driven on a rural road. Study the settings for the left and right camber and caster angles. State the direction the vehicle will steer. Explain your answer.

FIGURE 16-20. Illustration for review question 15.

FIGURE 16-21. Illustration for review question 16.

FINAL TEST

This examination is multiple choice. Only one answer will be accepted. Carefully read every statement.

1. Mechanic A says that positive caster is the rearward tilt of the top of the wheel. Mechanic B says that positive caster is the rearward tilt of the top of the steering axis centerline. Who is right? (A) Mechanic A, (B) Mechanic B, (C) Both A and B, (D) Neither A nor B.
2. Mechanic A says that positive caster is measured when the wheel is located in the straight-ahead position. Mechanic B says that caster is not considered adjustable and is calculated in millimeters. Who is right? (A) Mechanic A, (B) Mechanic B, (C) Both A and B, (D) Neither A nor B.
3. Mechanic A says that positive caster influences

a wheel to toe-in. Mechanic B says that negative caster influences a wheel to toe-in. Who is right? (A) Mechanic A, (B) Mechanic B, (C) Both A and B, (D) Neither A nor B.

4. Mechanic A says that positive caster aids the front wheel of a vehicle to return to the straight-ahead position. Mechanic B says that positive caster encourages the vehicle to steer in a straight direction. Who is right? (A) Mechanic A, (B) Mechanic B, (C) Both A and B, (D) Neither A nor B.
5. Mechanic A says that excessive positive caster increases the influence of road shock. Mechanic B says that incorrect positive caster tends to influence a wheel to shimmy at low speeds. Who is right? (A) Mechanic A, (B) Mechanic B, (C) Both A and B, (D) Neither A nor B.
6. Mechanic A says that positive caster may be used to offset for the effects of road crown. Mechanic B says that negative caster may be used to offset for the effects of road crown. Who is right? (A) Mechanic A, (B) Mechanic B, (C) Both A and B, (D) Neither A nor B.
7. Mechanic A says that positive caster has a pushing effect on a wheel. Mechanic B says that positive caster has a trailing effect on a wheel. Who is right? (A) Mechanic A, (B) Mechanic B, (C) Both A and B, (D) Neither A nor B.
8. Mechanic A says that when the rear of a vehicle is raised, the caster angles become less positive or more negative. Mechanic B says that when the front of a vehicle equipped with unequal control arms is raised, the caster angles become less positive. Who is right? (A) Mechanic A, (B) Mechanic B, (C) Both A and B, (D) Neither A nor B.
9. Mechanic A says that when negative caster is designed into the front end, the front wheels are influenced to return to the straight-ahead position through the effects of steering axis inclination. Mechanic B says that when negative caster is designed into the front end and the vehicle is turning left, the left front of the vehicle lifts and the right front drops. Who is right? (A) Mechanic A, (B) Mechanic B, (C) Both A and B, (D) Neither A nor B.
10. Mechanic A says that excessive positive caster causes rapid return of the steering wheel. Mechanic B says that excessive negative caster causes increased road feel. Who is right? (A) Mechanic A, (B) Mechanic B, (C) Both A and B, (D) Neither A nor B.
11. Mechanic A says that when positive caster is used to counteract the effects of road crown, more positive caster is placed on the left front wheel. Mechanic B says that when negative caster is used to counteract the effects of road crown, more negative caster is placed on the right front wheel. Who is right? (A) Mechanic A, (B) Mechanic B, (C) Both A and B, (D) Neither A nor B.
12. Mechanic A says that the caster angle differential should not vary more than 3/4° between the left and right sides. Mechanic B says that it is acceptable if the caster angle settings are: right front wheel, 1/4° positive; left front wheel, 1/4° negative. Who is right? (A) Mechanic A, (B) Mechanic B, (C) Both A and B, (D) Neither A nor B.

Note: The following caster and camber settings pertain to Question 13: Right front wheel caster is +1 1/8° and camber is -1°. Left front wheel caster is -3/4° and camber is +1°.

13. Mechanic A says that the vehicle will steer to the right because of the caster settings. Mechanic B says that the vehicle will steer to the left because of all the settings. Who is right? (A) Mechanic A, (B) Mechanic B, (C) Both A and B, (D) Neither A nor B.

CHAPTER 17

Toe

LEARNING OBJECTIVES

After studying this chapter, you should be able to:

- Explain three purposes of correct toe,
- Illustrate and describe excessive incorrect toe-in tire tread wear,
- Illustrate and describe excessive incorrect toe-out tire tread wear,
- Define *featheredging tire tread wear.*

INTRODUCTION

When you were a child, your parents bought you a brand new pair of shiny black shoes. They looked terrific, and when you ran, your feet were as light as feathers. To draw attention to your new shoes, you started to drag your heels. In a very short time, you caused the heels of the shoes to scuff, and your parents, to say the least, were not very happy.

A scuffed heel is similar to a tire's worn tread. A tire that is subject to an incorrect toe setting will be dragged or scuffed while the wheel is rolling on the surface of the road. This scuffing action greatly decreases the tread life of even a new tire. Before we discuss this factor, do you recall the definition of toe?

PURPOSES OF TOE

Correct toe has three purposes:

- To compensate for slight flexing in the steering linkage mechanism,
- To aid the vehicle's directional control and stability,
- To reduce tire tread scuff wear.

Purpose 1: Correct toe compensates for slight flexing in the steering linkage mechanism. When a vehicle is moving straight ahead, the two front wheels and the two rear wheels, at least in theory, should be parallel when viewed from above. In reality, this is seldom the case. Tie-rods, tie-rod ends,

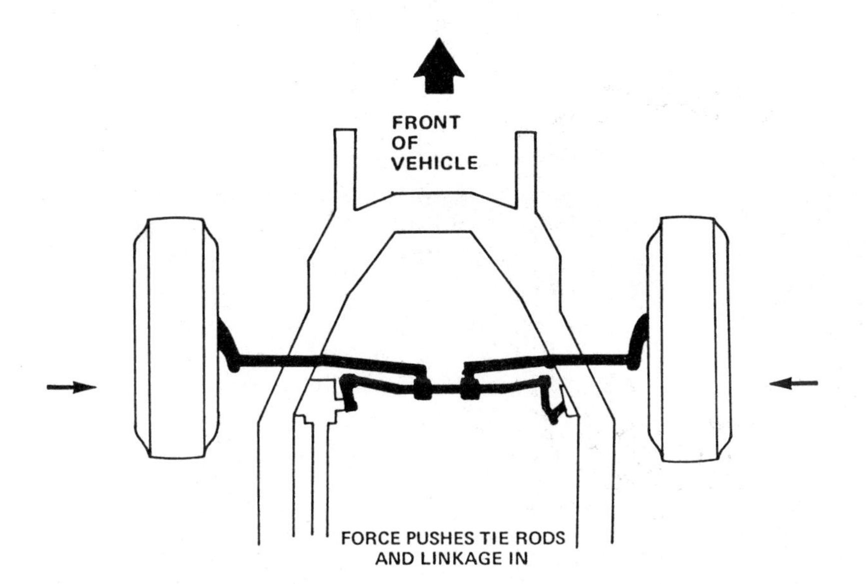

FIGURE 17-1. Toe-In Adjustment to Compensate for Outward Movement of Wheels while Driving. (Courtesy of Chrysler Corporation)

idler arms, and pitman arms, due to their design and construction, allow a slight amount of unavoidable flexing (slack) in the steering linkage mechanism.

To compensate for the flexing in the linkage, manufacturers specify a slight amount of toe, 5 millimeters out to 5 millimeters in (3/16 inch out to 3/16 inch in), when the front wheels are being aligned (Figure 17-1). When the vehicle is being driven, the front wheels (rear wheel drive vehicle) tend to move out, compressing the tie-rods. If the vehicle is front wheel drive, the front wheels tend to move inward when torque is applied, also compressing the tie-rods. The compression on the linkage mechanism moves the running toe measurement to approximately zero.

The term *approximately zero* has been used because it is impossible to set a toe specification that will be exactly right for all driving situations. Road surfaces (wet or dry), road surface temperatures, types of tires, the gross weight of the vehicle, and its speed are variable toe-changing factors. Also, a steering linkage system that has been subject to considerable road use will be looser than a vehicle with new parts. It doesn't take too much flexing in steering linkage parts to produce toe-out. Actually, slight toe-out will not wear the tread of a tire any more than an equal amount of toe-in, however, a *toe-out running condition* (that is, toe-out while the car is being driven) will contribute to some loss of the vehicle's directional control and stability.

Purpose 2: Correct toe is used to aid in establishing a straight, definite course, thus aiding the vehicle's directional control and stability. Driving a vehicle while its directional control is unstable is mentally and physically exhausting. Correct toe can offset a number of problems that produce unstable steering: worn ball joints, incorrect camber and caster settings, excessive play in the steering gearbox, loose or worn steering parts, and even defective tires.

Purpose 3: Correct toe reduces tire tread scuff wear. For some years, wheel alignment technicians believed that a cambered wheel had a strong effect on the toe of a wheel, however, vehicle and tire manufacturers have proved that the old theory is somewhat false. The two alignment factors are not that dependent on one another. *Note:* Modern automobiles are not designed with excessive positive camber; also, modern tires are low in sidewall profile height and hence smaller in diameter. A tire that is subject to excessive incorrect toe-in or toe-out will scuff the tread surface. Figure 17-2 illustrates how a tire subject to a toe setting error of 3 millimeters (1/8 inch) is equivalent to dragging the tire crossways 853 centimeters (28 feet) for each 1.6 kilometers (1 mile) the vehicle is driven.

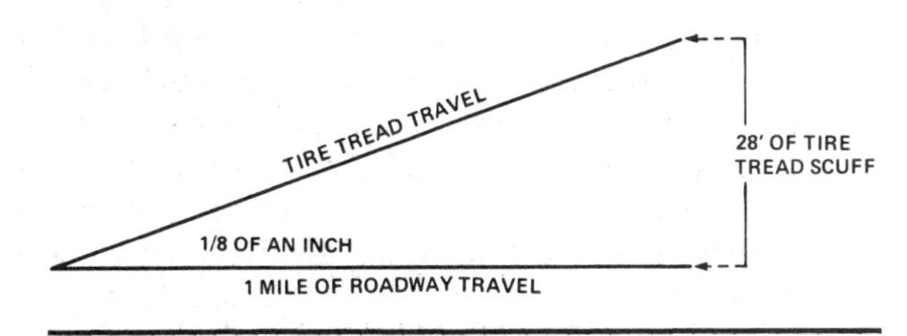

FIGURE 17-2. Diagram of Tire Wear Due to Toe-Setting Error. Only 3 millimeters (1/8 inch) out-of-adjustment toe will cause 853 centimeters (28 feet) of extra wear per mile.

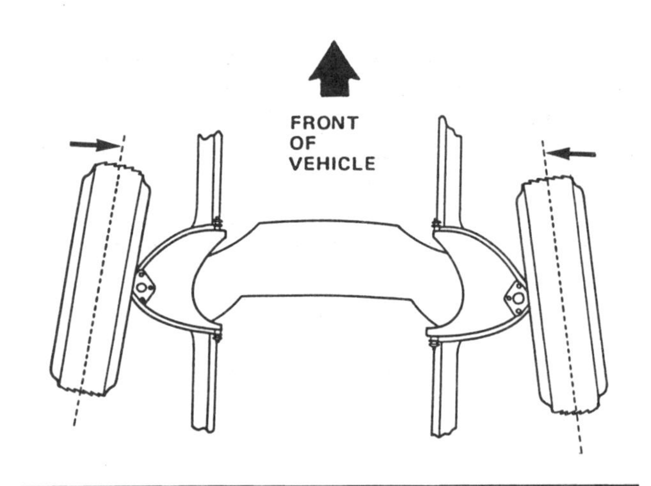

FIGURE 17-3. Tires Scuffing on Road Surface Due to Incorrect Toe-In.

TREAD WEAR AND INCORRECT TOE

Tire tread wear that has been caused by incorrect toe is distinctive and cannot be mistaken for any other type of tread wear pattern. Figures 17-3 and 17-4 illustrate the causes and the results of incorrect toe-in on the tires' tread surface.

Cause and Result of Incorrect Toe-In

High-speed driving, worn steering linkage parts, and sagging springs cause toe-in change. When toe-in is severe, the tire tread surfaces wear unevenly because the treads contact the pavement at an angle (Figure 17-3).

Incorrect toe-in causes each tread rib to wear at its outside edge, while the inner edge becomes sharp and ragged (Figure 17-4). This condition can be seen easily, or it can be felt by passing the hand across the tread face from the inside to the outside of the tread surface.

Cause and Result of Incorrect Toe-Out

Incorrect toe-out can result from the same causes as incorrect toe-in. When an incorrect toe-out condition exists, tread wear is reversed leaving a sharp featheredge pattern on the outside of the tread ribs (Figure 17-5).

Incorrect toe-out causes each tread rib to wear at its inside edge, while the outside edge becomes sharp and ragged. This condition can also be seen or felt by passing the hand across the tread face from the outside to the inside of the tread surface (Figure 17-6).

FACTS TO REMEMBER

1. The tread of a tire that is subject to excessive incorrect toe will indicate tread wear in less than 160 kilometers (100 miles) of driving. The tread wear that results is called *featheredging.*
2. Replacing a tire that indicates signs of toe tread wear does not correct the problem. Toe tread wear is only an indication of existing problems in the vehicle's steering or suspension systems. The condition that caused incorrect toe must be corrected before installing a replacement tire.

ADJUSTING TOE

After the wheel alignment technician has analyzed the operating condition of the steering and suspension system and has replaced any defective parts, the camber, caster, and toe factors are then adjusted to the manufacturer's specifications. It is important to remember that when the camber and caster factors are changed, the toe settings of the front wheels are automatically changed because of the geometric design and relationship of the parts in the suspension system. For example, let's assume that it is

FIGURE 17-4. Tread Wear Pattern Caused by Incorrect Toe-In. (Courtesy of Snap-on Tools of Canada Ltd.)

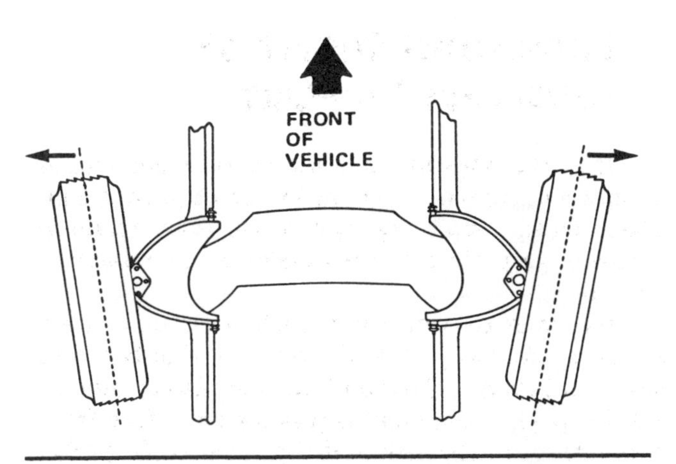

FIGURE 17-5. **Tires Scuffing on Road Surface Due to Incorrect Toe-Out.** Incorrect toe-out causes the tires to scuff on the road surface.

necessary to move the caster from 1° negative to 1° positive. That means that the top of the steering axis centerline must be moved toward the rear of the vehicle. Attached to the bottom of the steering knuckle is a steering arm, which is attached to the tie-rod. When the top of the axis line is moved back, the point at which the steering arm attaches to the tie-rod is also moved. Therefore, the tie-rod is no longer parallel with the lower control arm, forcing the front of the wheel to move to a toe-out position. Figure 9-22 will help you understand how toe is changed when caster and camber are altered. If the toe factor only is adjusted, the camber or caster settings will not change. Toe is measured when the front wheels are in the straight-ahead position. Toe is adjusted by rotating the threaded tie-rod adjusting sleeves that are part of the tie-rods (Figure 17-7).

FACTS TO REMEMBER

1. When camber and caster are adjusted, the toe measurement changes.
2. The toe factor is not used to counteract the influence of a crowned road.
3. Incorrect toe is a major cause of tire tread scuff wear.

REVIEW TEST

1. List the three purposes of toe.
2. Describe how a change to the caster and or camber alters the toe settings.
3. Define the term *featheredging.*
4. Sketch a tire tread surface that has worn due to an excessive toe-in condition. Label the outside and inside portions of the tire.
5. Prepare a list of possible causes in the steering and suspension system that would result in toe-in or toe-out tread wear.
6. Explain why the toe setting changes as the vehicle's speed increases.
7. Briefly describe how toe is adjusted.

FIGURE 17-6. **Featheredge Tread Wear Pattern Caused by Incorrect Toe-Out.** (Courtesy of Snap-on Tools of Canada Ltd.)

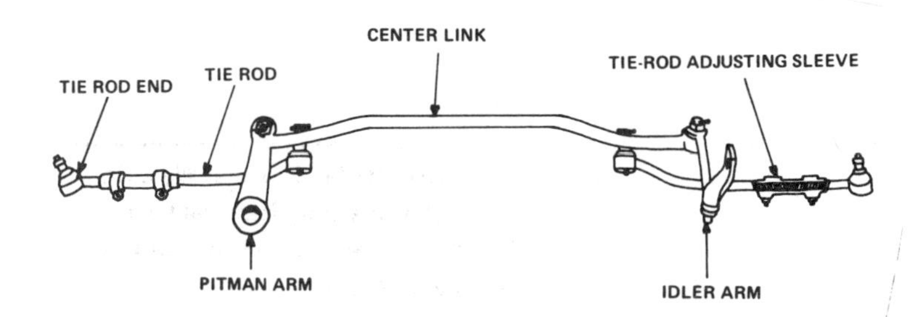

FIGURE 17-7. **Tie-Rod Adjusting Sleeves Used to Change Toe Setting.**

FINAL TEST

This examination is multiple choice. Only one answer will be accepted. Carefully read every statement.

1. Mechanic A says that the main purpose of toe is to compensate for worn parts. Mechanic B says that the main purpose of toe is to return the steering wheel to the straight-ahead position. Who is right? (A) Mechanic A, (B) Mechanic B, (C) Both A and B, (D) Neither A nor B.
2. Mechanic A says that the main purpose of toe is to assist the vehicle's directional control and stability. Mechanic B says that the main purpose of toe is to offset the influence of a crowned road. Who is right? (A) Mechanic A, (B) Mechanic B, (C) Both A and B, (D) Neither A nor B.
3. Mechanic A says that toe is measured in degrees. Mechanic B says that incorrect toe is the major cause of tread wear. Who is right? (A) Mechanic A, (B) Mechanic B, (C) Both A and B, (D) Neither A nor B.
4. Mechanic A says that toe is never adjusted before caster and/or camber. Mechanic B says that correct toe measurements are calculated at axle height. Who is right? (A) Mechanic A, (B) Mechanic B, (C) Both A and B, (D) Neither A nor B.
5. Mechanic A says that toe and camber are measured when the wheels are positioned straight ahead. Mechanic B says that toe is altered by adjusting the tie-rod sleeves. Who is right? (A) Mechanic A, (B) Mechanic B, (C) Both A and B, (D) Neither A nor B.
6. Mechanic A says that the front wheels of a rear wheel drive vehicle tend to toe-in as the vehicle's speed increases. Mechanic B says that if the feathering is on the outside of tread segments, the cause of the problem is excessive toe-out. Who is right? (A) Mechanic A, (B) Mechanic B, (C) Both A and B, (D) Neither A nor B.

Kingpin or Ball Joint Inclination

LEARNING OBJECTIVES

After studying this chapter, you should be able to:

- Explain three purposes of the kingpin or ball joint inclination factor,
- Explain the meanings of *scrub radius area*, *point of intersection*, and *included angle*,
- Explain how the kingpin or ball joint inclination angle is measured and the importance of the result of this measurement.

INTRODUCTION

Kingpin or ball joint inclination is a major contributor to a vehicle's directional control and stability. Yet its purposes and benefits in aiding the steering control of a vehicle are the least known by wheel alignment technicians. Appropriately, this important front end wheel alignment factor could be called the forgotten angle.

Do you recall the definition for kingpin or ball joint inclination? Let's review some of the points that have been discussed about this factor. You were introduced to the term and its definition in Chapter 14. Do you remember the association of the terms *kingpin inclination*, *ball joint inclination*, and *steering axis inclination*? These three terms have a common definition. When we speak of them, we are referring to the inward tilt of the top of the kingpin or ball joint pivot axis centerline through that particular steering-suspension part. *Note:* The top of that unseen centerline is always tilted inward at the top as compared to its bottom position when viewed from the front of a vehicle (Figure 18-1).

Do you recall the first time you drove an automobile? You drove along the road and turned a corner, and after the turn, the vehicle's front wheels

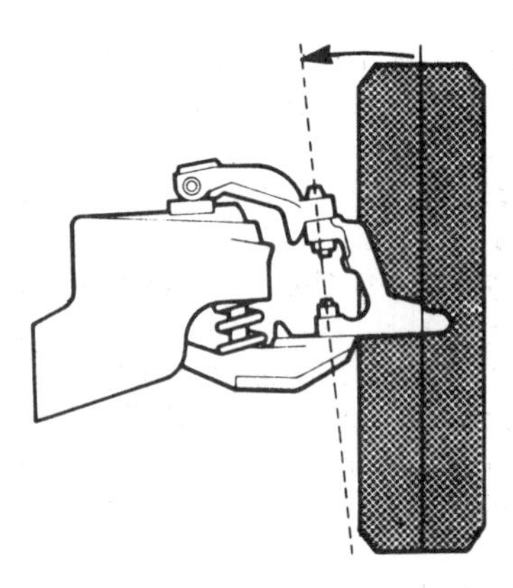

FIGURE 18-1. Inward Tilt of Kingpin or Ball Joint Pivot Axis Centerline. (Courtesy of Ford Motor Company)

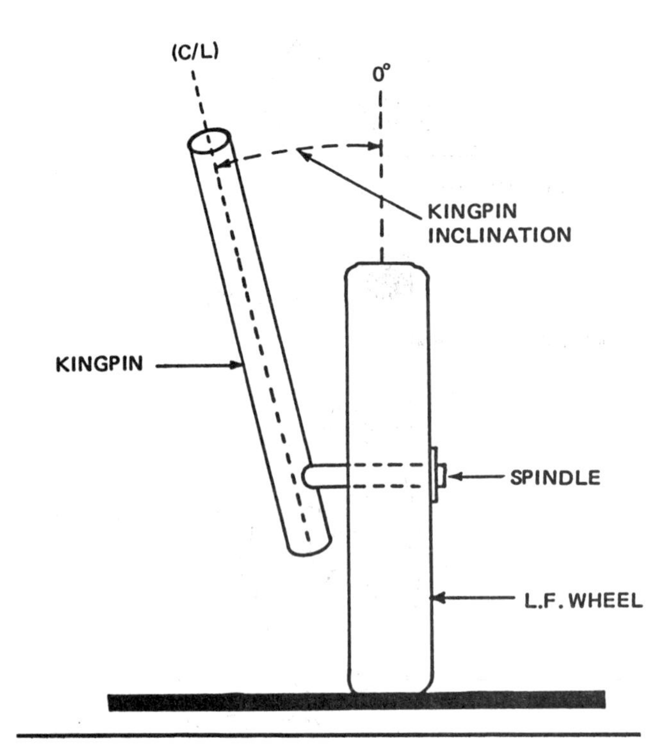

FIGURE 18-2. Front View of Kingpin Inclination.

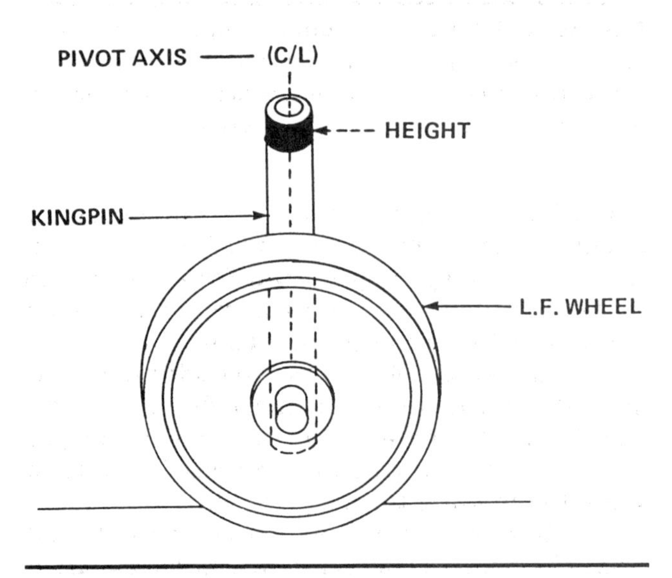

FIGURE 18-3. Side View of Kingpin Inclination.

and the steering wheel returned to the straight-ahead driving position. The reason the steering wheel re-centered was due primarily to the influence of king-pin or ball joint inclination. To understand how kingpin or ball joint inclination actually functions, let's study the three purposes of this factor.

PURPOSES OF KINGPIN OR BALL JOINT INCLINATION

The three purposes of kingpin or ball joint inclination are as follows:

1. Kingpin or ball joint inclination is a major factor in increasing a vehicle's directional control by aiding the front wheels to return automatically to their straight-ahead position.
2. Kingpin or ball joint inclination reduces the effort required to turn the front wheels of a vehicle, increasing the service life of the vehicle's steering and suspension systems.
3. Correct kingpin or ball joint inclination reduces the need for a wheel to be subject to excessive positive camber.

Returning Wheels To Straight-Ahead Position

Figure 18-2 illustrates a front view of a vehicle's left front wheel, spindle (axle), and kingpin. The Kingpin inclination angle has been exaggerated to assist you in understanding the first purpose of this important front end factor. The next diagram, Figure 18-3, illustrates a side view of the left front wheel and kingpin in the straight-ahead position. Notice the indicated height of the top of the kingpin.

Figure 18-4 shows a front view of the left front wheel in a left turn position. Note what has happened to the kingpin's height; the top of the kingpin has lifted. The reason that the kingpin has lifted is the inward tilt of the top of the kingpin. The weight

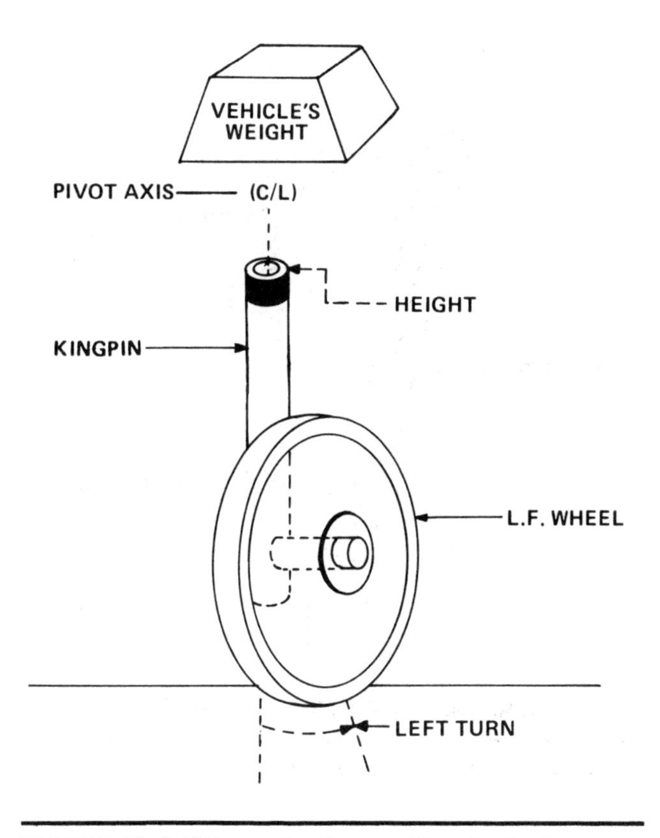

FIGURE 18-4. Effects of a Turn on Kingpin. In the course of a turn, the kingpin is raised; when the turn is completed, the weight of the vehicle forces the kingpin down and returns the wheels to the straight-ahead position.

that a front wheel supports is directed through the steering knuckle and kingpin assembly. The weight that the assembly supports will force the kingpin to rotate to its lowest position, returning the left front wheel to the straight-ahead position.

The same principle applies to the left front wheel in a right turn position. Once again, the kingpin lifts, and the weight of the vehicle will force the kingpin to rotate to its lowest position, moving the left front wheel to the straight-ahead position.

To this point, we have discussed only the left front wheel of a vehicle. Since a vehicle has two front wheels, let's discover what happens to the opposite wheel by performing the following assignment:

1. Obtain a hydraulic floor jack. Position it under the center of an automobile's front crossmember.
2. Turn the steering wheel and position the front wheels straight ahead.
3. Raise the front of the vehicle so that the tread of the tires is .25 centimeter (1/2 inch) above the surface of the garage floor. When the front wheels are straight ahead, the left and right front axles will be almost horizontal.
4. Turn the steering wheel of the vehicle to an extreme rotational position.
5. As you view the distance between the treads of the tires and the surface of the garage floor, you will observe that the distance has decreased. In all probability, the face of the treads may even be compressed against the surface of the garage floor.

Let's consider why this has happened. Rotating the steering wheel caused the spindles and the front wheels to move closer to the floor since the ends of the spindles move within arcs. The arcs are created by the inward tilt of the top of the ball joint pivot axis centerlines (Figure 18-5). It is obvious that the wheels of automobiles and trucks are firmly supported by the hard surface of the road. When the front wheels of a vehicle are turned to the left or right, the wheels cannot move lower than the road surface. Therefore, the suspension and the body-chassis of a vehicle must lift.

We now know what happens when the front wheels have been turned from their straight-ahead position. The kingpin or ball joint inclination factor, aided by vehicle weight, plus correct alignment, stable tire treads, and a wide track, contributes to the directional control and stability of a vehicle by resisting any force that may tend to turn the front wheels away from the straight-ahead position.

Reducing Turning Effort

Figure 18-6 illustrates the second principle related to the kingpin or ball joint inclination factor. Imagine two 454 kilogram (1000 pound) weights resting on two supports, A and B. You are required to rotate the two masses of weight and their supports. It is obvious that support A will require less physical effort to rotate than support B. You may prove this by placing the sharpened point of a lead pencil on a hard surface like a desk. Rotate the pencil. Now, invert the pencil so that the eraser is on the surface of the desk. Again, rotate the pencil. There is a noticeable difference between the two turning efforts. Manufacturers have adapted this principle into the front end design of a vehicle's steering and suspension systems.

This principle is easy to understand when you consider what will happen if the pivot axis center-

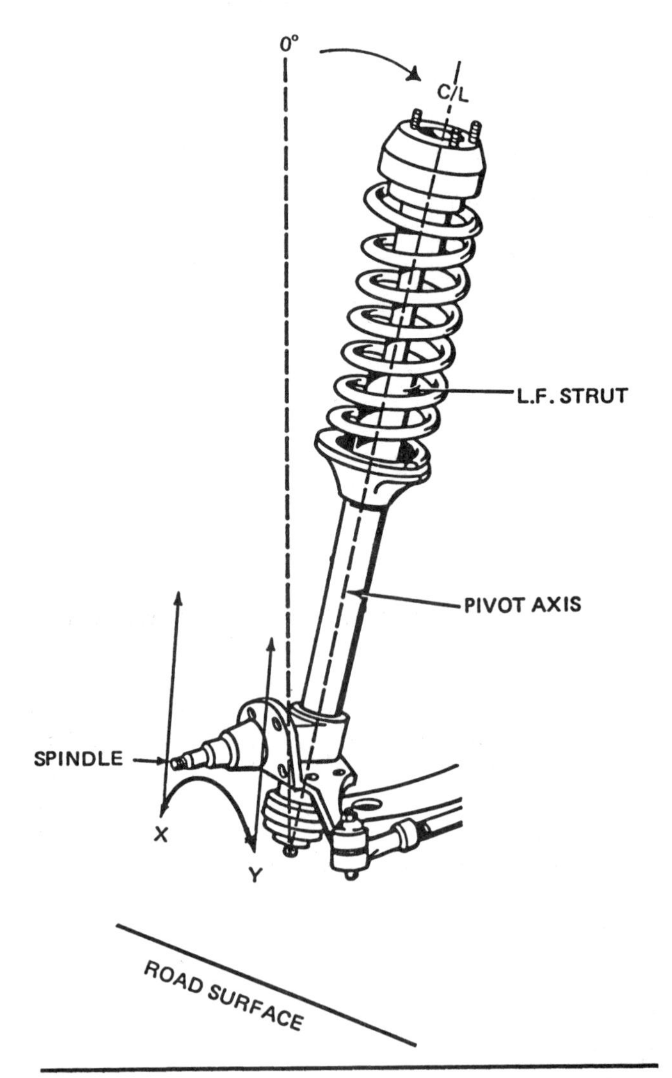

FIGURE 18-5. Effects of Steering Axis Inclination. Since the ball joint pivot axis centerline tilts inward when the spindle rotates to positions X and Y, the suspension and body-chassis must lift. (Courtesy of Moog-Canada Automotive, Ltd.)

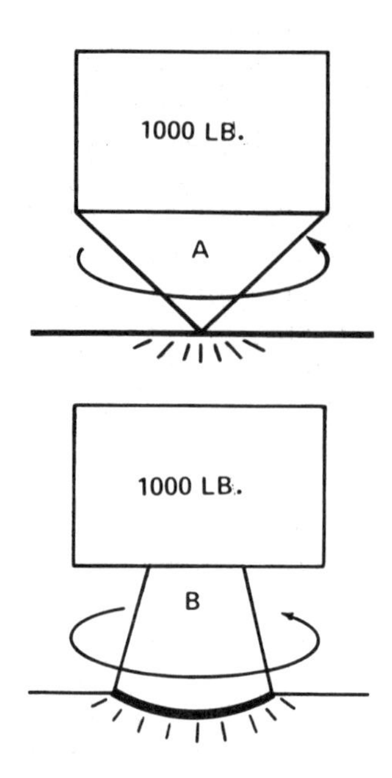

FIGURE 18-6. Illustration for Determining which Weight Will Rotate with the Least Effort: A or B?

line is not inclined inward at the top. If the pivot axis centerline were vertical (0°), the projected extension of the centerline would intersect the road surface some distance from the actual tire tread and road surface contact area. Every time the wheel turned left or right, the tire tread would scrub against the road surface. In other words, it would be forced to rotate around a pivot axis centerline considerably outside the tire tread and road surface contact area (Figure 18-7).

If a vehicle's steering and suspension systems were constructed in this way, you could readily understand some of the disadvantages of the design.

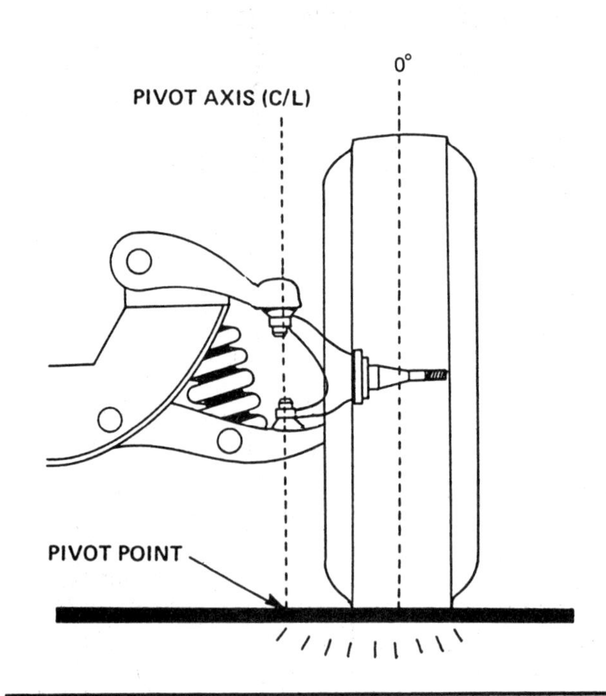

FIGURE 18-7. Effect of 0° Inclination. With 0° inclination, the tire's pivot point is far from the tread surface. (Courtesy of Hunter Engineering Company)

1. There would be excessive overload (stress) on all suspension parts—for example, ball joints, steering knuckle, bearings, and wheels. In time, the spindle would probably shear or break due to the stress.

2. Increased effort would be required to turn the front wheel of a vehicle, thereby increasing the stress on the steering mechanism.
3. Decreased tire tread life would result due to excessive road surface scrubbing.
4. Excessive road shock and kickback at the steering wheel would be experienced by the driver because of the extended leverage action on the spindles (the distance between the 0° position of camber and the parallel 0° line of the pivot axis).

Figure 18-8 illustrates the inward tilt of the top of the pivot axis centerline. The projected extension of the centerline is close to the tire tread and road surface contact area. The effort required to turn the front wheels of a vehicle is greatly reduced by locating the projected pivot axis centerline near the centerline of the wheel at the road surface contact area.

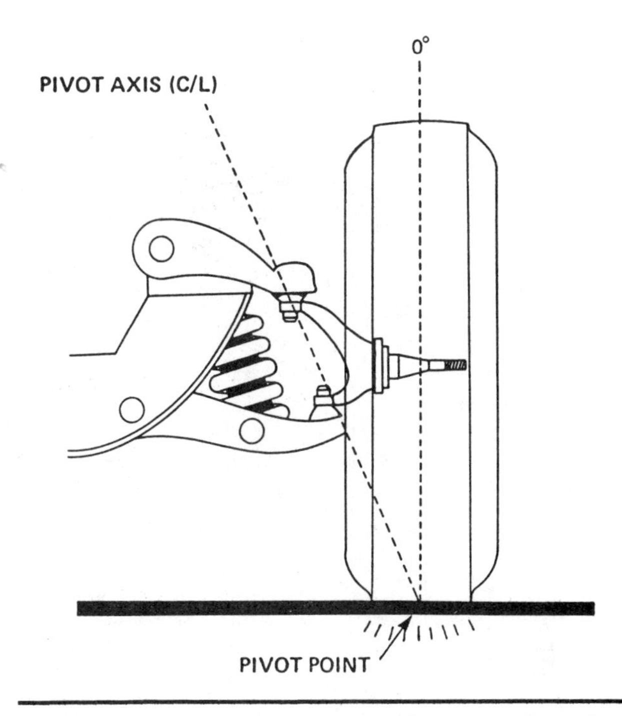

FIGURE 18-8. Inward Tilt of Top of Pivot Axis Centerline. The load projected through the pivot axis centerline and the weight supported by the wheel meet at a central point on the road surface, reducing wear and facilitating turning. (Courtesy of Hunter Engineering Company)

Reducing Need for Excessive Positive Camber

If manufacturers were to design the pivot axis centerline at the 0° position and the camber of a front wheel at a 0° setting, you now know the potential disastrous results of that design. To compensate for a vertical (0°) pivot axis centerline, manufacturers could design the camber of a front wheel excessively positive—for example, 15° positive; however, this would also produce a problem, as Figure 18-9 shows.

Chapter 15 discussed the problem of a tire that was subject to excessive positive camber. Do you remember what the text stated? The outside portion of the tire tread surface, being smaller in circumference, must endeavor to rotate faster than the inside portion of the tire tread. The outside part of the tread would slip on the surface of the road, producing tire tread scuff wear. The effort required to turn the wheels left or right when the vehicle would be moving at a low speed would be reduced, but the tires would not last long.

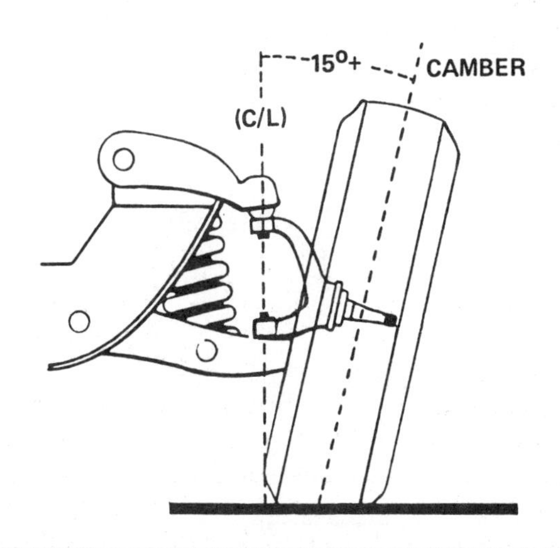

FIGURE 18-9. Excessively Positive Front Wheel Camber. Excess positive camber would offset a 0° ball joint inclination, but other problems would arise. (Courtesy of Hunter Engineering Company)

THREE RELATED TERMS

So far, you have been introduced to three important purposes of kingpin or ball joint inclination; however, that's only half of the story. To understand completely how this factor controls and in-

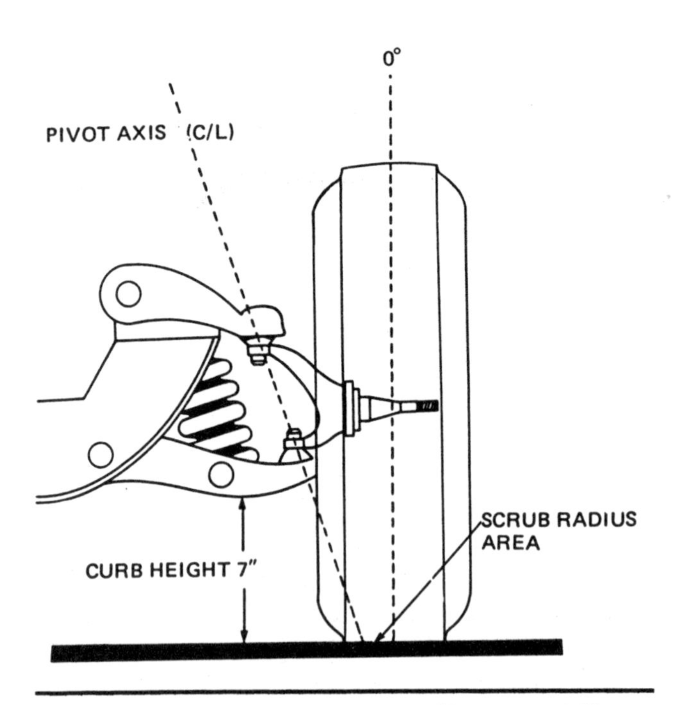

FIGURE 18-10. Scrub Radius Area. (Courtesy of Hunter Engineering Company)

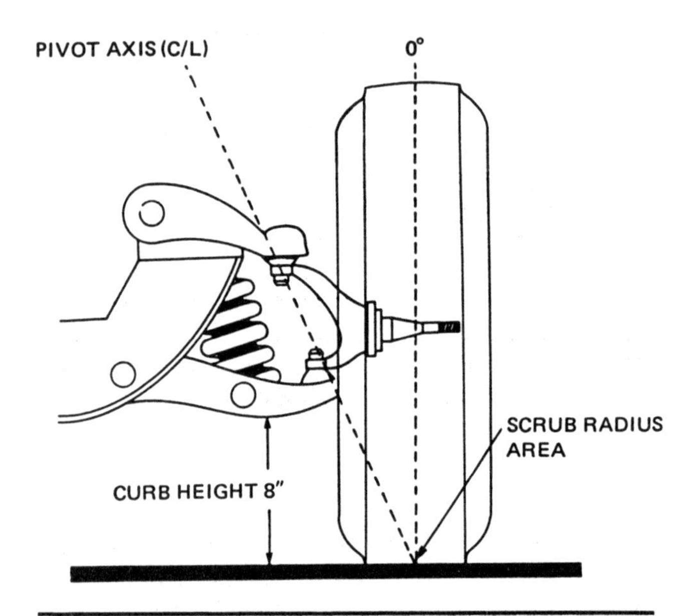

FIGURE 18-11. Effect of Excess Tire Size on Scrub Radius Area. Tires larger than those specified by the manufacturer will change the scrub radius area. Note how the riding height and scrub radius area have changed from Figure 18-10. (Courtesy of Hunter Engineering Company)

fluences a revolving wheel, we will now introduce three related terms and explain their meanings.

Scrub Radius Area

By modifying the design of a vehicle's steering and suspension components, it is possible to increase or decrease the steering effort required by a driver to turn the front wheels of a vehicle. In reality, a slight turning load or resistance is essential to the driver since he or she must sense how to control a vehicle when it is moving. This sense is partially accomplished through the **scrub radius area,** the distance between the extended centerline of the pivot axis and the centerline of the tire at the point where the tire touches the road. (See Figure 18-10.) A good way to remember this term is to recall Figure 18-6. It compared two frictional areas.

Ms. Allen has recently purchased a new set of tires for her vehicle. The tires are slightly larger in diameter (higher) than those specified by the manufacturer. Figure 18-11 illustrates that the height of the vehicle has been raised. Notice how the scrub radius area has changed from Figure 18-10.

How would an increase in tire diameter and curb riding height affect a vehicle? Ms. Allen has noticed that although she has new tires on her vehicle, it has lost some of its directional control and also requires less effort to maneuver the vehicle at low speeds. This has been caused by decreasing the scrub radius area at the two front wheels.

The scrub radius area also will be changed by installing reverse rims on the front wheels of a vehicle. It is not uncommon to see a modern van proceeding along the road equipped with extra-wide tires or reverse rims. These changes will result in moving the centerline of the wheels closer to the outside of the vehicle's body, thereby increasing the scrub radius area. Additional load will be placed on all suspension parts and to some extent will change the camber angles of the vehicle's two front wheels. Installing reverse wheel rims on a vehicle is not recommended by vehicle manufacturers.

Point of Intersection

When a vehicle's front suspension system is on the drawing board, the engineer designs the front wheels with a predictable tendency. You will recall that in Chapter 15, the text illustrated that a wheel rolls in the direction of its tilt. If a wheel is subject to positive camber, it will roll outward. However, there is an exception to the rule, and it is governed by the location of the **point of intersection**, the apex of the extended centerline of the pivot axis and the extended centerline of a front wheel (Figure 18-12).

We have shown how the scrub radius area is altered by changing tire diameter. When a larger-diameter tire is installed on a front wheel, the height is increased from the road surface to the midpoint of the spindle, causing the apex of the point of intersection also to be raised. How does a change in height affect a front wheel?

There are two important points to bear in mind:

1. When a front wheel is subject to positive camber and the point of intersection is located below the surface of the road, the wheel will be influenced to roll outward.
2. When a front wheel is subject to positive camber and the point of intersection is located above the surface of the road, the wheel will be influenced to roll inward.

To understand these important facts, let's examine some diagrams. Figure 18-13 illustrates two sticks that are supporting weight. The sticks are hinged near their base, and the weight that they are supporting is endeavoring to move them outward at the

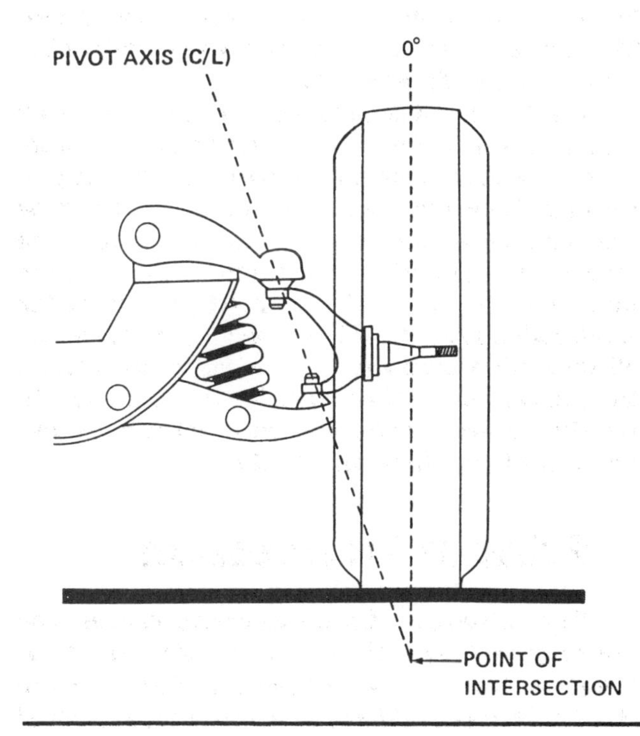

FIGURE 18-12. Point of Intersection Below Road Surface. (Courtesy of Hunter Engineering Company)

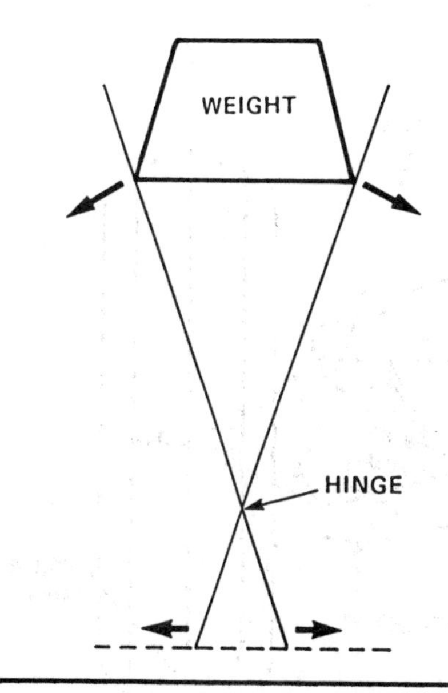

FIGURE 18-13. Illustration of Projected Force and Its Results. The weight moves the sticks in opposite directions.

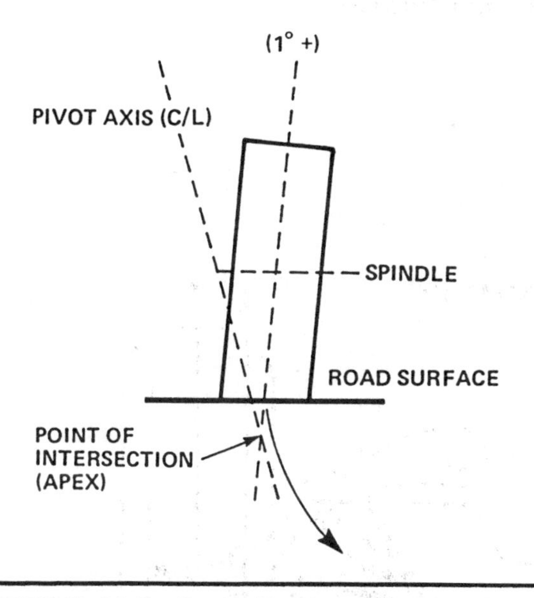

FIGURE 18-14. Predictable Movement of a Wheel Outward. With the point of intersection below the road, wheels roll outward.

top. Since they are hinged above their base, what is the direction of movement at the base? You're right; they move outward, in opposite directions.

Figure 18-14 illustrates the left front wheel of a vehicle that is subject to positive camber. The point of intersection is located below the surface of the road. Since the apex is below the road surface, the projected line of force through the wheel governed

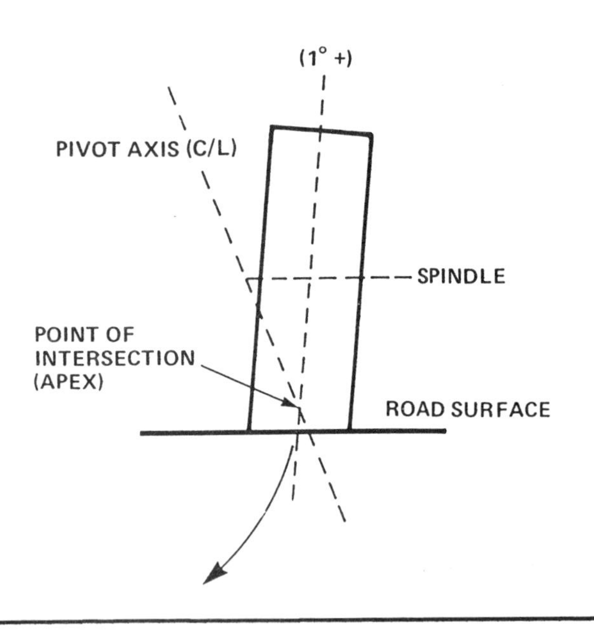

FIGURE 18-15. Predictable Movement of a Wheel Inward. When the point of intersection falls above the road, the wheels roll inward.

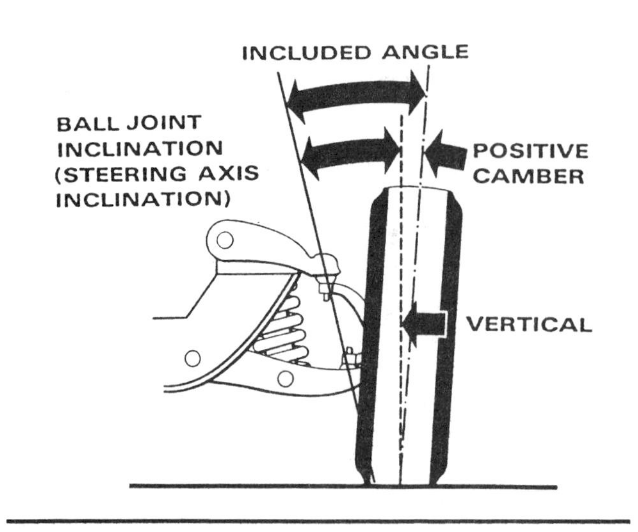

FIGURE 18-16. Included Angle. (Courtesy of Hunter Engineering Company)

by the location of the point of intersection will force the wheel to roll outward.

Figure 18-15 illustrates the left front wheel of the same vehicle. The point of intersection is located above the surface of the road due to an increase in the tire diameter. Since the apex is above the road surface, the projected line of force through the wheel governed by the location of the point of intersection will force the wheel to roll inward even though the wheel is subject to positive camber.

How will all this affect the directional control and stability of your vehicle? Let's assume that you purchased at a bargain a set of oversized-diameter tires for your vehicle. You have now had the opportunity to road test your vehicle, and to your displeasure, it has lost some of its stable steering characteristics. What has happened is that by increasing tire size, you have changed the point of intersection to a dangerous position. When the vehicle is being driven, the heights of the front wheel spindles are constantly changing. One moment they are low, and the next moment they are high because of bumps and other road surface deviations. In other words, the point of intersection is below the road and then above the road, causing the front wheels to roll outward and then inward. Your vehicle now has unstable directional control and very little stability.

How can this problem be corrected, since it is not the fault of incorrect wheel alignment? There is only one way: Replace the vehicle's tires with the correct size specified by the manufacturer, and in the future be careful of these special tire deals.

Included Angle

The third term to be introduced is **included angle**, defined as the sum of the kingpin or ball joint inclination angle and the camber angle of a vehicle's front wheel (Figure 18-16). Why is it necessary to know the importance of this particular angle? Since the centerline of a vehicle's front wheel is seldom at the 0° position, it is necessary to understand what the included angle represents to determine the kingpin or ball joint inclination angle within the vehicle's suspension system.

MEASURING THE KINGPIN OR BALL JOINT INCLINATION ANGLE

Since the kingpin or ball joint inclination angle is part of the included angle, let's discuss why and how this factor is measured. The kingpin or ball joint inclination is used to detect a bent steering knuckle. It was pointed out earlier that the inclination angle cannot be adjusted since it is designed into the steering knuckle when the part is manufactured. Generally, if the camber of a front wheel can be adjusted to manufacturer's specifications, it can be assumed that the steering knuckle is not damaged. Kingpin or ball joint inclination, similar to the caster angle,

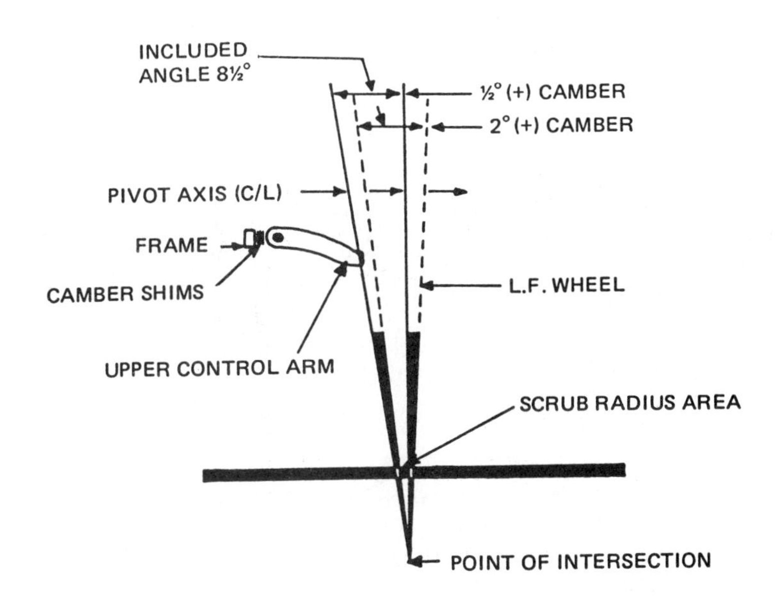

FIGURE 18-17. Effect of Adjusting Camber. Adjusting the camber angle will not change the kingpin or ball joint inclination angle.

is measured by determining the total camber angle change produced by turning the front wheels through an included angle for which the equipment is calibrated. If it is determined that the steering knuckle is damaged (bent), a new steering knuckle must be installed and camber, caster, and toe readjusted. The following procedure tells you how to measure the kingpin or ball joint inclination angle.

Step-by-Step Procedure

1. Determine the camber angle of the front wheel when the wheels have been positioned straight ahead. Record the reading.
2. Attach the hub gauge to the hub of the wheel. *Note:* Manufacturers recommend that when using the hub gauge, a brake pedal jack be installed to lock the wheel and brake mechanism as a single unit.
3. Generally, the wheel is turned in 20°.
4. Set the kingpin or ball joint inclination scale to zero.
5. Turn the wheel 40° in the opposite direction.
6. Record the number of degrees shown on the scale.
7. If the camber that you recorded in step 1 was positive, add the positive figure to the reading obtained on the kingpin or ball joint inclination scale of the gauge. If the camber that you recorded in step 1 was negative, subtract the negative figure from the reading obtained on the kingpin or ball joint inclination scale of the gauge.

When using a hub gauge (see Figure 16-19), remember that the gauge is capable only of calculating the angle from the center of the steering axis centerline to a true vertical position. If the camber were 0°, the steering axis inclination gauge would then calculate the actual steering axis inclination angle.

8. Compare the result with the manufacturer's alignment specification chart. Normally, you are allowed 1° variation from the specification. If the variation is greater than 1°, it denotes a distorted steering knuckle. *Note:* Always verify your analysis and conclusions before you replace the part.

An Important Point

The included angle and the kingpin inclination or ball joint inclination angle within the vehicle's right and left front suspension system will not be altered by changing the camber angle of a wheel. Figure 18-17 illustrates this point. You are looking at a front view of a vehicle's frame, its adjusting camber shims, and the upper control arm. The vertical solid lines represent the centerline positions of the pivot axis and the 1/2° positive camber position of the left front wheel. Let's assume that you have moved the wheel outward to a new camber position of 2° positive. This change has been accomplished by adding additional shims between the upper inner control arm and the vehicle's frame. Since the wheel has moved outward, the pivot axis line has also moved outward. As you study the illustration, it is obvious that the included angle is still the same and has not diminished the benefits of the kingpin inclination-ball joint inclination factor.

REVIEW TEST

1. Explain how the kingpin or ball joint inclination factor assists the front wheels of a vehicle to maintain a straight-ahead course.
2. Explain how the kingpin or ball joint inclination factor reduces the effort required to turn the front wheels of a vehicle.
3. Explain how the kingpin or ball joint inclination factor reduces tire tread wear.
4. Define (a) *scrub radius area*, (b) *point of intersection*, and (c) *included angle.*
5. Explain how a front wheel, subject to positive camber with the point of intersection located above the road, will influence a wheel to roll outward.
6. Describe how to measure the steering axis inclination factor.
7. Assume that you have placed the vehicle described here on an alignment rack. Since the automobile has sustained damage to the left front suspension, you are going to measure the ball joint inclination angle prior to aligning the front wheels. The vehicle is a late model Empress (four door); the ball joint inclination manufacturer's specification is 13° at 0° camber. You have measured and recorded a 1° negative camber reading at the left front wheel. You have measured and recorded a 14° ball joint inclination reading from the scale on the gauge. Based on the figures that have been given, determine if the left front steering knuckle is damaged.
8. You have again placed a vehicle on an alignment rack. The manufacturer's ball joint inclination specification is 8° at 0° camber. You are going to measure the ball joint inclination of the vehicle at its two front wheels since the front of the automobile has sustained major body-frame damage at its left and right sides. You have obtained the following readings: Recorded camber readings, right front wheel -2°, left front wheel -2½°. Recorded ball joint inclination readings, right front wheel 10°, left front wheel 5°. Based on the figures that have been given and the information in Chapters 12 and 13, you are to determine if the steering knuckles are damaged and what other problems may exist in the front of the vehicle's frame.

FINAL TEST

This examination is multiple choice. Only one answer will be accepted. Carefully read every statement.

1. Mechanic A says that steering axis inclination is the inward tilt of the top of the wheel. Mechanic B says that steering axis inclination is the inward tilt of the top of the pivot axis centerline. Who is right? (A) Mechanic A, (B) Mechanic B, (C) Both A and B, (D) Neither A nor B.
2. Mechanic A says the factors that return the front wheels to the straight-ahead position are negative camber and ball joint inclination. Mechanic B says the factors that return the front wheels to the straight-ahead position are positive caster and kingpin inclination. Who is right? (A) Mechanic A, (B) Mechanic B, (C) Both A and B, (D) Neither A nor B.
3. Mechanic A says that when the front wheels are rotated from the straight-ahead position, the front end of the vehicle must lift. Mechanic B says that when larger front diameter tires are installed, the scrub radius areas are altered. Who is right? (A) Mechanic A, (B) Mechanic B, (C) Both A and B, (D) Neither A nor B.
4. Mechanic A says that an incorrect steering axis inclination angle may be corrected by altering the camber angle. Mechanic B says that an incorrect steering axis inclination angle may be corrected by altering the caster angle. Who is right? (A) Mechanic A, (B) Mechanic B, (C) Both A and B, (D) Neither A nor B.
5. Mechanic A says that a major purpose of ball joint inclination is to reduce the need for the wheel to be subject to excessive camber. Mechanic B says that a major purpose of ball joint inclination is to assist the front wheels of a vehicle to maintain a straight-ahead course. Who is right? (A) Mechanic A, (B) Mechanic B, (C) Both A and B, (D) Neither A nor B.
6. Mechanic A says that the included angle is the sum of positive camber and steering axis inclination. Mechanic B says that the included angle is the sum of positive caster and steering axis inclination. Who is right? (A) Mechanic A, (B) Mechanic B, (C) Both A and B, (D) Neither A nor B.
7. Mechanic A says that if the camber of a wheel

is set at 0°, the steering axis inclination gauge will measure the true inclination angle. Mechanic B says that if the camber of a wheel is set at 1° positive, the positive camber reading must be subtracted from the reading obtained on the ball joint inclination scale. Who is right? (A) Mechanic A, (B) Mechanic B, (C) Both A and B, (D) Neither A nor B.

8. Mechanic A says that in order to measure the steering axis inclination angle accurately, a brake pedal jack must be applied to the brake pedal. Mechanic B says that in order to determine the steering axis inclination angle accurately, the parking brake need be applied only prior to measuring the caster angle. Who is right? (A) Mechanic A, (B) Mechanic B, (C) Both A and B, (D) Neither A nor B.
9. You have a vehicle on the alignment rack, and you suspect that the MacPherson strut has been damaged. The manufacturer's specification is 13° inclination angle at 1° positive camber. Mechanic A says that if the camber of the wheel is 0°, the scale of the kingpin-ball joint inclination gauge should indicate 14° inclination angle. Mechanic B says that if the camber of the wheel is 0°, the scale of the kingpin-ball joint inclination gauge should indicate 12° inclination angle. Who is right? (A) Mechanic A, (B) Mechanic B, (C) Both A and B, (D) Neither A nor B.

Toe-Out on Turns or Turning Radius

LEARNING OBJECTIVES

After studying this chapter, you should be able to:

- Explain the purpose for the toe-out on turns wheel alignment factor,
- Explain how the manufacturer enables the front wheels of a vehicle to toe-out when it is turning a corner,
- Explain how excessive cornering speed affects the tread of a tire,
- Describe how the toe-out on turns factor is measured.

INTRODUCTION

So far, we have considered four wheel alignment factors: camber, caster, toe, and kingpin or ball joint inclination. All of these geometrical wheel alignment factors are designed into a vehicle's suspension system to arrive at specifications for straight-ahead driving.

When the front wheels of a vehicle are turned, a fifth wheel alignment factor becomes important. This is commonly referred to as *toe-out on turns.* Like kingpin or ball joint inclination, it is a nonadjustable factor. Its prime purpose is to control the tracking of the front wheels when the vehicle is turning a corner, thereby minimizing side slippage and tire tread wear. Do you recall the definition of toe-out on turns?

In the very early years of automotive design, the front axle of the vehicle would pivot in its center, like a child's four-wheel wagon. When the entire axle was turned, each front wheel was automatically held at a right angle to its radius, and both front wheels would turn around on the same common center (Figure 19-1). Today's vehicles are not designed as they were at the turn of the century. When a modern vehicle turns a corner, the front end design is such that the front wheels pivot independently and at different distances from the

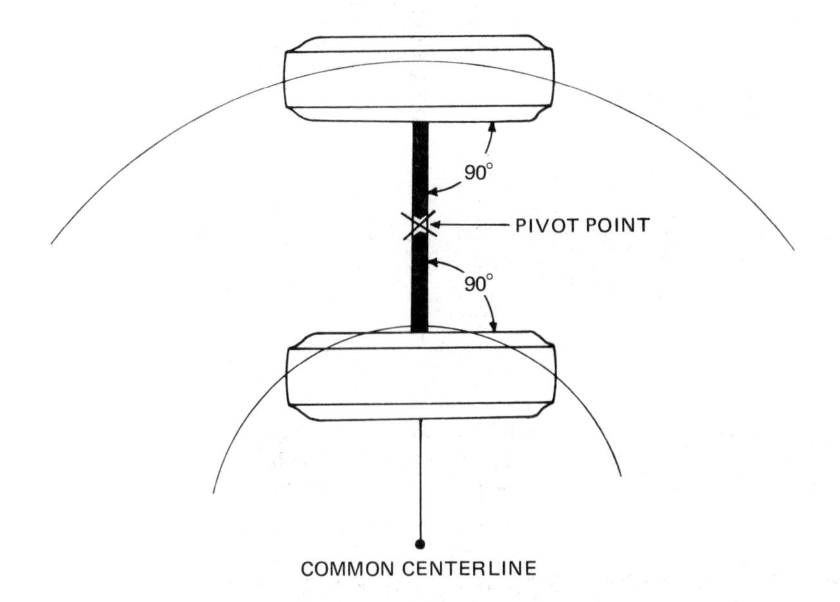

FIGURE 19-1. Illustration of an Early Motor Vehicle Steering System.

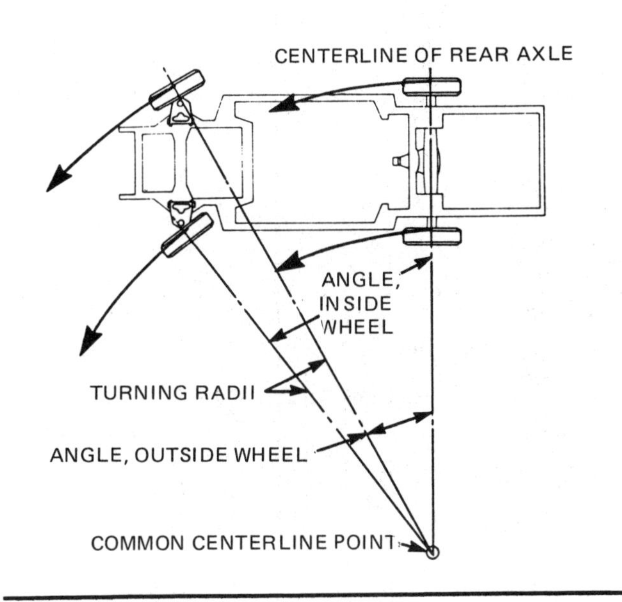

FIGURE 19-2. Illustration of Modern Steering Designs. (Courtesy of Ford Motor Company)

center of the turn. Therefore, the front wheels must turn at different angles, and the inside front wheel must be ahead of the outside wheel on a turn. As you study Figure 19-2, notice that the inside wheel must turn at a different angle from the opposite wheel. This enables the front wheels to remain perpendicular to their turning radius, avoiding undesirable scrubbing of the tires' tread surfaces.

PRINCIPLE OF DESIGN

How do manufacturers enable the front wheels of a vehicle to toe-out while turning a corner? To understand the principle of steering design and how the steering arms (levers) actually function, let's first consider the peculiarity of a straight lever moving in and through an arc of 90°. Figure 19-3 shows that as the lever moves from point A to B, it moves toward the right, a distance of X. As the lever continues to move in the arc, it will move from point B to point C, a distance of Y. *Note:* As the end of the lever moves farther through the arc of 90°, the distance that it must travel is increased. Compare distance X to distance Y.

With slight modification, vehicle manufacturers utilize the same principle in the steering arms (levers) of a vehicle's steering system. Figure 19-4 illustrates an overhead schematic view of two front wheels and the principal parts of the steering system. As you study the illustration, notice how the arms are designed inward where they attach to the tie-rods. The arms can increase or decrease the distance of movement, depending on their rotary position within their individual operating arcs.

Figure 19-5 shows the front wheels of a vehicle turning left. The left steering arm has traveled through the part of the arc represented by distance X. The right steering arm has traveled through that part of the arc represented by distance Z. Since distance X is greater than distance Z, the left steering arm will cause the left front wheel to toe-out more

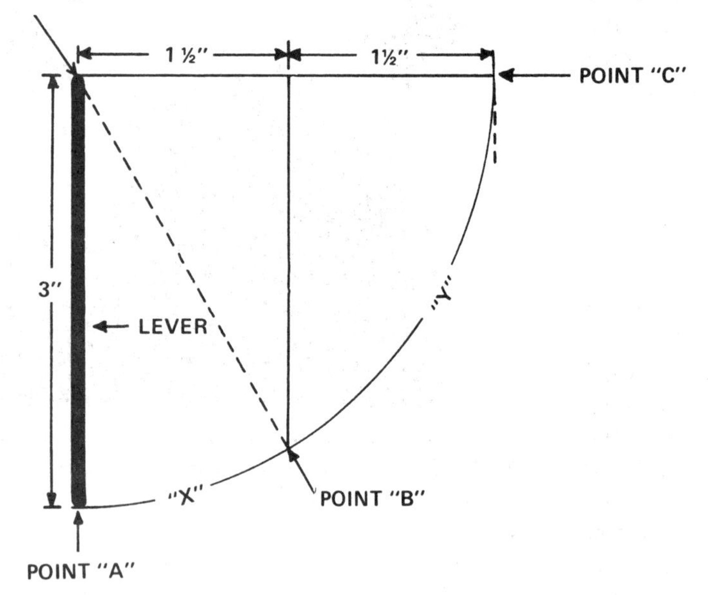

FIGURE 19-3. Illustration of the Principle of Modern Steering Design. (Courtesy of Ford Motor Company)

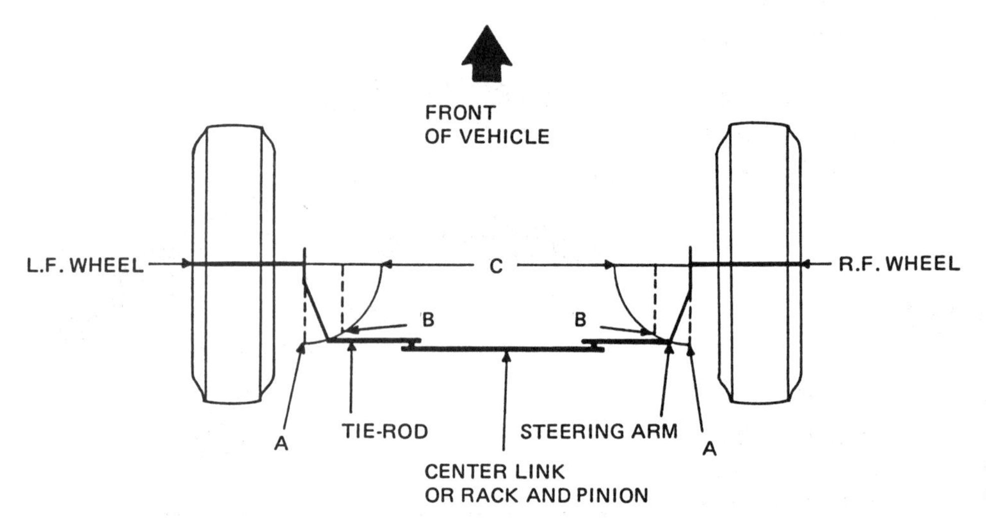

FIGURE 19-4. Steering Arms Acting as Levers Moving through an Arc.

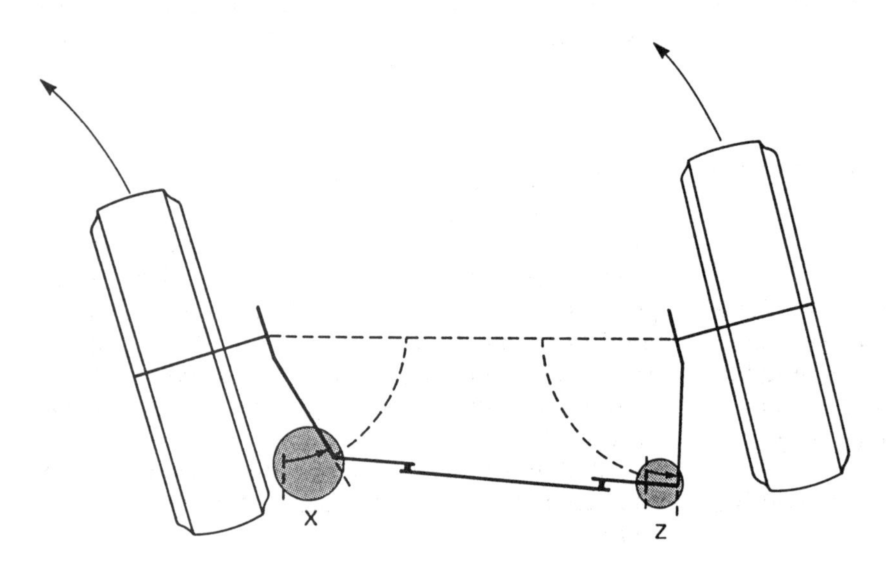

FIGURE 19-5. Result of Correct Toe-Out on Turns. Because arc X described by the left front wheel is greater than arc Z described by the right front wheel, the left front wheel toes-out more than the right front wheel toes-in.

than the opposite wheel toes-in. When the front wheels of a vehicle are turned to the right, the same geometrical relationship exists; the right front wheel will toe-out a greater number of degrees than the opposite wheel toes-in.

TIRE TREAD WEAR

Regardless of a vehicle's steering-suspension system design and the contribution of the wheel alignment factors, there is a practical limit to the tire's tread ability to grip the road's surface without slipping. When this limit has been exceeded, the tread of the tires will tend to slip diagonally, producing tread wear that can easily be mistaken for incorrect camber or underinflation. Figure 19-6 illustrates tread surface of a tire that has sustained wear due to excessive high-speed cornering. Whenever high-speed cornering tread wear is apparent, the owner should be shown the tires and the rounding of the inside and outside tread shoulders. It should be explained to the owner that he or she is actually grinding off the tread surfaces by excessive speeds on turns, thereby allowing the tires to slip (Figure 19-7).

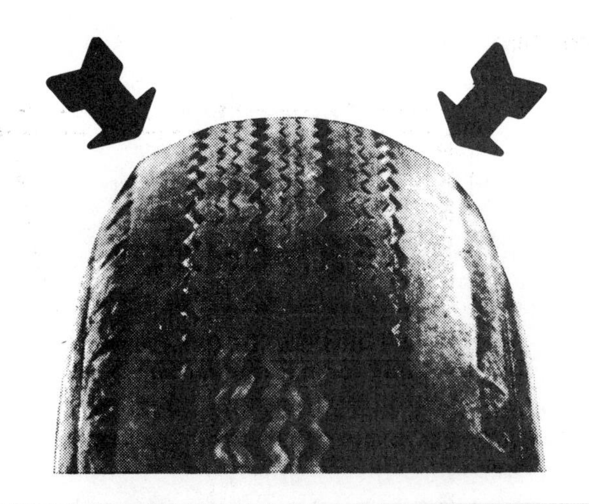

FIGURE 19-6. Tire Tread Wear Resulting from High-Speed Cornering. Continuous high-speed cornering will result in wear on both the inner and outer shoulders of all four tires. This is a greater problem with bias ply than with radial ply tires. (Courtesy of Hunter Engineering Company)

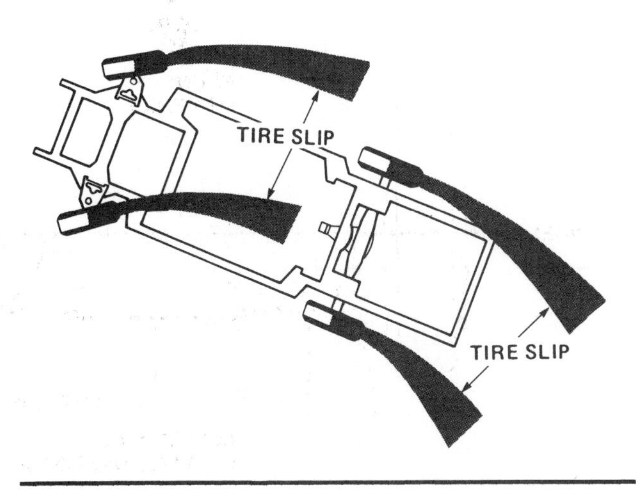

FIGURE 19-7. Excessive Tire Slippage Produced by High-Speed Cornering. (Courtesy of Hunter Engineering Company)

MEASURING TOE-OUT ON TURNS

All wheel alignment factors are interrelated. When one factor is altered, the other factors are affected. The following order of procedure should be used when checking or aligning the front wheels of a vehicle:

- Caster,
- Camber,
- Kingpin or ball joint inclination (if necessary),
- Toe,
- Toe-out on turns.

In reality, there is very little that can be done to correct tire tread wear caused by excessive high-speed cornering. Tell the driver what is causing the tread wear and suggest that he or she effect a cure by slowing down on corners, providing the toe-out on turns is correct.

Use the following procedure when measuring toe-out on turns:

Step-by-Step Procedure

1. Place the front wheels of the vehicle straight ahead on rotary turntable plates. Each plate has a scale that indicates the number of degrees the wheel is turned (either to the left or right) from the straight-ahead position. Because of the design of some alignment equipment, it is recommended that the brake pedal be depressed with a brake pedal jack.

2. Remove the lock pins. Adjust the dials on the turning radius gauges so that both pointers are on zero.

3. Turn the right front wheel in toward the center of the vehicle until the dial and the scale read 20°. Go to the opposite gauge and read the scale. This is the angle of toe-out on turns or turning radius of the left wheel. Record the reading.

4. For the turning radius on the right front wheel, turn the left front wheel in toward the center of the vehicle until the dial reads 20°. The reading obtained on the right dial is the turning radius of the right wheel.

5. Check the recorded readings against the manufacturer's specifications.

If the readings do not conform to the manufacturer's specifications (after the front wheels have been correctly aligned), the tie-rods, steering arms, or other linkage parts are bent. It is then necessary to check the steering mechanism and replace the parts that are distorted. Although this is not stated in manufacturer's specifications, you are generally allowed 1° variation from the specifications.

FACTS TO REMEMBER

1. Toe-out on turns is not an adjustable wheel alignment factor.
2. Before checking the toe-out on turns factor, other factors must be adjusted to specifications.
3. Defective steering linkage parts cannot be straightened; they must be replaced.
4. Incorrect toe-out on turns decreases tire tread life.
5. Tire squeal on a cool day is indicative of an incorrect toe-out on turns relationship between the front wheels of a vehicle.

REVIEW TEST

1. Explain the purpose of the toe-out on wheel alignment factor.
2. Explain how the manufacturer enables the front wheels of a vehicle to toe-out when it is turning a corner.
3. Explain how excessive cornering speed affects the tread of a tire.
4. Describe the correct procedure that must be followed when measuring the toe-out on turns factor.
5. List the five Facts to Remember that conclude this chapter.

FINAL TEST

This examination is multiple choice. Only one answer will be accepted. Carefully read every statement.

1. Mechanic A says that toe-out on turns is an adjustable front end wheel alignment factor. Mechanic B says that incorrect toe-out on turns may be corrected by straightening a bent steering arm after it has been heated. Who is right? (A) Mechanic A, (B) Mechanic B, (C) Both A and B, (D) Neither A nor B.
2. Mechanic A says that tire squeal on a cool day is indicative of an incorrect toe-out on turns relationship between the front wheels of a vehicle. Mechanic B says that adjusting the toe-out on turns changes the camber and ball joint inclination factor. Who is right? (A) Mechanic A, (B) Mechanic B, (C) Both A and B, (D) Neither A nor B.
3. Mechanic A says that correct toe-out on turns reduces tire tread wear. Mechanic B says that correct toe-out on turns reduces tire squeal. Who is right? (A) Mechanic A, (B) Mechanic B, (C) Both A and B, (D) Neither A nor B.
4. Mechanic A says that correct toe-out on turns is dependent on correct toe. Mechanic B says that correct toe-out on turns is established by the parallel relationship of the steering arms. Who is right? (A) Mechanic A, (B) Mechanic B, (C) Both A and B, (D) Neither A nor B.
5. Mechanic A says that when a vehicle turns a corner, both steering arms turn an equal distance in their prescribed arcs. Mechanic B says that since the front wheels of a vehicle pivot independently, they are at different turning radii from the common center point. Who is right? (A) Mechanic A, (B) Mechanic B, (C) Both A and B, (D) Neither A nor B.
6. Mechanic A says that the turning angles for both front wheels are not the same. Mechanic B says that the turning angles for both front wheels are calculated from a common line through the rear axle housing of the vehicle. Who is right? (A) Mechanic A, (B) Mechanic B, (C) Both A and B, (D) Neither A nor B.
7. Mechanic A says that since the front wheels

of a vehicle turn at different angles, the unit of measurement is calculated in millimeters. Mechanic B says that to reduce tire tread scuffing, the wheel must be perpendicular to its turning radius. Who is right? (A) Mechanic A, (B) Mechanic B, (C) Both A and B, (D) Neither A nor B.

8. Mechanic A says that to decrease turning radius tire tread wear, increase the toe-in setting. Mechanic B says that to decrease turning radius tire tread wear, inflate the tires to their correct tire inflation pressure and suggest that the driver slow down on curves. Who is right? (A) Mechanic A, (B) Mechanic B, (C) Both A and B, (D) Neither A nor B.

9. Mechanic A says that centrifugal force and the speed of the vehicle have no bearing on tire tread slippage. Mechanic B says that a worn idler or pitman arm will not affect the turning radius of a vehicle. Who is right? (A) Mechanic A, (B) Mechanic B, (C) Both A and B, (D) Neither A nor B.

10. Mechanic A says that to decrease turning radius tire tread wear, move the camber more positive. Mechanic B says that to decrease turning radius tire tread wear, move the camber more negative. Who is right? (A) Mechanic A, (B) Mechanic B, (C) Both A and B, (D) Neither A nor B.

Wheel Alignment

LEARNING OBJECTIVES

After studying this chapter and with the aid of wheel alignment equipment, a manufacturer's shop manual, and practical instruction from your shop instructor, you should be able to:

- Explain and demonstrate how to change the caster, camber, and toe factors to manufacturer's specifications,
- Explain the meaning of the term *toe change* and how it is measured,
- Explain and demonstrate how to measure the toe-out on turns factor,
- Explain and demonstrate how to center the vehicle's steering wheel to its correct riding position,
- Explain four different methods that are used by automobile manufacturers to adjust the caster and camber to the correct specifications,
- Explain how the caster and camber factors are set on a truck equipped with a single I-beam front axle.

INTRODUCTION

This chapter is the culmination of what you have been studying from the first chapter in the book. In this chapter, you will put together everything you have learned and apply that information to aligning the wheels of a vehicle.

Wheel alignment, as applied to automobiles and trucks, means the correct relative position of the front and rear wheels to obtain a true, free-rolling movement over the surface of the road. It is a system of interrelated angles of axles, wheels, and other body-chassis parts to permit ease of steering and greater safety and to prevent abnormal and wasteful

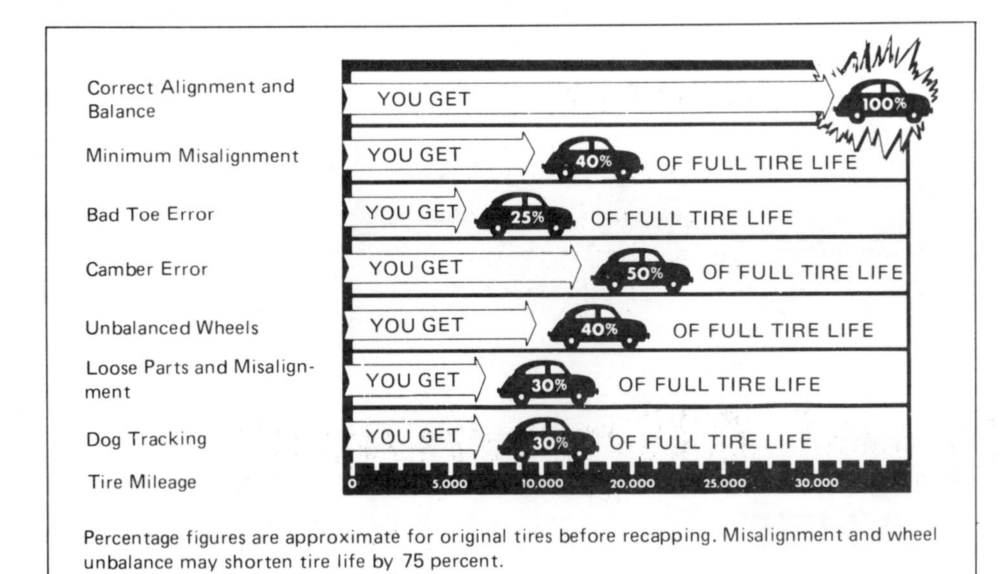

FIGURE 20-1. Tire Mileage Chart. (Courtesy of Bear Automotive, Inc.)

wear of tires and suspension parts. The modern concept of wheel alignment includes another important factor, wheel balancing: the science of placing weights on a wheel so that all portions will exert an equal pull from the center as the wheel assembly revolves.

The importance of wheel alignment, wheel balance, and body-chassis straightening in the motor vehicle field cannot be minimized. Its proper understanding by repair technicians is essential if motorists are to reduce tire tread wear and prolong the life of their vehicles. No other phase of vehicle operation, when neglected, can be so costly or so dangerous. A wheel badly out of balance may pound suspension parts unmercifully. With worn parts, wheel alignment factors become maladjusted, and soon a tire begins to show rapid tread wear, or perhaps a serious defect appears in the steering linkage. An emergency arises on the highway, and a crash results. Lives may be lost, and damage to the involved vehicles may run into half a million dollars.

Because alignment angles are constantly changing under ordinary everyday driving conditions, it is essential that every vehicle have the steering, suspension, and alignment inspected at least once a year, and, in some cases, even twice a year, depending on the total accumulated kilometers (mileage). Such an inspection under the direction of a competent alignment technician equipped with the necessary tools is an investment in the benefits of steering ease, tire tread life, riding comfort, directional control, steering stability, and, most important of all, safety.

Figure 20-1 presents a chart that shows how several unfavorable conditions will affect an automobile's tire tread mileage. As you study the chart, notice the results of misalignment and wheel unbalance. The combination of such conditions could shorten tire tread life by as much as 75 percent.

INSPECTION PROCEDURES

Inspection of the components of a vehicle's steering and suspension systems will vary according to the type of suspension. Both front and rear suspension components should be inspected prior to any alignment corrections. Check both the camber and caster alignment factors before making any adjustments. This information will allow you to obtain a complete picture of the various readings and what will be required to align the factors to the manufacturer's specifications. *Note:* Careless checking will not reveal existing or potential problems. A checklist such as the one shown in Figure 20-2 will help assure a thorough inspection.

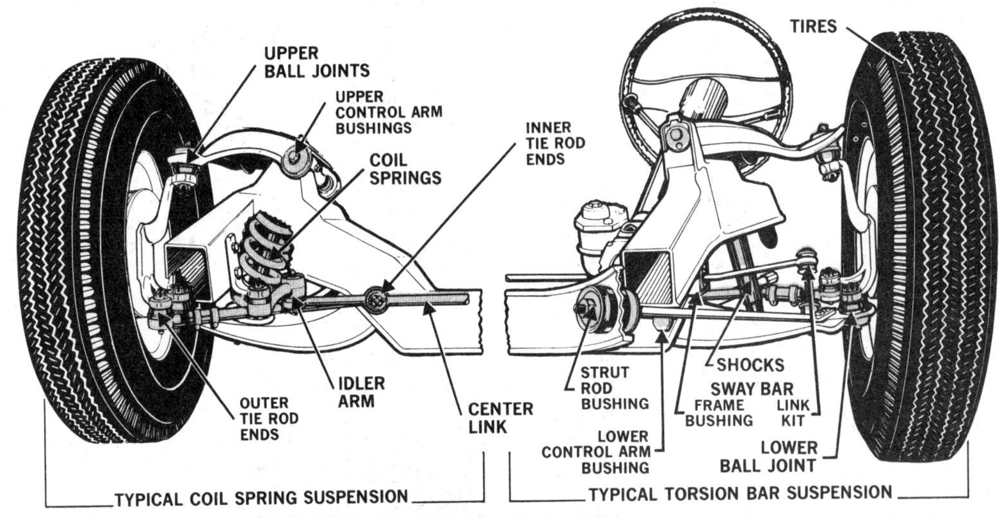

YOUR CAR DESERVES A

WHEEL TO WHEEL SECURITY CHECK

Owner__________ Date__________ Phone__________

Make__________ Model__________ Year__________

Mileage__________ License Number__________ Engine Size__________

PARTS DESCRIPTION	OK	COMMENTS	PARTS	LABOR
SPRINGS				
CONTROL ARM BUSHINGS				
POWER STEERING				
LOWER BALL JOINT				
UPPER BALL JOINT				
WHEEL BRGS.				
BALANCE				
TIRES				
TIE ROD ENDS				
IDLER ARM				
PITMAN ARM				
CENTER LINK				
SWAY BAR FRAME BUSHINGS				
SWAY BAR LINK KIT				
STRUT ROD BUSHINGS				
SHOCK ABSORBERS				
ALIGNMENT				
REMOVED PARTS REQUESTED	YES / NO		SUBTOTAL / **TOTAL**	

CAR HEIGHT			
	Left	Right	Specs.
FRONT			
REAR			
BALL JOINT READINGS			
Load Carrier	Left	Right	Specs.
AXIAL			
RADIAL			
ALIGNMENT			
	Left	Right	Specs.
CAMBER			
CASTER			
TOE			

REMARKS:

____
Inspector

FIGURE 20-2. **Sample Wheel Inspection Checklist.** (Courtesy of Moog-Canada Automotive, Ltd.)

Vehicle and Equipment Preparation

Whenever possible, wheel alignment inspection should be performed on permanent stationary alignment equipment that has been calibrated to ensure accurate readings. In the absence of such equipment, the vehicle may be positioned on portable stands that allow it to sit level. The floor area, or the stands, should be level within 3 millimeters (1/8 inch) from the front to the rear of the vehicle and within 3 millimeters (1/8 inch) from side to side.

With the permission of Snap-on Tools of Canada Ltd., this chapter will detail how the Snap-on wheel alignment equipment is used to determine camber, caster, and steering axis inclination readings. The WA 210 four wheel optical toe and tracking system is used to calculate the toe and the tracking of a vehicle. The turntable dial plates are used to measure the toe-out on turns factor. This equipment may be used with any stationary rack or portable wheel stand equipment.

FIGURE 20-3. Positioning Snap-on WA 40A Gauge onto Wheel Hub. (Courtesy of Snap-on Tools of Canada Ltd.)

Step-by-Step Procedure

1. Position the gauge.
2. Remove the hubcap or wheel cover and the dust cap from the front wheels.
3. Wipe off the machined end of the hub flange.
4. Remove the cotter pin. Tighten the wheel bearing adjusting nut to a torque of approximately 20 newton-meters (15 foot-pounds). This step is done to ensure that there is no lateral end play in the wheel bearings.
5. Holding the Snap-on gauge near the hub flange, set the self-centering plunger in the lathe center hole in the end of the front wheel spindle (Figure 20-3). Powerful magnets will attach the gauge to the machined hub flange on the wheel. Twist the gauge about a quarter turn several times to let it get a positive seat on the flange. If there is a rocking motion, remove the gauge and check the flange and the gauge for grit or foreign particles. The self-centering plunger makes sure that readings will be taken from the exact center of the spindle. For wheels with inaccessible hubs, use the universal rim adapters. These adapters will fit the rims of most wheels ranging from 30.5 to 46 centimeters (12 to 18 inches) (Figure 20-4). When the rim adapter has been properly located, the part that the gauge attaches to should be located opposite the center of the wheel.

FIGURE 20-4. Universal Wheel Rim Adapter. (Courtesy of Snap-on Tools of Canada Ltd.)

Checking for Wheel Runout with Rim Adapters

Before making any camber and caster checks on the vehicle, check for wheel runout, as follows.

Step-by-Step Procedure

1. Raise the car so the front wheels are free to revolve.
2. Place the magnetic gauge on the adapter as shown in Figure 20-4.
3. Set the bubble in the caster vial to zero by turning the adjusting nut on the caster gauge.
4. Turn the wheel 180°. Turn the gauge back 180° and read the caster. If there is a change in the caster reading, reduce the reading to one-half by adjusting the runout screw. For example, if this reading is 4, adjust it to 2.
5. Set the bubble in the caster vial to zero again by turning the adjusting nut on the caster gauge.
6. Turn the wheel 180° again. Turn the gauge back 180°. The bubble in the caster vial should still be on zero. If it is not, carefully repeat steps 3 through 6.
7. Lower the car on the turntables, while keeping the rim adapters in an upright position. Perform all wheel alignment checks except toe.

Checking Camber

Step-by-Step Procedure

1. With the gauge in position on the wheel, remove the locking pins that hold the turntable dial plates in a fixed position. The front wheels of the vehicle must be in the straight-ahead position with the turntable pointers located on zero.
2. Jounce the vehicle several times by lifting up and down on the front bumper. This step will allow the vehicle to be at its curb riding height.
3. Read the camber scale on the right-hand side of the gauge. The position of the center of the bubble will indicate the camber of the wheel being checked.
4. Write down the camber reading.
5. Check the other front wheel in the same manner as described.

Checking Caster

Step-by-Step Procedure

1. Obtain a brake pedal jack. Depress the brake pedal and hold the pedal down, as shown in Figure 20-5. In vehicles with power brakes, the engine must be operating when the brakes are set and locked.
2. Be sure the pointers on both turntables are on 0°. Turn the front of the wheel being checked out to 20° on the turntable dial (Figure 20-6). Be sure the turntable on the opposite side is in the nonlocking position.
3. Set the bubble in the caster vial to zero. For the right wheel, 0° is the extreme right of the scale. For the left wheel, 0° is at the extreme left.
4. Turn the front of the wheel in 40° from its present position on the turntable dial (Figure 20-7).
5. Write down the caster reading.
6. Check the other front wheel in the same manner as described.

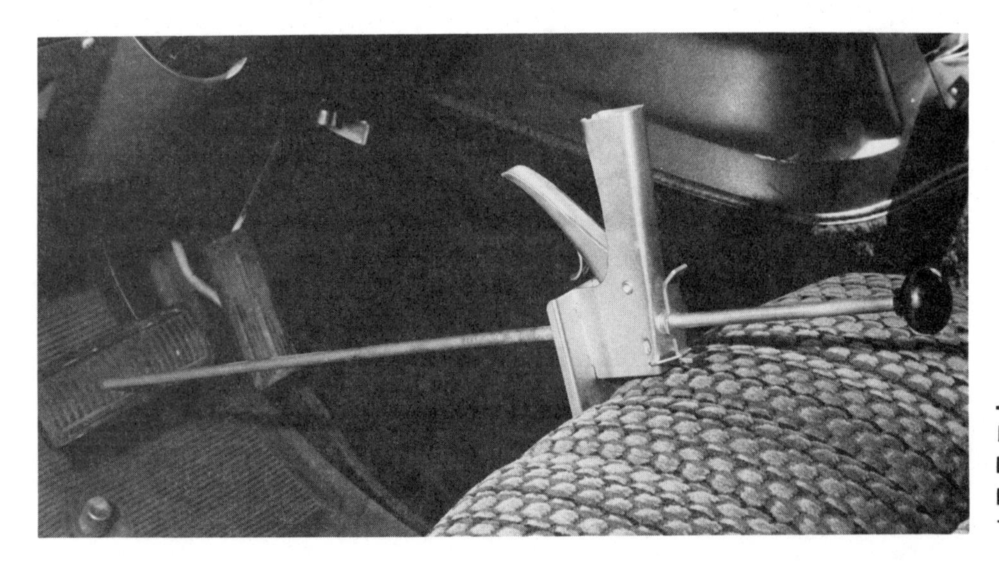

FIGURE 20-5. Using Brake Pedal Jack To Depress Brake Pedal. (Courtesy of Snap-on Tools of Canada Ltd.)

FIGURE 20-6. Checking Caster, 20° out. (Courtesy of Snap-on Tools of Canada Ltd.)

FIGURE 20-7. Checking Caster, 20° in. (Courtesy of Snap-on Tools of Canada Ltd.)

Checking Steering Axis Inclination

The steering axis inclination reading is not normally checked during the alignment process if the camber can be adjusted to the manufacturer's specifications. However, if you suspect that the steering knuckle, MacPherson strut, or kingpin has been damaged, determining the extent of the damage is very important. This can be determined by checking steering axis inclination. To review how steering axis inclination is measured, turn to Chapter 18.

CASTER-CAMBER CORRECTION PROCEDURES

You now have an automobile positioned on the alignment rack and have assembled all the necessary tools and equipment to align the wheels of the vehicle. You have inspected the vehicle's steering and suspension systems and have concluded that the component parts do not require replacement. You have measured the camber and caster factors and have recorded the following readings:

	Right Front Wheel	Left Front Wheel
Camber	+5/8°	+3/8°
Caster	+2°	+1 1/2°

Since you now know the caster and camber readings, you must also determine the manufacturer's preferred specifications from the shop manual. Now, determine the make of the vehicle, the model, and the production year. The vehicle is an Empress, two door, and was manufactured last year. With these facts, let's obtain the manufacturer's shop manual and determine the alignment specifications. This is what you will see in the manual.

Front Wheel Alignment Specifications for Empress 2 Door

Alignment Factors	Minimum	Preferred	Maximum
Caster	+1°	+2°	+3°
Camber (left)*	+1/8°	+3/8°	+5/8°
Camber (right)	-1/8°	+1/8°	+3/8°
Toe-In	3mm (1/8 in.)	6mm (1/4 in.)	9mm (3/8 in.)

Note: Cross caster variations should not exceed 1/2° between sides.

*Left and right side should be different at least 1/4° but no more than a 1/2°, with the left side having the greatest positive reading.

Questions and Answers

Now that you have studied the specification chart, you no doubt have some questions in mind.

1. Why does a manufacturer specify minimum and maximum caster settings? *Answer:* Chapter 16 discussed the advantages and disadvantages of positive and negative caster. If, for example, the driver was experiencing excessive road shock and everything else within the vehicle's steering and suspension systems was normal, the alignment technician could move the caster adjustment to the minimum allowable setting, 1° positive to lessen the influence of road shock. If the driver required increased directional control, the caster angle could be moved to the maximum permitted setting, 3° positive.

2. What does the term *preferred specification* mean? *Answer:* When the alignment factors are adjusted to the preferred specifications, the manufacturer is saying that a vehicle will sustain favorable directional control, steering stability, and potential tire tread life expectancy.

3. Are the settings the same for bias belted ply tires as compared to radial ply tires? *Answer:* Although it is not generally specified by vehicle manufacturers, tire manufacturers have stated that if a vehicle is equipped with radial tires, set the caster angle for the left and right side to the maximum allowable setting, 3° positive, to ensure the maximum steering control. Set the camber angles close to 0°, preferably 1/8° positive. Set the toe factor 1.6 millimeters (1/16 inch) in for both front wheels.

4. Since incorrect toe is a major contributor to tire tread wear, should I set both front wheels to the preferred toe specification when the vehicle is equipped with bias ply or bias belted ply tires? *Answer:* When setting the toe factor, an experienced alignment technician will divide the preferred toe specification by 2, thereby placing an equal setting at both front wheels.

Changing Caster and Camber Factors

Right Front Wheel. Figure 20-8 shows an overhead view of the vehicle's frame and upper control

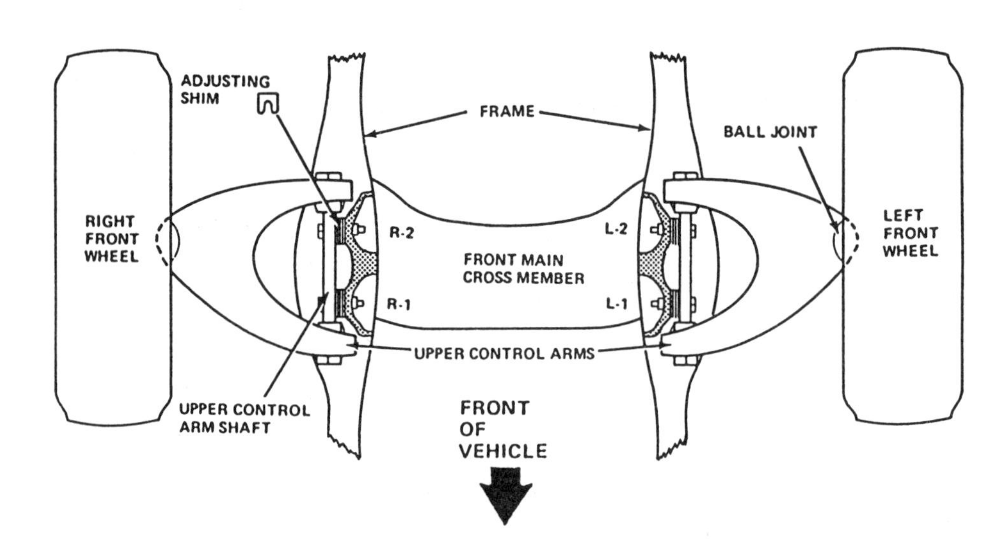

FIGURE 20-8. Overhead View of a Vehicle's Frame and Suspension.

arm assemblies. As you study the diagram, notice the location of the upper inner control arm shafts in relation to the vehicle's frame and the shims that are located between the frame and the control arm shafts. Since the vehicle is a shim-type suspension adjustment, it is recommended that the caster angle be adjusted first and then the camber. To assist in understanding what is required, the readings and the specifications are listed as follows:

	Right Front Wheel	Left Front Wheel
Caster reading	+2°	+1 1/2°
Caster specification	+2°	+2°
Camber reading	+5/8°	+3/8°
Camber specification	+1/8°	+3/8°

Let's analyze what is required by starting at the right front wheel. Since the caster reading and the specification are the same, there is no need to adjust the caster. What about the camber? The camber reading is 1/2° more positive than the specification, (+5/8°) — (+1/8°) = +1/2°. Therefore, the top of the wheel must be moved in. To set the caster and/or camber to specifications, manufacturers may use four different shim thicknesses at either the front or rear suspension (alignment adjustment) locations. The following shim thicknesses are equivalent to a fraction of a degree of camber angle change (the figures in millimeters are approximate):

Shim Thickness	Fraction of a Degree
0.4 millimeter (1/64 inch)	1/16°
0.8 millimeter (1/32 inch)	1/8°
1.6 millimeters (1/16 inch)	1/4°
3 millimeters (1/8 inch)	1/2°

We know that the top of a wheel must be moved in 1/2°. The list shows that a shim thickness of 3 millimeters (1/8 inch) is equal to 1/2° of camber angle change.

The next question is whether we need to add or remove shims at positions R-1 and R-2 in Figure 20-8 to change the camber of the wheel. Since only the camber angle at the right front wheel needs to be adjusted (the top of the wheel moved in), the upper control arm shaft must be positioned closer to the vehicle's frame. At this point, the alignment technician would loosen the two hex-head retaining nuts at R-1 and R-2. It may be necessary to use a pry bar to move the control arm shaft away from the frame brackets. Moving the control arm shaft will allow you the opportunity to remove a 3 millimeter (1/8 inch) shim at positions R-1 and R-2 (Figure 20-9). *Note:* When changing only the camber angle, whatever is done at R-1 must be done at R-2.

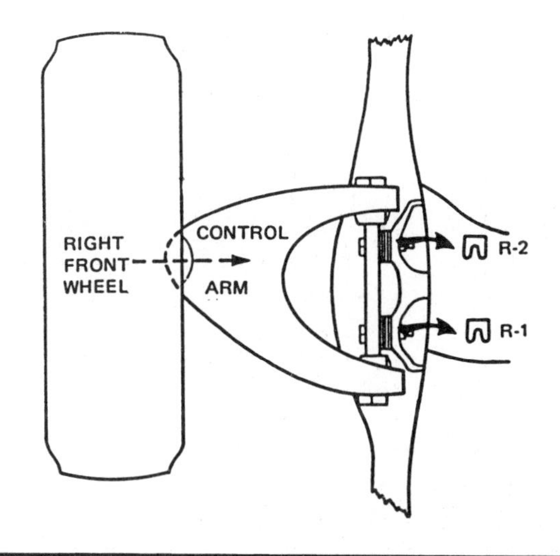

FIGURE 20-9. Shim Locations for Making Camber Adjustments.

When you have removed the two shims, correctly torque the two retaining nuts. Lift up and press down on the middle of the front bumper to settle the vehicle at its curb riding height. Measure the caster and camber again. If the readings correspond to the specifications, you may then direct your attention to adjusting the alignment factors at the opposite wheel.

Left Front Wheel. Let's analyze what is required at the left front wheel. Refer once again to the earlier table that listed the caster and camber readings and specifications. Since the camber reading and the specification are the same, there is no need to adjust the camber.

What about the caster? Since the caster reading is 1/2° less positive than the specification, (+2°) — (+1 1/2°) = +1/2° the top of the steering knuckle (pivot axis centerline) must be moved back toward the rear of the vehicle (Figure 20-10). To set the caster to the specification requires a bit more concentration on your part. Since you now know the required caster angle change is 1/2°, that amount of change must be divided by 2 (1/2° ÷ 2 = 1/4°). The reason for this is that if the caster angle change was not divided by 2, you would actually move the

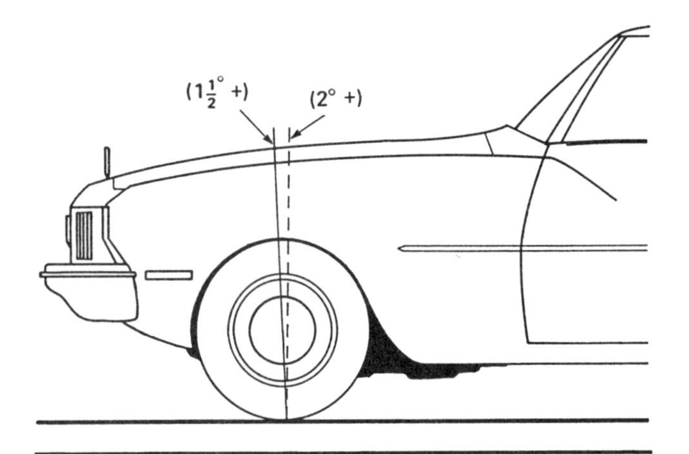

FIGURE 20-10. Increasing the Positive Caster Angle. The caster angle change equals 1/2°.

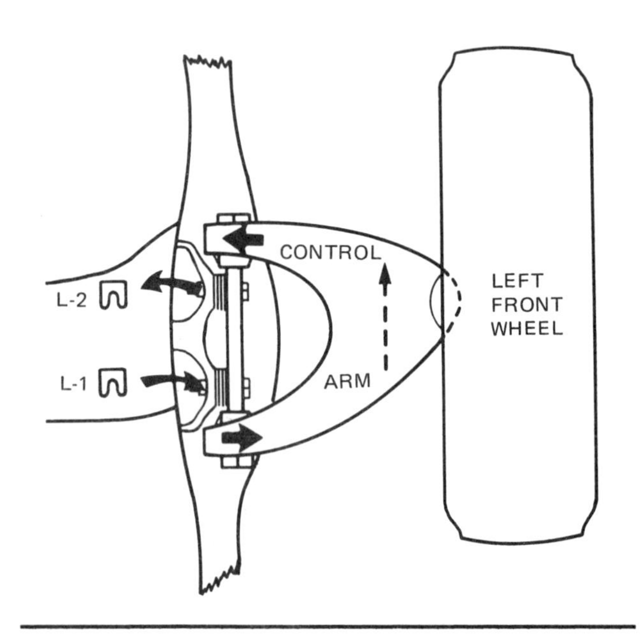

FIGURE 20-11. Shim Locations for Making Caster Adjustments.

caster, when adjusting, twice the required amount. You will understand this better as you continue to read the text.

The shim thickness equal to 1/4° is 1.6 millimeters (1/16 inch). The next question is, How will I change the caster angle? (See Figure 20-11.) Since only the caster angle at the left front wheel needs to be adjusted (the top of the steering knuckle moves rearward), the upper control arm at the front, L-1, must be moved outward, and the control arm at the rear, L-2, must be moved inward. At this point, the alignment technician would loosen the two hex-head retaining nuts at L-1 and L-2. It may be necessary to use a pry bar to move the control arm shaft away from the frame bracket. This will allow you the opportunity to add a 1.6 millimeter (1/16 inch) shim at position L-1 and remove a 1.6 millimeter (1/16 inch) shim at position L-2. When you have added and removed the two shims, correctly torque the two retaining nuts. Lift up and press down on the middle of the front bumper to settle the vehicle at its curb riding height. Measure the caster and camber again. If the readings correspond to the specifications, you can then direct your attention to the next alignment procedure, after completing the following assignment questions:

1. Figure 20-12 illustrates an overhead view of a vehicle's frame and upper control arm assembly. To the left of the diagram are the camber reading and the manufacturer's specification. Based on what the text has discussed, determine the shims that must be either added or removed from positions R-1 and R-2 to align the right side of the vehicle's suspension. Briefly state your answer.*
2. Figure 20-13 illustrates an overhead view of the vehicle's frame and upper control arm assembly. To the left of the diagram are the caster reading and the manufacturer's specification. Based on what the text has discussed, determine the shims that must be either added or removed from positions R-1 and R-2 to align the right side of the vehicle's suspension. Briefly state your answer.*

*Answers:

1. To move the camber from 1/2° negative to 1/4° positive equals a total camber angle change of 3/4°. *Note:* You must calculate from a 1/2° negative number to a 1/4° positive number. 3/4° of movement is equal to a 3 millimeter (1/8 inch) and 1.6 millimeter (1/16 inch) in shim sizes. Therefore, to set the camber, add 3 millimeter (1/8 inch) and 1.6 millimeter (1/16 inch) shim at positions R-1 and R-2.

2. To move the caster from 1/2° positive to 2° positive equals a total caster angle change of 1 1/2°. The caster angle must then be divided by 2: 1 1/2° - 2 = 3/4°. 3/4° is equal to 3 millimeter (1/8 inch) and 1.6 millimeter (1/16 inch) in shim sizes. Therefore, to set the caster, add 3 millimeter (1/8 inch) and 1.6 millimeter (1/16 inch) at position R-1. Remove 3 millimeter (1/8 inch) and 1.6 millimeter (1/16 inch) at position R-2.

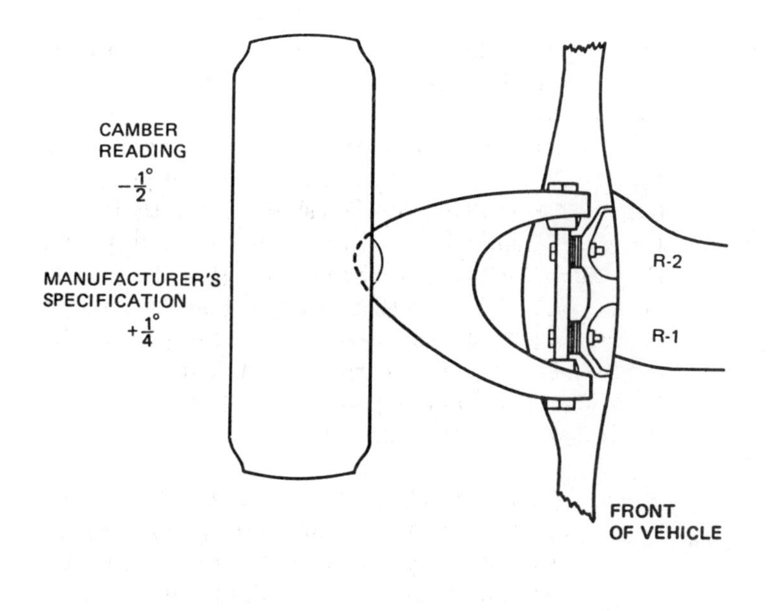

FIGURE 20-12.

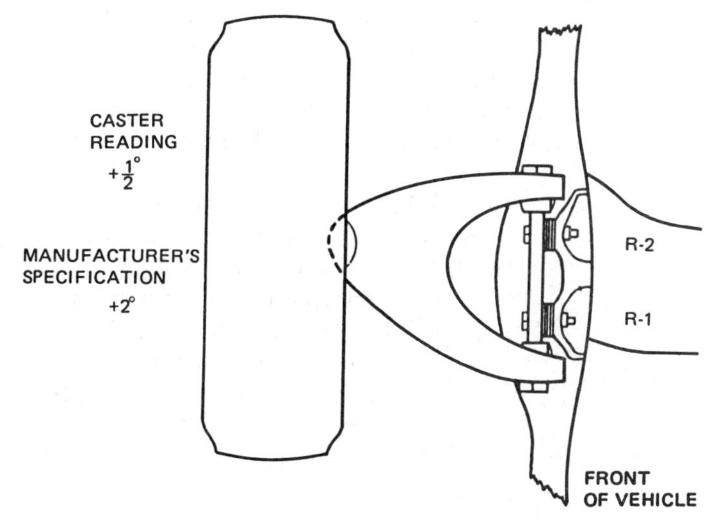

FIGURE 20-13.

SETTING TOE

Figure 20-14 illustrates the Snap-on WA 210 four wheel optical toe and tracking equipment. This system enables you to make the critical geometric measurements necessary for precision toe and track alignment settings. Correct toe and track settings are two of the important alignment factors that enable a vehicle to roll in a straight line with the steering wheel in a center-point position. On a vehicle with correct four wheel alignment, steering is easy, tires wear uniformly and last, and a minimum of strain is placed on the steering and suspension parts. More important, the vehicle is safe to operate.

The objective of toe alignment is to keep the wheels rolling parallel to each other when the vehicle is moving forward. Toe alignment is adjusted after caster and camber have been set because other adjustments will affect it. Because toe is the most serious tire wearing angle adjustment, care should be taken when setting this factor. Also, errors in toe setting will adversely affect track alignment.

Preliminary Procedure

Before using the four wheel optical toe and tracking system, the following items must be considered:

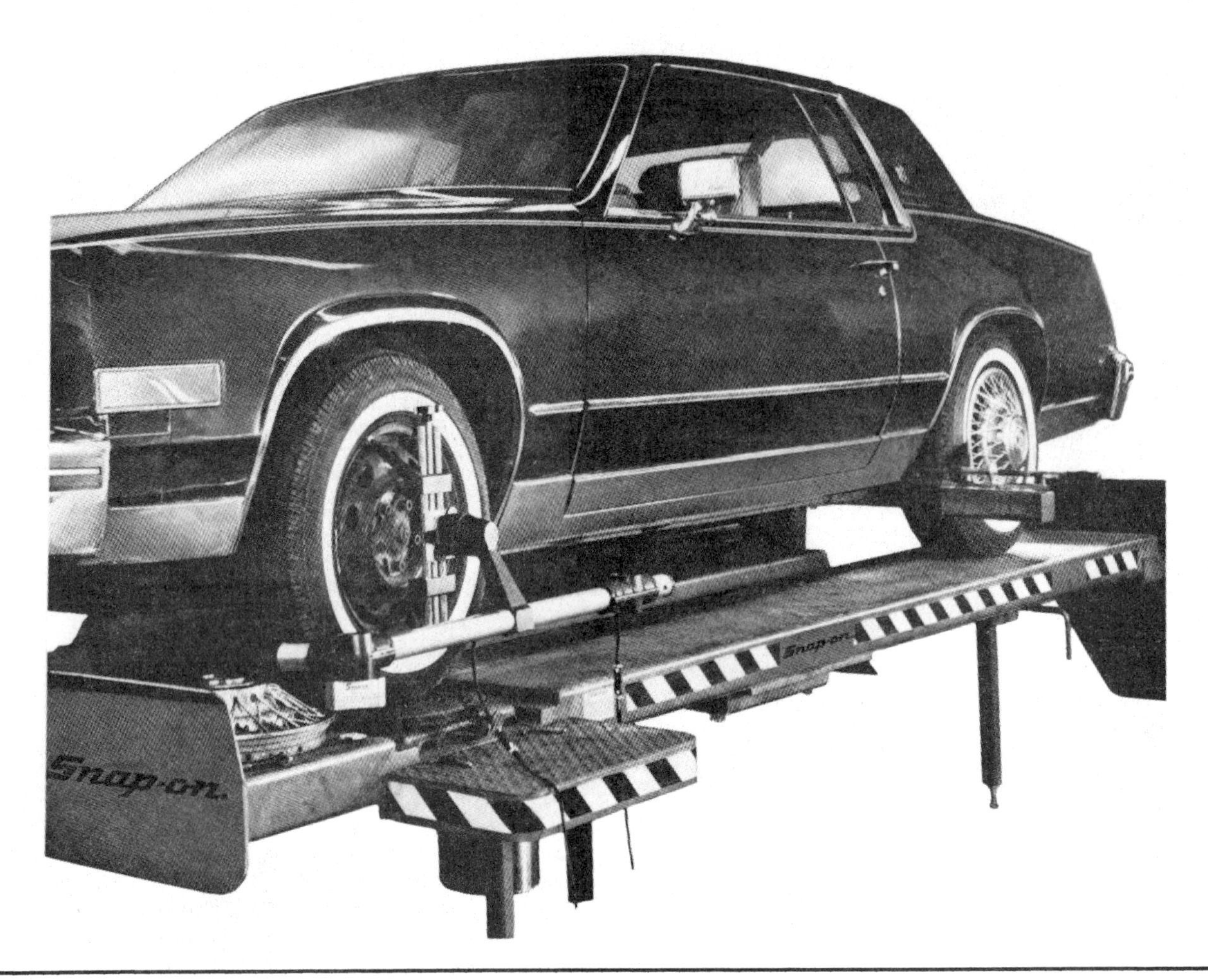

FIGURE 20-14. Snap-on WA 210 Four Wheel Optical Toe and Tracking Equipment. (Courtesy of Snap-on Tools of Canada Ltd.)

1. Make certain that caster and camber are correct.
2. The vehicle should be positioned on the alignment rack or stands with the front wheels centered on turntables. The turntables should be unlocked.
3. The steering wheel should be locked in the straight-ahead position (Figure 20-15). On vehicles with power steering, the engine must be operating when the steering wheel is locked in position.
4. The brakes must be set and locked (see caster checking instructions previously described in this chapter).

Reading and Setting Toe

Step-by-Step Procedure

1. Remove the wheel covers and grease caps. Clean the hub faces and magnet faces. Mating surfaces should be free of grease, dirt, and burrs.

2. Jounce the vehicle a few times and allow it to settle.

3. Mount gauges on the hub faces with the tow projector to the front of the wheel, as shown in Figure 20-14. The spring-loaded centering plunger must be located in the wheel spindle centering hole with the magnet centered on the hub face. *Note:* If the gauge assemblies do not fit on the hub face (because the hubs are aluminum or magnesium) or if the wheel rim protrudes beyond the hub and the gauge assembly will not adhere securely, use rim adapters for mounting.

4. Connect the battery lead clips to the vehicle's battery (12 volt only) or to an external 12 volt direct current (DC) power source.

5. Connect the power leads to the toe projector leads.

6. Using the leveling indicators located near the sleeves, level the gauge assemblies. After each adjustment or movement of the suspension, check to ensure that the gauges are level.

FIGURE 20-15. **Preparing To Set Toe.** Position the steering wheel in the straight-ahead location. Hold the wheel by using a wheel lock. (Courtesy of Hunter Engineering Company)

7. Note the projected image on the opposite toe scale and focus each toe projector until the image is sharp.
8. Write down the toe reading for each wheel. Add the two toe readings together. The sum will give you total toe.
9. Compare your total toe reading with the toe-in specification for the vehicle. If total toe is within specifications, toe check is completed. If total toe is not within specifications, complete step 10.
10. Loosen the bolts on both tie-rod sleeves. Adjust the tie-rod sleeve on the driver's side until the scale on the passenger's side reads one-half of the vehicle's toe-in specifications. Adjust the toe-rod sleeve on the passenger's side until the scale on the driver's side reads one-half of the vehicle's toe-in specifications.

Reading and Setting Rear Toe

Use this procedure for setting the rear toe on vehicles with four wheel independent suspension. To set rear toe, mount the gauges on the rear wheels, as shown in Figure 20-16. Use the same basic procedure as was used to set the front toe. *Note:* Reversing the gauge assemblies will cause toe-in to read as toe-out and toe-out to read as toe-in.

Reading and Setting Front Track

The correct tracking of the front wheels is vital to the directional control and stability of the vehicle. The track gauge can be used to assist the wheel alignment technician to determine if the front track of the vehicle is correct. Front track alignment is performed only after checking and/or setting front toe to make sure it is within specifications.

Step-by-Step Procedure

1. Attach the gauges to the front wheels, as shown in Figure 20-14.
2. Connect the leads to the track projector leads.
3. Attach the scales to the rear tires, as shown in Figure 20-14. Make sure that the clamp assembly is resting on the tire only (not against the brake drum, suspension members, or other part). The scale face should be vertical, with the center of the face positioned about 15 centimeters (6 inches) below the centerline of the wheel. Using the level indicators, level the scales.
4. Adjust the scale range (if necessary) to the range in which the projected image intercepts the scale face. To adjust the scale range, loosen the locking knob and slide the scale to one of the three positions indicated by the detents. Tighten the locking knob. Check the level indicators. Relevel if necessary. *Note:* Both scales must be set in the same range.
5. Note the projected image on the scales. Adjust the focus knob on each track projector until the image is sharp and clear.
6. Note the track reading for each wheel. If the track readings are the same, front track is properly set. If the readings are not equal, complete steps 7 through 9.
7. Add the two track readings together. The sum will give you total track for the vehicle.
8. Divide total track by 2. The result will give you the correct track setting for each wheel. It should be one-half of total track.
9. Loosen the bolts on both tie-rod sleeves. Adjust the sleeves until both wheels are set to one-half of the total track. Tighten the tie-rod sleeve bolts.

FIGURE 20-16. Using Snap-on WA 210 Equipment To Measure Toe of Rear Wheels. (Courtesy of Snap-on Tools of Canada Ltd.)

Reading and Setting Rear Track

Use this procedure for setting the rear track on vehicles with four wheel independent suspension. Rear track alignment is performed only after checking and/or setting rear toe to make sure it is within specifications.

To set rear track, mount the gauges on the rear wheels with the track projectors directed toward the front of the vehicle, as shown in Figure 20-16. Attach the scales to the front tires. Use the same basic procedure as is used to set front track.

TOE AND TRACK ALIGNMENT PROCEDURES

There are four basic toe and track alignment procedures. The procedures vary in accordance with the vehicle's suspension system and/or the conditions causing misalignment.

Vehicles with Four Wheel Independent Suspension

Step-by-Step Procedure 1

1. Mount the gauges on the rear wheels and the scales on front. Set rear toe to specification. *Note:* Keep in mind that toe-in will now read as toe-out.
2. Read total track from the scales and adjust until both scales read one-half of total track. Lock the adjustment.
3. Mount the gauges on the front wheels and the scales on the rear. Set front toe to specification.
4. Read total front track from the scales. Adjust the tie-rod sleeves until both scales read one-half of total track. Tighten the tie-rod sleeve bolts to lock adjustment.

Vehicles with Solid Rear Axle

Step-by-Step Procedure 2

Use this procedure for all solid axle vehicles, unless rear axle condition is known to be bad. If the vehicle has axle offset, use procedure 3. If the vecle has axle side-set, use procedure 4.

1. Mount the gauges on the front wheels and the scales on the rear. Set front toe to specification.
2. Read total track from the scales. Adjust the tie-rod sleeves until both scales read one-half of total track. Tighten the tie-rod sleeve bolts to lock adjustment.

Solid Axle Vehicles with Axle Offset

Step-by-Step Procedure 3

1. Mount the gauges on the front wheels and the scales on the rear wheels. Turn the steering wheel until equal track is read on both scales. Lock the steering wheel.
2. Mount the gauges on the rear wheels and the scales on the front wheels. Loosen the locking knob on the scales and slide the scales until track reads equal on both scales. Tighten the locking knobs.
3. Mount the gauges on the front wheels and the scales on the rear. Turn the steering wheel to center position and relock. Set front toe to specification.
4. Read total front track from the scales and adjust the tie-rod sleeves until both scales read one-half of total track. Tighten the tie-rod sleeve bolts to lock adjustment.

Solid Axle Vehicles with Axle Side-Set

Step-by-Step Procedure 4

This procedure is also used if proper track cannot be set using procedures 1 to 3.

1. During road test, note the position of the steering wheel when the vehicle is driving straight.
2. Mount the gauges on the front wheels and the scales on the rear.
3. Turn the steering wheel to the position noted during the road test. Note the track reading for each wheel.
4. Turn the steering wheel to the center-point position and lock.
5. Adjust the tie-rod sleeves until each wheel is set to the track reading noted in step 3. Tighten the tie-rod sleeve bolts to lock adjustment.

ADJUSTING TOE

Step-by-Step Procedure

1. Prior to taking a toe reading, spread the front of the front wheels by pushing lightly outward on both wheels simultaneously (this is done to remove any slack from the steering linkage parts).

2. Read and record the toe reading. Then turn off the engine. **Caution:** If you are working in a confined space and the engine is allowed continually to idle, carbon monoxide will induce sickness. Never run the engine longer than necessary.

3. Prior to adjusting the toe, use penetrating oil to lubricate the tie-rod adjusting sleeve and the clamp sleeve locking bolts and nuts. Loosen the locking bolts and nuts sufficiently to allow the clamps to rotate partially. The sleeves probably will be

seized partially on the threads of the tie-rods. With an air hammer and its special adapter, move horizontally across the sleeves, shaking them loose. When the sleeves are loose, rotate the left and right sleeves clockwise or counterclockwise to obtain the desired toe-in specification (Figure 20-17).

When endeavoring to encourage a seized, rusted adjusting sleeve to move, it is sometimes the practice to use an acetylene torch to expand the metal. This work practice is not recommended by vehicle manufacturers. If heat is used, the part should be replaced. **Caution:** Never use a pipe wrench on an adjusting sleeve. The pipe may crush the sleeve, causing the metal part to fracture and creating a potentially dangerous situation.

4. When the toe has been adjusted to the specification, tighten the clamp bolts and nuts to the correct torque specification. *Note:* If the vehicle is equipped with power steering, check the toe reading with the engine idling to eliminate the slack in the power steering gear mechanism. When tightening the clamps, care must be exercised not to change the setting. Also, tie-rods and tie-rod ends must be in their correct position to allow them to pivot. *Note:* The tie-rod clamps must not interfere with other steering or suspension parts. General Motors of Canada recommends for its models that the tie-rod adjusting sleeve clamp be positioned on the adjuster tube (Figure 20-18) because there is less danger of possible failure by the linkage part.

Toe Change

The term *toe change* means just what the words say. The toe measurement between the vehicle's two front wheels constantly changes as the body-chassis deflects upward and downward from its curb riding height. This condition is normally caused by sagged springs and/or an incorrectly aligned front end. When a manufacturer designs the steering and suspension systems, the operating position of the tie-

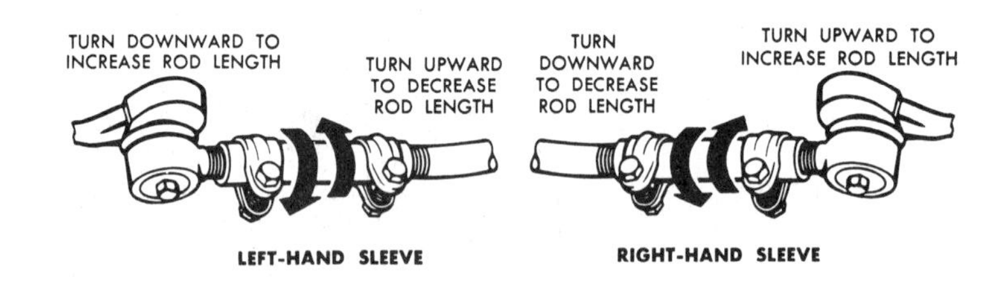

(a)

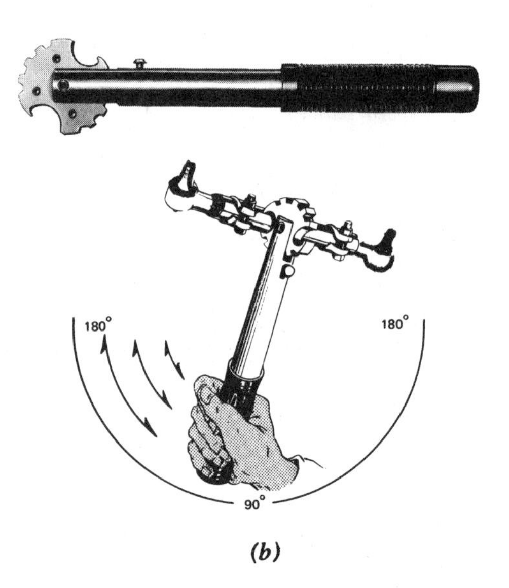

(b)

FIGURE 20-17. Adjusting Tie-Rod Sleeve. (a) Procedure; (b) special tool for rotating the tie-rod adjusting sleeves. (Part a Courtesy of Ford Motor Company; part b Courtesy of Bear Automotive, Inc.)

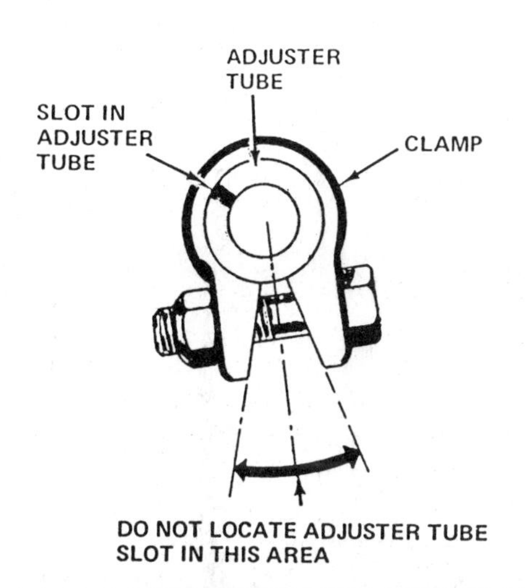

FIGURE 20-18. Correct Position of Clamp on Adjuster Tube. The adjuster tube must be set so that it will not interfere with other parts. (Courtesy of General Motors of Canada, Limited)

rods must be parallel to the positional planes of the right and left lower control arms. To determine if toe change exists after the toe has been correctly adjusted, lift up and press down on the middle of the front bumper while viewing the toe scales of the alignment equipment. If any appreciable change is evident from the original toe setting, this condition will influence tire tread wear.

In the past, some alignment technicians have tried to correct the problem by bending the spindle arms or the pitman and idler arms. This work procedure is not recommended by vehicle manufacturers. *Note:* If this condition exists, it is an indication of another problem within the vehicle's steering and suspension systems. When you have measured, adjusted, and checked for possible toe change, you may then remove the steering wheel holder.

Toe-Out on Turns

The next factor to examine is toe-out on turns. You will recall that it was discussed and illustrated in Chapter 19. If it is necessary to measure this factor, review the procedure described in Chapter 19.

Road Testing the Vehicle. Let's assume that our imaginary vehicle has been correctly aligned. Before replacing the dust caps, torque the wheel bearings to the manufacturer's specifications. Then install the correct-sized cotter pin. Chapter 4 describes the procedure.

Now let's discuss road testing the automobile. Since you are driving an unfamiliar vehicle, it is important that you drive defensively. Don't forget to check the brake pedal's reserve height (height of the pedal above the floor after the brakes have been applied). If the brake pedal is spongy or low, it may be best not to road test the vehicle because the vehicle may not be safe to drive. If this is the case, inform the customer and obtain permission to diagnose the cause of the problem.

It is best to perform a road test on a smooth asphalt pavement at a safe speed based on driving conditions. If the vehicle possesses directional control and stability, you will know this before the vehicle has been driven more than two city blocks from the shop. The center spoke of the steering wheel should be in its correct position, and the vehicle should not drift to the left or to the right. If the road is isolated from other traffic, remove your hands from the steering wheel and determine if steering control is needed. If the center spoke of the steering wheel is not in its correct position, additional service is required. Return the vehicle to the repair shop to recenter the steering wheel.

Centering the Steering Wheel. If the center spoke of the steering wheel is not centered, it is not difficult to correct this minor problem. To recenter the steering wheel, use the following procedure:

Step-by-Step Procedure

1. Obtain a hydraulic floor jack and two safety support stands.
2. Correctly place the jack under the front of the automobile, raise the vehicle, and position it on two safety stands. Position the front wheels in their straight-ahead location.
3. With a piece of chalk, mark the position of the left and right tie-rod adjusting sleeves in relation to the tie-rods (Figure 20-19).
4. Position the spoke of the steering wheel where it was located when the vehicle was being road tested. Turn the steering wheel to its correct centered position and notice the directional movement of the front wheels. Figure 20-20 illustrates how the problem is corrected. *Note:* Both tie-rod sleeves must be rotated an equal amount from their chalk-marked position. For example, if the left tie-rod sleeve is lengthened one-quarter turn, the right tie-rod sleeve must be shortened one-quarter turn.
5. When the sleeves have been rotated the de-

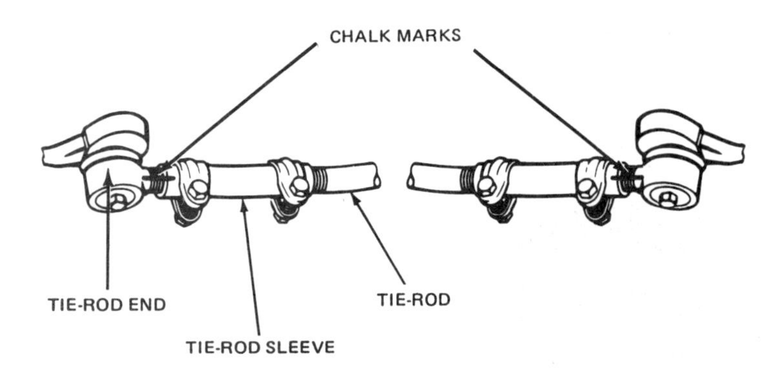

FIGURE 20-19. How to Determine Amount of Rotation. (Courtesy of Ford Motor Company)

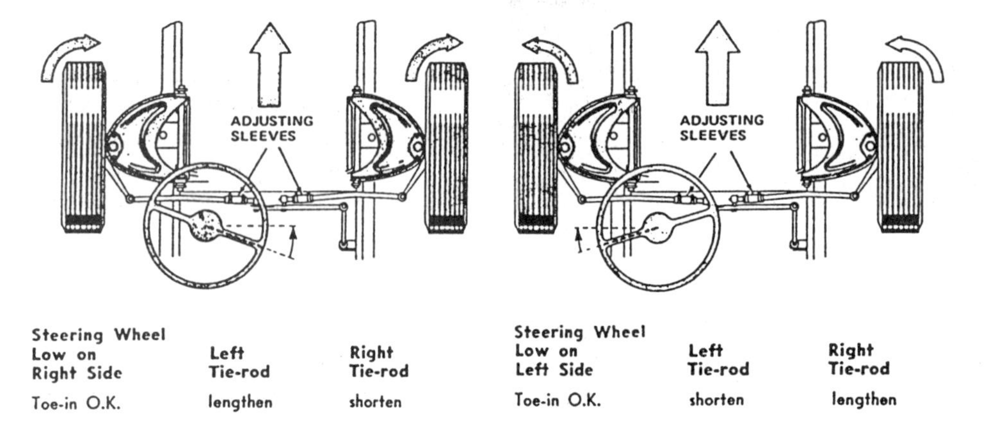

FIGURE 20-20. Proper Method of Steering Wheel Adjustment. The tie-rod sleeves must receive equal but opposite adjustment. (Courtesy of Bear Automotive, Inc.)

sired fraction of a turn, tighten the clamp bolts and nuts to the correct torque. Lower the vehicle and road test again.

WHEEL ALIGNMENT EQUIPMENT

There are many makes and models of wheel alignment equipment. Because of the variety, it would be impractical to discuss and illustrate how each of the various makes of equipment is used by alignment technicians. However, it is important to realize that all alignment equipment is precision instrumentation and must be operated according to the manufacturer's instructions. Before a garage or service station installs alignment equipment, the owner should assess the dollar return on the amount of money invested in the equipment and the utilization of service floor space. Another factor that is often overlooked by a business is the staff required to operate the equipment on a profit margin basis. Will the equipment be required to handle only alignment service, or will the equipment be required to handle major alignment, frame, and brake service?

Figure 20-21 illustrates a full rack alignment installation. This type of installation is compatible with both automobile and truck alignment service. Many alignment technicians prefer this equipment because it provides accessibility to the lower suspension of the vehicle. The horizontal front crossbeam allows the technician to move adjusting clamps and hydraulic jacks for straightening crossmembers or front axles. The rack provides an open, uncluttered work area and has a strong merchandising appeal in promoting alignment service.

The Hunter A111 system provides the most up-to-date computer technology in wheel alignment. The system uses sensors on all four wheels to detect and compute the angles of each wheel relative to the geometric centerline of the vehicle. As the operator adjusts toe, the system automatically references the front wheels to the rear wheel thrust line. Setback and axle offset are automatically taken into account. The A111 provides a continuous display of alignment data, from the initial check of the vehicle, through caster, camber, and toe changes, to the final check. The system simultaneously displays all wheel alignment factors on a television-type screen.

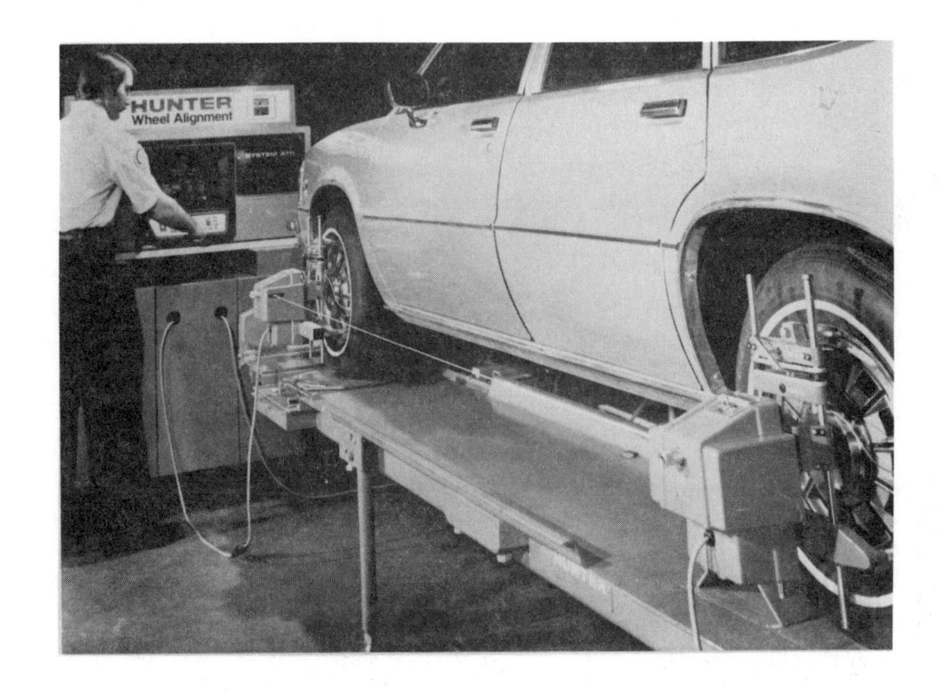

FIGURE 20-21. Hunter A111 Microprocessor Computer Four Wheel Aligner. (Courtesy of Hunter Engineering Company)

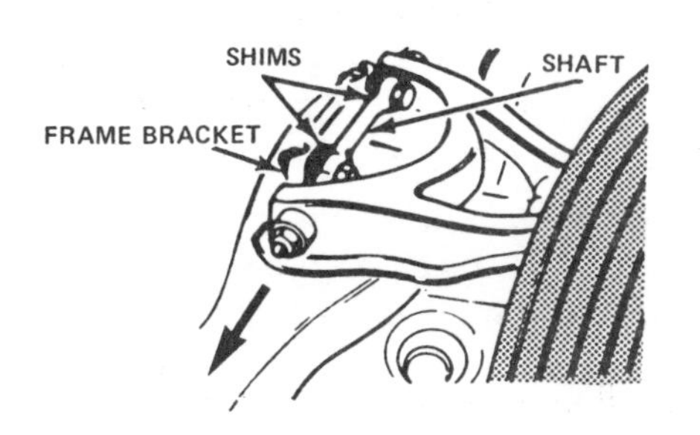

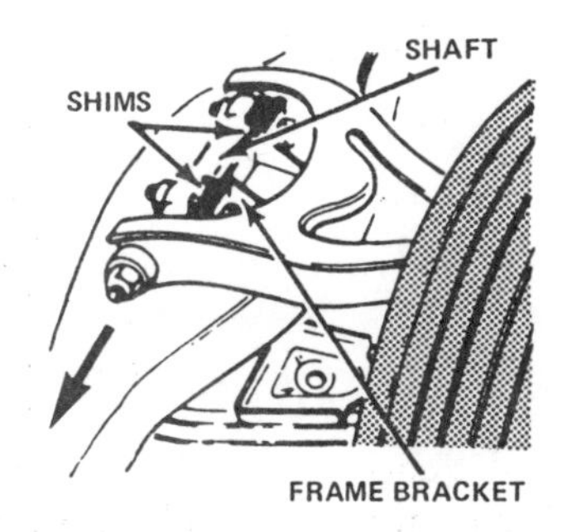

FIGURE 20-22. Two Methods of Caster and Camber Shim Adjustment. (Courtesy of FMC Corporation, Automatic Service Equipment Division)

METHODS OF ADJUSTING CASTER AND CAMBER

All vehicle manufacturers do not use the same method for adjusting caster and camber. The purpose of the following illustrations is to assist you in understanding some of the methods and how the two major directional control factors are changed.

Shim Method

The following illustrations, although they appear to be the same, are different. In Figure 20-22, the positions of the inner control arm shafts in relation to the frame brackets are reversed. For shim-type adjustment, adjust the caster angle first and then the camber. The number of shims necessary to correct the caster should be shifted from the rear to the front, or vice-versa. Then, remove an equal number of shims from both the front and the rear positions to correct the camber.

MacPherson Strut Suspension

Since there are now many cars with MacPherson strut suspension, some manufacturers do not provide a means of changing caster, only camber. To alter the camber, the rivet is removed, as shown in Figure 20-23. Loosen the front nut one-half turn. If the camber is to be moved more positive, move the top of the MacPherson strut outward.

To assist you in this procedure, it is often helpful to place a jack underneath the frame and raise the body of the vehicle 50 to 76 millimeters (2 to 3 inches). Loosen the two remaining nuts, and move the top of the tower outward. Once the tower has been moved outward, tighten the three retainer nuts and check the camber reading.

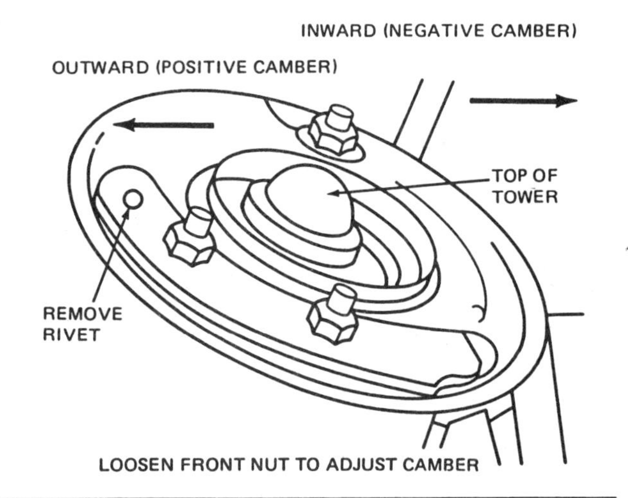

FIGURE 20-23. Changing Camber on MacPherson Strut Suspension. (Courtesy of Hunter Engineering Company)

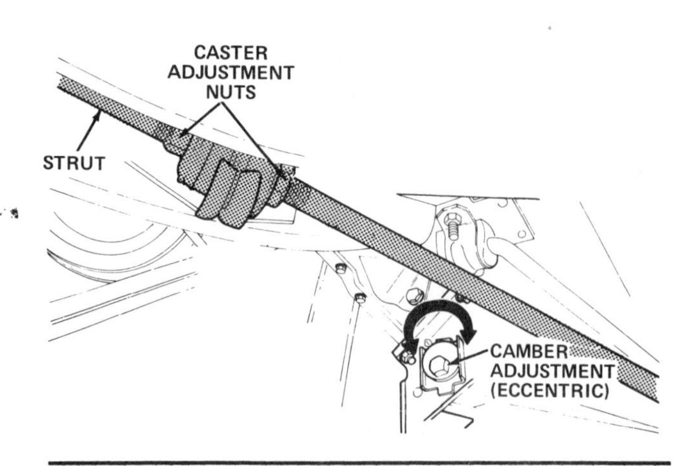

FIGURE 20-24. Adjusting Caster and Camber on Vehicle with Lower Eccentric and Strut Rod. (Courtesy of Ford Motor Company)

Lower Eccentric and Strut Rod

For automobiles designed with a camber adjustment and adjustable strut rods, to obtain the caster settings adjust the camber angle first and then the caster. A change in camber will affect the caster (Figure 20-24).

Elongated or Slotted Holes

Some automobile manufacturers adjust the caster and camber by sliding the upper control arm within elongated or slotted holes (Figure 20-25). This allows the upper arm to move inward and outward and to swing forward and backward when the attaching bolts are loosened. The two factors (caster and camber) are obviously interrelated since changing one changes the other. Special tools are used to move and position the upper arm while adjusting caster and camber.

A PRACTICAL SERVICE TIP

When the factors have been adjusted to the manufacturer's specifications, always recheck the caster and camber after tightening the adjusting bolts and locking nuts to the manufacturer's specification.

TRUCK AXLE SUSPENSION

Although this chapter has primarily discussed automotive steering and suspension systems, a text that deals with the important subject of alignment would not be complete unless it had discussed and illustrated the method used to align a truck's front and rear axles.

Preliminary Inspection

Before attempting to straighten a truck's axle, check the vehicle's frame to determine if it is bent. If the frame is bent, this condition must be corrected first. Second, examine spring shackles and hangers for possible distortion. Replace any sagged springs, broken leaves, and faulty center tie bolts. Examine and replace (if necessary) any worn or defective parts such as spindles, kingpins, and bushings.

Removing Twist from an I-Beam Axle

A twisted axle must be fixed before remedying an incorrect camber condition. To determine if an axle has sustained a twist, take caster readings and compare the left and right sides. If the readings are excessively different, the axle is twisted. To determine where the twist is located, lay a straightedge along the horizontal face of the axle. Figure 20-26

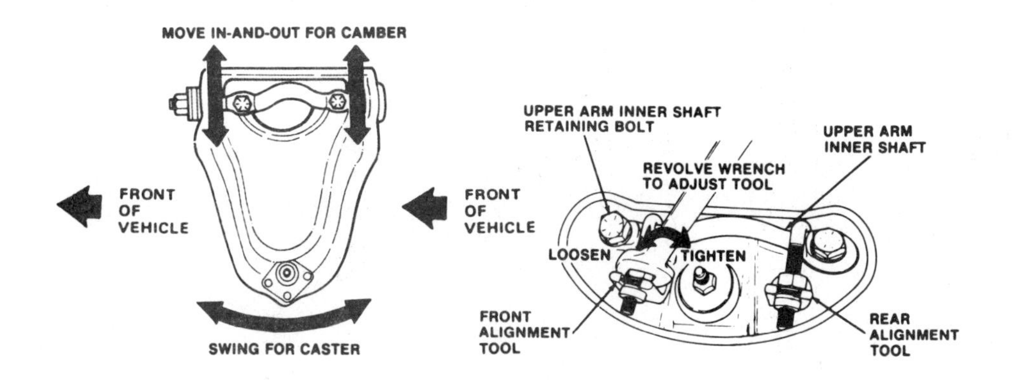

FIGURE 20-25. Setting Caster and Camber when Upper Control Arm Slides in Elongated or Slotted Holes. (Courtesy of Ford Motor Company)

FIGURE 20-26. Removing a Twist from I-Beam Axle. The arrows indicate the direction of force. (Courtesy of Bear Automotive, Inc.)

illustrates the procedure for removing a twist condition in the I-beam axle. A twist condition is corrected by using a series of tie yokes, connectors, and hydraulic jacks. Through leverage, a force of 180 tonnes (200 tons) may be produced, as shown by the arrows in Figure 20-26.

Changing Camber

Camber corrections are made after fore and aft bends and twists have been removed. Figure 20-27 illustrates how the camber angle may be moved to a more or less positive position.

Changing Caster

On a single I-beam front axle, the caster angle is adjusted by means of tapered axle shims installed

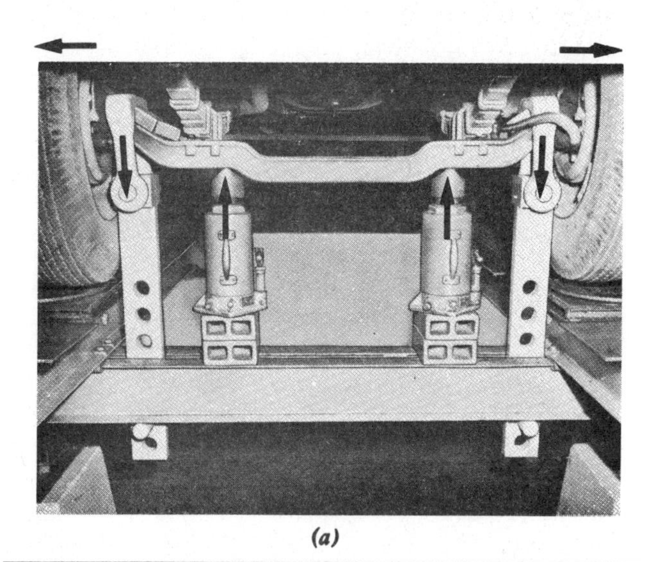

(a)

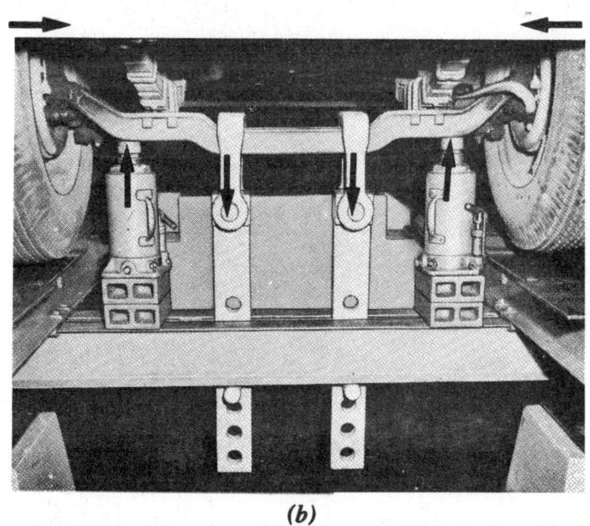

(b)

FIGURE 20-27. Adjusting Camber. (a) Moving camber to a more positive position; (b) moving camber to a more negative position. (Courtesy of Bear Automotive, Inc.)

between the springs and the axle, as illustrated in Figure 20-28. When there is a shim already under the leaf spring but the caster is not correct, remove the shim and replace it with one having a greater or lesser degree of taper. Never install two shims on one side of the axle.

Truck rear axle housings may also sustain misalignment. The rear wheels of a rigid rear axle housing should have 0° camber and a zero toe setting. Figure 20-29 illustrates the procedures to correct an incorrect camber and toe setting condition. To correct housing misalignment, place the housing in the position over a press beam so pressure can be applied against the bend. Heavy cables and connectors are used to hold down the housing to the alignment equipment.

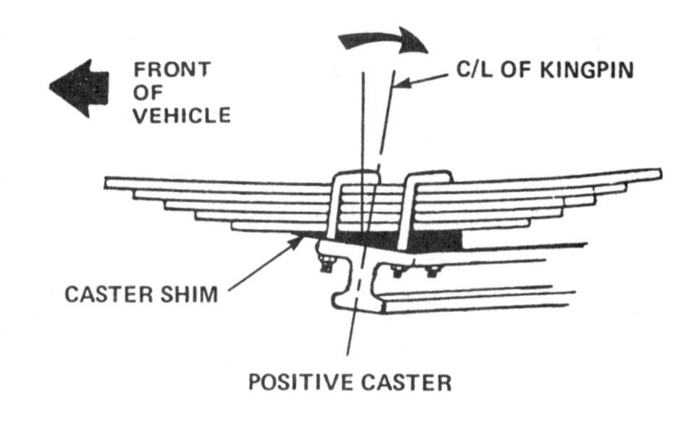

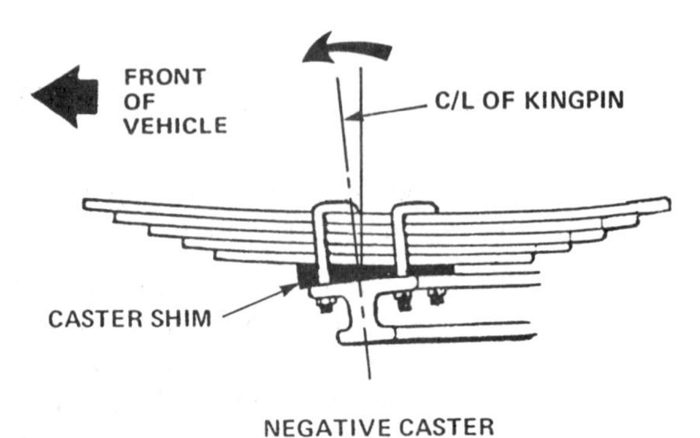

FIGURE 20-28. Adjusting Caster Angle Using Caster Shims. (Courtesy of Ford Motor Company)

PRACTICAL SERVICE TIPS

1. When correcting a truck's front or rear axle, make sure that all correcting equipment is properly installed as the straightening process proceeds.
2. Always recheck the position of all parts to prevent possible damage to the vehicle or potential injury to the operator.

REVIEW TEST

1. Prepare a complete list, in point form, of all the items to be inspected prior to checking the caster, camber, and toe factors.
2. Describe how to measure camber.
3. Describe how to measure caster.
4. When is it necessary to measure steering axis inclination?
5. Describe how to measure steering axis inclination.

(a)

(b)

FIGURE 20-29. Procedure for Correcting Camber and Caster in Rigid Rear Axle Housing. (a) Correcting camber and (b) correcting toe. (Courtesy of Bear Automotive, Inc.)

6. Why does a manufacturer specify minimum and maximum caster settings?
7. What does the term *preferred specification* mean?
8. The right upper inner control arm shaft is located on the outside of the vehicle's frame, and the camber and caster readings are adjusted by the shim method. The readings for the two factors are: camber, +1°; caster, 0°. The specifications are: camber, +1/2°; caster, +2°. Describe how to align the right front wheel of the vehicle.
9. Describe how to measure toe.
10. How is it possible to determine that the track of a vehicle is correct?
11. How is it possible to determine if the front wheels of a vehicle are subject to toe change?
12. Describe how to measure toe-out on turns.
13. Describe the correct procedure for recentering a steering wheel.

FINAL TEST

This examination is multiple choice. Only one answer will be accepted. Carefully read every statement.

1. Mechanic A says that only curb riding height needs to be checked before a vehicle is aligned. Mechanic B says that only tire pressure needs to be checked before aligning a vehicle. Who is right? (A) Mechanic A, (B) Mechanic B, (C) Both A and B, (D) Neither A nor B.
2. Mechanic A says that positive camber is the outward tilt of the top of the wheel. Mechanic B says that negative caster is the forward tilt of the top of the wheel. Who is right? (A) Mechanic A, (B) Mechanic B, (C) Both A and B, (D) Neither A nor B.
3. Mechanic A says that when reading caster, the front wheel must be placed in the straight-ahead position. Mechanic B says that when reading camber, the front wheel must be placed in the straight-ahead position. Who is right? (A) Mechanic A, (B) Mechanic B, (C) Both A and B, (D) Neither A nor B.
4. Mechanic A says that if it is necessary to use a wheel adapter, determine lateral runout before measuring camber. Mechanic B says that if it is necessary to use a wheel adapter, torque the front bearing to the manufacturer's torque before measuring camber. Who is right? (A) Mechanic A, (B) Mechanic B, (C) Both A and B, (D) Neither A nor B.
5. Mechanic A says to use a brake jack and depress the pedal before measuring caster. Mechanic B says that if the vehicle has power brakes, use a brake pedal jack while the vehicle's engine is operating. Who is right? (A) Mechanic A, (B) Mechanic B, (C) Both A and B, (D) Neither A nor B.
6. Mechanic A says that if the camber is positive, the positive camber reading must be added to the reading obtained from the steering axis inclination scale. Mechanic B says that if the camber reading is 0°, the gauge cannot read the true steering axis inclination angle. Who is right? (A) Mechanic A, (B) Mechanic B, (C) Both A and B, (D) Neither A nor B.
7. Mechanic A says that if the camber reading is 1/4° negative and the specification is 3/4° positive, the total angle change is 1/2°. Mechanic B says that if the caster reading is 1° negative and the specification is 1° positive, the total angle change is 2°. Who is right? (A) Mechanic A, (B) Mechanic B, (C) Both A and B, (D) Neither A nor B.
8. Mechanic A says that if the minimum caster setting is 1° positive and the maximum setting is 3° positive, the preferred setting is 2° positive. Mechanic B says that based on the text, an experienced alignment technician will divide the preferred toe specification by 2. Who is right? (A) Mechanic A, (B) Mechanic B, (C) Both A and B, (D) Neither A nor B.

Questions 9, 10, and 11 pertain to shim-type adjustments.

9. Mechanic A says that based on the text, a 1/2° of camber movement is equal to a 1.6 millimeter (1/16 inch) shim. Mechanic B says that based on the text, a 1/2° of caster movement is equal to a 3 millimeter (1/8 inch) shim. Who is right? (A) Mechanic A, (B) Mechanic B, (C) Both A and B, (D) Neither A nor B.
10. Mechanic A says that when changing caster, it is necessary to divide the total angle change by 2. Mechanic B says that when changing camber, it is also necessary to divide the total angle

change by 2. Who is right? (A) Mechanic A, (B) Mechanic B, (C) Both A and B, (D) Neither A nor B.

11. Mechanic A says that in order to change only camber, equal shims must be added to or removed from the adjustment positions. Mechanic B says that in order to change only caster, the shims removed from one adjustment position must be added to the opposite position. Who is right? (A) Mechanic A, (B) Mechanic B, (C) Both A and B, (D) Neither A nor B.
12. Mechanic A says that the toe setting will not adversely affect track alignment. Mechanic B says that altering the caster and camber settings will not change the toe setting. Who is right? (A) Mechanic A, (B) Mechanic B, (C) Both A and B, (D) Neither A nor B.
13. Mechanic A says that if camber is used exclusively to offset road crown, the caster settings must be the same. Mechanic B says that if caster is used exclusively to offset for road crown, the camber settings must be the same. Who is right? (A) Mechanic A, (B) Mechanic B, (C) Both A and B, (D) Neither A nor B.
14. Mechanic A says that toe change is caused by incorrect toe settings. Mechanic B says that toe change is caused by incorrect curb riding height. Who is right? (A) Mechanic A, (B) Mechanic B, (C) Both A and B, (D) Neither A nor B.
15. Mechanic A says that it is recommended that the caster angle be changed prior to altering the camber factor. Mechanic B says that if the caster is changed through the strut rod, lengthening the rod will decrease the positive caster setting. Who is right? (A) Mechanic A, (B) Mechanic B, (C) Both A and B, (D) Neither A nor B.

Introduction to Brakes

LEARNING OBJECTIVES

After studying this chapter, you should be able to:

- Name the component parts of a vehicle's hydraulic brake system,
- Describe the external design of the component parts of a vehicle's hydraulic brake systems and explain their purpose,
- Define three terms related to operation of a brake system: *kinetic friction*, *static friction*, and *coefficient of friction*,
- Illustrate and explain how force applied to a vehicle's brake pedal is multiplied upon brake application through the mechanical advantage of a brake pedal's lever,
- Define Pascal's law,
- Explain the term *hydraulic pressure*,
- State three variations of a formula pertaining to Pascal's law,
- Illustrate and explain how force and motion are transmitted through the application of hydraulics,
- Explain how force is increased and decreased through the application of hydraulics.

INTRODUCTION

One of the world's greatest inventions is the wheel. The introduction of the wheel, however, created a problem: how to control its speed and stop it from revolving.

When the first four-wheel carriage came on the scene, the driver would attempt to control its speed and stop its forward progress by dragging a log behind the carriage. This method of stopping a vehicle was totally ineffective and sent the engineers of that day back to their drawing boards.

During the evolution of primitive brake systems, the second method for controlling a moving vehicle came with the introduction of a type of skid called a *shoe*, which was pressed against the outside rim of the wheel (Figure 21-1). A crude device, the shoe was usually a piece of curved wood operated by mechanical linkage as the driver pulled on a long lever. To stop a slow-moving carriage, this design was adequate; however, even for early motorized vehicles, this type of brake mechanism needed to be improved.

In the mid-1920s, hydraulic brakes were introduced to the motoring public by vehicle manufacturers. This brake mechanism was a vast improvement over the mechanical system of rods and operating levers. The driver had more control over the vehicle and, as the saying goes, could stop on a dime.

Let's begin our study of vehicle brake systems by becoming familiar with the names of the component parts, their external design and purpose, frictional facts and terms, mechanical advantage through leverage, and principles of hydraulics as they relate to the operation of brakes.

BRAKE SYSTEMS

When you are required to stop a vehicle, you place your foot on the brake pedal and exert force. (See Figure 21-2.) The downward movement of the pedal is transferred through a lever and rod to an optional power assist unit. The purpose of this unit is to reduce the effort required by the driver to stop the vehicle. From the power assist unit, an internal push rod (not shown in Figure 21-2) operates a tandem (dual) master cylinder, the main controlling component part of a brake system. As you study Figure 21-2, you will notice that a modern vehicle actually has two brake systems: one for the front wheels, and one for the rear wheels. In some vehicles, the brake systems are diagonal: left front to right rear.

As the brakes are applied, the master cylinder, which is a hydraulic pump, pressurizes brake fluid and transmits the liquid through two hollow steel tubes to the pressure differential valve and switch. The purpose of the valve and switch is to indicate if the front or rear brake systems have developed a loss of fluid and are therefore unable to produce hydraulic pressure. If a leak has developed, a red warning light will appear on the instrument panel of the dash.

Front Brake System

From the pressure differential valve, fluid is directed to a metering valve, which is a part of some front disc brake systems. The purpose of this valve is to reduce line pressure to the front brakes during initial brake application. In other words, the rear

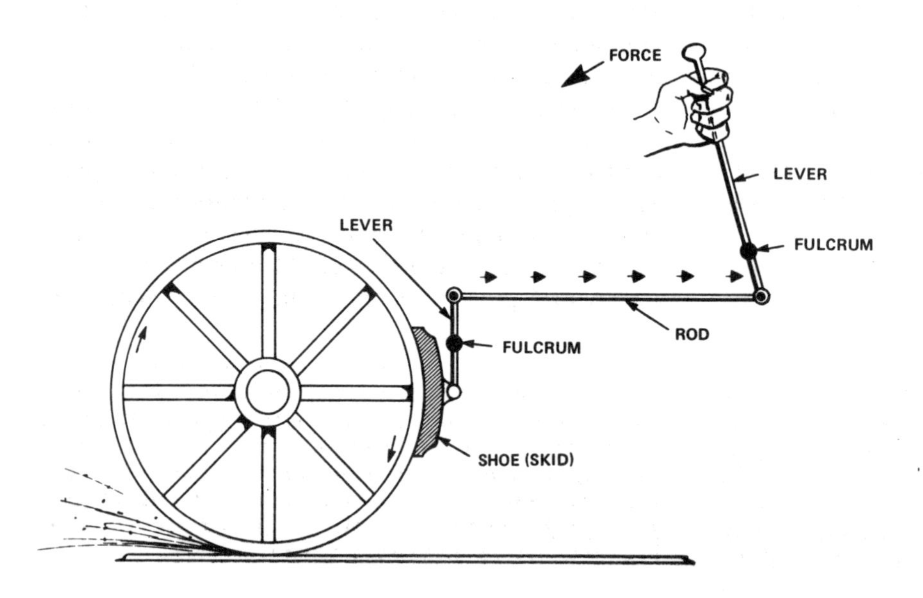

FIGURE 21-1. Primitive Brake System. (Courtesy of Ford Motor Company)

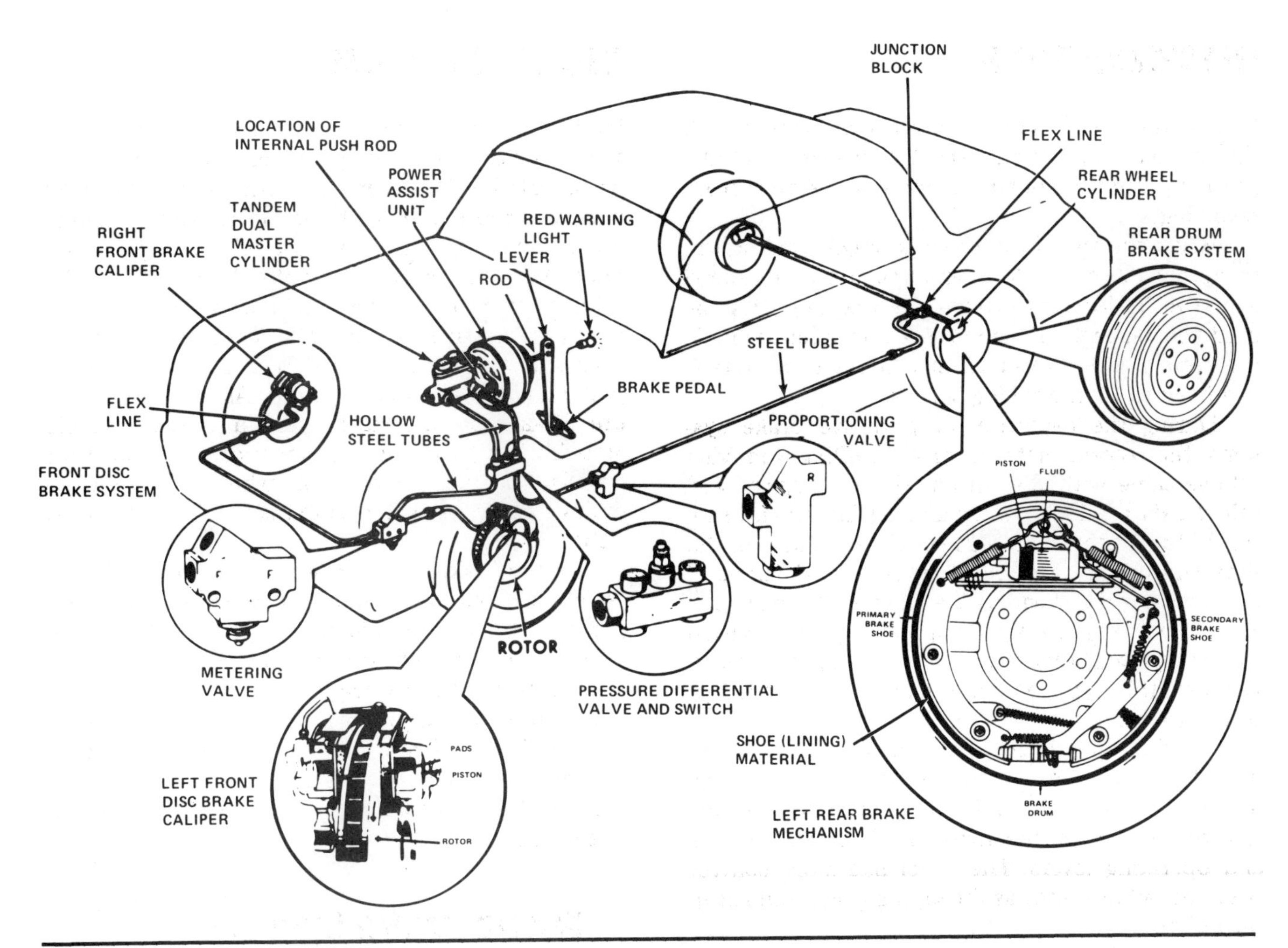

FIGURE 21-2. Diagram of a Modern Brake System. (Courtesy of Wagner Brake Company, Ltd.)

brakes begin to function before the front. Fluid is then transmitted through steel tubing and flex lines to the front brake calipers. Since hydraulic brake systems are confined (sealed) systems, the fluid under pressure, acting on pistons in calipers, causes brake pads made from asbestos fibers to press and rub against revolving rotors attached to the front wheels of the vehicle. The friction created between the pads and the rotors retards and stops the movement of the front wheels.

Rear Brake System

For a vehicle with a rear drum brake system, a line from the pressure differential valve and switch directs fluid to a proportioning valve.* The purpose of this valve is to reduce line pressure to the rear wheel cylinders during severe or panic stop brake application. In an emergency stop, the front end of the vehicle dips down due to a transfer of weight from the rear of the vehicle to the front.

After fluid has passed through the proportioning valve, it is transmitted through a steel tube and flex line to the rear wheel cylinders. During brake application, pistons (inside the wheel cylinder) press primary and secondary brake shoes against the inside surface of the brake drums. Although this is not shown in Figure 21-2, the drums are connected to

*When learning technical information, it is often a help to use some connecting idea as an aid to remembering the name of a part and its location in the vehicle. To help you remember the location of the proportioning valve, let's look at how the word is spelled. *Proportioning* has two r's. Since the *proportioning* valve is an important component of the *rear* brake system, the two r's in rear connect with the two r's in proportioning. Therefore, the proportioning valve is located in the rear drum brake system.

the rear wheels by the rear axles. The friction created between the brake shoes and the drums retards and stops the movement of the rear wheels.

While you were being introduced to the component parts of the vehicle's front and rear brake systems, the word *friction* was mentioned. What is friction? Friction is resistance to motion. Any time an object slides across another object, friction will tend to resist the motion of the moving object or objects, and, as we know, friction generates heat. When your hands are cold, you rub your palms together to warm them. Through sliding or moving friction, you are producing heat. Heat is also generated when the brakes of your vehicle are applied. The brakes retard and stop the motion of the wheels of a vehicle by developing controlled friction between the shoes and drums, or the disc pads and the rotors. This type of friction, between a more or less stationary surface and a moving surface, is called *kinetic friction.*

Tire Traction

The friction created between the lining material and the drums and/or rotors is essential to braking; however, that is only part of the story. Traction (friction) between the tire treads and the road surface is also important.

When a wheel is revolving on the surface of the road, the tire's tread, regardless of the speed of the vehicle, momentarily stops for a split second. This type of friction between two stationary surfaces is called *static friction.* Without static friction between the tire's tread and the road, the vehicle could not go forward or backward, nor could it stop on brake application. This subject was discussed in Chapter 6.

Characteristics of Brake Lining Material

The material used in a brake system comes in contact with the drums or rotors—called *lining* or *pads*—and therefore must provide efficient safe braking regardless of the vehicle's speed or weight. This means that the heat developed in the braking mechanism during braking must be dissipated into the air as quickly as possible.

Special material, consisting of asbestos and resins, is used for linings and pads (see Figure 21-2). To reduce the possibility of excessive heat buildup, the linings and pads must have total contact with

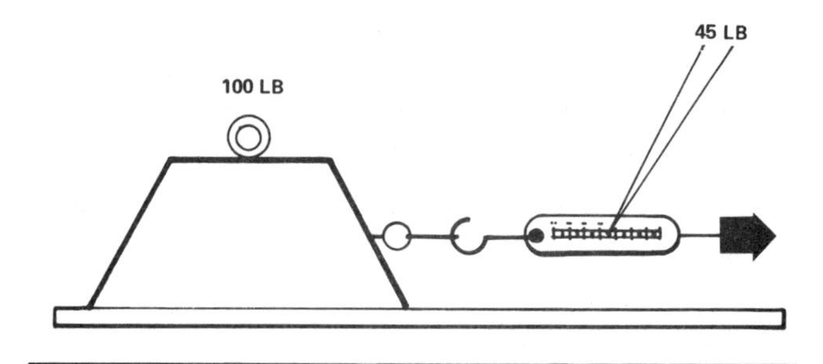

FIGURE 21-3. Illustration of Coefficient of Friction. (Courtesy of Wagner Brake Company, Ltd.)

the drums and/or rotors. Also, the drums and rotors must have sufficient metal thickness to be able to absorb and disperse the heat. When brake lining material and the drums and/or rotors become excessively hot due to severe braking applications, drums and rotors alter their shape, and the frictional characteristics of the lining change. When this situation occurs, the lining material is unable to retard the motion of the drums and/or rotors and produces a condition known as **brake fade.** When brakes fade, the maximum force applied to the brake pedal will not stop the vehicle.

Brake manufacturers, when producing linings, must determine through testing the coefficient of friction of the lining's material to assess durability and dependability. The coefficient of friction can be described as a ratio: the force required to slide an object over a surface in proportion to the weight of the object. Figure 21-3 illustrates 44 kilograms (100 pounds) of mass (weight) being pulled by 20.4 kilograms (45 pounds) of force. The ratio of force to weight is the coefficient of friction. In this example, it is 0.45.

If a vehicle's brakes were to work at an excessively high coefficient of friction, abrasion would very quickly wear down the brake linings, drums, and rotors. Since brake pads and shoes must be durable, lining material is generally manufactured with a low and medium coefficient of friction. Brake design, vehicle weight, wheel sizes, and the force required to apply the brakes are important factors taken into consideration in the basic engineering of brake systems.

FACTS TO REMEMBER

1. Providing that there are no hydraulic and mechanical deficiencies with a vehicle's brake systems, the brakes, when applied, will effectively stop a vehicle when:

there is kinetic friction between the linings and drums and/or rotors.
there is static friction between the tire treads and the surface of the road.

2. When you slam on the brakes, the wheels cease to roll on the road's surface, and you produce:

   static friction between the linings and the drums and/or rotors.
   kinetic friction between the tire treads and the surface of the road.

3. During a panic stop, the distance required to brake and stop a vehicle will be increased, there will probably be a loss of directional control, and tire tread wear will be sustained.
4. Since friction produces heat and wear, never drive with your foot lightly resting on the vehicle's brake pedal. Premature wear of the linings and drums and/or rotors will result.

BRAKING FORCE

To this point, we have discussed the external design of a modern hydraulic brake system and the facts pertaining to friction; however, friction is also dependent on force. *Force* is the amount of push or pull on an object. If you pull a cart, you use force; if you push a brake pedal, you use force. Even if you push against an immovable stone wall, you use force.

How does force work in a vehicle's brake systems? Even more important, how can the force from a human foot be multiplied to slow or stop a vehicle weighing thousands of pounds and traveling at high speeds? When the driver presses the brake pedal, he or she is using physical force to stop the vehicle. The force applied to the brake pedal can vary depending on the distance required for braking. To understand the following illustration (Figure 21-4), let's assume that the force applied to a brake pedal is 22.6 kilograms (50 pounds).

Since the brake pedal mechanism is a lever, the mechanical advantage (M.A.) of the lever can be calculated and expressed as a ratio:

$$\frac{d-2}{d-1} = \frac{25.40 \text{ cm (10 in.)}}{5.08 \text{ cm (2 in.)}} = \text{M.A. } 5:1$$

In this example, the answer is 5:1. In other words, since the M.A. is 5:1, force (push) applied to the brake pedal will be multiplied 5 times. As you study Figure 21-4, you will notice that the force of 22.6 kilograms (50 pounds) has been increased on the master cylinder's push rod to 113 kilograms (250 pounds). Now that we know how force applied to a brake pedal is multiplied, let's become familiar with the basic principles of hydraulics.

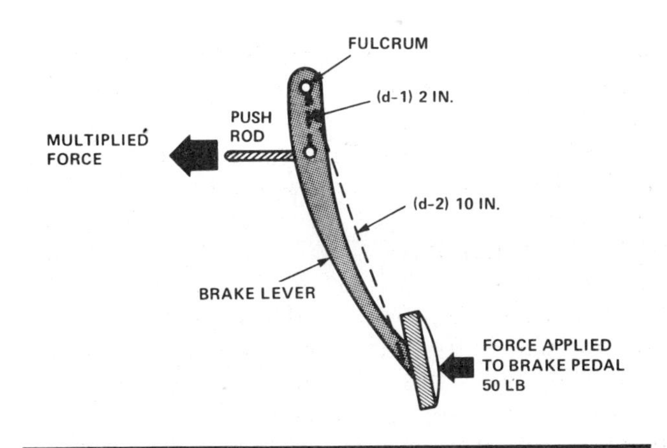

FIGURE 21-4. Illustration of Braking Force.

PRINCIPLES OF HYDRAULICS

In 1650, a French physicist, Blaise Pascal, discovered the fundamental law of physics on which all modern hydraulics is based. Hydraulics is basically the study of the potential energy of liquid under pressure and its use to perform work. Examples are the hydraulic floor jack used to raise a vehicle, and the brake system used to control and stop the momentum of a moving vehicle. Essentially, Pascal's law states, "Pressure exerted on a confined liquid is transmitted undiminished in all directions and acts with equal force on all equal areas."

Hydraulic Pressure

What is **hydraulic pressure** and how is it created? To begin, let's assume for practical purposes that liquids are essentially incompressible. (Although liquid is said to be incompressible, in a high pressure hydraulic system, oil does compress slightly. For example, at 21,000 kPa [3000 psi], oil loses approximately 1.2 percent of its volume. This small volume depreciation is not significant in a vehicle's hydraulic brake system.) Figure 21-5 illustrates an enclosed cylinder that contains a liquid and is sealed by a piston. Exerted on top of the piston is a force of 45.34 kilograms (100 pounds). The area of the piston is 2.54 square centimeters (1 square inch).

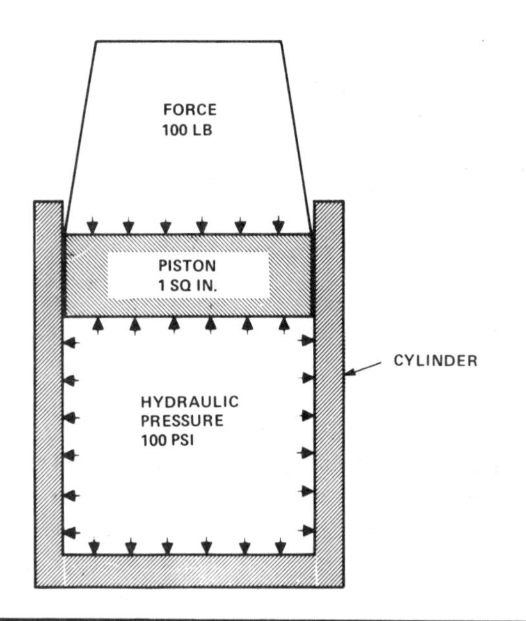

FIGURE 21-5. Illustration of Hydraulic Pressure.

Since the force on the piston is endeavoring to compress a liquid that is noncompressible, hydraulic pressure is created on the entire inside surface of the cylinder, as well as the bottom surface of the piston. The pressure within the cylinder is approximately 700 kPa (100 psi).

Calculating Force, Area, and Pressure

Through the use of simple mathematics and the device shown in Figure 21-6 and based on the formula for Pascal's law, force, area, and pressure can be calculated. Although there appear to be three separate formulas, there is only one formula expressed in three variations: f refers to force, a indicates area, and p represents pressure. Thus, the variations of the formula are written:

$$f = a \times p$$

$$p = \frac{f}{a}$$

$$a = \frac{f}{p}$$

Any letter in the triangle may be expressed as the product or the quotient of the other two, depending on its position within the triangle. For example, to calculate the hydraulic pressure on the entire inside surface of the cylinder and the bottom surface of the piston, you would consider the letter

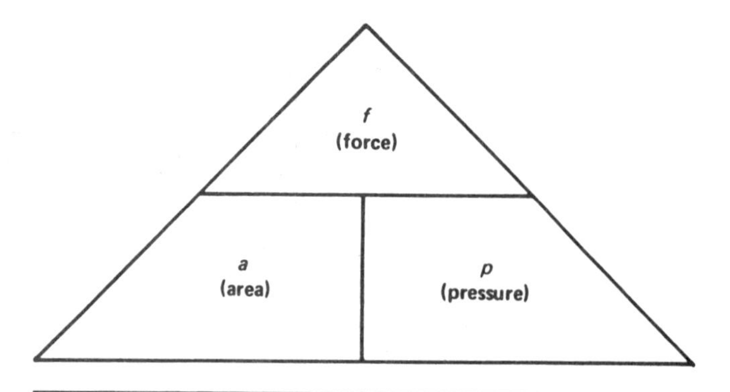

FIGURE 21-6. Simple Device for Remembering Variations of the Formula to Calculate Force, Area, and Pressure.

p as being set off by itself. Look at the other two letters. The letter f is the letter a; therefore,

$$\frac{f}{a} = p$$

Figure 21-5 illustrated an enclosed cylinder containing a liquid that was sealed by a piston and subjected to a downward force. The pressure within the cylinder was calculated by using the formula:

$$\frac{f}{a} = p$$

and substituting:

$$\frac{100 \text{ lb } (45.35 \text{ kg})}{1 \text{ sq in. } (2.54 \text{ cm}^2)} = 100 \text{ psi } (700 \text{ kPa})$$

Force and Motion

During his early experiments with hydraulics, Pascal also discovered that liquids can be used to transmit force and motion. When a driver presses down on the brake pedal, force and motion occur. Through the operation of the master cylinder, the action on the pedal is transmitted through fluid in the brake lines to the pistons located in the front disc brake calipers and the rear wheel cylinders. The movement of the pistons forces the pads against the rotors and the shoes against the brake drums. In summary, the force and motion exerted on the brake pedal has been transmitted through the vehicle's brake systems and has produced motion and force in the brake mechanisms at the wheels.

Figure 21-7 illustrates two cylinders, A and B. The pistons within the two cylinders are identical in area. The piston within cylinder A is subject to

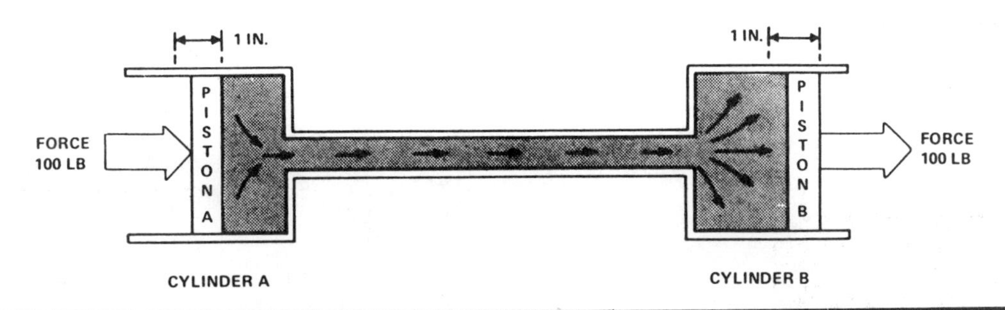

FIGURE 21-7. Illustration of Transmitted Force and Motion.

a force of 45.35 kilograms (100 pounds) and has moved a distance of 2.54 centimeters (1 inch) within its cylinder's bore. The piston in cylinder B is producing a force of 45.35 kilograms (100 pounds) and has moved 2.54 centimeters (1 inch) within its cylinder's bore. Since both pistons are identical in area, whatever force is exerted on piston A, the same force will be produced by piston B. Also, the movement of both pistons will be the same. In reality, piston A has displaced the fluid from cylinder A through a line to cylinder B, thus transmitting force and motion. As you study Figure 21-7, reread the definition for Pascal's law. It will now have more significant meaning.

Increasing Force and Decreasing Force

It is not only possible to transmit force and motion. It is also possible to increase force, as well as decrease force.

In Chapters 24 and 25, you will be requested to dismantle, inspect, service and reassemble an automobile's brake systems and their component parts. Upon inspection of the front and rear systems, you will find that the pistons in the front calipers (if the vehicle is equipped with disc brakes) are larger than those in the rear wheel cylinders.

Earlier in this chapter, we saw that weight is transferred to the front of a vehicle upon brake application. Since there is a transfer of weight, the pistons in the front hydraulic cylinders must be capable of producing and applying more force to the brake pads or shoes than the rear brake cylinders. This is accomplished by having larger pistons in the front brake cylinders and smaller pistons in the rear cylinders.

Figure 21-8 provides a schematic view of a hydraulic brake system. Cylinder A is the master cylinder and has a force of 90.71 kilograms (200 pounds) applied to the piston. The surface area of the piston is 2.54 square centimeters (1 square inch).

Since we know the applied force and the piston's area, we can calculate the pressure created within the hydraulic system at this time of brake application. To calculate pressure, you would use the formula:

$$\frac{f}{a} = p$$

and substitute:

$$\frac{90.71 \text{ kg } (200 \text{ lb})}{2.54 \text{ cm}^2 \ (1 \text{ in.}^2)} = 1400 \text{ kPa } (200 \text{ psi})$$

Now that we know the hydraulic pressure within the brake system, let's direct our attention to cylinder B. Cylinder B represents the front brake cylinder in our illustration; the area of the front piston is 5.08 square centimeters (2 square inches). Since we have determined the hydraulic pressure and know the area size of the piston in the front brake cylinder, we can calculate the force being produced by the piston. To calculate force, you would use the formula:

$$a \times p = f$$

and substitute:

$$5.08 \text{ cm}^2 \ (2 \text{ in.}^2) \times 1400 \text{ kPa } (200 \text{ psi}) = 181.40 \text{ kg } (400 \text{ lb})$$

At this point, it is interesting to note that the force applied to the master cylinder's piston has been doubled, through hydraulics, by doubling the area size of the piston in the front brake cylinder.

Cylinder C represents the rear brake cylinder in our schematic diagram. The area of the rear piston is 1.27 square centimeters (0.5 square inch). Since we have previously calculated the hydraulic pressure

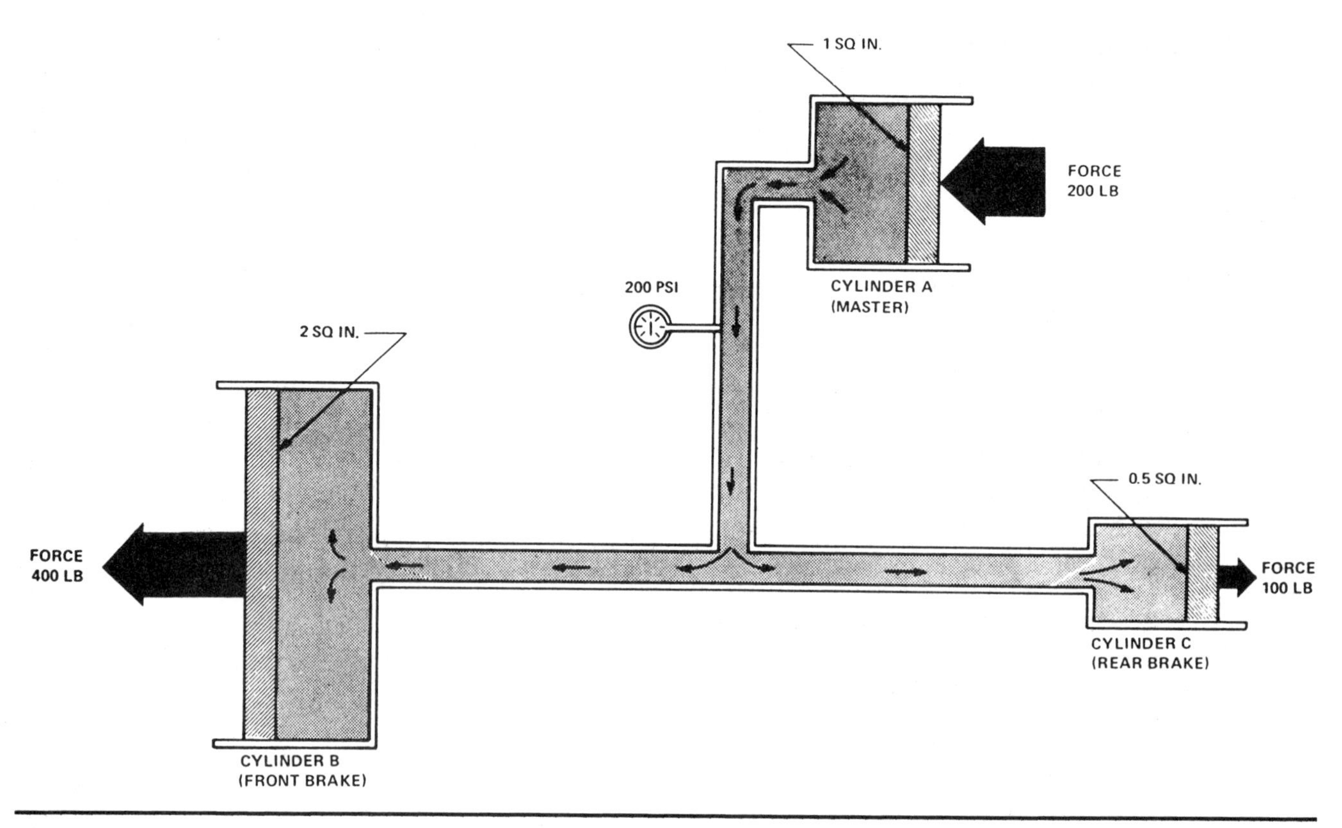

FIGURE 21-8. Schematic View of Hydraulic Brake System.

in the brake system and know the area size of the piston in the rear brake cylinder, we can calculate the force being produced by the piston. To calculate force, you would use the formula:

$$a \times p = f$$

and substitute:

$$1.27 \text{ cm}^2 \ (0.5 \text{ in.}^2) \times 1400 \text{ kPa } (200 \text{ psi}) = 45.35 \text{ kg } (100 \text{ lb})$$

As we compared the force applied to the master cylinder's piston, we can see that through the application of hydraulics, it is also possible to decrease force by decreasing the area size of the piston in the rear brake cylinder. In this case, the force has been halved by making the rear brake piston half the size of the master cylinder piston. An engineer designing a vehicle's brake system utilizes Pascal's law and the application of hydraulics to produce the necessary stopping forces that provide the driver with safe, efficient braking.

FACTS TO REMEMBER

1. Upon brake application, hydraulic pressure within the brake system is equal in every part of the system unless controlled by a metering and/or a proportioning valve.
2. Through the application of hydraulics, liquids transmit force and motion.
3. Through the application of hydraulics, force may be increased and decreased.

REVIEW TEST

1. List the names and briefly explain the purpose of the major component parts of a vehicle's hydraulic brake systems.
2. Define the term *kinetic friction.*
3. Explain how static friction is utilized during brake application.
4. Briefly list four desirable characteristics of brake lining material.

5. Define the term *coefficient of friction.*
6. Describe how force applied to a vehicle's brake pedal is multiplied through the mechanical advantage of the lever.
7. Explain the term *hydraulic pressure.*
8. State the variations of the formula to calculate force, area, and pressure.
9. Illustrate and explain how force and motion are transmitted through the application of hydraulics.
10. Briefly explain why the pistons located in a vehicle's front brake cylinders are larger in area size than the pistons in the rear brake cylinders.

FINAL TEST

This examination is multiple choice. Only one answer will be accepted. Carefully read every statement.

1. Mechanic A says that when you apply the brake pedal, you exert force, not pressure. Mechanic B says that the primary purpose of a dual master cylinder is to produce flow and pressurize the fluid. Who is right? (A) Mechanic A, (B) Mechanic B, (C) Both A and B, (D) Neither A nor B.
2. Mechanic A says that fluid from the master cylinder is directed through hollow steel tubes to a pressure differential valve. Mechanic B says that the purpose of the pressure differential valve is to restrict fluid flow to the rear brakes. Who is right? (A) Mechanic A, (B) Mechanic B, (C) Both A and B, (D) Neither A nor B.
3. Mechanic A says that the metering valve is part of the rear brake system. Mechanic B says that the proportioning valve is part of the front brake system. Who is right? (A) Mechanic A, (B) Mechanic B, (C) Both A and B, (D) Neither A nor B.
4. Mechanic A says that in a disc brake mechanism, brake pads made from asbestos fibers press against a brake drum. Mechanic B says that in a drum brake mechanism, primary and secondary brake shoes press against the surfaces of a rotor. Who is right? (A) Mechanic A, (B) Mechanic B, (C) Both A and B, (D) Neither A nor B.
5. Mechanic A says that when the wheels are revolving and the brakes are applied, there is kinetic friction between the pads/shoes and the rotors/drums. Mechanic B says that when the wheels are not revolving and the brakes are applied, there is static friction between the tire's tread and the road. Who is right? (A) Mechanic A, (B) Mechanic B, (C) Both A and B, (D) Neither A nor B.
6. Mechanic A says that brake fade is the result of a loss of friction during brake application. Mechanic B says that when brake fade occurs, maximum force applied to the brake pedal will not stop the vehicle. Who is right? (A) Mechanic A, (B) Mechanic B, (C) Both A and B, (D) Neither A nor B.
7. Mechanic A says that the mechanical advantage of a lever cannot be calculated until the force applied to the pedal is stated. Mechanic B says that since the pedal mechanism is a lever, the mechanical advantage of the lever can be expressed as a ratio. Who is right? (A) Mechanic A, (B) Mechanic B, (C) Both A and B, (D) Neither A nor B.
8. Mechanic A says that hydraulics is basically the study of the potential energy of liquid under pressure and its use to perform work. Mechanic B says that Pascal's law states that all liquids, regardless of type, are essentially compressible. Who is right? (A) Mechanic A, (B) Mechanic B, (C) Both A and B, (D) Neither A nor B.
9. Mechanic A says that through the use of liquids, it is possible to transmit motion. Mechanic B says that through the use of liquids, it is possible to transmit force. Who is right? (A) Mechanic A, (B) Mechanic B, (C) Both A and B, (D) Neither A nor B.
10. Mechanic A says that the formula used to calculate the area of a piston within a cylinder is $a = p/f$. Mechanic B says that the formula used to calculate the force produced by a piston within a cylinder is $f = a \times p$. Who is right? (A) Mechanic A, (B) Mechanic B, (C) Both A and B, (D) Neither A nor B.
11. Mechanic A says that an automobile has larger pistons in the front brake system than in the rear brake system. Mechanic B says that hydraulic pressure must be greater in the front brake system than in the rear brake system during initial brake application. Who is right? (A) Mechanic A, (B) Mechanic B, (C) Both A and B, (D) Neither A nor B.

Brake Fluid and Master Cylinders

LEARNING OBJECTIVES

After studying this chapter, you should be able to:

- Describe the unique characteristics of brake fluid,
- Illustrate and describe the design of a single piston master cylinder,
- Describe the operation of a single piston master cylinder,
- Describe the operation of a dual master cylinder,
- Explain how to diagnose problems associated with the operation of a dual master cylinder,
- Explain and demonstrate how to service a dual master cylinder,
- Explain and demonstrate how to test the operation of a dual master cylinder.

INTRODUCTION

In Chapter 21, you were introduced to the principles of hydraulics as they apply to the operation of a vehicle's brake systems. Since brake fluid is the means by which force and motion are transmitted through the hydraulic systems during braking, let's first discuss the unique characteristics of this special liquid and then the internal operation of a single and dual master cylinder.

CHARACTERISTICS OF BRAKE FLUID

Brake fluid is composed primarily of glycol. Because of the internal design of the parts within the brake systems and the need to counter the effects of changes in temperature based on climate and operating conditions, special additives are necessary. These additives give brake fluid a number of characteristics.

Lubricates. When a driver applies and releases the brake pedal, internal parts within the hydraulic master cylinder—controlling valves and wheel cylinders—move. Brake fluid manufacturers blend castor oil into the liquid in order to provide lubrication to these internal moving parts, thereby preventing them from seizing.

Is Noncorrosive. The various components of front and rear brake systems are constructed of different types of metals and materials. To prevent brake fluid from attacking or corroding these parts, suitable chemicals are incorporated into the formulation of the fluid, thereby neutralizing acids or alkalines contained in it.

Maintains Viscosity. Throughout the world, vehicles are driven in geographical areas of extreme heat and intense cold. To ensure safe and efficient operation of the brake systems, brake fluid is manufactured so that it flows consistently in cold weather as well as hot. Brake fluid must have a stable rate of viscosity. It must flow freely regardless of operational conditions and changes in temperature.

Absorbs Moisture. Modern brake systems are sealed; however, when the top lid or cover is removed from the master cylinder, fluid is then subject to moisture in the air. If the fluid could not absorb and retain moisture, vital parts within the brake systems would seize when temperatures drop below freezing. To enable brake fluid to absorb moisture and to reduce the possibility of its congealing, an additive similar to methol-hydrate is blended into the fluid during the manufacturing process.

Has High Boiling Point. During repeated severe brake applications, blistering heat is generated in the brake mechanisms as a result of friction between the pads and rotors and between the brake shoes and the drums. Temperatures in excess of 816°C (1500°F) have been recorded during brake application. When high temperatures have been reached, brake fluid does not remain in a liquid state; it becomes vaporous and therefore compressible. As a result, the brake pedal becomes very spongy when it is depressed, indicating that the brake system is inoperative and unpredictable.

To ensure the operational efficiency of brake fluid, manufacturers have developed fluids that have a very high boiling point. To maintain that high boiling point, brake fluid must not be exposed to moisture.

In a recent test, a manufacturer of brake fluid concluded that when a container of brake fluid absorbs only 3 percent of moisture into its volume, it will drastically lower its boiling point by 45 percent.

Caution: To prevent brake fluid from becoming contaminated with moisture, never leave the top off a vehicle's master cylinder or off a container of brake fluid.

Limits Swelling of Rubber Parts. Inside the component parts of a vehicle's brake systems are O-rings and piston seals (cups) made of rubber. The purpose of these parts is to seal the fluid, thereby preventing internal and external leakage. Since these parts perform an important function during the operation of the brake systems, manufacturers have developed brake fluids that limit the swelling of these vital parts. If the wrong type of liquid were to be dispensed into a brake system, the rubber parts would expand excessively beyond engineering design, and the brakes would not operate.

Since the quality of brake fluid is so essential to the safe operation of a vehicle's brake systems, the federal governments of Canada and the United States have set specifications for three types of brake fluid for automotive use. The three types are designated DOT 3, DOT 4, and DOT 5. Each type must be a specified color if manufactured after September 1978. DOT 3 (formerly SAE 70R2 or SAE J1703) and DOT 4 are amber to clear in color. The major difference between the two is that DOT 4 has a higher wet **boiling point** and absorbs moisture more slowly than DOT 3 fluid. DOT 5 brake fluid is silicone based, is purple, and has a higher boiling point than DOT 3 or 4. DOT 3 and 4 may be mixed with each other, but DOT 5 should not be mixed with DOT 3 or DOT 4. The reason is that DOT 3 and DOT 4 have a chemical base different from DOT 5, and the chemical bases are not compatible.

FACTS TO REMEMBER

1. Always store brake fluid in a clean, dry place.
2. Always keep the lid or cover on a container of brake fluid.
3. Never dispense brake fluid into a vehicle's mas-

ter cylinder from a container that has held a mineral base ingredient (gasoline, varsol, engine oil, transmission oil, etc.). *Note:* The slightest amount of mineral base oil will cause the rubber seals and cups within the brake systems to swell excessively.
4. Never clean hydraulic parts in a mineral base solvent.
5. Never reuse brake fluid.
6. To ensure safe, dependable braking, dispense only heavy-duty brake fluid into a vehicle's master cylinder.
7. When dispensing fluid into an imported vehicle's brake system, use only the fluid specified by that manufacturer.

MASTER CYLINDERS

The *master cylinder* of a modern vehicle is actually two hydraulic pumps designed to operate within a single cylinder and a metal housing. Originally, master cylinders were made of cast iron and were heavy. Master cylinders used in today's vehicles are often referred to as *composite master cylinders.* These types of master cylinders use an aluminum housing with a plastic nylon reservoir attached to the metal housing. Composites offer reduced weight and lower material cost and have a greater application flexibility.

To assist us in understanding how a dual master cylinder functions during the application and release of the brake systems, let's first study the design of a single piston master cylinder and its operation. Then we can discuss the dual master cylinder, its operation, and how it may be serviced.

Single Piston Master Cylinder Design

Figure 22-1 illustrates a cross-sectional view of a typical single piston master cylinder. Within its housing are two compartments: (1) a circular chamber in which a piston operates and (2) a reservoir that contains a limited supply of brake fluid. A removable filler cap, containing a vent, enables fluid to be dispensed into the reservoir.

The brake fluid reservoir not only stores fluid but also allows for expansion and contraction of fluid based on the climate and the operating temperatures of the brake system. If there is a slight external leak of fluid at the master or wheel cylinders, the reservoir compensates for this small loss by storing fluid.

Located at the end of the cylinder (left side of Figure 22-1) is a piston stop. The purpose of the

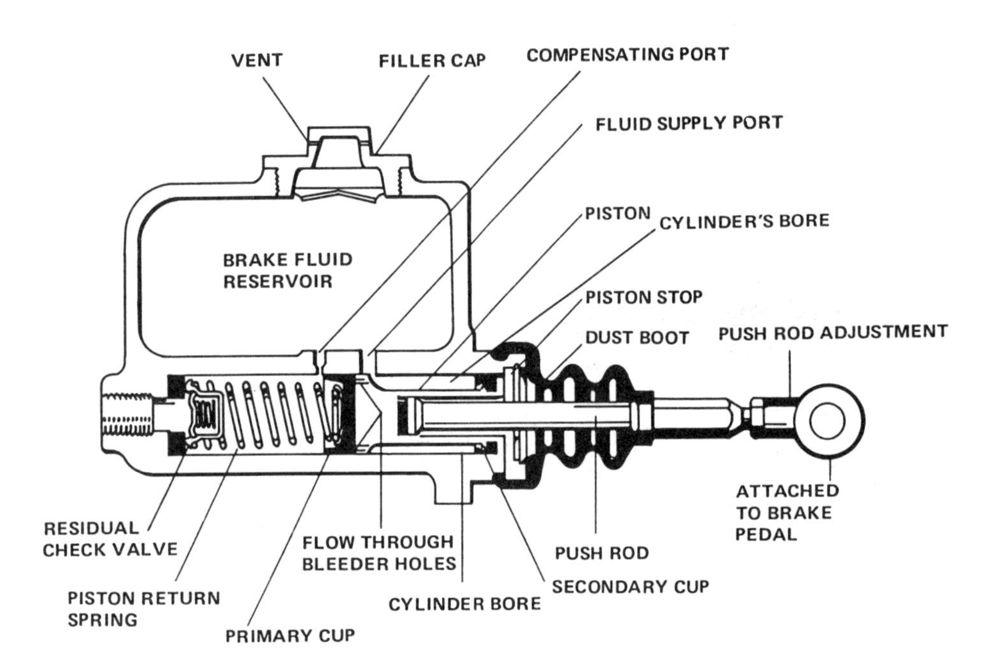

FIGURE 22-1. Typical Single Piston Master Cylinder.

stop is to limit the travel of the piston since it is under the influence of the piston return spring. Positioned and seated around the piston is a seal made of neoprene and rubber; the seal is a secondary cup. The purpose of the secondary cup is to prevent external leakage of fluid at the open end of the cylinder's bore.

To prevent dust or dirt from marring the smooth surface of the cylinder's bore, a rubber dust boot is tightly fitted over the end of the bore and the push rod. The push rod is attached to the brake pedal's lever and activates the piston within the cylinder during brake pedal application.

Connecting the cylinder to the reservoir are two ports: the fluid supply port and the compensating port. The purpose of the fluid supply port is to supply brake fluid to the cylinder's chambers during the application and release of the piston. The compensating port supplies brake fluid to the compression side of the cylinder's chamber (the area between the primary cup and the residual check valve). The compensating port allows brake fluid to return to the reservoir after the piston has gone back to its original nonoperating position. Also, the compensating port allows fluid to return to the storage reservoir when there is an expansion or contraction of fluid in the hydraulic lines and wheel cylinders.

Drilled through the head of the piston, next to the primary cup, are typically three to six flow-through bleed holes. The purpose of these small passages is to allow fluid to flow through the head of the piston and then over and around the primary cup to the area in front of the cup, when the piston is returning to its original nonoperating position. This operation will be explained further when we discuss the master cylinder.

The primary cup, located next to the piston, is also made of a composition of neoprene and rubber. The purpose of this important part is to seal the fluid within the cylinder when it has moved past the compensating port, thereby pressurizing the fluid in the master cylinder, lines and wheel cylinders.

The piston return spring keeps the primary cup against the head of the piston, as well as positioning the residual check valve against its seat, at the end of the cylinder's bore. The residual check valve, controlled by the piston return spring, is a one-way check valve. Its primary purpose is to retain fluid under slight pressure—56-105 kPa (8-15 psi)—thereby assisting in the application of the brake system. The valve also aids the wheel cylinder cups inside the cylinders to seal, thereby reducing the possibility of external leakage.

Single Piston Master Cylinder Operation

Brakes Applied. As the driver depresses the brake pedal, force directed through the pedal's lever and push rod causes the piston in the master cylinder to move into the cylinder's bore (Figure 22-2). When the piston has moved the primary cup past the compensating port, brake fluid is trapped in the chamber so that further piston movement builds up hydraulic pressure. Fluid is then directed through the residual check valve, brake lines, and flex lines. This fluid, under pressure, enters the wheel cylinders between the opposed pistons and their rubber sealing cups, forcing the cylinders outward. This movement of the pistons is transferred through piston

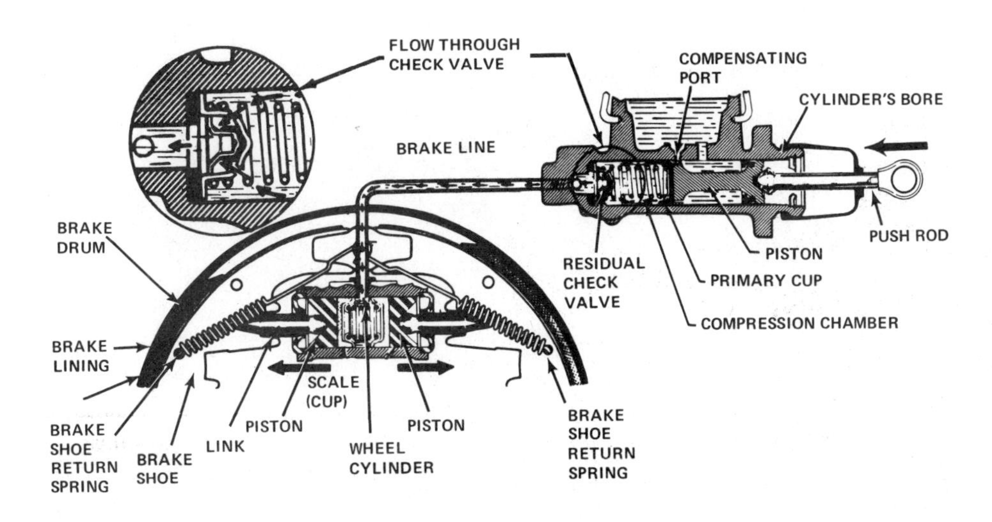

FIGURE 22-2. Master Cylinder and Brake Drum Mechanism. (Courtesy of General Motors of Canada, Limited)

link pins to the brake shoes and linings, causing the brake shoe return springs to lengthen and forcing the shoes into contact with the revolving brake drums. The linings, unable to revolve with the brake drums, create kinetic friction between the linings and the drums. This friction retards and stops the revolving drums, which are attached to the wheels.

Brakes Released: Initial Stage. When the brake pedal is released initially, strong brake shoe return springs contract, pulling the brake shoes and their linings away from the brake drums. The movement of the brake shoes is once again transferred through piston link pins, forcing the wheel cylinder pistons and rubber sealing cups to their original nonoperating position. Fluid is thus forced to return back through the hydraulic brake lines.

Inside the master cylinder, the primary cup and the piston are forced back to their nonoperating position by the piston return spring. While the piston assembly is being forced to return to its original position, the brake fluid, returning through the lines, does not immediately return and fill the cylinder's area between the primary cup and the residual check valve. During this phase of the master cylinder's operation, a low pressure area (partial vacuum) is created in the cylinder's compression chamber.

Since brake fluid in the reservoir is under atmospheric pressure, fluid is forced to flow through the fluid supply port and through the horizontal bleeder holes in the head of the piston (Figure 22-3). Because the fluid is subject to atmospheric pressure, the fluid is forced to flow over and around the primary cup, thereby filling the cylinder's compression chamber. This additional charge of fluid, entering the chamber during the release of the brake pedal, allows the driver to pump up the brake pedal when it is low or has insufficient operating height.

Brakes Released: Final Stage. The low pressure in the compression chamber and fluid returning through the hydraulic brake lines combine to move the residual check valve away from its seat at the end of the cylinder's chamber (Figure 22-4). Now there is an excessive volume of fluid in the chamber.

Since the piston and the primary cup have returned to their original nonoperating position, the excess volume of fluid in the compression chamber is returned to the reservoir through the compensating port. When hydraulic pressure is reduced in the compression chamber and brake lines, the check valve, influenced by the compression of the piston return spring, causes the valve to close and rest against its seat. The fluid that remains in the lines

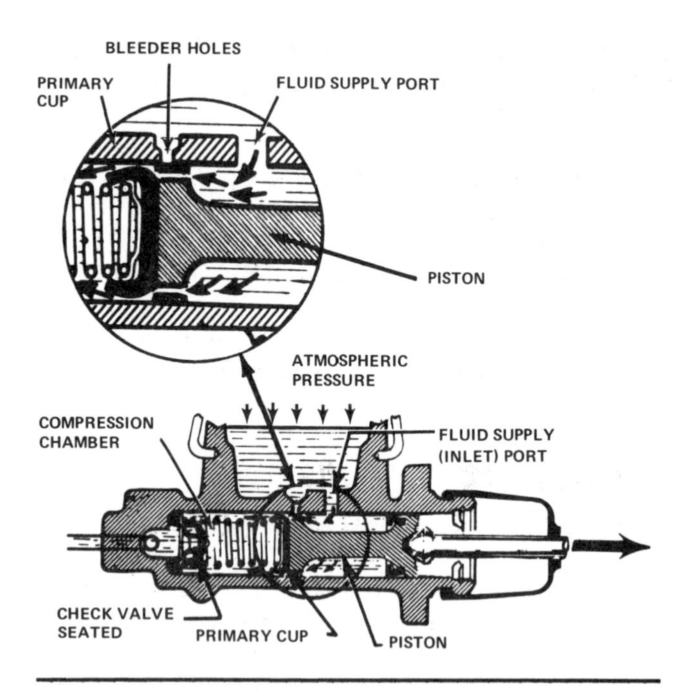

FIGURE 22-3. Initial Stage of Brake Release. During the initial release of the brake pedal, fluid flows through the fluid supply port and bleeder holes in the piston, deflecting the primary cup and filling the compression chamber. (Courtesy of General Motors of Canada, Limited)

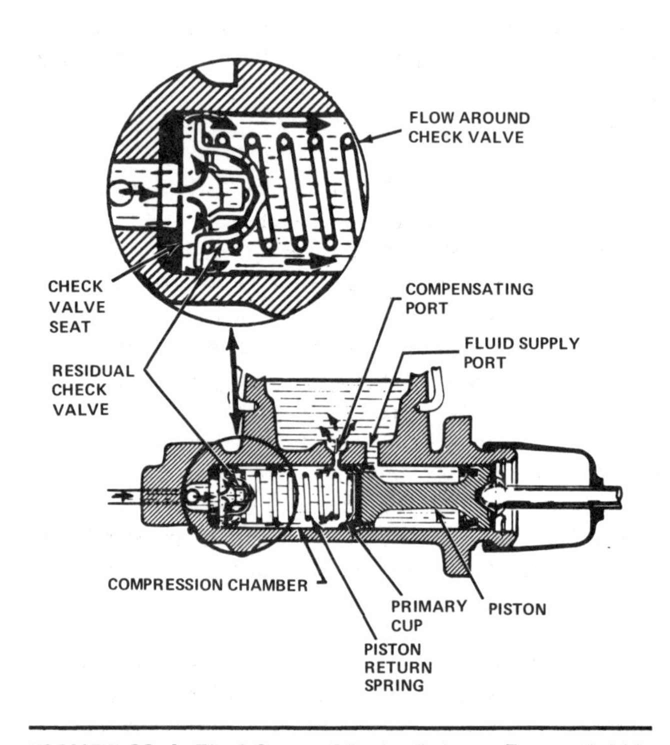

FIGURE 22-4. Final Stage of Brake Release. Excess fluid is returned to the reservoir through the compensating port. Note the operation of the residual check valve in the insert. (Courtesy of General Motors of Canada, Limited)

and wheel cylinders is then subject to slight residual static pressure. The residual static pressure aids the operation of the brakes by:

- Preventing air from entering the hydraulic system at the wheel cylinders,
- Reducing the possibility of external fluid leakage at the wheel cylinder's rubber sealing cups,
- Enabling the hydraulic system to respond immediately to brake pedal application.

Dual (Double Piston) Master Cylinder Design

Since 1967, all vehicles manufactured in North America have been designed with front and rear brake systems that utilize the **dual (double piston) master cylinder.** The advantage of this dual system is that in case of failure in one of the systems, the system still operating will be able to bring the vehicle to a safe, controlled stop.

Figure 22-5 illustrates a typical dual master cylinder. As you study the illustrations, you will observe that the cylinders' housings contain two fluid reservoirs and two pistons: primary and secondary. Located in the middle of the cylinders' bore are two secondary seals (cups) that divide the bore into two separate compression chambers. Each compression chamber is connected to its reservoir by a fluid supply port and a compensating port. A vented cover and flexible rubber diaphragm, at the top of the master cylinders' reservoirs, seal the hydraulic systems from possible entrance of contamination while at the same time permitting expansion and contraction of fluid within the reservoirs.

In the pressure chambers, coil springs hold rubber primary seals (cups) against the pistons. These seals, and the rubber secondary seals on the opposite end of the pistons, prevent escape of fluid past the pistons. The secondary piston is retained in its section of the bore by the push rod retainer. A rubber boot (Figure 22-5A) excludes foreign matter from the end of the cylinder's bore.

An Observation. When you have the opportunity to inspect a dual master cylinder on a vehicle, you will observe that the reservoirs may be equal or unequal in their size and shape. If the vehicle has front and rear drum brakes, the size of the reservoirs will be equal. If the vehicle has front disc brakes and rear drum brakes, the reservoir that contains the fluid for the front disc system will be the larger of the two. The reason is due to the unique design of the disc brakes. As a disc pad wears, the piston that forces the pad against the rotor will (because of wear to the pad) move outward in its caliper cylinder's bore (see Figure 24-2). This results in a larger fluid displacement area behind the piston. The only way to compensate for this increased area is to store additional fluid in the master cylinder's reservoir. Drum brakes, unlike disc brakes, have a self-adjusting mechanism that compensates for brake shoe lining wear. Hence, the pistons in a wheel cylinder operate within a relatively fixed area of the cylinder's bore. We'll talk about this further when we discuss disc and drum brake designs.

Dual (Double Piston) Master Cylinder Operation

The operation of a dual master cylinder is similar to that of the single piston type. When force, applied to the brake pedal, is transmitted through the pedal's lever and push rod, the primary and secondary pistons move deeper into the cylinder's bore. When both pistons have moved the primary seals past the compensating ports, hydraulic pressure is transmitted through the residual check valves to the front and rear brake systems. *Note:* If the vehicle is equipped with disc brakes, the part of the master cylinder that supplies fluid to the disc brakes will not have a residual check valve at the line outlet (Figure 22-5B).

When the brake pedal is released, the pistons in the cylinder's bore return to their nonoperational positions. The excess fluid in the systems is directed to the reservoirs through the compensating ports.

Rear Brake System Failure. When there is an external loss of hydraulic fluid—due to a ruptured rear brake line, for example—the primary piston will move into the cylinder's bore upon application (Figure 22-6). The piston, however, will not be able to produce pressure since that system is no longer sealed or confined. When this situation occurs, a slight force is transferred to the secondary piston through the compression of the primary piston spring and then through the piston's extension. When the extension contacts the secondary piston, the force exerted by the push rod is transmitted directly to the secondary piston. Sufficient pressure is then created to operate the front brake system.

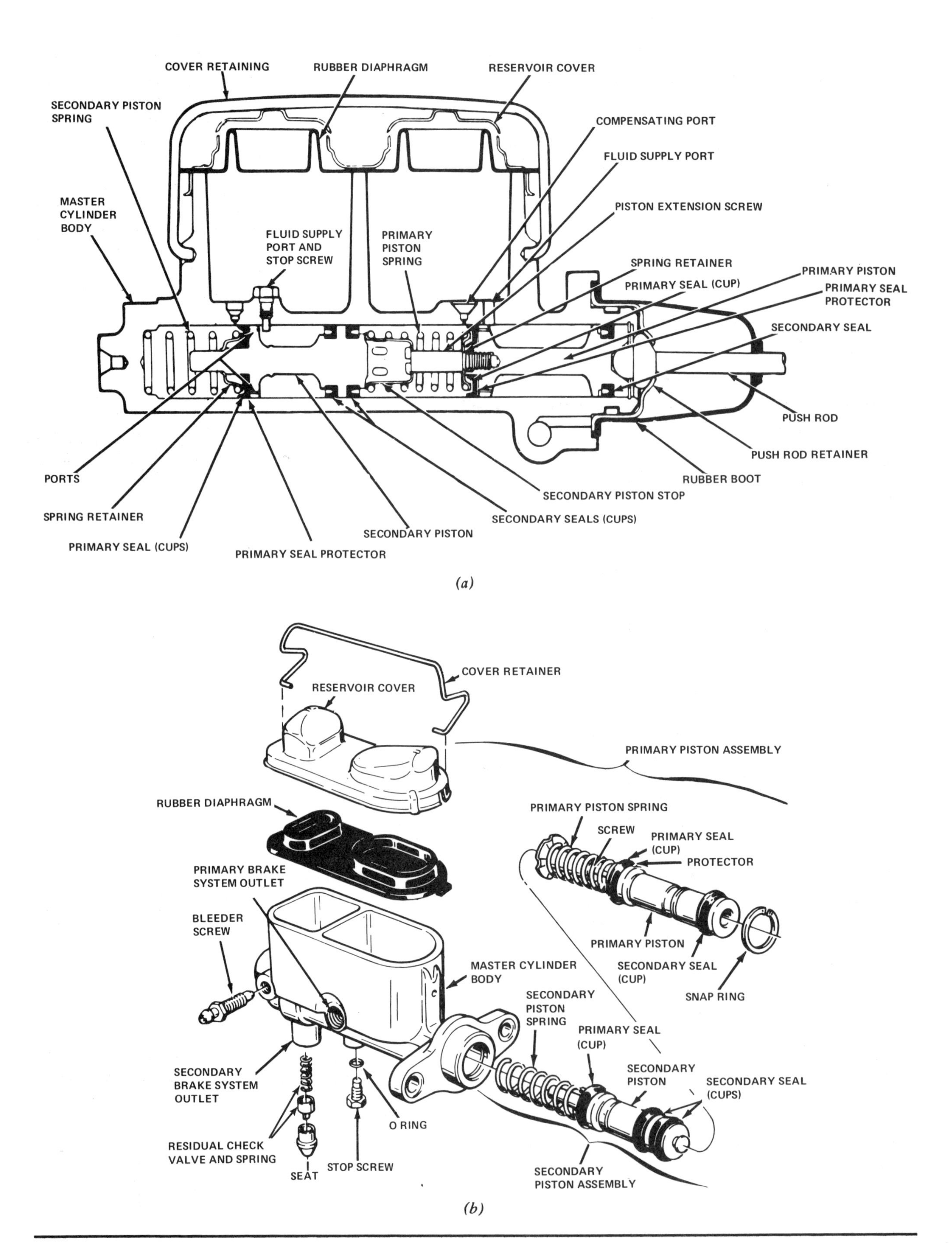

FIGURE 22-5. Typical Dual Master Cylinder. (a) Profile view; (b) three-dimensional view of components. (Part a courtesy of General Motors of Canada, Limited; part b courtesy of Ford Motor Company)

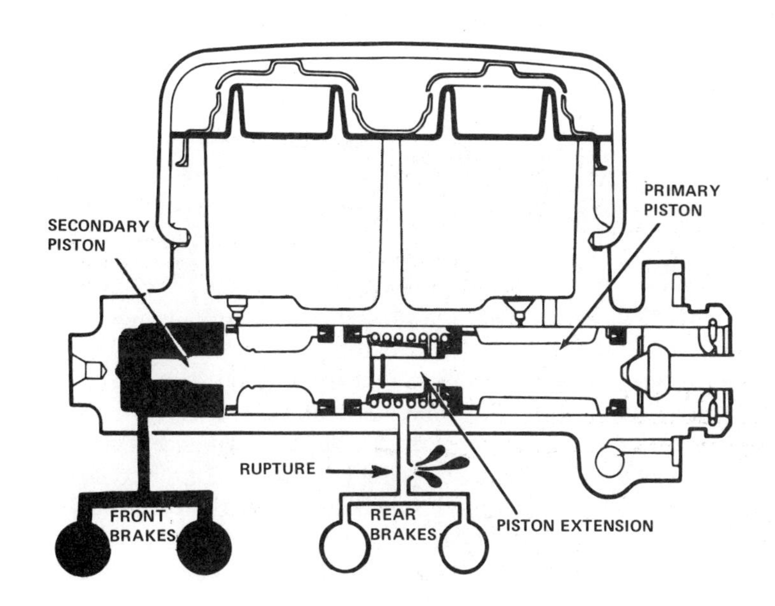

FIGURE 22-6. Ruptured Rear Brake Line. (Courtesy of General Motors of Canada, Limited)

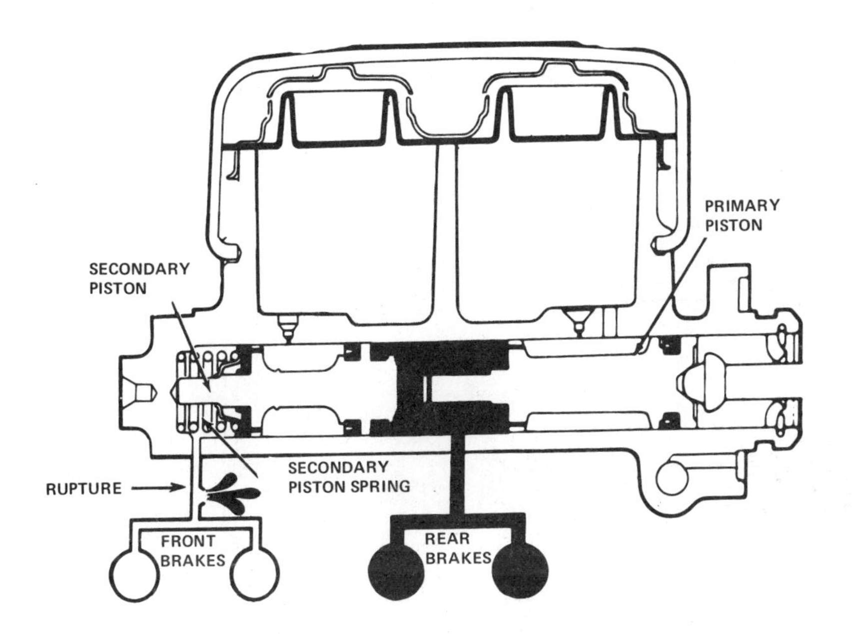

FIGURE 22-7. Ruptured Front Brake Line. (Courtesy of General Motors of Canada, Limited)

Front Brake System Failure. If there is an external loss of hydraulic fluid in the front brake system (Figure 22-7), both pistons will move into the cylinder's bore when the brakes are applied under normal operating conditions. However, due to the rupture in the line, there is nothing to resist piston travel except the secondary piston spring. This permits the primary piston to produce only negligible pressure until the secondary piston bottoms at the end of the cylinder's bore. Then, sufficient hydraulic pressure will be created to operate the rear brake system.

When there is a loss of fluid in the rear or front brake systems, additional pedal travel will be evident, as well as a need for increased applied force to the brake pedal when stopping the vehicle. The reason the driver will need to increase the applied force to the pedal is that only half the brake system is operational.

Both of the effects—additional pedal travel plus the need for increased force—should be noticeable to the driver. As an added safety feature, a warning light operated by the pressure differential valve and switch has been incorporated into the vehicle's brake systems. The dash-mounted red warning light will light up when there is insufficient hydraulic pressure in the brake systems. On some vehicles, the light will flash off and on; on others, it will stay lit as long as the ignition key is in the on position. In both cases, the light is a signal to the driver to have the brakes repaired immediately.

Quick Take-up Master Cylinder Design

In recent years, North American vehicle manufacturers have introduced a new type of master cylinder, the *quick take-up master cylinder* (Figure 22-8). This unit is designed for a diagonally split brake system (the left front brake and right rear brake are connected) and incorporates dual proportioning valves plus the brake warning switch system. In use, this cylinder has a quick take-up feature that provides a large volume of low pressure fluid at the beginning of the stroke to take up the pad and lining clearances at the wheels. High pressure is then generated to create the needed braking forces.

In this brake system, the caliper seals cause the pistons to retract farther than in prior designs. Retraction relieves the brakes of any slight drag that could adversely affect fuel economy.

Quick Take-up Master Cylinder Operation

In the initial brake application stage, more fluid is displaced in the primary low pressure chamber than in the *primary* high pressure chamber because of the larger diameter of the low pressure chamber and piston. The excess volume of fluid must go somewhere, so it goes around the primary piston lip seal into the primary high pressure chamber. From there, it flows to two wheels. Since the pressure and displacement is equal in both the primary and secondary systems, the secondary piston must move farther because its bore is smaller. Flow and pressure to the two wheels it serves is the same as that to the two wheels served by the primary piston. This large initial displacement has now taken up all clearances in the brake system and applied a light pressure throughout the system (see Figure 22-9A).

The quick take-up valve contains a spring-loaded check valve that holds pressure in the primary low pressure chamber up to a specified value. When that

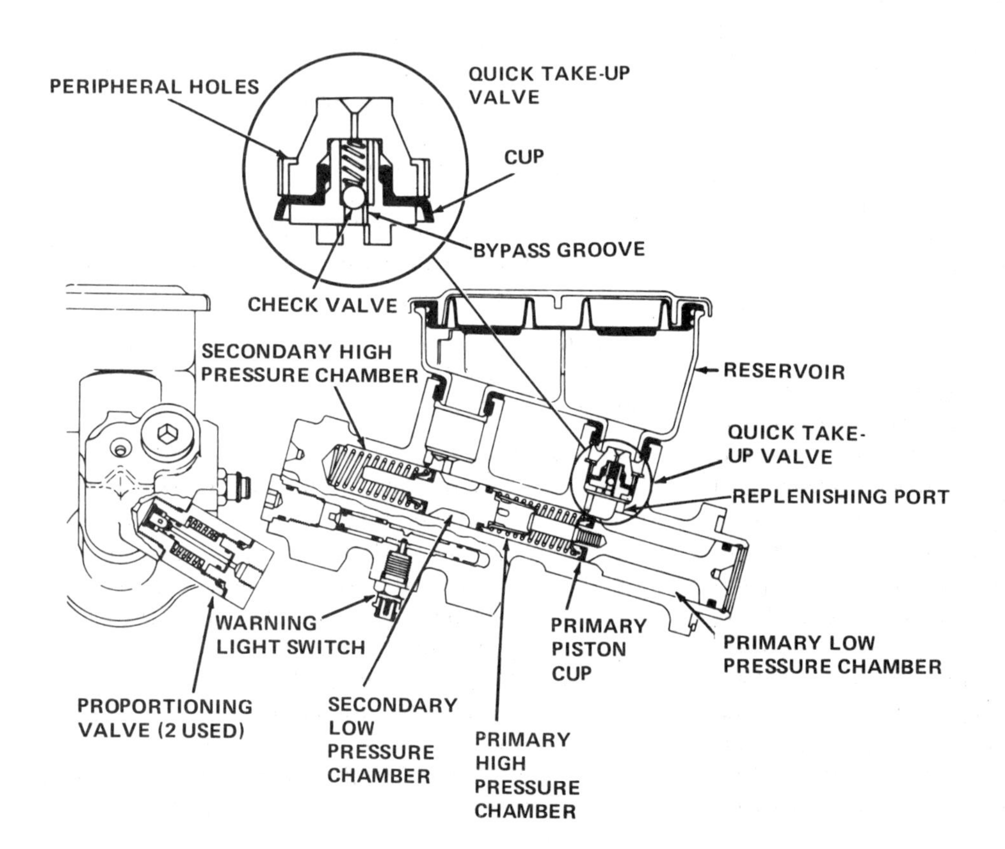

FIGURE 22-8. Quick Take-Up Master Cylinder. (Courtesy of Wagner Brake Company, Ltd.)

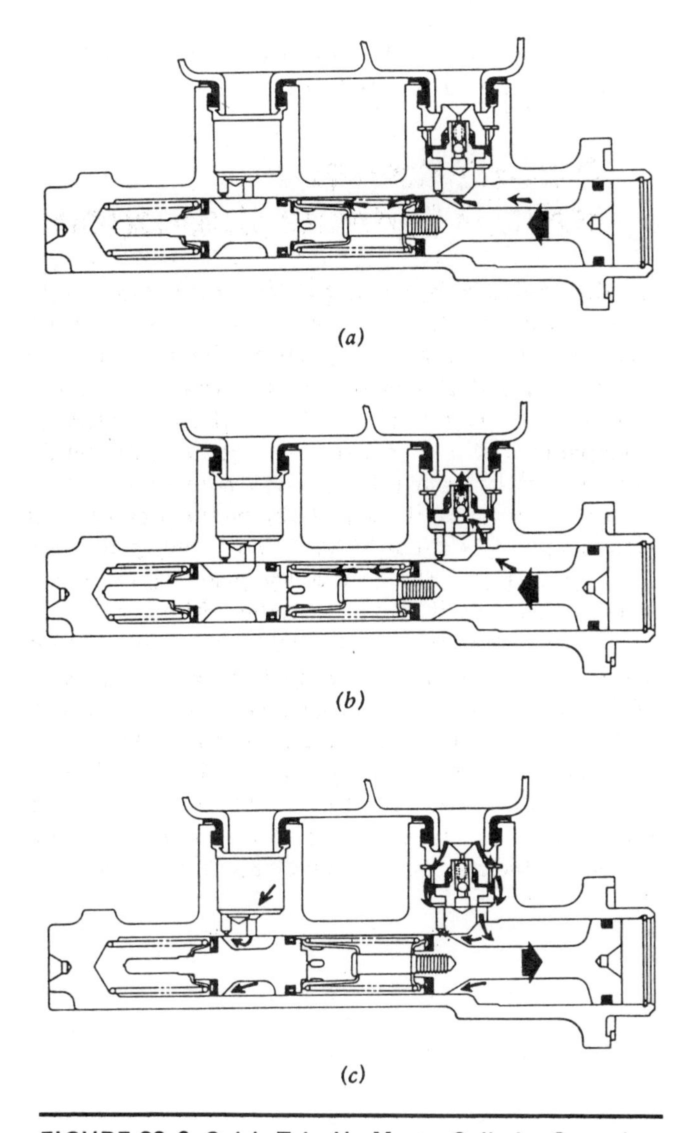

FIGURE 22-9. Quick Take-Up Master Cylinder Operation. (a) Low pressure application; (b) application above transition pressure; (c) brake release. (Courtesy of Wagner Brake Company, Ltd.)

pressure is exceeded in the cylinder, this spring-loaded valve opens. Then, fluid in the primary low pressure chamber will bypass to the reservoir, and the primary and secondary pistons function as a normal dual master cylinder. These pistons operate in the smaller bore to develop the higher brake operating pressures required (see Figure 22-9B). When the brakes are released, fluid replenishment occurs through a normal replenishing port for the secondary piston. *Note:* The piston moves back faster than the fluid returns, and fluid must flow around the cup to the front of the piston to balance the pressures. Fluid makeup for the primary piston occurs through the peripheral holes in the quick take-up valve, around the quick take-up cup, and through the replenishing port to the low pressure chamber of the primary piston. The fluid then flows around the primary piston cup into the high pressure chamber (see Figure 22-9C).

A bypass groove in the quick take-up valve accommodates fluid expansion in the primary chambers through the vent and replenishing ports. The secondary chambers handle expansion and contraction through the normal vent and replenishing ports directly connected to the front reservoir.

DIAGNOSING OPERATIONAL PROBLEMS

Modern dual master cylinders seldom provide problems for the driver of a vehicle. However, vehicle repair technicians must be capable of inspecting and diagnosing problems associated with the performance of a vehicle's master cylinder. When you suspect that the operation of a master cylinder may be faulty, use the following procedure.

Step-by-Step Procedure

1. Visually inspect the external surface of the cylinder's housing for signs of excessive leakage.

   a. External leakage may be caused by insufficient sealing of the rubber diaphragm due to a loss of tension of the cover's spring clip retainer.

   b. If, after cleaning the outside of the cylinder and tightening the clip, there is still a seepage of fluid, examine the housing for a small hole caused by a defect when the housing was manufactured.

2. Remove the cover of the master cylinder and examine the levels for fluid in the primary and secondary reservoirs.

   a. If the primary reservoir is almost empty of fluid and if the secondary reservoir contains too much fluid, the cause of the problem may be a defective secondary seal located in the middle of the compression chamber. Extract some of the excess fluid from the secondary reservoir. (The level should be approximately 0.5 millimeters (3/8 inch) from the reservoir's brim). Fill the primary reservoir with brake fluid to its correct level. Install the cover and repeatedly pump the pedal 10 to 15 times. Remove the cylinder's cover. Once again examine the level of the fluid. If the same situation occurs, the master cylinder will need to be serviced.

b. If the primary reservoir is almost empty of fluid and if the secondary reservoir contains the correct amount of fluid, the cause of the problem may be a defective secondary seal located on the primary piston near the open end of the cylinder's bore. To determine if this is the cause, visually inspect the inside of the dust boot for a collection of fluid. If the vehicle is equipped with a power brake assist unit, the master cylinder will need to be removed from the power unit to determine if fluid has leaked into that part. If this situation has occurred, the master cylinder will need to be serviced.

c. If the primary and/or secondary reservoirs are almost empty of fluid, and you have concluded that the master cylinder is not the cause of the problem, visually examine the valves, lines, and the front and/or rear wheel cylinders for signs of external leakage. These parts will be discussed in later chapters.

3. Obtain the feel of the brake pedal by applying a light force to the pedal for a period of 45 to 60 seconds. If the vehicle is equipped with a power brake assist unit, the engine must be operating.

a. If the brake pedal, when lightly depressed, sinks toward the floor, a primary seal within the compression chamber is bypassing fluid in the cylinder. When the primary seals are performing their function, the height of the pedal will remain constant. If the pedal sinks during this test, the master cylinder will need to be serviced.

b. If the pedal is hard to push at the very top of its compression stroke, the cause of the problem may be a blocked compensating port due to contamination of the brake fluid in the reservoirs, or in the brake systems. To determine if the ports are blocked, have another person pump the brake pedal several times; then, keep the pedal depressed. While the pedal is still depressed, remove the cylinder's cover. Observe the fluid and have the pedal released. When the pedal is released, the fluid returning to the reservoirs should disturb the surface of the fluid still in the reservoirs. If there is no disturbance, the compensating ports may be blocked due to some foreign matter. If the ports are blocked, the dual master cylinder will have to be removed from the vehicle, dismantled, and repaired or replaced. If the fluid inside the reservoirs is contaminated (contains grit), the front and rear brake systems will have to be flushed and bled with new brake fluid. Flushing and bleeding procedures are discussed in detail in Chapters 24 and 25. When the master cylinder has been reinstalled on the vehicle, check again for the disturbance of the fluid in the reservoirs after repeated pedal applications.

c. If the pedal is hard to push, there may be no operating clearance between the end of the push rod and the primary piston. If there is a lack of operating clearance—0.5 millimeters (0.020 inch)—brake fluid will not return to the reservoirs through the compensating ports. Keep in mind that after repeated brake applications while driving, the fluid in the brake mechanisms will be heated and find its way through the hydraulic lines back to the master cylinder. This causes loss of operating clearance between the piston and the push rod. The brakes will remain applied, and the wheels will not revolve. To correct this problem, manufacturers maintain the clearance between the push rod and piston either by changing the length of the push rod or by placing a shim between the master cylinder and its mount position (the firewall and the master cylinder housing). Correct clearance at the end of the push rod can be determined by measuring brake pedal free play (Figure 22-10).

4. Pump the vehicle's brake pedal to determine if it has a spongy feeling.

If the pedal is spongy, probably a small pocket of air is trapped in the master cylinder, brake lines, and/or wheel cylinders. If the pedal becomes firm after the brakes are bled and then becomes spongy again, the cause of the problem will be a loss of fluid from the brake system. If this situation exists, the only remedy is to repair or replace the master cylinder.

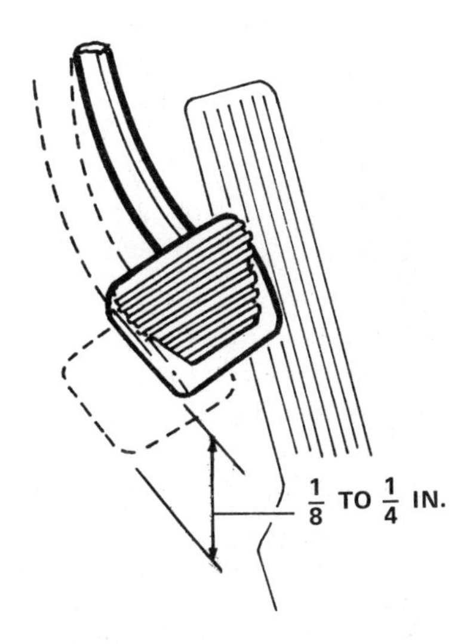

FIGURE 22-10. Normal Free Play in Brake Pedal.

Quick take-up master cylinders may evidence a problem unique to them. You have bled the brakes and have expelled all the air from the brake system. The brake shoes and pads are operating correctly, and the brake pedal is still very low. If the pedal is low, the cause of the problem is probably a quick take-up valve inside the master cylinder. This valve can be replaced on many vehicles after the nylon reservoir has been removed from the top of the master cylinder. After you remove the reservoir, a snap ring retains the quick take-up valve inside the metal housing. A sheet metal screw may be threaded into the center of the valve. Then, using pliers, you may pull on the screw and extract the valve from the metal housing. To install a new valve, place an appropriate-sized socket on top of the valve. Using a plastic-tipped hammer, install the valve in the metal casting.

FACTS TO REMEMBER

You can more easily diagnose problems associated with the operation of a vehicle's master cylinder when:

1. You have a clear mental picture of the design and internal operation of the master cylinder,
2. You follow a logical and systematic diagnostic approach to determine the cause of the problem.

SERVICING THE MASTER CYLINDER

Although manufacturers use a common principle in the construction of vehicle brake systems, there are minor differences in design. Since this is the case, it would be impractical to list the service procedures for all vehicles. Therefore, during the following practical assignment, let's rely on the manufacturer's shop manual for detailed service instruction.

If you are a student in an automotive technical school or college or serving an apprenticeship in a garage business, ask your instructor or shop foreman if you may overhaul a dual master cylinder. In order to complete this assignment, you will require the following tools and equipment:

- Fender cover,
- Manufacturer's shop service manual,
- Tool box,
- Flare nut brake line wrench,
- Lock ring pliers,
- Needlenose pliers,
- Denatured alcohol or brake cleaning fluid,
- Roll of masking tape,
- Small electric drill,
- Master cylinder hone,
- Set of allen wrenches (if necessary),
- Parts tray,
- Set of 6.3 millimeter (¼ inch) wide feeler gauges, ranging from 0.025 to 0.254 millimeter (0.001 to 0.010 inch),
- Container of heavy-duty brake fluid,
- Air line hose and blow gun.

Since we can only discuss general procedures, you will find the following information very useful to the commencement of your assignment. *Note:* Brake fluid will remove paint. Before you actually start disconnecting or removing any part from the vehicle, place a cover over the vehicle's fender to protect its paint finish.

Removing and Disassembling the Master Cylinder

Step-by-Step Procedure

1. Before removing the cylinder from the vehicle's firewall or power assist unit, carefully clear

and, if possible, remove other parts that may cause interference.

2. When disconnecting brake lines from the cylinder, use a flare nut brake line wrench to prevent the possibility of damage to the line nuts.

3. When the brake lines have been disconnected from the master cylinder, cover the ends of the lines and nuts with masking tape to prevent dirt from entering their open ends.

4. When the cylinder has been removed from the vehicle, remove the reservoir cover and pour the fluid into a drain can. Hold the cylinder in a work vise, being careful not to distort the housing. To remove all the fluid, activate the pistons in the cylinder by pushing and releasing the push rod several times. Catch the fluid in an appropriate container and discard the fluid.

Now, obtain from the manufacturer's shop manual the descriptive procedure for the removal and disassembly of the master cylinder. Read the procedure and follow its instructions. When the master cylinder has been disassembled, place all the parts in the parts tray.

Cleaning and Inspecting the Parts

Wash the exterior and the interior surfaces of the cylinder's housing with denatured alcohol or a brake cleaning fluid. Then, inspect the bore for possible scoring or pitting. If the cylinder's wall has light scratches or shows signs of slight corrosion, the wall can usually be cleaned with crocus cloth. When using crocus cloth, coat the surface of the material with brake fluid and slide the cloth into the bore by using a circular, rather than a lengthwise, motion, If the wall has deep scratches or score marks, the wall of the cylinder will require honing (Figure 22-11).

Honing is an acceptable method of restoring the surface of the cylinder wall as long as the diameter of the bore is not increased more than 0.076 millimeter (0.003 inch). After the cylinder has been honed, inspect it for excessive piston clearance and for burrs formed on the edges of the fluid supply ports and the compensating ports. Use the air line and the blow gun to clear the compensating port passages.

To check maximum piston clearance, place a feeler gauge 0.102 millimeter (0.004 inch) lengthwise into the cylinder's bore and determine if the piston can be inserted with the shim in place (Figure 22-12). If it can, the cylinder's diameter is oversized. The master cylinder unit will need to be replaced.

FIGURE 22-11. Honing the Cylinder Wall. When using a hone, lubricate the stones by immersing them in brake fluid. (Courtesy of Wagner Brake Company, Ltd.)

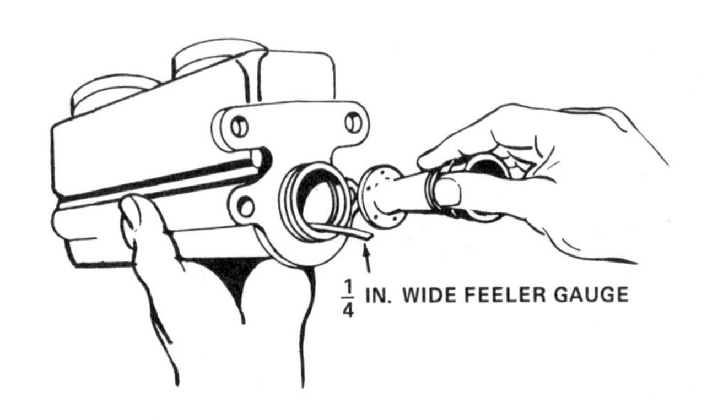

FIGURE 22-12. Using a Feeler Shim to Check Cylinder Diameter. (Courtesy of Wagner Brake Company, Ltd.)

If the master cylinder pistons are badly scored or corroded, replace them. Also, piston cups and seals must always be replaced when reconditioning the master cylinder. Now, clean, inspect, and hone the cylinder. When you have completed this portion of the workshop assignment, your shop instructor or shop foreman will inspect your work.

Aluminum Bore Clean-up

Inspect the bore for scoring, corrosion, and pitting. If the bore is scored or badly pitted and corroded, the assembly should be replaced. Do not use an abrasive material to clean the bore. The material will remove the wear and corrosion resistant anodized surface. Clean the bore with a clean piece of cloth around a wooden dowel and wash thoroughly with alcohol. Do not confuse bore discoloration or staining with corrosion.

Reassembling the Master Cylinder

Step-by-Step Procedure

1. Before assembling the master cylinder, dip all the internal component parts in a container of clean brake fluid. Place the parts on a clean, lint-free shop towel.

2. When installing the rubber seals into the cylinder's bore, be extra careful not to damage the lips of the seals.

3. Following the instructional procedures described in the shop manual, assemble the dual master cylinder.

Bleeding the Master Cylinder

Before installing the master cylinder on the vehicle, it must be bench bled (Figure 22-13). To do this, fill the reservoirs with new brake fluid. Install bleeding tubes at the openings for the brake lines and bend them into the reservoirs. Activate the pistons in the cylinder by stroking the push rod several times. When all the air has been expelled from the cylinder's compression chambers, the fluid being discharged into the reservoirs will not contain air bubbles. You may then install the reservoir cover and proceed to install the master cylinder.

Installing the Master Cylinder

When installing the master cylinder, you will find it easier to thread the brake line nuts into the cylinder's housing if the unit is not solidly attached to its mounting position. Now, install the master cylinder as described in the shop manual and torque the brake line nuts and the cylinder's mounting bolts to the manufacturer's specifications.

Testing the Master Cylinder

Step-by-Step Procedure

1. When the master cylinder has been installed, pump the vehicle's brake pedal several times. If no fluid has leaked from the lines, and provided that there is no air in the brake systems, the brake pedal should be firm when depressed and held.

2. The internal operation of the cylinder may also be tested by pumping the pedal several times, holding the pedal down and then releasing it. When the pedal is released, fluid returning through the compensating ports will disturb the surface levels of the fluid in the reservoirs. If there is no disturbance, the push rod's clearance will need to be examined and adjusted, if necessary. Adjusting probably will be necessary on vehicles with drum brakes and probably will not be necessary if the master cylinder supplies fluid to an old-style disc brake system since the pads are resting against the rotors. In the new-style disc brake mechanism on cars equipped with a quick take-up master cylinder, fluid

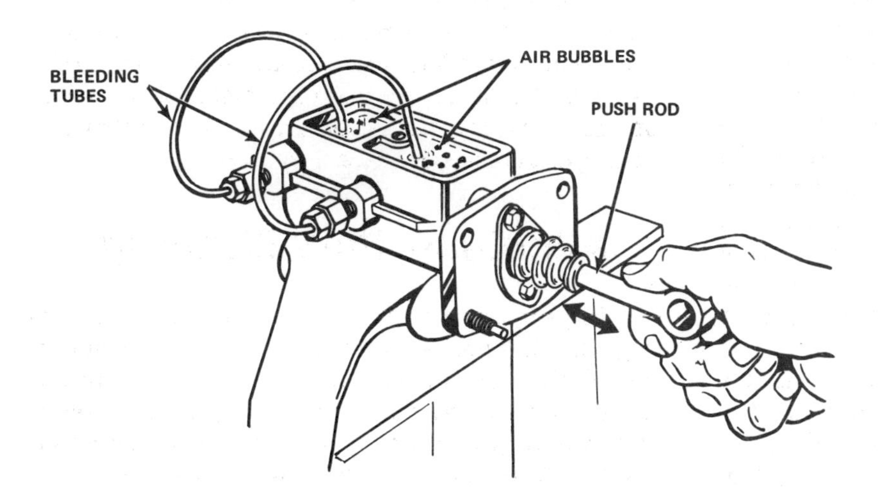

FIGURE 22-13. Bench Bleeding the Master Cylinder. (Courtesy of Wagner Brake Company, Ltd.)

is discharged from the compensating ports upon brake release.

3. Firmly apply the brake pedal with the ignition switch turned on. If the vehicle has a power brake assist unit, the engine must be operating. If only one of the cylinder's compression chambers is producing hydraulic pressure, the brake warning light on the dash will light up.

When you have performed the tests, request that the shop instructor or shop foreman examine and assess your work. After you have successfully completed this workshop assignment, return the tools and equipment to their correct place.

REVIEW TEST

1. List the unique characteristics of brake fluid.
2. What are the distinct differences among DOT 3, DOT 4, and DOT 5 brake fluids?
3. Write a brief summary, in point form, describing the seven facts to remember about brake fluid.
4. List and explain the purpose of all the parts and passages contained within a dual master cylinder.
5. Describe the operation of a dual master cylinder during the applied stage and released initial stage.
6. Describe the operation of a quick take-up master cylinder when the brake pedal is applied and released.
7. Describe how to diagnose the operational problems associated with a dual master cylinder.
8. Describe how to overhaul a dual master cylinder.
9. Describe how to bench bleed a dual master cylinder.
10. You have just installed a master cylinder on a vehicle. Describe how to test the operation of the master cylinder.

FINAL TEST

This examination is multiple choice. Only one answer will be accepted. Carefully read every statement.

1. Mechanic A says that brake fluid is designed to have a constant viscosity. Mechanic B says that brake fluid possesses a very high boiling point and a low freezing point. Who is right? (A) Mechanic A, (B) Mechanic B, (C) Both A and B, (D) Neither A nor B.
2. Mechanic A says that all types of brake fluids may be mixed together regardless of color. Mechanic B says that DOT 3 and DOT 4 brake fluids may not be mixed with DOT 5. Who is right? (A) Mechanic A, (B) Mechanic B, (C) Both A and B, (D) Neither A nor B.
3. Mechanic A says that when brake fluid absorbs moisture, its boiling point is raised. Mechanic B says that brake fluid exposed to air for some period of time will not have altered efficiency. Who is right? (A) Mechanic A, (B) Mechanic B, (C) Both A and B, (D) Neither A nor B.
4. Mechanic A says that the reservoir not only holds brake fluid but allows for expansion and contraction of the fluid. Mechanic B says that if there is internal leakage of fluid within the master cylinder, the problem is always a defective primary seal. Who is right? (A) Mechanic A, (B) Mechanic B, (C) Both A and B, (D) Neither A nor B.
5. Mechanic A says that flow-through bleeder holes are designed to expel air from the compression chambers. Mechanic B says that compensating ports are designed to allow fluid to return to the reservoirs. Who is right? (A) Mechanic A, (B) Mechanic B, (C) Both A and B, (D) Neither A nor B.
6. Mechanic A says that if the fluid leaks from the bore of the master cylinder, the problem is always a defective secondary seal. Mechanic B says that if there is internal leakage of fluid within the master cylinder the problem is always a defective primary seal. Who is right? (A) Mechanic A, (B) Mechanic B, (C) Both A and B, (D) Neither A nor B.
7. Mechanic A says that additional fluid enters the compression chamber of a master cylinder during the release of the brake pedal. Mechanic B says that the reason a disc brake reservoir is larger than a drum brake reservoir is to compensate for pad wear. Who is right? (A) Mechanic A, (B) Mechanic B, (C) Both A and B, (D) Neither A nor B.
8. Mechanic A says that a cast-iron dual master cylinder has three primary seals and two secondary seals. Mechanic B says that a cast-iron dual master cylinder has two reservoirs, two pistons, and one piston return spring. Who is

right? (A) Mechanic A, (B) Mechanic B, (C) Both A and B, (D) Neither A nor B.

9. Mechanic A says that when there is a loss of fluid from one-half of the master cylinder, the pressure differential valve and switch should activate a warning light. Mechanic B says that when there is a loss of fluid from one-half of the master cylinder, additional pedal travel will be evident, and increased stopping distance is required. Who is right? (A) Mechanic A, (B) Mechanic B, (C) Both A and B, (D) Neither A nor B.
10. Mechanic A says that the best way to test the operation of a master cylinder is to apply maximum force to the pedal while driving the vehicle. Mechanic B says that the best way to test the operation of a master cylinder is to apply light force to the pedal while the vehicle is stationary. Who is right? (A) Mechanic A, (B) Mechanic B, (C) Both A and B, (D) Neither A nor B.
11. Mechanic A says that the only way to determine clearance between the push rod and the primary piston is to use a feeler gauge. Mechanic B says that the best way to determine clearance between the push rod and the primary piston is to use your hand and activate the pedal. Who is right? (A) Mechanic A, (B) Mechanic B, (C) Both A and B, (D) Neither A nor B.
12. Mechanic A says that the quick take-up master cylinder has a straight bore and a special valve. Mechanic B says that the quick take-up master cylinder has two high and two low pressure chambers. Who is right? (A) Mechanic A, (B) Mechanic B, (C) Both A and B, (D) Neither A nor B.
13. Mechanic A says that it is permissible to hone the bore of an aluminum-type master cylinder. Mechanic B says that after honing the bore of a master cylinder, 0.18 millimeter (0.007 inch) clearance between the wall and piston is acceptable. Who is right? (A) Mechanic A, (B) Mechanic B, (C) Both A and B, (D) Neither A nor B.
14. Mechanic A says that the primary purpose of a master cylinder is to produce flow. Mechanic B says that the primary purpose of a master cylinder is to produce pressure. Who is right? (A) Mechanic A, (B) Mechanic B, (C) Both A and B, (D) Neither A nor B.

Switches, Valves, and Brake Lines

LEARNING OBJECTIVES

After studying this chapter, you should be able to:

- Draw a schematic diagram and label the parts of a vehicle's stop light switch, pressure differential safety valve and switch, and their electrical circuits,
- Demonstrate and explain how to determine brake pedal free play,
- Illustrate and describe the internal design and operation of the pressure differential safety valve and switch,
- Describe how to test the operation of the pressure differential safety valve and switch,
- Describe how to test the operation of the metering valve,
- Describe how to test the operation of the proportioning valve,
- Describe how to inspect a vehicle's brake lines and hoses,
- Describe how to test a vehicle's brake lines and hoses to ensure their dependability to withstand maximum hydraulic pressure,
- Describe how to flare steel brake tubing.

INTRODUCTION

In this chapter, we will study the stop light switch and its electrical circuit, the control valves, and the hydraulic brake lines through which the fluid flows from the dual master cylinder to the wheel cylinder. We will begin by discussing how the stop lights of a vehicle are activated.

STOP LIGHT SWITCH AND ITS ELECTRICAL CIRCUIT

When the dual master cylinder was introduced by manufacturers, the *stop light switch*, was relocated from the hydraulic brake system and mounted under the vehicle's dash. This relocation of the switch was necessary since a vehicle was designed with two separate brake systems. The brake pedal, when depressed, operates a mechanical stop light switch that completes an electrical circuit to the rear stop light lamps. Figure 23-1 is a hydraulic and an electrical schematic view of a vehicle's stop light switch, pressure differential safety valve and switch (discussed later in this text), and their electrical circuits.

Test Procedures

To ensure the correct operation of the dual master cylinder and the stop light switch, free play—approximately 6.35 millimeters (¼ inch) of brake pedal movement—is essential (see Figure 22-9). To determine if a brake pedal has the necessary free play on a vehicle not equipped with power brakes, with your hand slowly depress the pedal and physically feel and measure for that movement and specified distance.

To determine the free play at the pedal when the vehicle is equipped with a power assist unit, have the vehicle's engine operating at idle speed. Then, remove the cover from the master cylinder's reservoir. Next, have another person move the brake pedal the specified distance. If the pedal has the necessary free play, there will be a slight discharge of brake fluid from the compression chambers, through the compensating ports, disturbing the level of fluid in the reservoirs. If the brake pedal does not have the necessary free play, consult the manufacturer's shop manual about the way to obtain the correct distance.

If the rear stop lights fail to operate when the brake pedal is depressed, the stop light switch may be defective. To test the switch, remove the wires from the switch and bypass the switch by using a jumper wire to connect the wires together. If the rear stop lights then operate, replace the switch by following the instructions that are detailed in the shop manual. If the rear stop lights do not function when a jumper wire has been used, check for a burned-out fuse, a faulty electrical connection, a broken wire, or burned-out bulbs.

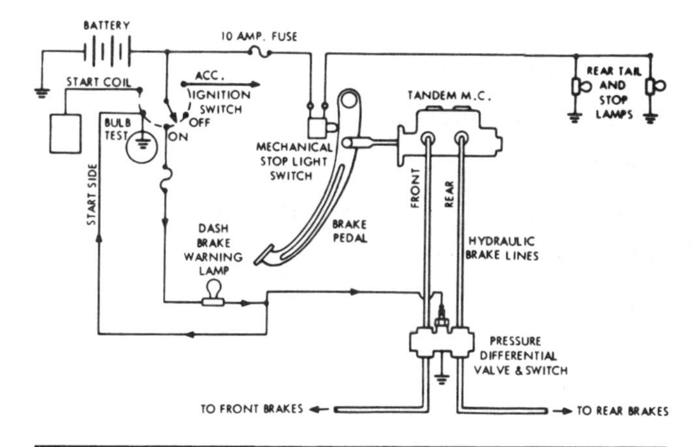

FIGURE 23-1. Schematic View of Electrical Circuits in a Brake System. (Courtesy of Wagner Brake Company, Ltd.)

FACTS TO REMEMBER

1. When inspecting a vehicle's brake systems, always examine the brake pedal's free play.
2. Too little free play can cause the brakes to drag, thereby producing premature wear to the brake linings, drums, and/or rotors.
3. Excessive free play of the brake pedal will result in a very low pedal reserve (the distance that the pedal must travel before the pedal is firm and above the floor).
4. When inspecting a vehicle's brake systems, always examine the operation of the rear stop lights.

PRESSURE DIFFERENTIAL SAFETY VALVE AND SWITCH

When you studied Chapter 21, you were introduced to the names of three valves that are component parts of a vehicle's brake systems. The first valve mentioned was the *pressure differential (warning)*

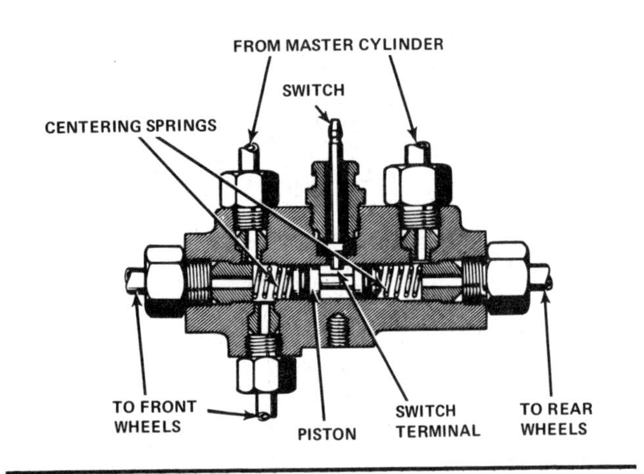

FIGURE 23-2. Cross-Section of Valve Housing Showing the Horizontal Piston Centered. (Courtesy of General Motors of Canada, Limited)

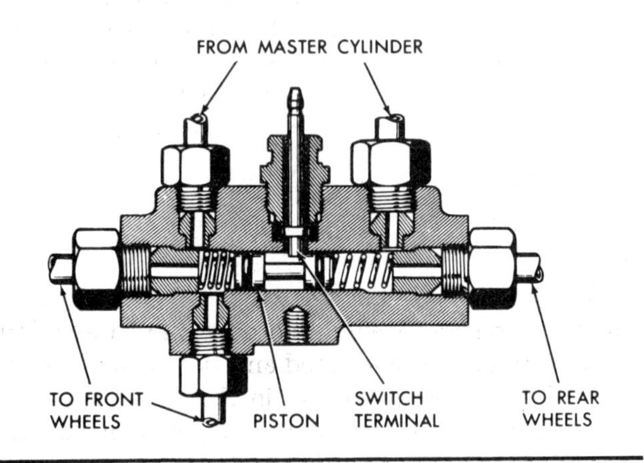

FIGURE 23-3. Cross-Section of Valve Housing Showing Piston Off Center. In this position, the piston comes in contact with the switch terminal, providing a ground for the brake's warning light circuit. (Courtesy of General Motors of Canada, Limited)

valve and switch. The purpose of this valve is to indicate to the driver if there has been a loss of brake fluid, and consequently of hydraulic pressure, in either the front or rear brake systems.

Figure 23-2 is a cross-sectional view of a valve housing that contains the pressure differential (warning) valve and switch. When hydraulic pressure is equal in both the front and rear brake systems, the horizontal piston remains centered and does not contact the switch terminal. On either side of the piston are centering springs that hold the piston in a neutral position.

Figure 23-3 illustrates that the piston has moved to the left, indicating a loss of pressure in the line leading to the front brakes. When the piston moves to the inoperative side, the piston then contacts the switch terminal, providing a ground for the brake's warning light circuit. Since the piston is centered by springs, the brake's warning light on the vehicle's dash will be illuminated only when the brakes are applied. The light will not remain lit when the brakes are released.

You will recall that in Chapter 22, you were introduced to the design of the quick take-up master cylinder. Figure 22-8 illustrated the design of the pressure differential valve and switch that is located in the housing of the master cylinder. Although the valve does not have centering springs at either end of the piston, it operates in the same manner as the one shown in Figure 23-4.

Test Procedures

To enable the driver to test the dependability of the brake warning light, on some makes of vehicles the bulb is lit when the ignition key is turned to the start position. On other makes of vehicles, the bulb is lit when the parking brake is activated. If the light does not operate, remove the bulb from the dash and inspect it to determine if it has burned out. If, after replacing the bulb, it still does not light, check the fuse in the fuse block, normally located under the dash, to determine if it is defective.

If the fuse and the bulb seem all right, check the wiring. To check the continuity of the wire that leads from the warning switch to the bulb, turn the ignition on. Then, carefully remove the terminal that attaches to the top of the switch. With the aid of a jumper wire, ground the wire inside the terminal. If the electrical circuit has continuity, the warning light on the vehicle's dash will be illuminated.

To test the operational dependability of the pressure differential valve, you will require assistance. First, turn the ignition key to the on position (not the start position). Then, have your assistant apply a constant force to the brake pedal. While the force is being applied to the pedal, take the appropriate brake line wrench and loosen the brake line nut that threads into the master cylinder's primary compression chamber. Approximately three complete turns should be enough. Since the force is being applied to the pedal, fluid will escape from the line at the master cylinder. When pressure within the brake system is almost 0 kPa (0 psi), the brake warning light should be activated.

If the brake light operates, tighten the line nut and pump the brake pedal several times. Have your assistant continue to apply force to the brake pedal. Then, loosen the opposite brake line nut. If the

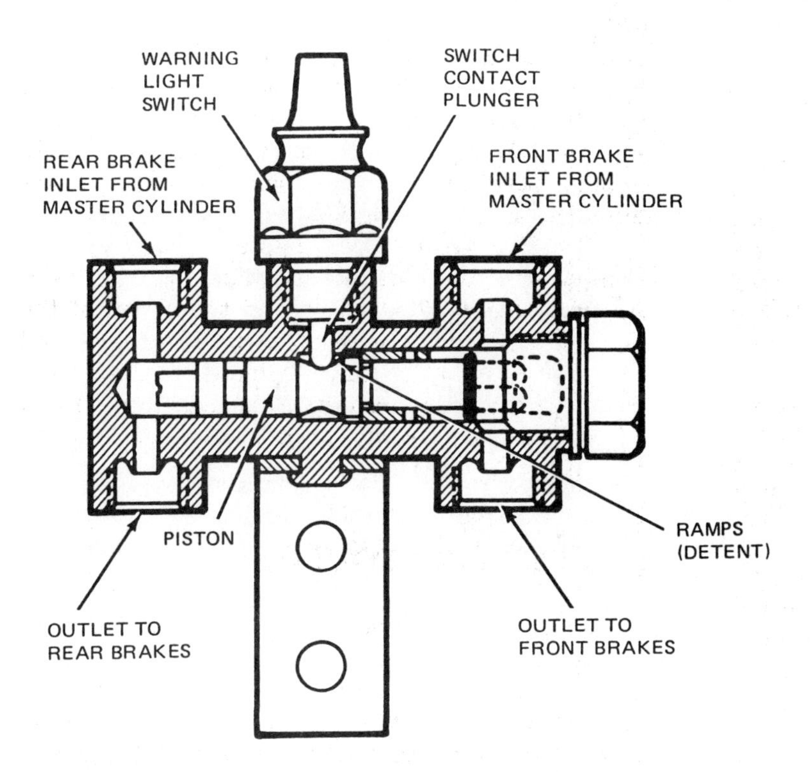

FIGURE 23-4. Different Type of Pressure Differential Safety Valve and Switch. (Courtesy of Wagner Brake Company, Ltd.)

warning light is once again illuminated, tighten the line and pump the brake pedal. Remove the master cylinder's cover and fill the reservoirs to the required level. Then, install the cover. If the brake warning light failed to be illuminated when pressure was lost from the front and/or rear brake systems, the entire valve must be replaced as a unit since it is a nonserviceable part.

Figure 23-4 illustrates a pressure differential safety valve and switch slightly different in design from that generally encountered. This valve is used on the Pacer, Gremlin, Hornet, and late model Matadors. The valve consists of a housing that contains a piston and a plunger-type switch. The switch contact plunger is actuated by ramps on the valve's piston to open and close the electrical circuit between the switch and warning light. Both the front and rear brake hydraulic systems are connected to the valve.

If there is a loss of pressure in either the front or rear systems, a pressure differential of 490 to 2100 kPa (70 to 300 psi) will cause the valve piston to shuttle toward the low pressure side. The piston ramps will then push the switch contact plunger upward. This action closes the electrical circuit between the switch and warning light, causing the bulb to illuminate.

With this design, unless the steering column ignition lock is turned to the off position, the warning light will remain illuminated until the cause of the loss of pressure is corrected and the piston is recentered. To center the piston in this particular valve, and deactivate the warning light, the manufacturer has recommended the following procedure.

Step-by-Step Procedure

1. Correct the cause of the loss of pressure.
2. Turn the steering column ignition lock to the on position.
3. Check the master cylinder reservoirs' fluid levels and fill to within 6.35 millimeters (¼ inch) of the rim, if necessary.
4. Apply the brake pedal several times. This operation will cause the piston in the valve to recenter itself.
5. Turn the ignition lock to the off or lock position.

The procedure for testing this valve and switch is similar to the steps that have been previously discussed, with one exception. To test the electrical switch, once it has been removed from the valve's housing, reattach the wire terminal to the top of the switch and turn on the ignition key. Then, depress and ground the switch contact plunger. If the warning light is lit, the switch does not need to be replaced.

METERING VALVE

The second valve that we briefly discussed in Chapter 21 was the *metering valve.* The metering valve, sometimes called the hold-off valve, is a component part of a vehicle's front brake system on some vehicles equipped with front disc brakes and rear drum brakes. Vehicles equipped with only drum brakes do not require the metering valve. But since we are discussing the metering valve, can you remember its purpose? The purpose of this valve is to reduce line pressure to the front brakes during initial brake application. Although we have not discussed disc and drum brake mechanisms, if a vehicle equipped with front disc and rear drum brakes were to have the front brakes applied at the same time as the rear brakes on an icy road, the front wheels would lock up (stop revolving), thereby affecting the steering control of the vehicle.

The metering valve, due to its unique design, allows a free flow of brake fluid through the valve when the brakes are not applied. When the brakes are initially applied, a pressure of 28 to 210 kPa (40 to 130 psi) will cause the metering valve stem to move to the left. This action within the valve prevents the flow of brake fluid to the front disc brake cylinders. As the master cylinder continues to produce and increase hydraulic pressure, the metering valve then allows fluid to flow to the front brake cylinders at increased controlled pressure. When the pressure created by the master cylinder reaches approximately 875 kPa (125 psi), the metering valve then allows fluid flow and pressure to the front disc brake cylinders.

Proportioning Valve

Not all vehicles have *proportioning valves.* If the vehicle does have this valve, it is located in the rear brake system. The purpose of this valve is to reduce line pressure to the rear brake cylinders during a panic stop, thereby preventing the loss of directional control. Some vehicles that have quick take-up master cylinders have the proportioning valve threaded into the master cylinder. (See Figure 22-8.)

COMBINATION VALVE

In recent years, brake engineers have incorporated the design features of the pressure differential safety valve and switch and the metering valve into one housing, the *combination valve.* This valve (Figure 23-5) is used on many vehicles designed with front disc and rear drum brake systems. The purpose and design of the combination valve are similar to that of the types already discussed.

Test Procedures for Metering Valve

To test the operation of the metering valve in the combination valve, first remove the boot that covers the end of the metering valve stem. Then, have another person slowly depress the brake pedal. While the pedal is being depressed, the valve stem will move either slightly outward or slightly inward, depending on the make of the valve. If you are unable to acquire the assistance of another person, you may determine the operation of the valve by slowly depressing the brake pedal. As the pedal is depressed, you should feel a slight bump in the movement of the pedal. *Note:* If the vehicle is equipped with a power assist unit, the engine must be operating during this test.

If during the test, you observe a loss of fluid from the area around the metering valve stem or if the metering valve stem does not move, the entire combination valve must be replaced since it is a nonrepairable part.

The proportioning valve in the combination valve is designed not to operate during normal braking. However, during an emergency stop, hydraulic pressure produced by the tandem master cylinder will be reduced to the rear brake cylinders. For example, in a panic stop situation, the pressure in the brake line leading to the combination valve may be 6300 kPa (900 psi), whereas the pressure in the line leading to the rear brake cylinder will be considerably less, 4200 kPa (600 psi). This difference in pressure will greatly reduce the probability of rear wheel lock-up, thereby enabling the vehicle to retain directional control during braking.

Test Procedures for Proportioning Valve

To test the operation of the proportioning valve in the combination valve, two pressure gauges that are calibrated to 14,000 kPa (2000 psi) are required. The first gauge must be installed by using a T fitting in the line leading from the master cylinder to the combination valve. The second gauge, again using a

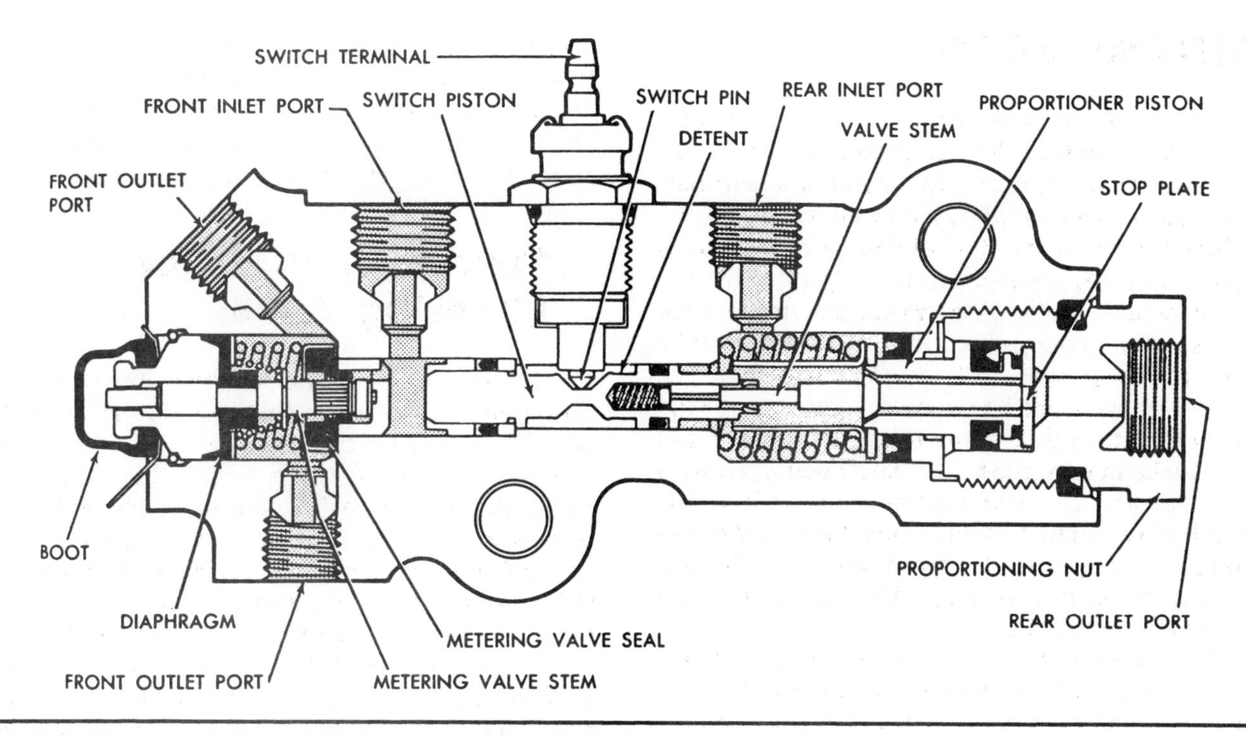

FIGURE 23-5. Cross-Section of Modern Combination Valve. (Courtesy of General Motors of Canada, Limited)

T fitting, is installed in the line leading to the rear brake cylinders. Have an assistant apply the brake pedal and determine if the pressures are proportional. *Note:* If the vehicle is equipped with a power assist unit, the engine must be operating during this test. If there is no difference between the readings of the two gauges, the combination valve is defective and must be replaced since it is a nonrepairable part.

HYDRAULIC BRAKE LINES

Hollow steel brake tubes and flexible rubber-coated hoses transmit brake fluid at various pressures from the vehicle's master cylinder to the control valves and to the wheel cylinders. Since brake lines consist of steel tubing and rubber hoses, let's become familiar with how they are made.

When brake tubing is manufactured, it is subjected to a heat treating process called *annealing.* This process removes the brittleness from the metal, leaving it bendable. The metal is then coated to resist rust and corrosion.

The ends of the brake line are double flared or have an International Standards Organization (ISO)-type flare to guard against leakage. Where the line

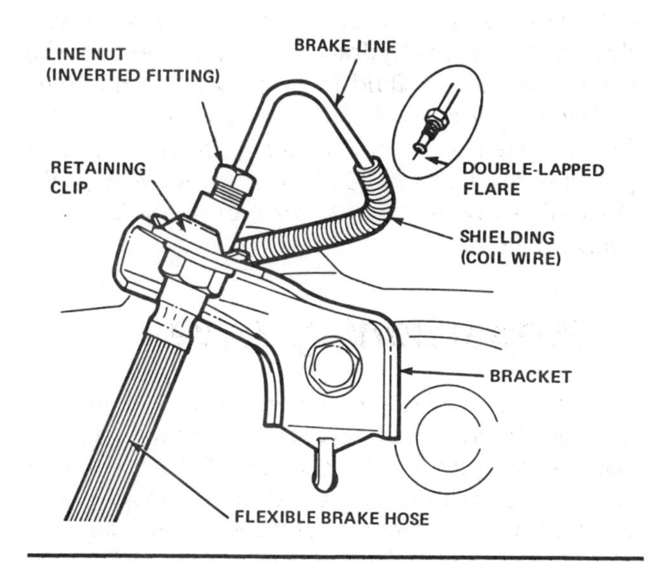

FIGURE 23-6. Component Parts of a Hydraulic Brake Line. (Courtesy of Ford Motor Company)

attaches to another part, the tubing is flared after it has been double lapped (see Figure 23-6). This double lapping allows the line nut, called an *inverted flare fitting,* to produce a strong, leakproof joint. Manufacturers also protect brake lines from potential damage by placing the line, where it is exposed

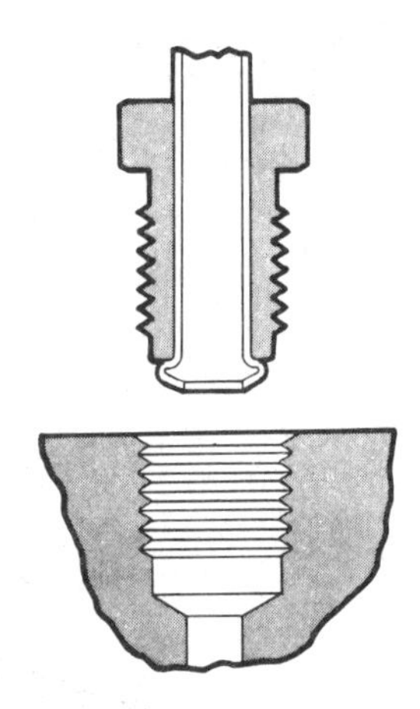

FIGURE 23-7. International Standards Organization (ISO) Design Brake Line Flare. (Courtesy of Wagner Brake Company, Ltd.)

to stones and other road objects, inside an outer shielded layer of coiled wire.

In recent years, manufacturers have adopted an ISO design brake line flare (Figure 23-7). This flare has several advantages:

1. The fitting is easier to machine than a double flare fitting.
2. When tightened, the shoulder of the nut bottoms out in the body of the part, creating a uniform pressure on the tube flare.
3. The design is not subject to operator overtightening. The service technician merely tightens the nut firmly on the seat, and the correct sealing pressure occurs automatically.

When you have the opportunity to inspect brake lines, you will see that the lines are attached to a vehicle's frame in such a way as to be routed away from the exhaust system and movable suspension parts.

Flexible brake hoses are located in a brake system between the vehicle's frame and its front and rear suspension. Hoses are made from several layers of bonded rubber reinforced with a fine steel mesh between each layer. Since a hose must move with the suspension, threaded steel fittings on each end of the hose enable it to be attached to the wheel cylinder or brake caliper and to the steel brake tube. A bracket that attaches to the frame provides a support mounting for the flex hose and steel brake tube. A retaining clip fits over the hose, locking it into the bracket.

Inspection Procedures

Brake lines and their fittings should be examined at least once a year or whenever the brake mechanisms at the wheel are inspected. The service technician should inspect the steel lines and their fittings for stains that indicate a very minute loss of fluid, for excessive external rusting, and for indentations or bruises caused by stones and other road objects. If the brake line fitting cannot be tightened or if the line is rusted and bruised, it must be replaced.

Flexible brake hoses should be examined for chafing, usually caused by stress or unnecessary twist. Hoses that have functioned under stress will often separate internally. Stress causes the inside lining of the hose to separate, thereby restricting fluid flow, acting as a check valve, and possibly producing a dragging brake (Figure 23-8). Hoses should also be inspected for possible ballooning, caused by a separation and rupture of its various rubber layers. Other conditions that hoses are subject to are cracking of the exterior surface or blistering near the metal fittings. If such conditions are evident or if you suspect a hose of being faulty, replace the part.

Test Procedures

Although brake tubes, fittings, and flexible hoses appear to be in good condition, potential problems may still exist. Since a vehicle's brakes are seldom required to be operated at maximum pressure, the driver of a vehicle should periodically test the dependability of the lines and the other hydraulic parts of the systems. This test is conducted when

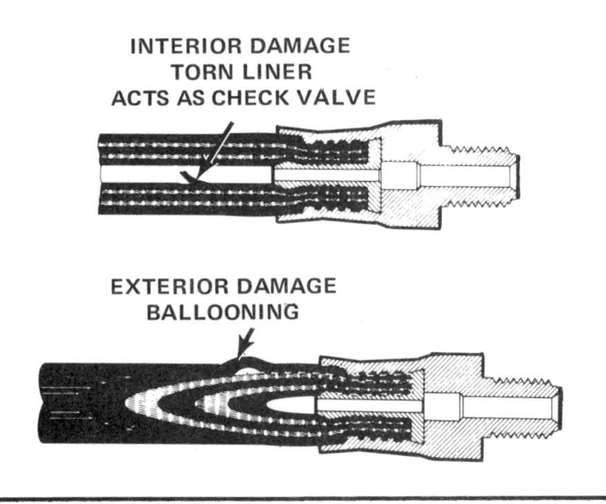

FIGURE 23-8. Interior and Exterior Damage to Brake Hoses. (Courtesy of Wagner Brake Company, Ltd.)

the vehicle is stationary, preferably parked in a driveway or a parking lot. With the use of both feet, apply maximum force to the brake pedal. If the vehicle is equipped with a power assist unit, the engine must be operating. If the pedal is spongy or suddenly sinks close to the floor, the brake's hydraulic systems are in need of repair.

A FACT TO REMEMBER

The test for determining the dependability of a vehicle's brake lines and other hydraulic parts should always be conducted after any service or repairs to the hydraulic systems.

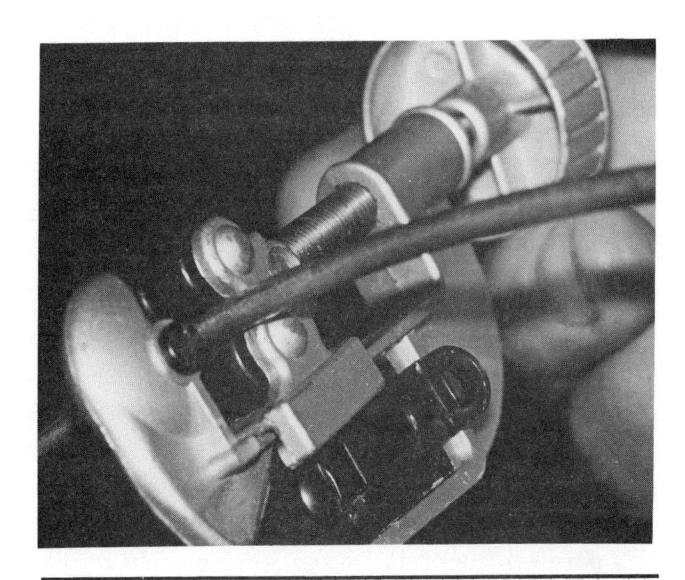

FIGURE 23-9A. Cutting Tubing with Brake Line Cutting Tool.

SERVICING BRAKE LINES AND HOSES

When you have carefully inspected and tested the brake lines and hoses and have determined that a part of a line must be replaced, chances are that other sections of that hydraulic line are also defective. Brake safety experts and vehicle manufacturers strongly recommend that a service technician replace the entire line, not just a section of it.

When replacing a brake line, it is often necessary to cut the line into several smaller pieces. Save each piece of the line and, with the aid of a string, determine the total length of the entire line. Then, measure the length of the string with a ruler. Since each end of the line must have a double-lap flare, add 6.35 millimeters (¼ inch) to the length of the line, plus another 12.7 millimeters (½ inch) for good measure. *Note:* It makes sense to have the line longer than required since you may improperly double-lap flare the line. When you have determined the length for the replacement line, use the following steps as your guide.

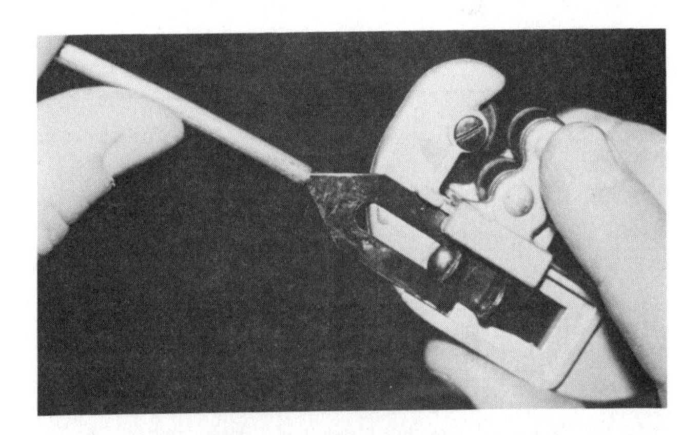

FIGURE 23-9B. Using Cutting Tool Reamer to Remove Sharp Edges from Inside of Tubing.

Step-by-Step Procedure

1. Cut the tubing to the desired length by using a brake line cutting tool (Figure 23-9A).
2. With the aid of a fine flat file, square off the ends of the tube and ream the sharp edges from the inside of the line by using the reamer provided on the cutting tool (Figure 23-9B).
3. Brake lines have various shapes. Place a tube bender over the line. Shape the line to the one that was previously removed from the vehicle (Figure 23-9C). Remove the tool from the line.

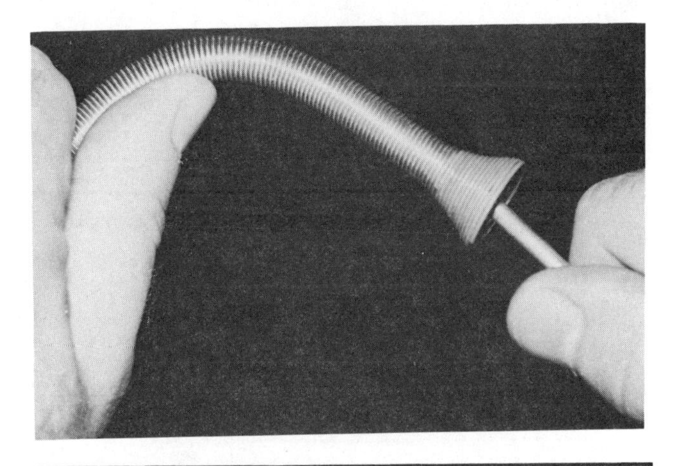

FIGURE 23-9C. Using a Tube Bender to Shape Tubing.

4. Install an inverted fitting on the line. Dip the end of the tubing to be flared in brake fluid. This lubrication will assist in a better formation of the flare.

5. Position the end of the tube in a vise. Then, unscrew the wing nuts. Position the end of the brake tube in the correct hole in the holder. Since the end of the line must have a double-lap flare, position the end of the tube 3.175 millimeters (1/8 inch) above the horizontal bars of the holder and tighten the two wing nuts. *Note:* To ensure that the end of the brake tube is the correct distance above the holder, use the double-lap adapter as illustrated in Figure 23-9D. If the tube is not at the correct height, reposition the tube in the holder.

6. Place the stem of the adapter into the tube and position the bridge over the adapter. Turn the handle of the threaded bridge screw until the adapter is flush with the bars of the holder. This procedure will compress the line (Figure 23-9E). When the end of the line has been formed to this shape, it is similar to an ISO flare.

7. Unscrew the bridge handle. Remove the adapter from the tube.

8. Once again, retighten the bridge handle and compress the lap into a flare (Figure 23-9F).

9. Remove the tube from the holder and inspect the finished product. There should be no cracks or splits in the flare. If the brake tubing has been correctly flared as illustrated in Figure 23-9G, repeat the same procedure for the opposite end of the brake line.

10. Blow the tubing out with compressed air to remove any foreign objects.

FIGURE 23-9D. Positioning Tubing and Double-Lap Adapter in Vise.

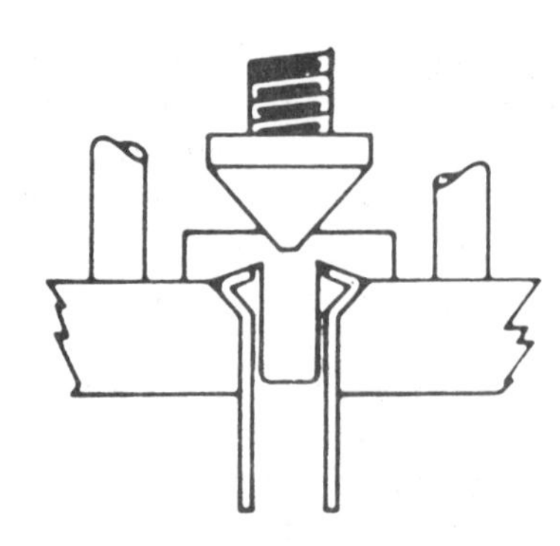

FIGURE 23-9E. Compressing the Adapter to Form Lap on Tube End.

FIGURE 23-9F. Compressing the Lapped End of Tube to Form ISO Flare.

FIGURE 23-9G. Correctly Flared Brake Line Tubing.

When replacing a flexible brake hose, the inverted line nut on the end of the steel line (see Figure 23-7) must first be unthreaded from the hose. Then, remove the retaining clip that locks the hose into the bracket and unscrew the hose from the wheel cylinder, brake caliper, or junction block terminal. When replacing a flexible hose, always install a new copper washer between the steel fitting and the part that the line is threaded into. Remove any dirt from the threaded portions of the hose and steel line. Tighten the various fittings to the manufacturer's torque specification, as detailed in the shop manual.

After the lines have been correctly installed, they will need to be bled to remove any air. This service procedure will be discussed and illustrated in the following chapters.

FACTS TO REMEMBER

1. Ordinary copper tubing must never be used as a substitute for steel brake lines. Copper tubing is subject to fatigue and will eventually break.
2. After the installation of a flexible hose, there should be no unnecessary twist in that particular part.

REVIEW TEST

1. Draw a schematic view and label the parts of a vehicle's stop light switch and pressure differential valve and switch and their electrical circuits.
2. List the reasons why the stop lights may not operate when the brake pedal is depressed.
3. Briefly describe how to determine brake pedal free play when the vehicle is not equipped with a power brake unit.
4. Describe how to test the operation of a pressure differential safety valve and switch.
5. Describe how to test the operation of a metering valve and a proportioning valve.
6. Describe how to inspect a vehicle's brake line and hoses and test their dependability to withstand maximum hydraulic pressure.
7. Describe how to double lap flare a brake line.

FINAL TEST

This examination is multiple choice. Only one answer will be accepted. Carefully read every statement.

1. Mechanic A says that the best way to test the fuse in the brake light circuit is to ground the positive terminal at the stop light switch. Mechanic B says that to test the operation of the switch, use a jumper wire to bypass the switch. Who is right? (A) Mechanic A, (B) Mechanic B, (C) Both A and B, (D) Neither A nor B.
2. Mechanic A says that when inspecting a vehicle's brake systems, always examine the brake pedal's free play. Mechanic B says that too little free play will not cause the brakes to drag. Who is right? (A) Mechanic A, (B) Mechanic B, (C) Both A and B, (D) Neither A nor B.
3. Mechanic A says that the only way to check the operation of the pressure differential valve and switch is to loosen one brake line fitting and then apply the brake pedal. Mechanic B says that the only way to check the operation of the pressure differential valve and switch is to loosen all brake line fittings that attach to the master cylinder and then apply maximum force to the brake pedal. Who is right? (A) Mechanic A, (B) Mechanic B, (C) Both A and B, (D) Neither A nor B.
4. Mechanic A says that the purpose of the metering valve is to reduce line pressure to the front brakes during initial application. Mechanic B says that the purpose of the proportioning valve is to reduce line pressure to the rear brakes during severe brake application. Who is right? (A) Mechanic A, (B) Mechanic B, (C) Both A and B, (D) Neither A nor B.
5. Mechanic A says that all vehicles are equipped with metering valves. Mechanic B says that all vehicles are equipped with proportioning valves. Who is right? (A) Mechanic A, (B) Mechanic B, (C) Both A and B, (D) Neither A nor B.
6. Mechanic A says that the test to determine the dependability of a vehicle's brake lines is conducted while the vehicle is stationary. Mechanic B says that to test the dependability of a vehicle's brake lines, drive the vehicle at 40 km/h (25 mph) and apply maximum force to the pedal. Who is right? (A) Mechanic A, (B) Me-

chanic B, (C) Both A and B, (D) Neither A nor B.

7. Mechanic A says that before you flare a brake tube, dip the ends of the line in engine oil. Mechanic B says that copper tubing is not an acceptable substitute for steel brake line. Who is right? (A) Mechanic A, (B) Mechanic B, (C) Both A and B, (D) Neither A nor B.
8. Mechanic A says that signs of surface cracking in a flex line are acceptable and do not mean that the line needs to be replaced. Mechanic B says that if a flex line shows signs of blistering near the metal fittings, this condition is not serious and the line does not need to be replaced. Who is right? (A) Mechanic A, (B) Mechanic B, (C) Both A and B, (D) Neither A nor B.

Disc Brake Service

LEARNING OBJECTIVES

After studying this chapter, you should be able to:

- List the names of the component parts of a sliding caliper disc brake assembly,
- Describe the operation of a floating caliper disc brake mechanism,
- Explain how to diagnose disc brake performance problems related to vibration, pull, squeal and noise, excessive pedal effort, pedal travel, and pedal feel,
- Describe four external conditions that affect brake performance,
- Describe how to perform a preliminary inspection of a disc brake assembly,
- Describe how to remove, disassemble, and service the caliper and disc and bleed a disc brake mechanism,
- Describe how to torque wheel lug bolts or nuts.

INTRODUCTION

Automobiles manufactured in North America are now equipped with sliding caliper front disc brake mechanisms. Although there are minor differences in design, disc brakes are similar in their operation and provide major advantages over drum brakes. Extensive testing has proved that disc brakes are less susceptible to brake fade after repeated applications at high speed. Disc brakes also decrease the possibility of a loss of directional control (pull to one side) during application and exhibit a minimum difference in braking performance whether the pads and discs are dry or wet. Since disc brakes are utilized by vehicle manufacturers, let's become familiar with the sliding caliper design, its operation, preliminary diagnostic tips, road testing the vehicle, brake pad inspection, repair service procedures, servicing the disc, and bleeding the brakes.

DESIGN AND OPERATION

When you look at an assembled sliding caliper disc brake unit, it is somewhat difficult to determine what parts move and what parts are stationary. Figure 24-1 illustrates a view of a disc brake mechanism. The major components are the revolving ventilated disc (often named the *rotor*) that attaches to the hub and the caliper that is positioned over the disc. The caliper contains two brake linings (called *inboard* and *outboard pads*), one internal cylinder, and a piston (not shown in this illustration). The caliper and its bushings float on two guide pins that are threaded into the anchor. The anchor is a stationary part that supports the caliper and is attached to the steering knuckle. When the brakes are applied, the anchor prevents the caliper from rotating with the revolving rotor. The caliper is machined and mated with the anchor so that during brake operations, it is able to slide laterally within the anchor. A fixed splash shield is positioned on the inboard side of the disc to protect the disc from water and the elements. The wheel itself protects the disc on the outside.

Front Disc Brake Operation

When the brake pedal is applied, fluid discharged from the dual master cylinder is transmitted through hydraulic lines and valves to the front disc brake calipers. When additional fluid enters the cylinder, it acts on the piston, thereby forcing the piston outward (to the left of Figure 24-2) in its cylinder's bore. The force produced by the piston (moving outward) causes the inboard pad to contact the disc.

What causes the outboard pad to make contact against the disc? We know that the law of physics states, For every action, there is an opposite and equal reaction. As you view Figure 24-2, you will see that hydraulic pressure is also reacting on the closed end of the cylinder's bore. This reaction moves the entire caliper inward (to the right of Figure 24-2), causing the outboard brake pad also to contact the disc. The caliper, prevented from rotating with the disc by the stationary adapter, transmits the retarding force to the vehicle. In other words, during brake application, the friction created while the pads are rubbing against the revolving discs retards and stops the front wheels from rotating.

How are the pads released? When the driver applies the brakes, a tight-fitting, square-shaped seal, located between the cylinder wall and the piston, tends to bend outward with the piston (Figure 24-3). When the brakes are released, the spring action of the distorted seal pulls the piston back into its cylinder's bore (Figure 24-4). This relieves the force on the brake pads, thereby establishing a slight running clearance between the linings and the disc. A dust boot, fitted around the piston and into the cylinder's bore, keeps out moisture and dirt, reducing the possibility of the piston's seizing in its cylinder.

In Chapter 21, we discussed how kinetic fric-

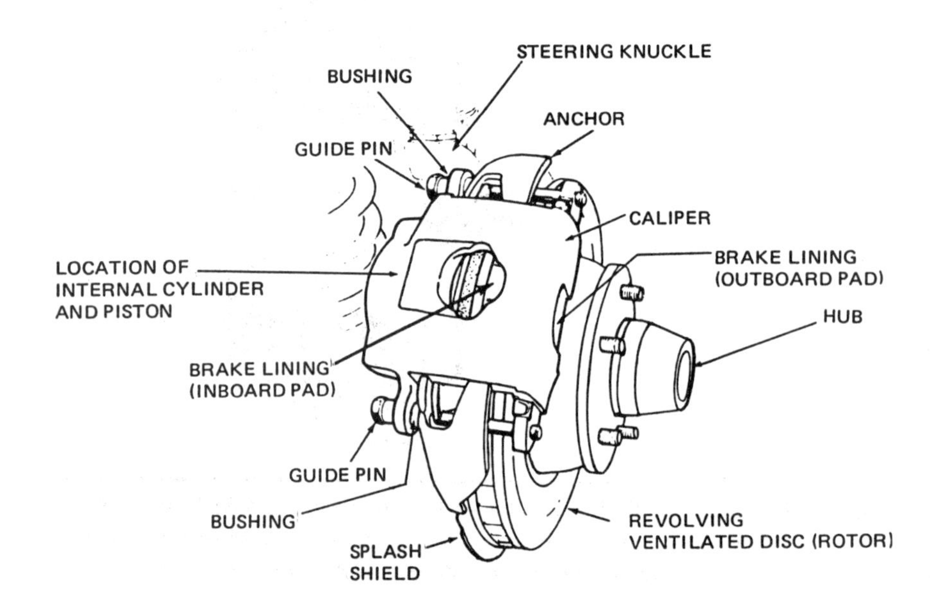

FIGURE 24-1. Disc Brake Mechanism. (Courtesy of Wagner Brake Company, Ltd.)

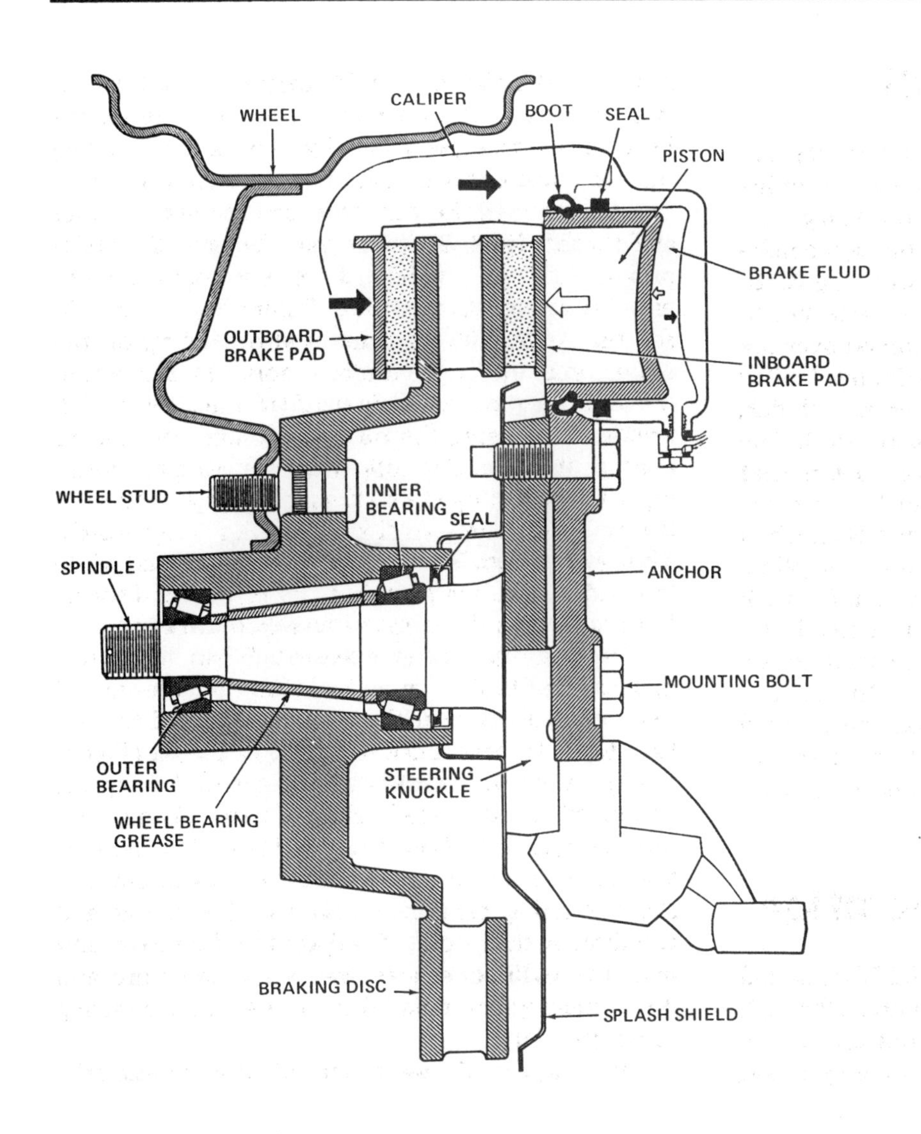

FIGURE 24-2. Cross-Section of a Typical Front Disc Brake Mechanism. (Courtesy of Chrysler Corporation)

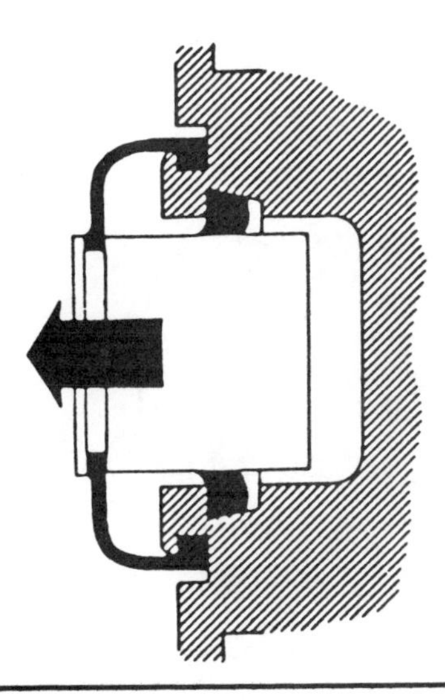

FIGURE 24-3. Operation of Piston and Seal during Brake Application. The seal bends outward when brakes are applied. (Courtesy of Wagner Brake Company, Ltd.)

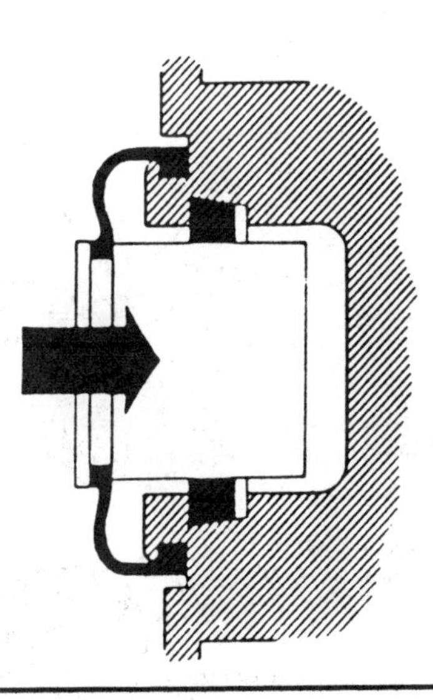

FIGURE 24-4. Operation of Piston and Seal during Brake Release. The seal pulls the piston back when brakes are released. (Courtesy of Wagner Brake Company, Ltd.)

tion is used to retard the motion of a revolving disc. Friction generates heat and causes the brake pads to wear. When the thickness of the brake pads is decreased, the piston inside the caliper's cylinder must change its operational position and move farther out in the bore. The caliper must also change its position within the adapter. Due to this situation, more fluid must be contained behind the piston in the cylinder. To allow for this additional fluid, the master cylinder's reservoir for a front disc brake system is larger than the reservoir for the rear drum brakes. Because of the unique design of a disc brake, the pads are always rubbing (wiping) the disc and are truly self-adjusting.

Rear Disc Brake Operation

Rear disc brakes have been used in a selected number of North American-built vehicles. The Kelsey-Hayes design has been used in Ford Motor vecles (Figure 24-5). This rear disc caliper assembly is similar to the front brake caliper except for an added parking brake and adjuster mechanism, which is actuated by the parking brake cable.

When the parking brake is applied, the lever rotates an operating shaft. There are three balls located in pockets between the operating shaft head and thrust screw. Shaft rotation forces the balls up ramps in the pockets. This action forces the thrust screw outward, driving the pistons and pads against the rotor. An automatic adjuster in the piston assembly moves on the thrust screw to compensate for pad lining wear and to maintain proper internal clearances. Special tools are required to service this type of disc brake mechanism.

DIAGNOSING PROBLEMS

In diagnosing problems associated with a vehicle's braking performance, valuable time can be saved if a logical sequence of analysis is followed. The three steps in making a diagnosis are asking the driver questions; road testing the vehicle, if necessary; and inspecting the brake pads.

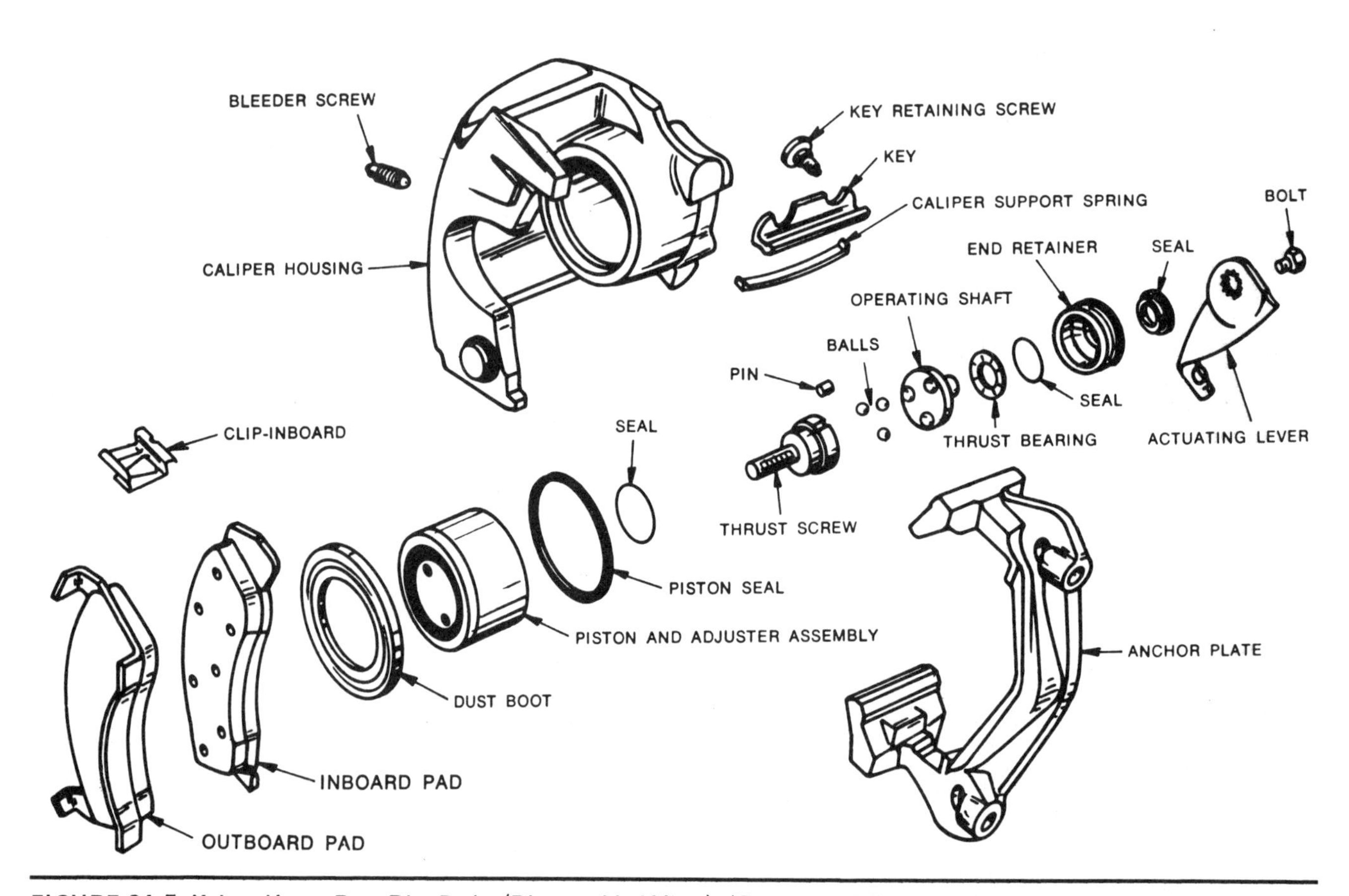

FIGURE 24-5. Kelsey-Hayes Rear Disc Brake (Disassembled View). (Courtesy of Wagner Brake Company, Ltd.)

Questioning the Driver

The easiest way to diagnose a vehicle's brake problems is to ask the driver questions. The pertinent questions to ask can be grouped into six categories: general, vibration, pull, brake squeal and noise, excessive pedal effort, and pedal travel and feel.

General

- How long ago (in kilometers or miles) did the problem first appear?
- Did the problem appear gradually or suddenly?

Analysis: Gradual appearance of the problem may indicate a partially seized disc brake piston or caliper. Sudden appearance of the problem suggests a severe application of the brake systems (such as an emergency stop). This may have resulted in damage to a major component part of the brake and/or suspension systems.

Vibration

- When the brakes are applied, is there a vibration? If so, at what speed is it most noticeable?
- Is there a vibration in the steering wheel or a pulsation in the brake pedal when force is applied?

Analysis: If, when the brakes are applied, a vibration is present above speeds of 80 km/h (50 mph), it is possible that there is a variation in the thickness of a disc. If the vibration is noticeable at a lower speed, it is possible that a rear brake drum is out-of-round. If the vibration is most noticeable in the steering wheel and/or the pedal pulsates, the problem may be related to the front brakes, wheels, or tires.

Pull

- With the driver's hands on the steering wheel and the brakes applied, is there a loss of directional control and steering stability to the left or right?

Analysis: This problem may be the result of a malfunctioning caliper assembly. An examination of the operation of the calipers and their piston will indicate if these are the source of the problem.

Note: If, after a careful inspection, there appears to be no cause for the complaint, an extensive examination of the vehicle's front suspension system is necessary. Pay particular attention to all control arm and strut rod rubber bushings to determine if they are fatigued. Also, examine the condition of the front crossmember where the strut rods enter the frame, or brackets, for rust and/or metal fatigue. If damage is evident, the caster factor, since it will be altered, will definitely cause a brake pull. (Chapter 16 described how positive caster influences the wheel to toe-in and negative caster influences the wheel to toe-out.)

Brake Squeal and Noise

- When the brakes are applied, is there a squeal or noise at all times or only sometimes—for instance, during the first few stops in the morning?

Analysis: If the brakes make a continuous noise on application, the problem is related directly to the pads and/or discs. If the noise is noticeable only during the first few stops in the morning, the surfaces of the pads are glazed or contaminated. Another probable cause may be damage to the insulation that adheres to the back of the pads.

Excessive Pedal Effort

- When the brake pedal is applied, is excessive force required to stop the vehicle?

Analysis: If the customer has this complaint, first test the operation of the power brake booster. The test procedures for the operation of this unit are described in Chapter 26 of this text. If the power unit is operating correctly, examine the condition of the front disc and the rear drum brake mechanisms, plus the operating condition of the various control valves and lines.

Pedal Travel and Feel

- When the brake pedal is applied, is there sufficient pedal reserve height? If not, how far down does the pedal travel? Does the pedal feel spongy?

Analysis: If the pedal height is low upon application, inspect the operating condition of the front disc and the rear drum brake mechanisms. *Note:* Pay particular attention to the adjustments for the rear brakes. There may be excessive clearance between the brake shoes and the drums.

If the pedal is spongy, examine the brake systems to determine if there has been a loss of fluid. Start by examining the fluid contained within the master cylinder's reservoirs. If there has been a loss of fluid and one of the compensating ports is exposed, air will have entered the brake system. Correct the cause of the leak and bleed the air from the system. This service procedure is described later in this chapter.

Road Testing the Vehicle

When a customer brings a vehicle to a repair shop for a diagnostic examination, problems associated with the brake system can often be assessed by asking the driver questions and without actually driving the automobile or truck. Sometimes, however, the vehicle will need to be road tested.

When you road test a vehicle's brake system, drive on a dry, clean, reasonably smooth, level roadway. A true test of stopping capabilities cannot be assessed if the roadway is wet, greasy, or covered with loose dirt since all tires will not be able to grip the road equally. Testing will also be adversely affected if the road is crowned because the vehicle's weight will shift toward the low side of the road. Also, if the roadway is rough, the wheels will tend to bounce. *Remember:* When you are required to perform a road test, you are driving an unfamiliar vehicle. It is more important than ever that you drive defensively.

Test the brakes at different speeds with both light and heavy pedal applications; however, you should avoid locking the brakes and skidding the wheels. Locked brakes and skidding tires do not indicate braking efficiency; on the contrary, heavily braked but turning wheels will stop a car in less distance than locked brakes. The following four points are external conditions that affect braking performance. Keep these points in mind when diagnosing brake problems.

1. *Tires:* Tires having unequal contact and grip on the road will cause unequal braking. Before road testing a vehicle, inspect the tread surface of each tire and be sure that their inflated pressures are correct.
2. *Vehicle loading:* When a vehicle has unequal loading, the wheels supporting the heaviest part of the load require more braking power than the others.
3. *Front end alignment:* Misalignment of the front end, particularly in regard to limits on camber and caster, as well as a defective front and rear suspension system, will cause the vehicle's brakes to pull to one side when applied.
4. *Front wheel bearings:* A loose front wheel bearing permits the front wheels to tilt and have a spotty or irregular contact with the pads and rotor. This condition will cause a low brake pedal and an erratic brake operation.

Inspecting Brake Pads

Let's assume that you have discussed the vehicle's braking problem with the customer and have road tested the automobile. Based on your preliminary diagnosis, you have concluded that further inspection of the brake system is required in the repair shop.

You now have the automobile in a workstall and have obtained the necessary tools and equipment to enable you to perform the inspection. Upon removing the hubcaps, you notice that the front wheels have balancing weights attached to the rims. To ensure that you do not disturb the balance of the front wheels, with a piece of crayon or chalk, index each wheel to its hub, as illustrated in Figure 24-6.

With the appropriate wrench, loosen each wheel nut approximately two revolutions. Keep in mind when doing this initial step that you will need first to determine if the left front wheel lug bolts are right- or left-hand thread. After loosening each wheel nut, raise the front of the car with a hydraulic floor jack. Position safety stands under the vehicle's frame between the front fenders and the doors (Figure 24-7). Lower the vehicle onto the stands and remove the jack from under the vehicle so that it does not become a potential safety hazard to anyone walking in front of it. Unthread the wheel nuts and remove the wheels from their hubs.

Inspect the brake pads by visually examining the outboard lining pad at each end of the caliper where the most wear occurs. The inboard lining pad is checked through the inspection hole (see Figure 24-1) on top of the caliper. If the linings are worn to less than 1.6 millimeters (1/16 inch) thickness, they must be replaced. If the condition of the pads is doubtful, it will be necessary to remove the caliper from the adapter and measure for minimum pad thickness.

Some manufacturers use wear indicators that are designed to rub on the rotor when the linings have become thin. If these wear indicators are less than

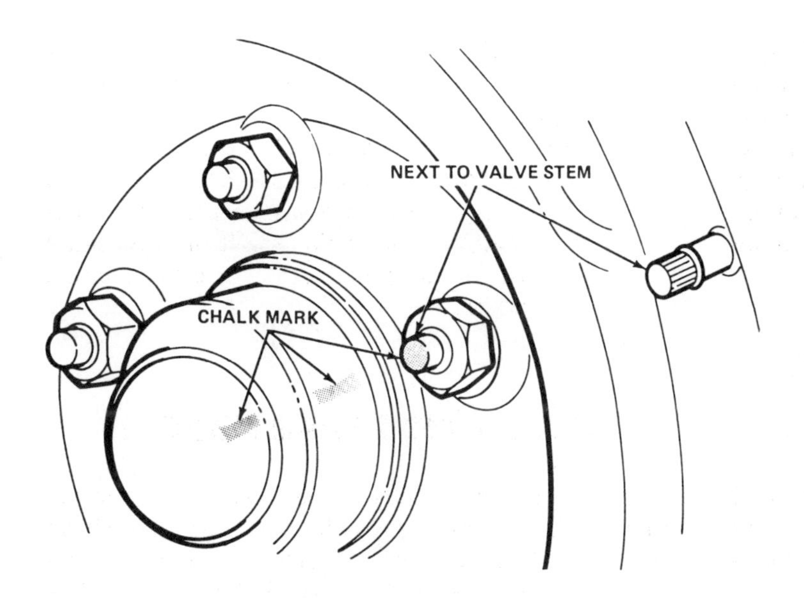

FIGURE 24-6. Indexing a Wheel by Using a Chalk Mark. (Courtesy of Ford Motor Company)

1.6 millimeters (1/16 inch) away from the rotor, it is strongly recommended that the pads be replaced.

Let's assume that you have examined the pads and have found that the outboard lining of the left caliper is tapered (thinner at the front than at the rear). When this condition exists, it is more than probable that the caliper is partially seized in the anchor and is causing a slight brake pull to the right. The disc brakes for both front wheels need to be overhauled and serviced.

CAUTIONS TO REMEMBER

Because studies have indicated that exposure to excessive amounts of asbestos dust may be a potential health hazard, the Occupational Safety and Health Administration (OSHA) has set maximum limits of levels of airborne asbestos dust to which workers may be exposed. Since most automotive friction materials normally contain a sizable amount of asbestos, it is important that people who handle brake linings and clutch facings understand the nature of the problem and know the precautions to be taken.

Areas where brake work is done should be set aside from the general work area if possible. Entrances should be posted with an asbestos exposure sign, as follows:

Asbestos Dust Hazard
Avoid breathing dust.
Wear assigned protective equipment.
Do not remain in area unless your work requires it.
Breathing asbestos dust may be hazardous to your health.

The amount of asbestos in the dust from brake lining wear is normally at an extremely low level because of chemical breakdown during use. If machining of friction material does not take place, simple procedures will minimize exposure. During brake servicing, the mechanic should wear an air-purifying respirator, either a throwaway type or one with replaceable particulate filter(s), as approved by the Mining Enforcement and Safety Administration or the National Institute for Occupational Safety and Health. The respirator should be worn during all procedures, starting with the removal of the wheels and including reassembly.

During disassembly, all parts should be carefully placed on the floor to minimize the possibility of creating airborne dust. Dust should first be cleaned from the brake drums, brake backing plates, and brake assemblies using an industrial-type vacuum cleaner equipped with a high-efficiency filter sys-

FIGURE 24-7. Correct Position of Safety Stands.

tem. After vacuum cleaning, any remaining dust should be removed by using a rag soaked in water and wrung until nearly dry. Under no circumstances should compressed air or dry brushing be used for cleaning.

Of extreme importance are the precautions that must be taken during machining of friction material. This is the operation in brake servicing when exposure to asbestos dust is at its highest. In addition to the approved respirator, there must be local exhaust ventilation such that the worker exposures are maintained at least below the 1976 OSHA asbestos standards. If there is any question as to the efficiency of asbestos dust removal by the machine, the manufacturer should be contacted.

Industrial vacuum cleaner bags containing asbestos dust and cloths used for wiping brake assemblies should be sealed in plastic bags and labeled with the following warning label printed in letters of sufficient size and contrast to be readily visible and legible:

Caution
Contains asbestos fibers.
Avoid creating dust.
Breathing asbestos dust may cause serious bodily harm.

All asbestos waste should be disposed of in accordance with the OSHA asbestos regulation. During removal of vacuum bags, an approved respirator should be worn.

All floor cleaning in areas where brakes are repaired should be done with the high-efficiency industrial vacuum cleaner. Under no circumstances should dry sweeping take place. Grinding (arcing) machines should also be cleaned with such a vacuum cleaner and any remaining dust wiped with a damp cloth. An approved respirator should be used during this cleaning.

Although adherence to these procedures should minimize any contamination of work clothing, it is necessary that the appropriate sections of the OSHA regulations on asbestos be followed when ceiling levels on asbestos dust exceed current standards. For example, special clothing and changing rooms may be necessary. The work clothes should not be taken home but should remain at work for industrial laundering.

Proper hygiene practices will help minimize exposure to asbestos dust. Wash thoroughly before eating, and do not eat in the work area. *Remember:* Smoking is harmful to the health. Smoking coupled with breathing asbestos dust is extremely dangerous

Caution: Do not breathe asbestos dust. Work on one wheel at a time to avoid popping the pistons out of the other caliper. Keep grease, oil, and brake fluid off the caliper, rotor, and pad and lining assemblies. Don't push the brake pedal when the pads are out or when the caliper is off the rotor.

SERVICING PROCEDURES

Since the vehicle has been raised and correctly positioned on safety stands, let's discuss how to remove and service the front disc brake calipers. *Remember:* Follow all safety guidelines.

Removing and Disassembling the Caliper

Step-by-Step Procedure

1. Place a protective cover over the front left fender. Any brake fluid that spilled on the fender would remove the paint.

2. Clean all dirt from and around the master cylinder's reservoir cover.

3. Remove the cover from the master cylinder. Using a syringe, siphon approximately half of the fluid from the reservoir that supplies fluid to the disc brakes (Figure 24-8). Removing the fluid now will prevent fluid overflow when the piston is pushed into its cylinder's bore prior to lifting the caliper from the disc.

4. With a large C clamp positioned over the caliper, tighten the handle of the screw and compress the piston into its cylinder (Figure 24-9). This simple step will enable you to lift the caliper from the disc easily.

FIGURE 24-8. Removing Fluid from Reservoir. (Courtesy of Wagner Brake Company, Ltd.)

5. Disconnect the hydraulic inverted line fitting from the brake flex hose at the frame bracket. Then, cap the end of the steel line. With pliers, pull the retainer lock from the bracket (Figure 24-10).

6. Using the necessary socket and ratchet, unscrew and remove the two caliper guide pins that attach the caliper to the adapter (Figure 24-11).

7. Lift the caliper off the adapter and away from the disc.

8. Place the caliper on the counter of a workbench.

9. Remove the inboard and outboard pads from the caliper.

10. With a screwdriver, carefully remove the outer and inner rubber bushings from the adapter. Discard the bushings.

11. Disconnect and remove the flex hose from the caliper.

12. With an appropriate deep socket and ratchet, unscrew and remove the bleeder valve. Clean the valve's exterior and, using an air gun and compressed air, blow out its internal passage. Drain and discard the remaining fluid from the caliper. Install and tighten the bleeder valve.

13. If the bleeder valve is seized in the caliper, apply penetrating oil to the threads of the valve. Allow time for the oil to do its job. After a few minutes, gently tap the end of the valve. If the bleeder valve is still seized, complete step 14. Then, using an acetylene torch, apply controlled heat to the caliper around the area of the valve. While the caliper is hot, you should be able to remove the valve.

14. Place the piston face down on a block of wood or rags on the workbench. Using an air gun, force compressed air through the caliper's inlet port (Figure 24-12). This will force the piston out of its bore and onto the block of wood or rags.

Caution: Keep your fingers away from the piston area to avoid personal injury.

Remove the block and the piston. *Note:* If the piston is seized in its bore, you can often force the piston out of its bore by attaching the caliper and the flex line to the steel brake line and by applying the brake pedal. Special tools are available that will assist you to remove a seized piston.

15. Mount the caliper in a vise that is equipped with protective jaws. *Remember:* If excessive vise pressure is used, you will distort the cylinder bore and cause the piston to bind.

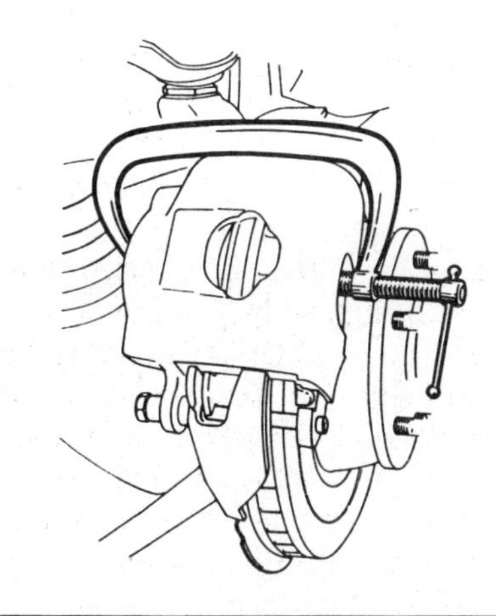

FIGURE 24-9. Compressing Piston into Its Cylinder Using a C Clamp. (Courtesy of Bear Automotive, Inc.)

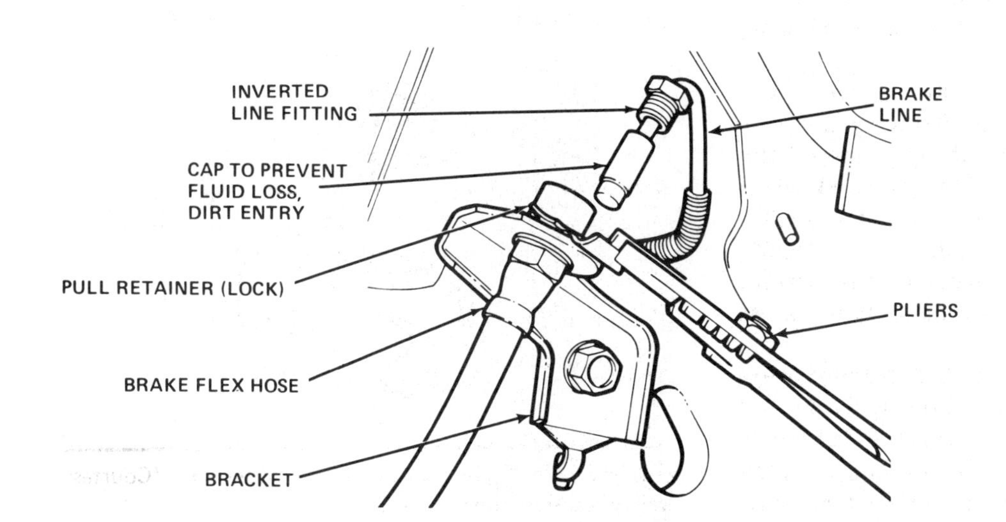

FIGURE 24-10. Disconnecting the Hydraulic Line. (Courtesy of Ford Motor Company)

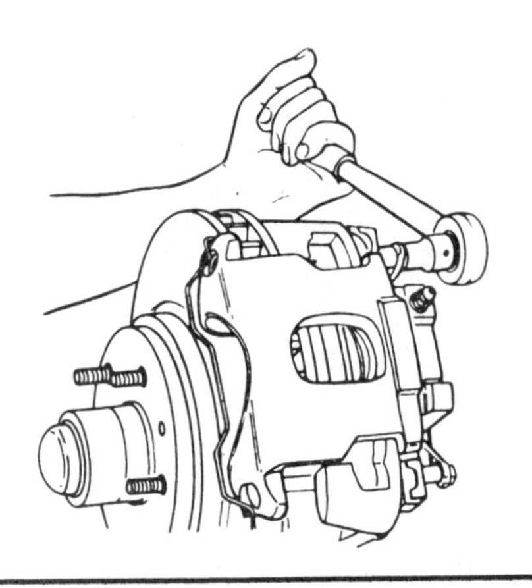

FIGURE 24-11. **Removing Guide Pins.** (Courtesy of Wagner Brake Company, Ltd.)

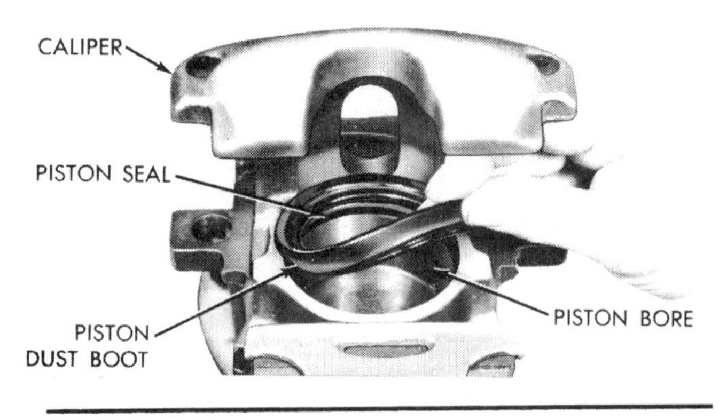

FIGURE 24-13. **Removing or Installing Piston Dust Boot.** (Courtesy of Chrysler Corporation)

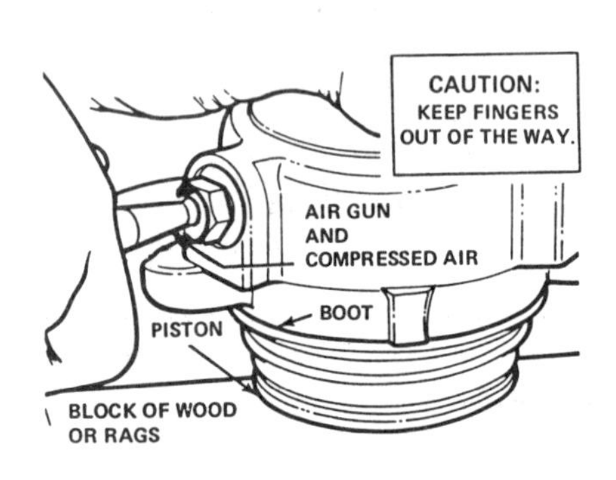

FIGURE 24-12. **Blowing Piston out of Bore.** (Courtesy of Ford Motor Company)

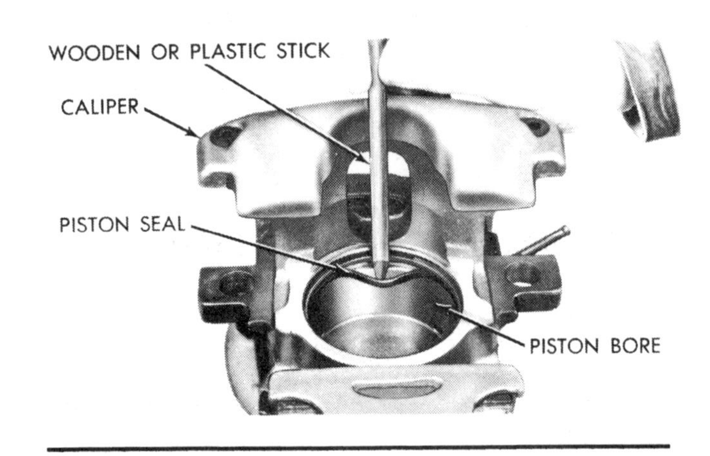

FIGURE 24-14. **Removing Piston Seal.** (Courtesy of Chrysler Corporation)

16. Using only your hands, remove the dust boot from its groove in the caliper (Figure 24-13).

17. Using a small, pointed, wooden or plastic stick, as illustrated in Figure 24-14, remove the piston seal out of its groove in the cylinder bore. If a metal object is used for this step, there is a danger of scratching the recessed groove in the bore.

18. Remove the bleeder valve from the caliper.

Cleaning and Inspecting the Parts

Vehicle manufacturers specifically state that only denatured alcohol or a suitable solvent (not varsol) be used to clean all the caliper's parts. After the parts have been washed and cleaned, use compressed air to blow out all the drilled passages and the bore. Inspect the piston and its bore for scoring and pitting. Surfaces that show light scratches or corrosion usually can be polished with crocus cloth, using a circular motion.

Some pistons are made of plastic. Do not hesitate to discard a piston that is pitted or cracked. If a piston is made of metal and has signs of plating wear, replace the piston.

Maximum piston clearance is specified by the vehicle's manufacturer and may be determined by placing a 6.3 millimeter (¼ inch) wide feeler shim of a specified thickness (refer to the manufacturer's shop manual) lengthwise in the cylinder bore (Figure 24-15). If the piston can be inserted with the shim in place, the piston clearance is oversize. If by trying a new piston the clearance has not been reduced, the caliper must be replaced.

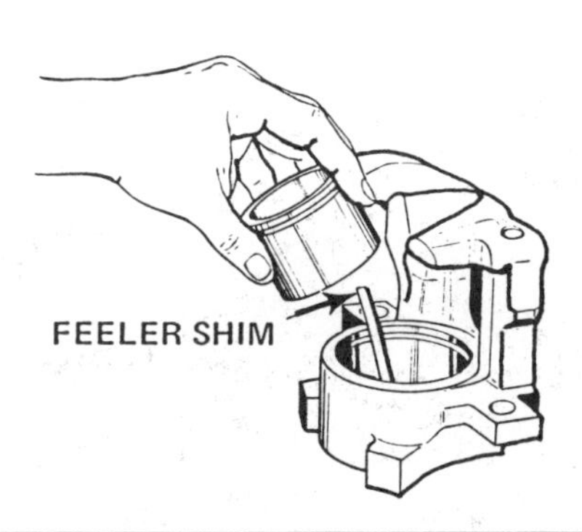

FIGURE 24-15. Testing for Piston Clearance. (Courtesy of Wagner Brake Company, Ltd.)

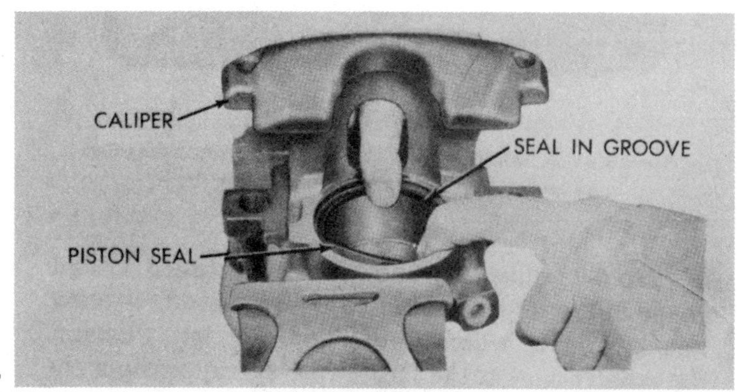

FIGURE 24-16. Installing Piston Seal. (Courtesy of Chrysler Corporation)

A PRACTICAL SERVICE TIP

When you are servicing a disc brake caliper, extreme cleanliness is absolutely essential.

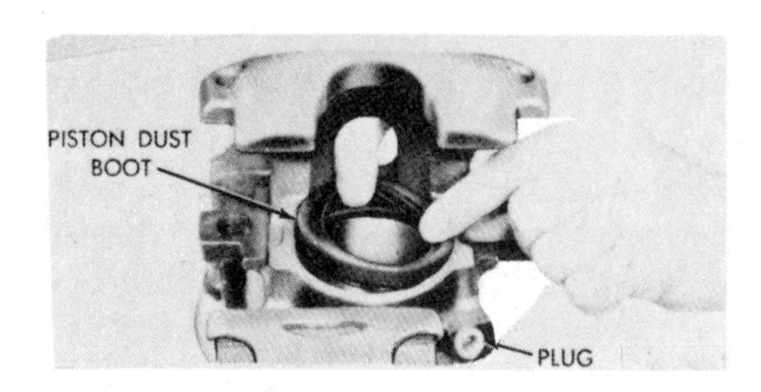

FIGURE 24-17. Installing Piston Dust Boot. (Courtesy of Chrysler Corporation)

Assembling the Caliper

Step-by-Step Procedure

1. Position and hold the caliper in the vise with the open end of the bore facing up.

2. Making sure that your hands are absolutely clean, dip the new piston seal in clean brake fluid and gently install it in the groove in the cylinder bore (Figure 24-16).

3. Coat the new dust boot with clean brake fluid, leaving a generous amount of fluid inside the boot. Install the lip of the boot in the caliper by working it into the outer groove using only your fingers. To ensure that the boot is properly seated, slide your forefinger around the inside of the boot (Figure 24-17).

4. Install the bleeder in the caliper and tighten with a wrench.

5. To assist you in this step, the manufacturer suggests that you obtain a plastic or brass plug that is the same size and thread as the inlet fitting of the flex line. Thread the plug into the inlet and tighten.

6. With your fingers spreading the boot, work the piston into the boot and press down on the piston (Figure 24-18). The entrapped air below the piston will force the boot around the piston and into its groove as the piston is depressed. Remove the plug. Carefully push the piston down in the bore until it is bottomed.

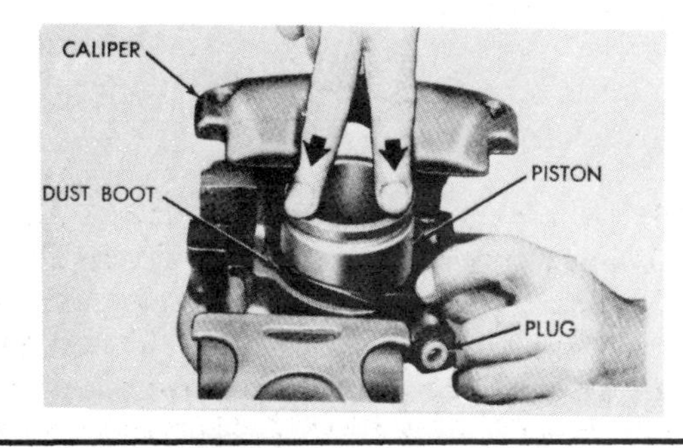

FIGURE 24-18. Installing Piston through Boot. (Courtesy of Chrysler Corporation)

7. Install new inner and outer rubber guide bushings in the caliper. Make certain that the bushings are correctly seated and that their flanges extend over the caliper casting evenly on both sides.

8. Examine the flex hose for cracks and chafing. If the hose may be used, install it in the caliper using a new copper washer.

SERVICING DISC BRAKES

When disc brake mechanisms are serviced, disc rotors must first be visually examined and then measured by using precise gauges and micrometers. Tolerances on depth of scoring, lateral runout, parallelism and overall thickness, and the refacing of disc braking surfaces are critical in relation to a vehicle's braking performance. Since a disc is a major component part of a brake mechanism, let's become familiar with the sequence that must be followed for servicing a disc.

Examining and Measuring the Discs

Visually examine discs for cracks, scoring, heat checking, and rusting. If the rotor is cracked, it is not repairable and must be replaced. If the rotor is scored, it may be machined providing the depth of the scores is not more than 0.38 millimeter (0.015 inch) deep following the refacing operation. Some vehicles manufactured by General Motors have discs that are intentionally grooved. This modification in disc design reduces noise and radial lift out of the pads.

Heat checking and disc pad surface areas may normally be restored during resurfacing. However, the results of an excessive accumulation of rust on the fins, between the two faces of the disc, may be determined only after the rust has been removed from the fins. A deterioration of the metal will result in a separation of the outboard faces of the disc.

Lateral Runout. Prior to measuring disc runout, the wheel bearings must be torqued and adjusted to the manufacturer's specification (Figure 24-19A). A torque wrench must be used for this step after the removal of the grease cup and the cotter pin (Figure 24-19B).

To determine the runout of a disc, fasten a dial gauge indicator securely to the splash shield, steering knuckle, or the wheel bearing adjusting nut. Position the pointer against the face of the disc approximately 2.5 to 3.1 centimeters (1 to 1¼ inch) from the outside diameter of the disc (Figure 24-20). Slowly turn the disc by hand and record the amount of lateral runout or wobble. If the reading exceeds specifications, the disc must be machined or replaced.

Parallelism and Overall Thickness. The term *parallelism* is used to refer to the even thickness of the disc. The outer edges of the disc must run parallel to one another or be at the same distance from one another all the way around. To determine if the disc is parallel, measure the thickness of the disc at eight different points and around its faces with a micrometer (Figure 24-21). *Note:* Manufacturers specify that the parallelism of the disc not vary more than 0.0125 millimeter (0.0005 inch). If the parallelism of the disc is not within specifications, the disc must be refaced. If the overall thickness of

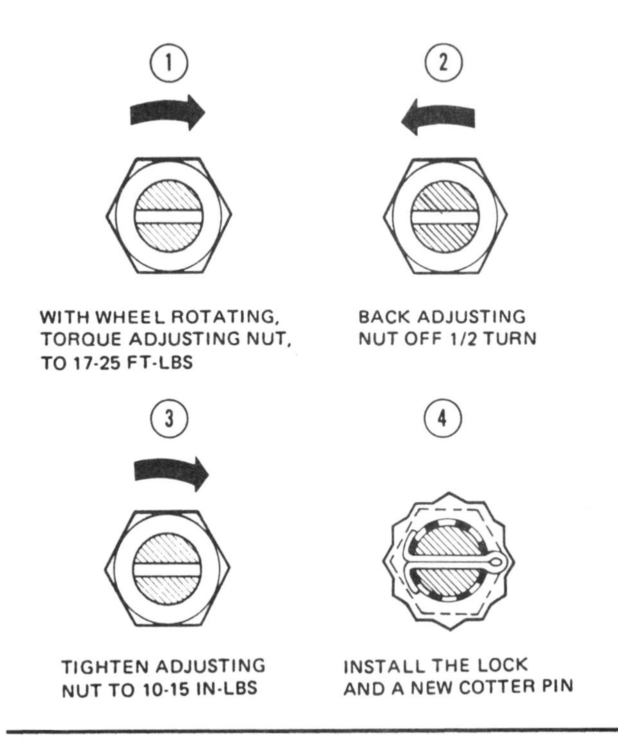

FIGURE 24-19A. Manufacturer's Specifications for Adjusting Wheel Bearings. (Courtesy of Ford Motor Company)

FIGURE 24-19B. Using Torque Wrench to Adjust Front Wheel Bearings. (Courtesy of Bear Automotive, Inc.)

FIGURE 24-20. Using a Dial Gauge Indicator to Measure Lateral Runout. (Courtesy of Bear Automotive, Inc.)

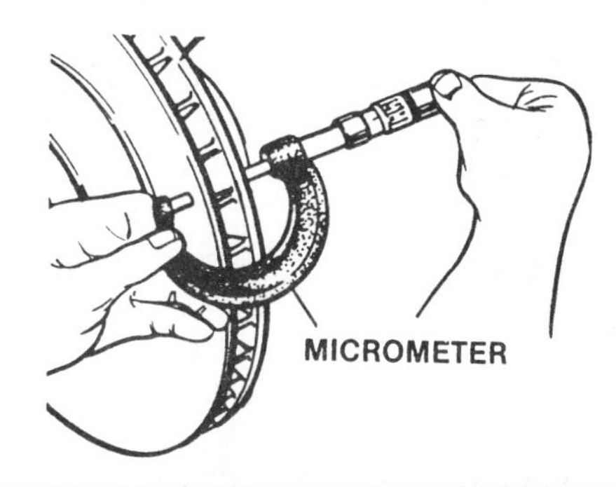

FIGURE 24-21. Measuring Disc for Parallelism with a Micrometer. (Courtesy of Ammco Tools, Inc.)

the disc is less than the manufacturer's specifications, the disc must be replaced.

If after a careful inspection and gauging of a rotor you determine that it does not need to be resurfaced, sand the braking surfaces of the disc with number 80 grit paper to remove previous lining contamination. Then wash the disc with denatured alcohol.

Machining the Discs

It is not always necessary to machine discs. However, let's assume that you have gauged the rotor and are required to reface its braking surfaces.

Step-by-Step Procedure

1. Using the appropriate size socket and ratchet, unthread the wheel bearing adjusting nut and remove the disc from the spindle. Place the outer bearing, washer, and nut in the hubcap.

2. Remove the inner bearing and seal from the hub. This step may be done by using a piece of round, wooden dowling, 30 millimeters (12 inches) in length, inserted through the outside bearing cup. Place one end of the dowling against the inner wheel bearing (the dowling will be positioned on a slight angle). Moving the dowling alternately from side to side, gently tap the opposite end of the dowling with a hammer and remove the seal and bearing from its recessed position in the hub. Discard the seal and place the inner bearing in the hubcap. With lint-free cloths dampened with varsol, thoroughly remove the grease from the inside of the hub so that it is clean.

3. To ensure that the disc will be accurately refaced, the shaft, hub and adapters, and bearing cups inside the rotor must be clean and free from any metal chips or contamination. Follow the manufacturer's instructions for the lathe that is in your shop and mount the disc on the lathe using the correct adapters. *Note:* If the disc is not mounted properly on the lathe, the runout will be worse after refacing than before.

4. After installing a vibration damper around the center of the rotor, use a double straddle cutter to reface both sides of the disc at the same time (Figure 24-22). *Note:* Make certain that the tool bits

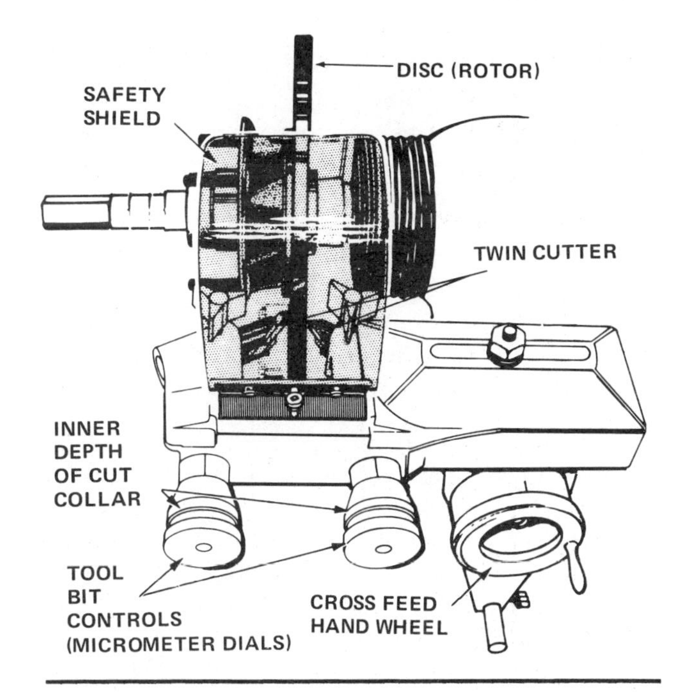

FIGURE 24-22. Refacing Both Sides of the Disc at the Same Time. (Courtesy of Ammco Tools, Inc.)

have rounded cutting points and are not chipped. Also, make sure that the safety shield is in place before you start machining the disc.

5. After the initial refacing of the disc, use the lathe to provide a finishing (slow cut) of 0.10 millimeter (0.004 inch) to both sides of the disc braking surfaces.

6. Using a micrometer, check the thickness of the rotor. Compare the reading to the manufacturer's specification embossed on the face of the rotor (Figure 24-23). If the measurement is less than the specified thickness, the disc must be replaced.

7. After refacing the rotor, operate the lathe again. Using number 80 grit paper, sand the surfaces. Wash both surfaces with denatured alcohol. When the rotor has been cleaned, do not touch the machined surfaces. Remove the disc from the brake lathe.

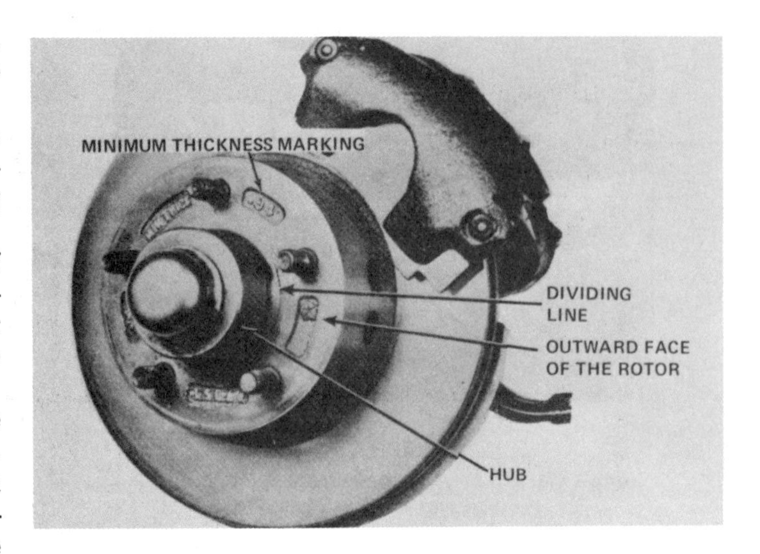

FIGURE 24-23. Manufacturer's Specifications Embossed on the Face of the Rotor. (Courtesy of Chrysler Corporation)

A FACT TO REMEMBER

Most rotors are now cast as a single component part. Always check a new rotor for lateral runout and parallelism before installing on the vehicle. Reface the rotor if necessary.

Lubricating the Bearings

To this point, you have studied procedures that will enable you to be better informed as to how to service a disc brake mechanism. The correct lubrication of the front wheel bearings is another part of the total service procedure. Chapter 4 discussed how wheel bearings are serviced and lubricated.

Installing the Caliper

The next important step in the total service of a disc brake mechanism is the installation of the caliper to the anchor (adapter).

Step-by-Step Procedure

1. With the aid of 80 grit paper, remove any surface rust from the anchor and the caliper where there is metal-to-metal contact. After the rust has been removed, coat only those surface areas with a light film of **molylube** or molybdenum grease.

2. Install the inboard and outboard brake pads in the caliper. Slowly slide the caliper down into position in the anchor and over the disc. Align the pin holes of the caliper, adapter, and inboard and outboard pads.

3. Install the guide pins through the inner bushings in the caliper, through the anchor, the inner and outer brake pads, into the outer bushings in the caliper and the antirattle spring (Figure 24-24).

4. Press in on the end of the guide pins. With extreme care (so you do not cross the threads of the pins and the threads in the caliper), use the correct-sized socket or allen wrench and ratchet to screw the guide pins into the adapter. After the guide pins have been installed, torque the pins to 34 to 55 newton-meters (25 to 40 foot-pounds).

Flushing the Front Hydraulic System

One important step in brake service that is often overlooked by repair technicians is the flushing of the hydraulic system before the calipers and their hoses are connected to the steel brake tubes. Let's consider why the flushing of a hydraulic system is important.

Brake fluid in a hydraulic system tends to pick up small particles of impurities, which contaminate the fluid. Since it is vital that the hydraulic system be clean, all particles must be expelled from the lines leading to the calipers before the flexible hoses are attached to the steel tubes. This step is accomplished by filling the master cylinder's disc brake

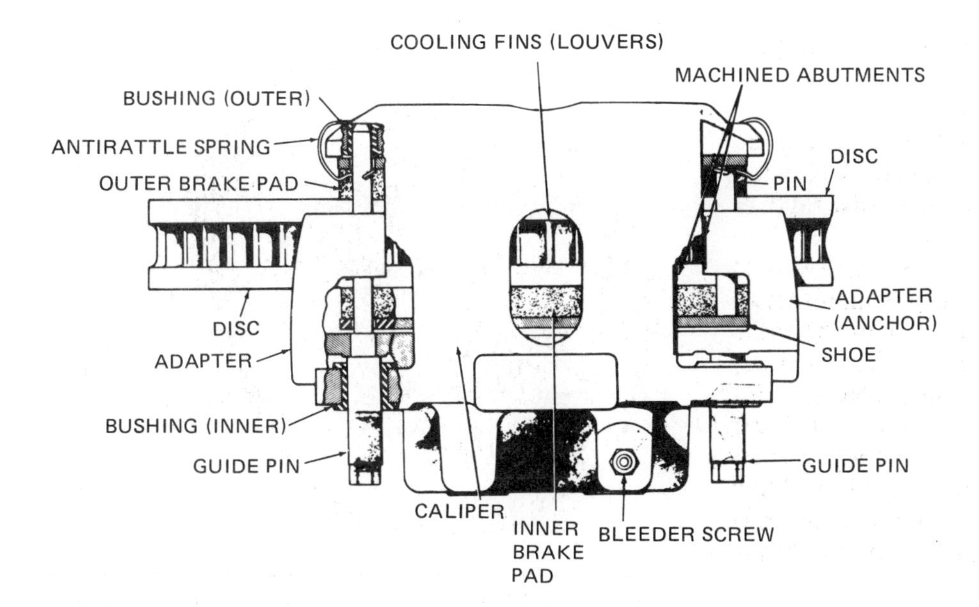

FIGURE 24-24. Floating Caliper Assembly (Sectional). (Courtesy of Chrysler Corporation)

reservoir and pumping the brake pedal until clean fluid is discharged from the steel brake tubes.

When it is evident that new, clean fluid is being discharged from the end of the steel brake line, position the end of the flexible hose through the frame bracket and thread the inverted line nut into the flex line. With the correct-sized wrenches, tighten the brake line fittings and insert the retainer lock.

BLEEDING FRONT DISC BRAKES

Once the front calipers have been serviced, their cylinders will need to be filled with new brake fluid and the air purged from the hydraulic system. Hydraulic brake systems can be bled manually or with the aid of pressure bleeding equipment.

Manual Bleeding

This step will require the assistance of another student or worker who will pump the brake pedal and then hold it down under force. A glass bottle, partially filled with new brake fluid, a length of rubber or plastic hose, and a bleeder valve wrench will be needed.

Step-by-Step Procedure

1. Remove the cap from the master cylinder. Fill the reservoir that supplies fluid to the front brakes. *Remember:* Keep the reservoir constantly full of fluid while bleeding the brakes.

2. Have your assistant pump the pedal several times and then apply a constant force by holding the pedal down.

3. Attach one end of the hose snugly around the bleeder valve that is located in the caliper farthest from the master cylinder. Immerse the opposite end of the hose in the brake fluid container in the glass bottle. *Note:* Keeping the hose immersed in the fluid will prevent air from reentering the system during the bleeding process.

4. Using the bleeder valve wrench, loosen (open), only one turn, the bleeder valve. When the valve is open, the air trapped inside the hydraulic system will be discharged through the valve, the hose, and the fluid contained in the bottle. When the air bubbles have ceased to appear, tighten (close) the valve and have the pedal pumped again and held. Continue this process to purge air from the caliper until only fluid, containing no air, is emitted from the hose.

5. Bleed the caliper at the opposite wheel. As the air is expelled from the system, the operational height of the brake pedal will be high and firm rather than spongy.

6. Fill the master cylinder reservoirs to their correct levels. Inspect the condition and placement of the gap's flexible rubber diaphragm. If it is correct, install the cover cap.

7. Test the dependability of the vehicle's hydraulic brake systems as previously described in Chapter 23.

GM Quick Take-Up Master Cylinder. Special procedures are required to manually bleed the quick take-up brake system used on some General Motors cars built since 1980. It is recommended that you bleed the master cylinder first (Figure 24-25). Then disconnect the left front brake line at the master cylinder, and fill the master cylinder until fluid flows from the port.

Caution: Catch fluid in a rag and don't allow fluid or rag to contact car finish. Connect the line and tighten fitting.

Depress the brake pedal one time slowly and hold. Loosen the same brake line fitting to purge air from the system. Retighten the fitting and release the brake pedal slowly. Wait 15 seconds. Then repeat the sequence, including the 15 second wait, until all air is removed. Bleed the right front connection in the same way as the left front.

Bleed the wheel cylinders and calipers only after you are sure that all the air has been removed from the master cylinder. Follow the specified right rear-left front-left rear-right front sequence and depress the brake pedal slowly one time before opening bleeder screw to release air. Tighten screw, slowly release pedal, and wait 15 seconds. Repeat all steps, including the 15 second delay until all air has been removed from the system.

Caution: Rapid pumping of this system moves the secondary master cylinder piston down the bore in a manner that makes it difficult to bleed the left front-right rear part of the system.

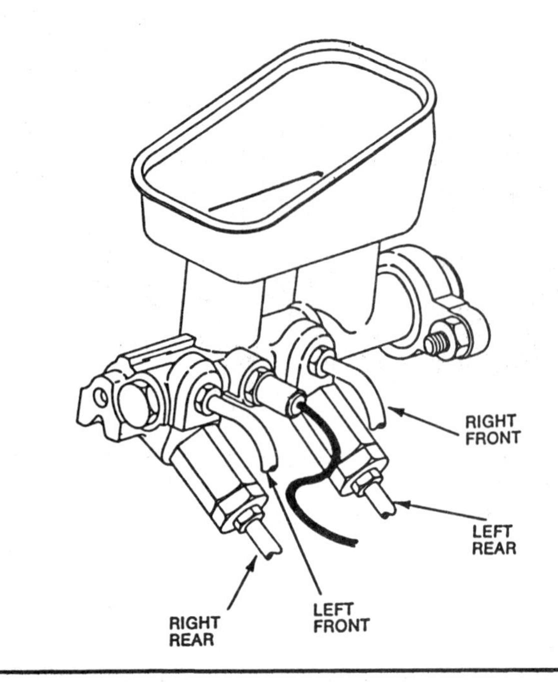

FIGURE 24-25. **GM Quick Take-Up Master Cylinder.** (Courtesy of Wagner Brake Company, Ltd.)

Pressure Bleeding

An alternate method for purging air from a brake system is the use of a pressure bleeder unit (Figure 24-26). Modern bleeders are internally designed with a flexible airtight diaphragm that divides the tank into two compartments. One compartment contains fluid; the other half of the equipment holds compressed air. Fluid, under pressure, is delivered to the master cylinder through a flexible rubber hose and a special adapter that fits over the openings for the reservoirs.

Prior to bleeding the brakes, the fluid compartment is filled with brake fluid, and the air compartment is charged with a maximum pressure of 175 kPa (25 psi).

When a vehicle's front brake system is equipped with a metering valve (part of the combination valve), the metering valve must be made inoperative before bleeding the front brakes, either by pushing in or pulling out on the stem of the metering valve (Figure 24-27).

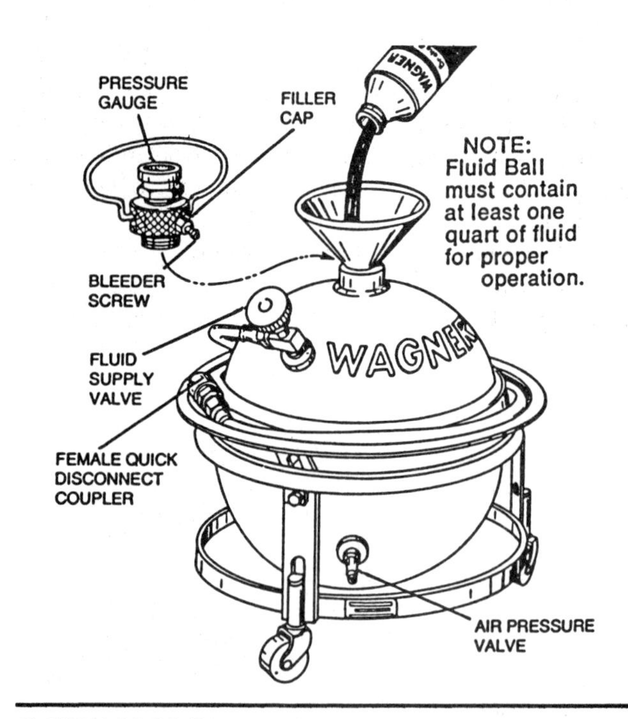

FIGURE 24-26. **Pressure Bleeder and Bleeder Adapters.** Fluid ball must contain at least 1 liter (1 quart) of fluid for proper operation. (Courtesy of Wagner Brake Company, Ltd.)

Step-by-Step Procedure

1. Remove the master cylinder's cap and attach the special adapter to the top of the cylinder.
2. Connect the hose to the adapter and to the pressure bleeder equipment.

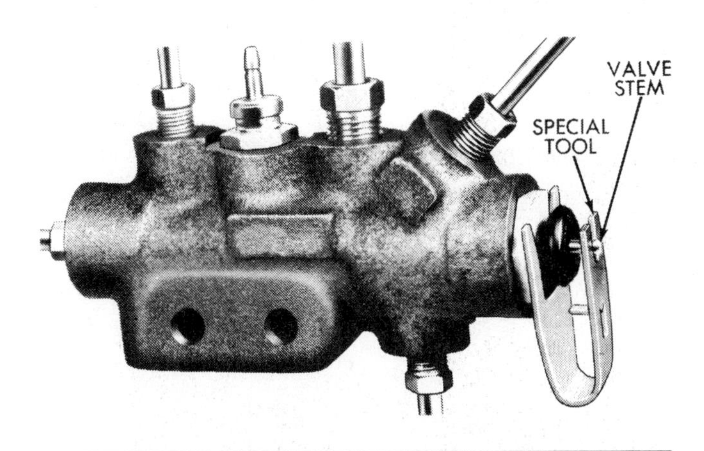

FIGURE 24-27. Holding Valve Open with Special Tool. (Courtesy of Chrysler Corporation)

3. Open the valve on the bleeder tank to admit pressurized fluid to the master cylinder.

4. Attach one end of the bleeder hose to the caliper's bleeder valve. Immerse the opposite end of the hose in the brake fluid contained in the glass bottle.

5. Using the bleeder valve wrench, loosen (open) the bleeder valve in the front caliper farthest from the master cylinder. With the valve open, allow fluid to be discharged into the bottle from the hose until air bubbles have ceased to appear. When there are no more bubbles, close the valve.

6. Bleed the caliper at the opposite wheel.

7. When the bleeding operation is completed, close the bleeder tank valve and disconnect the hose from the adapter. Remove the adapter from the cylinder.

8. Fill the master cylinder's reservoirs to their correct levels and install the cover cap.

9. Remove the special tool from the metering valve stem.

Caution: If the tool is not removed after bleeding, the diaphragm inside the valve will rupture when the brakes are applied, resulting in a loss of fluid.

10. Test the dependability of the vehicle's hydraulic brake systems as previously described in Chapter 23.

11. Using a hydraulic floor jack, raise the automobile and remove the safety stands from under the vehicle. Lower the vehicle's front wheels to the floor.

12. Retighten the lug bolts or nuts in sequence and torque to a final specification per the manufacturer's shop manual. This procedure will eliminate wheel rim, brake rotor, or brake drum distortion.

Cars with Plastic Reservoirs. The recent emphasis on vehicle weight reduction has resulted in the wide use of plastic master cylinder reservoirs. These require a special adapter for pressure bleeding, since pressurizing the plastic reservoirs would damage or burst them. The adapter connects directly with the master cylinder fluid ports. Figure 24-28 shows a universal adapter in place on a typical General Motors-type master cylinder with a plastic reservoir.

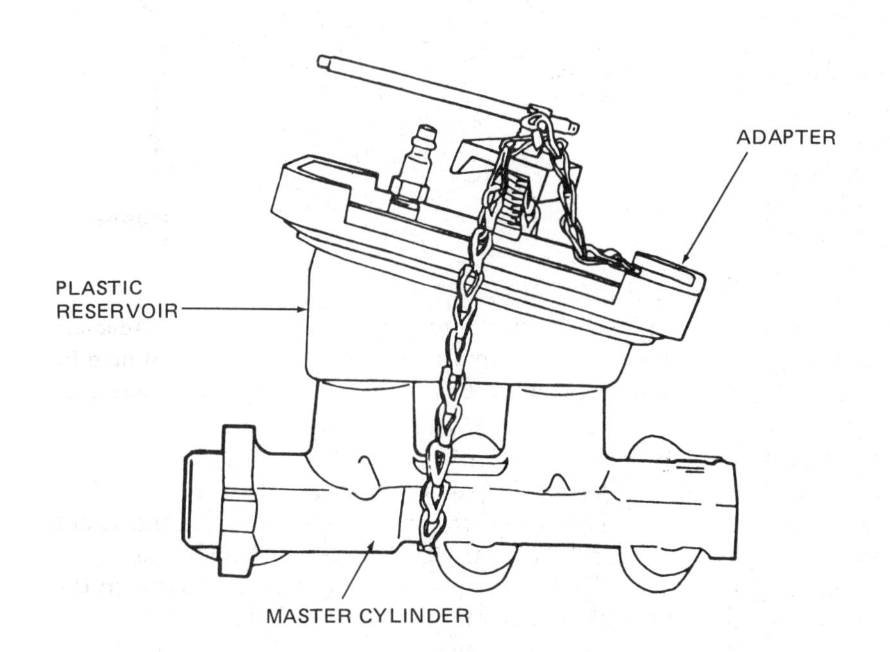

FIGURE 24-28. Universal Adapter in Place on Plastic Reservoir. (Courtesy of Wagner Brake Company, Ltd.)

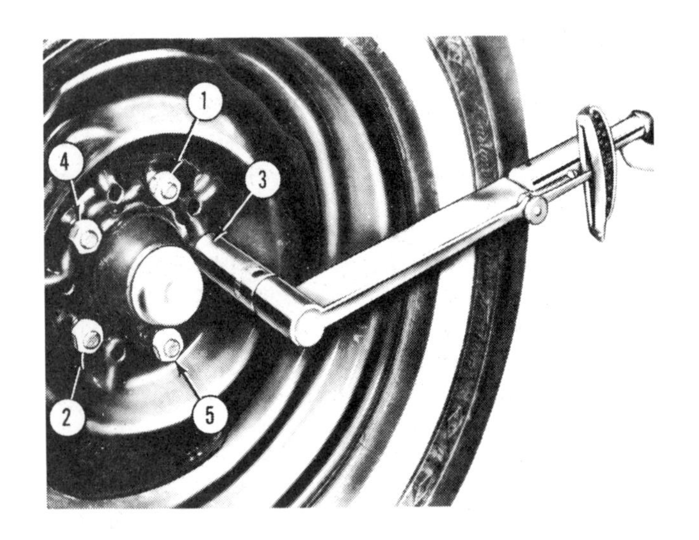

FACTS TO REMEMBER

Brake fluid, when bled from a vehicle's brake system, may appear to be clear and not contain any particles. However, manufacturers emphatically state: Never reuse brake fluid that has been drained from the hydraulic system or has been allowed to stand in an open container for an extended period of time.

Installing the Wheels

Once you have tested the hydraulic systems, install the front wheels to their indexed hubs. Tighten the wheel lug bolts or nuts in the correct sequence to an initial torque specification of 20 newton-meters (15 foot-pounds) (Figure 24-29).

REVIEW TEST

1. List the names of the component parts of a sliding caliper disc brake assembly.
2. Briefly describe the operation of a sliding caliper disc brake mechanism.
3. Explain why disc brakes are considered self-adjusting.
4. List four distinct advantages of a disc brake mechanism.
5. Briefly describe how to diagnose brake problems associated with the terms *vibration, pull, brake squeal and noise, excessive pedal effort,* and *pedal travel and feel.*
6. List four external conditions that affect braking performance. Then describe how to inspect a vehicle to determine if they are the cause of a brake problem.
7. In point form, summarize how to inspect the pads of a vehicle's disc brake mechanism.
8. In point form, describe how to remove and disassemble a sliding caliper disc brake mechanism.
9. Describe how to clean and inspect the component parts of a disc brake.
10. Describe how to assemble the caliper.
11. Briefly describe how to examine a disc visually for defects.
12. Describe how to determine if a disc has lateral runout.
13. Describe how to measure the parallelism and the overall thickness of a disc.
14. In point form, describe how to reface a disc.
15. Briefly describe how to lubricate the front wheel bearings when servicing a disc brake.
16. Explain why it is necessary to flush brake lines prior to connecting the calipers and their flexible hoses to the steel lines.
17. Briefly describe how to manually bleed, or pressure bleed, a vehicle's systems.

FINAL TEST

This examination is multiple choice. Only one answer will be accepted. Carefully read every statement.

1. Mechanic A says that disc brakes are less susceptible to brake fade than are drum brakes. Mechanic B says that disc brakes, due to their unique design, are self-adjusting. Who is right? (A) Mechanic A, (B) Mechanic B, (C) Both A and B, (D) Neither A nor B.
2. Mechanic A says that when the disc brake is applied, the anchor slides laterally and the caliper remains stationary. Mechanic B says that when the disc brake is applied, the outboard pad is moved against the disc by the direct action of the piston. Who is right? (A) Mechanic A, (B) Mechanic B, (C) Both A and B, (D) Neither A nor B.
3. Mechanic A says that the force produced by the piston (moving outward) causes the inboard

pad to contact the disc. Mechanic B says that disc brakes operate on the principle of physics: For every action, there is an opposite and equal reaction. Who is right? (A) Mechanic A, (B) Mechanic B, (C) Both A and B, (D) Neither A nor B.

4. Mechanic A says that as the pads wear, the pistons are required to move farther out of their bore. Mechanic B says that the pistons in the calipers are larger than the pistons in a master cylinder since the pressure in a disc brake is greater than the pressure in a drum brake system. Who is right? (A) Mechanic A, (B) Mechanic B, (C) Both A and B, (D) Neither A nor B.
5. Mechanic A says that when you are required to diagnose problems, you should first read the manufacturer's manual. Mechanic B says that when you are required to diagnose problems, you should first talk to the customer regarding the complaint. Who is right? (A) Mechanic A, (B) Mechanic B, (C) Both A and B, (D) Neither A nor B.
6. Mechanic A says that if, when the brakes are applied, there is a vibration above the speed of 80 km/h (50 mph), the master cylinder must be replaced. Mechanic B says that if, when the brakes are applied, there is a vibration above the speed of 80 km/h (50 mph), there is a variation in the thickness of the rotor. Who is right? (A) Mechanic A, (B) Mechanic B, (C) Both A and B, (D) Neither A nor B.
7. Mechanic A says that it is impossible for a defective suspension to contribute to a brake pull. Mechanic B says that if the outer pad is worn on a taper, the probable cause is a partially seized piston. Who is right? (A) Mechanic A, (B) Mechanic B, (C) Both A and B, (D) Neither A nor B.
8. Mechanic A says that it is acceptable to use varsol when washing a caliper. Mechanic B says that crocus cloth should not be used to remove light scratches. Who is right? (A) Mechanic A, (B) Mechanic B, (C) Both A and B, (D) Neither A nor B.
9. Mechanic A says that the first item to be removed from a caliper is the bleeder valve (screw). Mechanic B says that the best way to remove the piston seal from the caliper is by using a small metal screwdriver. Who is right? (A) Mechanic A, (B) Mechanic B, (C) Both A and B, (D) Neither A nor B.
10. Mechanic A says that the parallelism of a disc should not vary more than 0.125 millimeter (0.0005 inch). Mechanic B says that when measuring the lateral runout of a disc, the pointer of the dial gauge should be 2.5 centimeters (1 inch) from the outside of the rotor. Who is right? (A) Mechanic A, (B) Mechanic B, (C) Both A and B, (D) Neither A nor B.
11. Mechanic A says that before you machine a rotor, it is not necessary to remove the grease thoroughly from the inside of the hub. Mechanic B says that before you machine a rotor, the tool bits must have sharp points, not round ones. Who is right? (A) Mechanic A, (B) Mechanic B, (C) Both A and B, (D) Neither A nor B.
12. Mechanic A says that if the figure on the side of the rotor is 23.88 millimeters (0.940 inch), the minimum thickness is actually 23.27 millimeters (0.920 inch). Mechanic B says that if the figure on the side of the rotor is 23.88 millimeters (0.940 inch), the disc may be machined, but the part should not be used. Who is right? (A) Mechanic A, (B) Mechanic B, (C) Both A and B, (D) Neither A nor B.
13. Mechanic A says that you should wait 15 seconds between pedal applications when bleeding a quick take-up brake system. Mechanic B says that when a brake system is being pressure bled, the air compartment of the pressure bleeder is charged with a maximum pressure of 175 kPa (25 psi). Who is right? (A) Mechanic A, (B) Mechanic B, (C) Both A and B, (D) Neither A nor B.

Drum Brake Service

LEARNING OBJECTIVES

After studying this chapter, you should be able to:

- List the names of the component parts of a duo-servo drum brake assembly,
- Describe the operation of a duo-servo drum brake mechanism,
- Describe the operation of a parking brake mechanism,
- Explain how to diagnose drum brake performance problems related to pedal pulsation, pull, grab, brake squeal and noise, excessive pedal effort, and pedal travel and feel,
- Describe how to remove a brake drum, inspect its condition and wear, gauge its diameter, and machine its braking surface,
- Describe how to diagnose normal and abnormal brake shoe wear, remove the shoes, service the support plate, service a wheel cylinder, grind a brake shoe to fit the drum, and assemble and bleed a drum brake mechanism.

INTRODUCTION

In Chapter 21, we described a primitive brake mechanism, which consisted of a skid called a *shoe* pressing against the outside rim of a wheel. The engineers of those pioneering days used the term *external contracting* to describe its design and principle of operation. Although the design was a mechanical marvel, the entire mechanism was exposed to the elements.

Changes were made whereby the shoes, instead of being exposed, were protected by being placed inside a circular metal housing. The housing, called a *drum*, was attached to the wheel. Instead of moving in against the rim, brake shoes could now be moved outward against the inside of the drum. This improvement in design was called the *internal expanding* brake.

Drum brakes, like all other component parts of a modern vehicle, have had their design modified and improved. Since most automobiles now manufactured in North America are equipped with rear self-adjusting drum brakes (a term we'll discuss in a moment), we should know how the brake is constructed, its principle of operation, how it is inspected, and how it must be serviced and repaired.

DESIGN

The foundation of all drum brakes, regardless of their variations of design, is the metal support plate (often named the *backing plate*). This part is bolted to the rear axle housing by four hex head bolts (not shown). As you view the illustration of the brake's components (Figure 25-1), you will notice that the plate not only provides a mounting for the brake parts but also protects the mechanism from the elements.

Located above the double-ended hydraulic wheel cylinder is a fixed anchor and an anchor plate. Hooked around a groove in the anchor are two long primary and secondary coil return springs. The opposite end of each spring is hooked into the primary (front) and secondary (rear) brake shoes.

As you will observe, each shoe is held against the support plate by a shoe retainer and spring and pin assembly. Positioned at the lower ends of the two brake shoes and linking them together is an adjuster screw assembly. This part is actuated by an adjuster lever and a steel cable that is attached to the anchor. The adjuster screw (link) is held in place by the tension of a long shoe-to-shoe coil spring. Located just below the wheel cylinder is a strut (part of the parking brake mechanism). The strut is slotted into the primary shoe and the parking brake lever. The lever is slotted into or pinned to the secondary shoe and is operated by a steel cable (not shown) when the parking brake pedal is applied.

So far, you have been introduced to a number of parts that are attached to the support plate. There is still one more part that you must become ac-

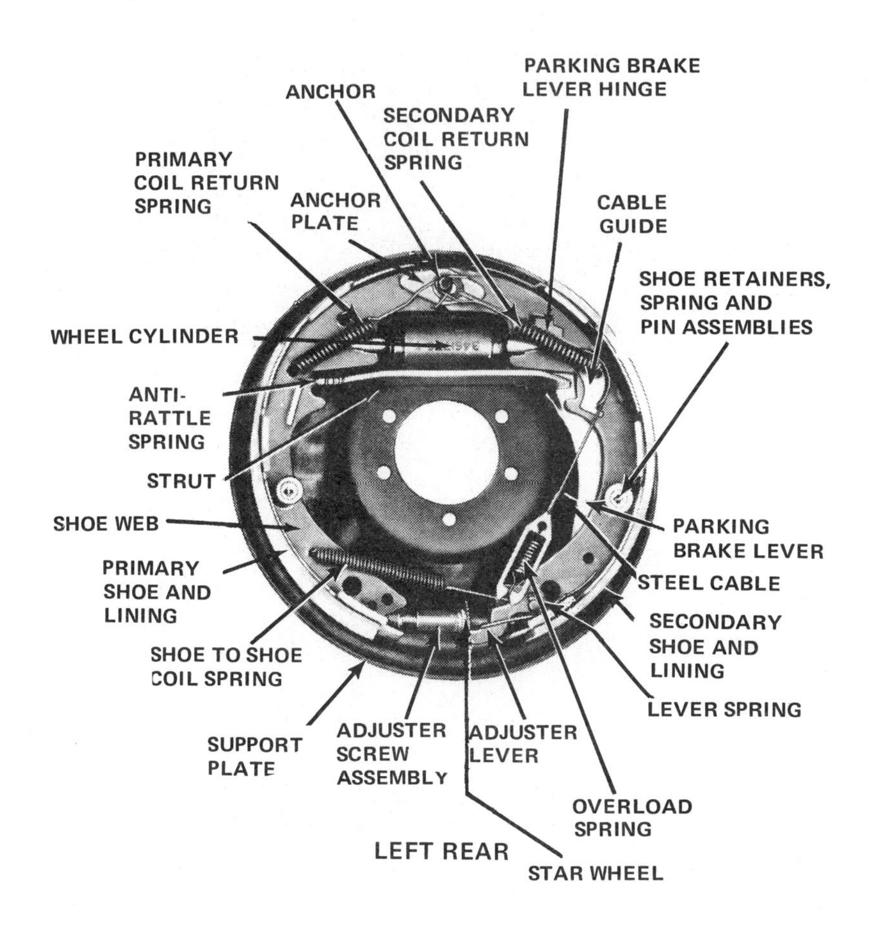

FIGURE 25-1. Drum Brake Assembly. (Courtesy of Chrysler Corporation)

quainted with: the brake drum. This part, usually made of cast iron, fits over the end of the rear axle, just behind the wheel (Figure 25-2). One of the purposes of a drum is to provide a surface for the shoes to rub against when the brakes are applied.

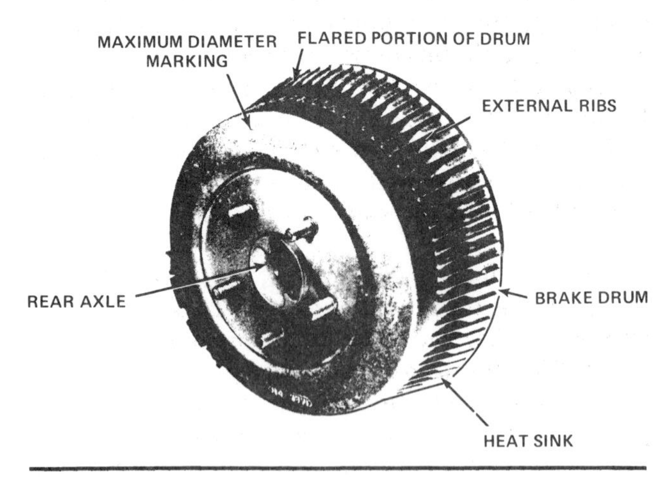

FIGURE 25-2. Brake Drum. (Courtesy of Chrysler Corporation)

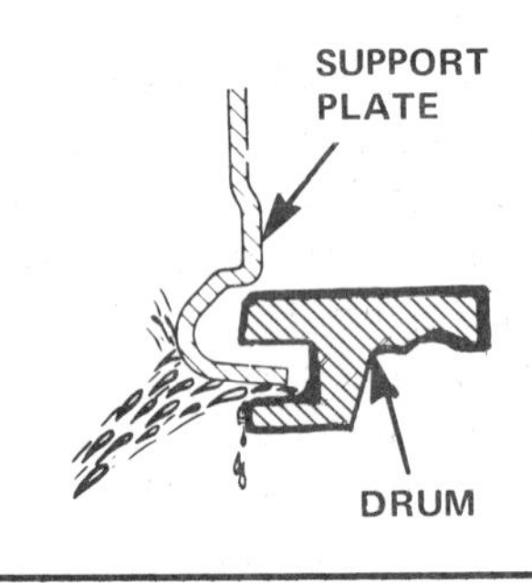

FIGURE 25-3. Labyrinth Seal. (Courtesy of Chrysler Corporation)

Later in this chapter, you will be required to remove a rear brake drum and inspect the brake components to determine their condition. When you have the opportunity to examine these parts, you will see that a labyrinth seal is formed by the flanged rim of the support plate (Figure 25-3). This part projects into a circular groove in the edge of the drum's rim. Because of this unique design feature, water splashed from the road is unable to enter the brake mechanism.

REAR DRUM BRAKE OPERATION

Duo-Servo Brake Mechanism

When the brake pedal is applied, fluid discharged from the dual master cylinder is transmitted through hydraulic lines to the rear drum wheel cylinders. This fluid acts on the two pistons within the cylinder's bore, forcing them to move outward. The movement of the pistons causes the brake shoes to move away from the anchor. During this stage of operation, the shoes are floating (free to turn slightly).

As the linings of the shoes touch the revolving drum, friction causes both shoes to revolve in the direction that the drum is turning for a short period of time. This rotational movement causes the top of the primary shoe to move away from the anchor (Figure 25-4).

The rotating force of the primary shoe is then transmitted through the adjuster screw to the bottom of the secondary shoe and pushes the rear shoe, which is stopped by the anchor, against the drum with considerable force. Because of this unique op-

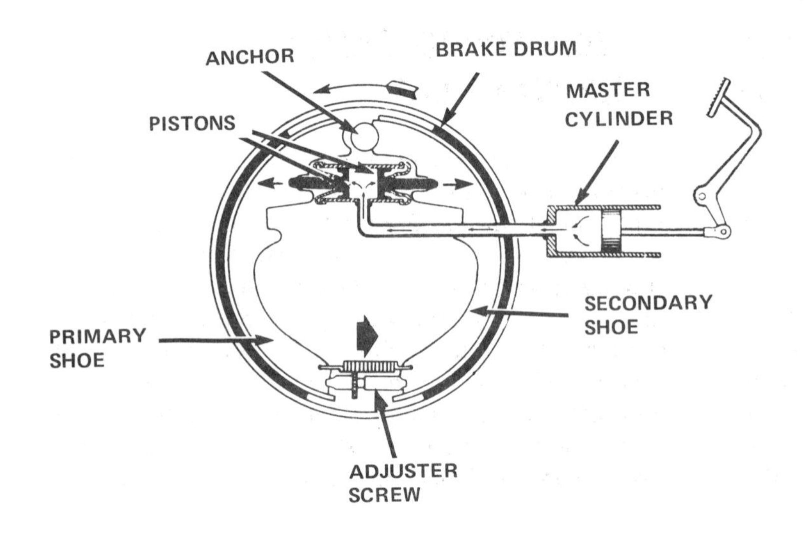

FIGURE 25-4. Operation of Brake Mechanism: Application Stage. (Courtesy of Wagner Brake Company, Ltd.)

erational feature, less pedal force is required to slow and stop a moving vehicle. This is a distinct advantage of a drum brake design as compared to a disc brake.

Since the secondary shoe is pressed against the drum with considerably more force than the primary shoe, the secondary shoe actually does most of the braking. Therefore, the lining of the primary shoe is composed of a high-friction material, giving it a greater tendency to turn with the drum. The lining of the secondary shoe is composed of a lower friction material to support the required braking load.

Now that we know how the drum brake operates, we must also understand that the same effect occurs when the vehicle is moving in reverse. The primary shoe takes over the function of the secondary shoe and in reverse does most of the braking. When a brake mechanism is designed to operate in this manner, it is called a *duo-servo* (meaning "two") *brake mechanism.*

When force is no longer applied to the brake pedal, the coil return spring pulls the shoe that has moved away from the anchor back to its original position. The additional fluid inside the cylinder, which caused the pistons to engage the shoes against the drum, will now be returned to the master cylinder and will be held in the reservoir until the next application of the brake pedal.

Self-Adjusting Mechanism. Earlier in this chapter, the adjuster screw assembly was mentioned and briefly described (see Figure 25-1). On most vehicles, the adjuster assembly mechanism is installed on the secondary shoe and thus operates only as the brakes are applied when the vehicle is traveling in reverse. When this happens, the top of the secondary shoe moves away from the anchor. As the shoe moves, the cable guide mounted to the secondary shoe moves and causes the cable, through the overload spring, to lift the adjuster lever. As the lever is raised, it will engage a tooth on the star wheel, which will lengthen the adjuster screw assembly. The screw will be lengthened only when the linings wear enough to permit sufficient shoe movement. The purpose of the overload spring is to prevent the unnecessary adjustment of the linings should there be a need for a panic stop.

A sudden stop will produce excessive heat, which may cause a brake drum to expand and deflect, thus changing its shape. To reduce the possibility of drum deflection, brake engineers have modified the external design of the drum. You will recall that in the illustration of a brake drum (Figure 25-2), the drum was flared and constructed with a number of external ribs. Engineers call this portion of the drum a heat sink since it acts with heat in much the same way as your kitchen sink acts with water. When you dump a pan of water in the sink too fast for all the water to go down the drain immediately, the sink holds the excess water until eventually it all runs down the drain.

Similarly, when the brakes have been applied for an extended operational period, a drum will absorb more heat than it can dissipate. The flared portion of the drum will hold the heat momentarily until it has been dispersed into the air.

Parking Brake Mechanism. To conclude our study of the operation of a drum brake, we must also be familiar with the parking brake mechanism. Figure 25-5 illustrates a basic design.

When a driver presses down the parking brake pedal, the front cable, which is partially enclosed within an outer covering, transmits the force and motion to an intermediate cable. The intermediate cable transfers the motion to the right and left rear cables. Each rear cable is attached to the parking brake lever, which pivots on the secondary shoe (see Figure 25-1).

When the parking brake lever is activated, the lower part of the lever moves toward the primary shoe, and, at the same time, the motion is transmitted to the strut. As the strut moves and forces the primary shoe against the drum, the lever must now pivot on the strut, thereby forcing the secondary shoe to move outward against the drum. The movement of the lever against the strut causes the rearward movement of the secondary shoe. Because of this mechanical design, there is action and reaction when the parking brake is applied.

Non-Servo Brake Mechanism

Since the inception of many down-sized vehicles, manufacturers have modified their design of drum brake mechanisms. Figure 25-6 illustrates a *non-servo drum brake mechanism,* which has a leading/trailing shoe design. The leading shoe does the majority of the braking when going in the forward direction; the trailing shoe does the majority of braking while going in the reverse direction. The automatic adjustment mechanism (which consists of the adjuster screw, washer, adjuster socket, adjuster lever, and parking brake lever pin) uses an

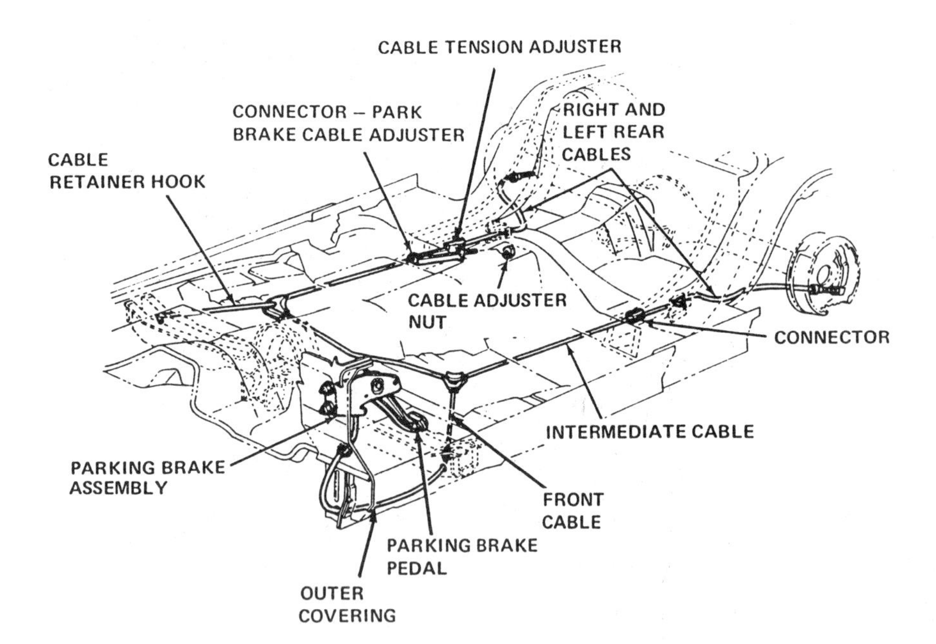

FIGURE 25-5. Operation of Parking Brake Mechanism: Application Stage. (Courtesy of Chrysler Corporation)

incremental adjuster (the adjuster lever) that adjusts during braking whenever the wear gap appears sufficient to actuate the adjuster screw. Brake adjustment occurs in forward and rearward braking. Since the brakes are self-adjusting, manual adjusting is required only after brake shoes have been replaced or when the length of the adjuster screw has been changed while performing some other service operation. These type of drum brake mechanisms perform approximately 15 to 20 percent of the total requirement.

DIAGNOSING PROBLEMS

In the previous chapter, we discussed the steps for diagnosing and analyzing potential brake problems. Once again, let's consider the step of questioning the driver. The pertinent questions to ask can be grouped into seven categories: general, pedal pulsation, pull, grab, brake squeal and noise, excessive pedal effort, and pedal travel and feel.

General

- How long ago (in kilometers or miles) did the problem first appear?
- Did the problem appear gradually or suddenly?

Analysis: Gradual appearance of the problem may indicate partially seized pistons within the rear hydraulic cylinders, a minute loss of fluid leaking past the cylinder's seals onto the linings, slight seepage of gear lubricant onto the linings and drums caused by a defective axle seal, or a partially seized front and/or rear parking brake cable. If the problem has occurred suddenly after a severe brake application, a major component part of the rear brake system has been damaged. Damage will be evident on inspection after the removal of the drums.

Pedal Pulsation

- When the brakes are applied, is there a pulsation felt on the pedal. If so, at what speed is it most noticeable?

Analysis: If the pulsation is noticeable at lower speed, it is probable that a rear brake drum is out-of-round (oval or eccentric) due to a severe brake application.

Pull

- On brake application, does the vehicle tend to steer to the left or to the right?

Analysis: This problem is seldom noticeable when

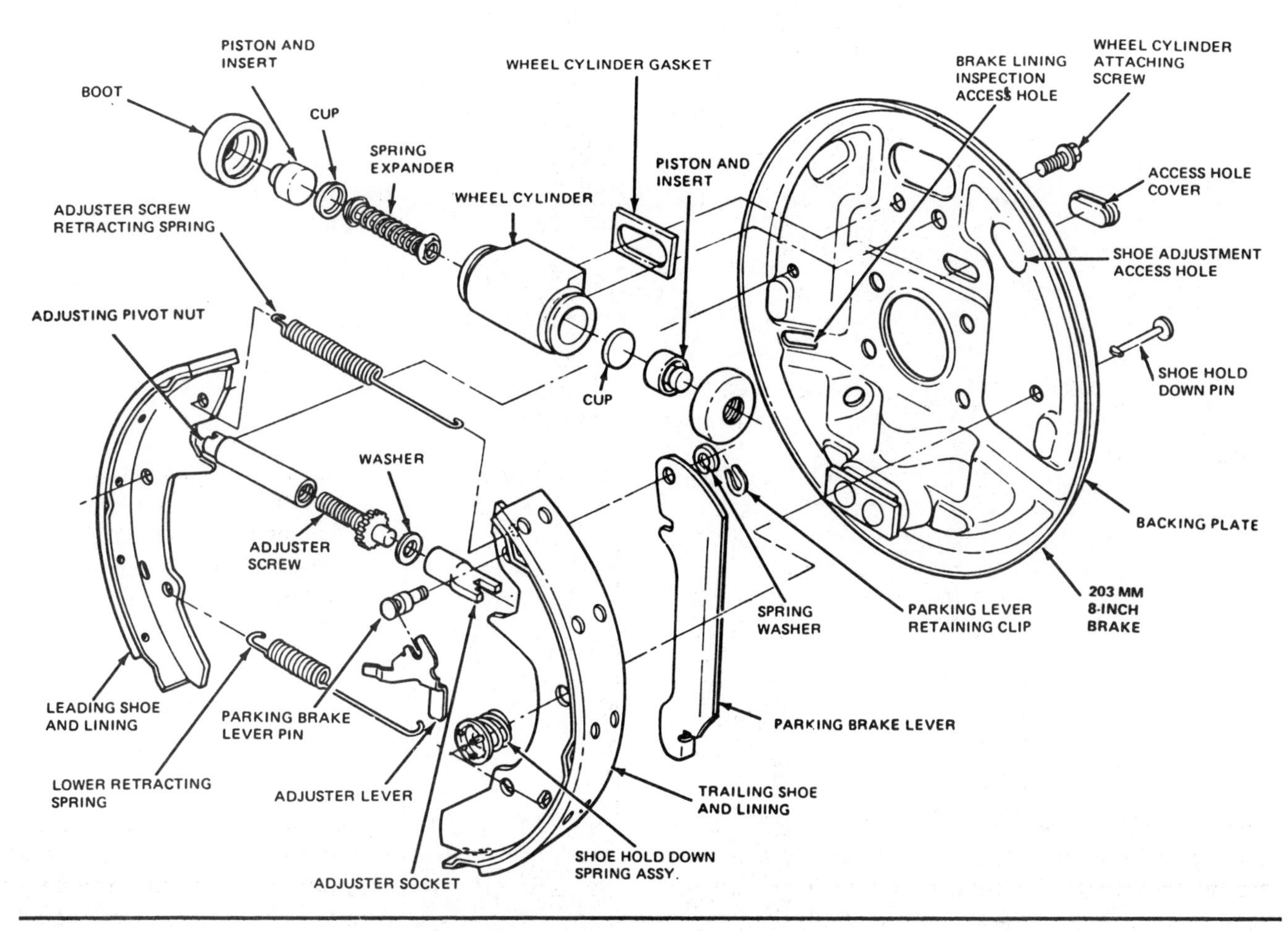

FIGURE 25-6. Non-Servo Rear Drum Brake Mechanism. (Courtesy of Ford Motor Company)

a rear brake malfunctions. If, however, the problem does exist, suspect seized pistons within a cylinder. Also, linings and the braking surface of one rear drum may be saturated with brake fluid or gear lubricant, thereby allowing effective braking at only one rear wheel. A distorted rear suspension control arm or a defective rear leaf spring should be suspected if there is no apparent brake cause.

Grab

- When the brakes are first applied, does the brake tend to lock one or both rear wheels, thereby preventing the wheel(s) from revolving?

Analysis: If this problem exists, it will probably be caused by one of the following conditions: a defective rear axle seal or a leaking wheel cylinder that is allowing a slight seepage of fluid onto the linings or drum, a self-adjusting mechanism overly expanded due to an eccentric drum, or a partially seized rear parking brake cable that has allowed the shoes to be partially expanded prior to brake application.

Brake Squeal and Noise

- Do the brakes, when applied, produce a squealing sound or a noise?

Analysis: If the rear brakes squeal when the vehicle is traveling in reverse, the surface of the linings is probably glazed. This condition is not detrimental to the brake's operation. If there is a grinding sound (similar to metal-to-metal contact), the lining's frictional material has become depleted through wear. The linings will need to be replaced and the drums machined.

Excessive Pedal Effort

- When the brake pedal is applied, are you required to exert excessive force on the pedal to stop the vehicle?

Analysis: If this problem is evident, suspect a faulty power assist unit. Other sources for this complaint may be seized pistons inside a wheel cylinder or a change in the frictional characteristics of the lining's material.

Pedal Travel and Feel

- When the brake pedal is applied, is there sufficient pedal reserve height? If not, how far down does the pedal travel? Does the pedal feel firm or spongy?

Analysis: If the pedal is low yet firm on initial application and then raised after repeated pumping, the self-adjusting mechanisms have seized. If the pedal is low and spongy on application, examine the brake's hydraulic system to determine if there has been a loss of fluid. Wheel cylinders that allow a seepage of fluid will also induce air into the system, and, as we know, air is compressible. Another source of the problem may be a defective residual check valve inside the master cylinder. You will recall that only a drum brake system utilizes a check valve. To determine if the master cylinder for that vehicle has a check valve, refer to the manufacturer's shop manual.

If the valve is the cause of the problem, make sure that the linings are correctly adjusted in relation to the drums. The brake pedal will be low and then raised after continued pumping. When you have removed your foot from the pedal and waited for approximately 30 seconds, on the next application, the pedal will once again be low. If this is the case, the master cylinder will need to be serviced or replaced.

SERVICING BRAKE COMPONENTS

Let's assume that you have discussed the vehicle's braking problems with the customer. Based on your diagnosis, further inspection of the brake system is required in the repair shop. After removing the hubcaps, index each wheel to its rear axle with a piece of crayon or chalk so you can return the wheels to their balanced position.

Before loosening the rear wheel nuts, determine if the left rear wheel lug nuts are right- or left-hand thread. Then, using the appropriate wrench, loosen each nut with only two revolutions. With the aid of a hydraulic floor jack positioned under the center of the axle housing, carefully raise the vehicle. Position safety stands under the automobile's frame, just in front of the rear suspension. Lower the car onto the stands and remove the jack from under the vehicle. Then unthread the wheel nuts and remove the wheels from their axles.

Removing the Drums

Removing a drum from its axle can be simple. First, the steel speed nuts are unthreaded from the studs. Then, with the parking brake fully released and with no tension on the rear cable, the drum is pulled straight out to separate it from the flange of the axle. The basic technique is easy enough, but the actual removal of a drum from an axle may at times try the patience of a saint. The following suggestions will assist you if you encounter problems during this procedure:

Step-by-Step Procedure

1. Using an oil can containing penetrating oil, liberally apply the oil around the area of the flange and the mounting face of the drum (Figure 25-7A).

2. After allowing time for the oil to counteract the rust, using a medium-sized hammer, tap the area marked by the positions X between the studs (Figure 25-7B).

Caution: At no time should you strike the round, ribbed area of a drum. Since a drum is made of cast metal, the drum will crack, and its operational performance will be impaired.

3. If the drum will not separate from the axle, you may, with the aid of an oxyacetylene torch, apply controlled heat at positions X. While the drum is hot, gently tap the drum at the flange area. Using work gloves to protect your hands, pull on the drum to remove it from the axle.

4. You may be able to move the drum only a short distance from the axle flange and then find that the drum will not move any farther. Further removal of the drum will be prevented by a ring of metal and rust on the drum's braking surface next to the support plate. To remove the drum completely, remove the plug and insert an adjusting tool through the opening at the rear of the support plate to contact the star wheel (Figure 25-8). Move the handle of the tool upward until a slight drag is felt when the wheel is rotated. Then insert a thin screwdriver or a piece of welding rod into the brake adjusting hole and push the adjusting lever out of engagement with the star wheel. While holding the

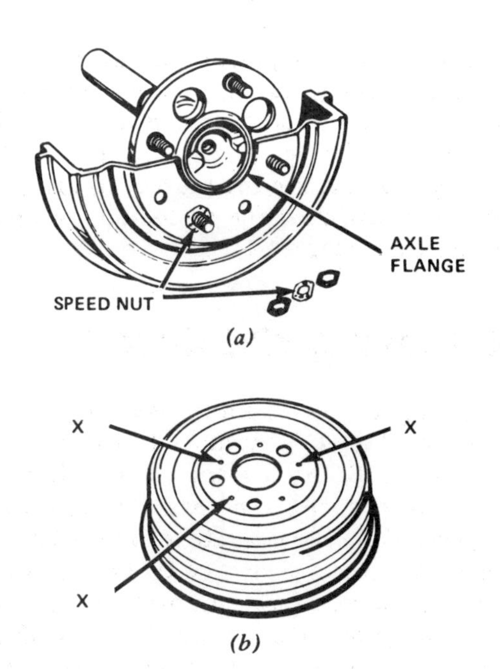

FIGURE 25-7. Brake Drum Removal. (a) Remove speed nuts from the studs and apply penetrating oil to the axle flange; (b) tap the drum face in the areas between the studs with a hammer. (Courtesy of Wagner Brake Company, Ltd.)

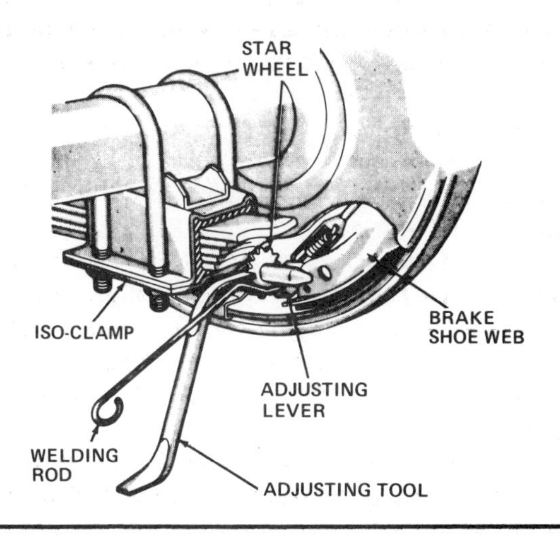

FIGURE 25-8. Using Adjusting Tool and Piece of Welding Rod to Back Off (Shorten) Adjuster Screw Mechanism. (Courtesy of Chrysler Corporation)

adjusting lever out of engagement, back off (by rotating) the star wheel using the adjusting tool. This will ensure a free revolving drum with no brake shoe interference. You may then remove the drum from the axle.

Caution: In the previous chapter, you were introduced to a number of safety tips concerning the handling of asbestos materials. Those safety rules also apply to drum brake servicing. To protect yourself from possible health hazards, wear an approved respirator. Vacuum loose dust from the brake mechanism. Wash the brake mechanism with an approved washer (Figure 25-9). After you have finished disassembling and cleaning the brake parts, change your coveralls and thoroughly wash your hands. At no time should compressed air ever be used to remove asbestos dust from any part of the brake mechanism or your clothing.

A PRACTICAL WORK TIP

If you have limited experience in performing brake repairs, work on one wheel at a time and use the other wheel for reference.

Inspecting the Lining and Drums

Since the drums have now been removed, let's discuss how to inspect the lining and drums. If the pistons inside the cylinders have been operational, the linings, since they have contacted a drum's braking surface, will visually indicate most wear conditions. The conditions that will be evident, if they exist, are scoring, bell mouth, and barrel shape (concave). Three more conditions will be evident only with careful examination of the drum: hard spots, cracks, and out-of-round.

FIGURE 25-9. Brake Assembly Washer. (Courtesy of Ammco Tools, Inc.)

Scoring. This condition has been caused by grit or sand entering the brake mechanism or, in some instances, caused by inferior hard lining material. If the braking surface of the drum is not machined (made smooth), new linings will sustain the same surface condition as the drum (Figure 25-10). If the grooves are not too deep, the drum may be machined in a lathe.

Bell Mouth. This condition is caused by excessive heat during periods of severe brake application and by insufficient metal support at the drum's exterior. As you study Figure 25-11, you will notice that the lateral face of the lining is tapered. In some cases, providing the condition is minor, the drum may be machined. However, the drum will be susceptible to brake fade.

Barrel Shape. A barrel-shaped (concave) drum is the direct result of excessive force produced by the shoes and linings to the center braking area of the drum. Heat from friction has caused the drum to distort, as shown in Figure 25-12. The drum may be machined and used again if it is not oversized after being resurfaced.

Hard Spots. Under this condition, hard spots will appear as raised metal areas on the drum's braking surface (Figure 25-13). The spots are caused by a change of metallurgy as a direct result of heat. Since these spots are raised, they will produce a multitude of brake problems, such as grab, pull, rapid wear, and noise. These spots can be smoothed only after the drum's surface has been ground with a revolving stone (part of the machining process). However, even though the drum has been ground, the hard spots will reappear since they are harder than the surrounding metal. The only permanent way to correct this condition is to replace the drum.

Cracks. A brake drum may sustain two types of cracks. When a drum has been subject to exces-

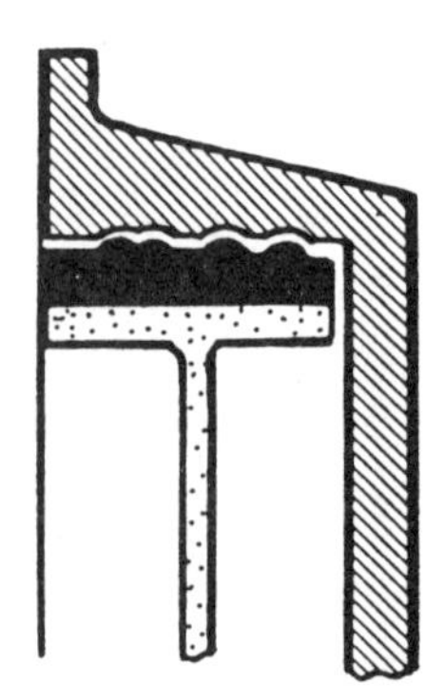

FIGURE 25-10. Scored Drum. (Courtesy of Wagner Brake Company, Ltd.)

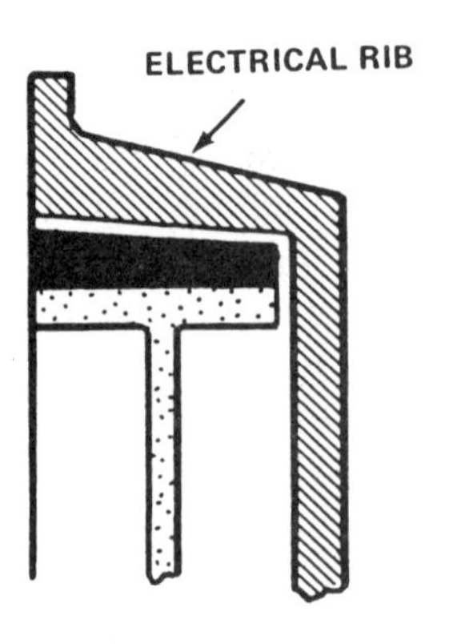

FIGURE 25-11. Bell-Mouthed Drum. (Courtesy of Wagner Brake Company, Ltd.)

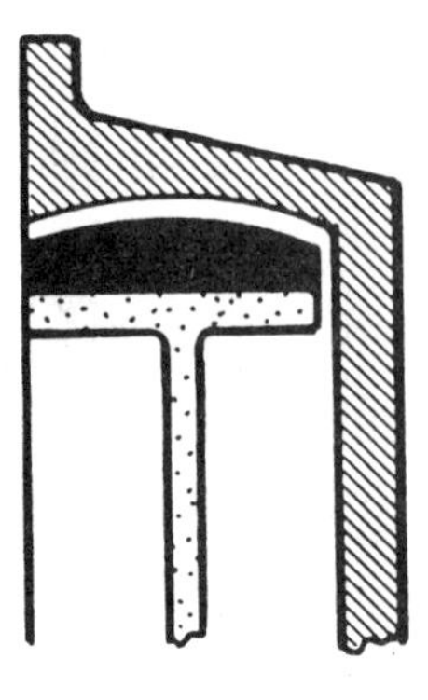

FIGURE 25-12. Barrel-Shaped Drum. (Courtesy of Wagner Brake Company, Ltd.)

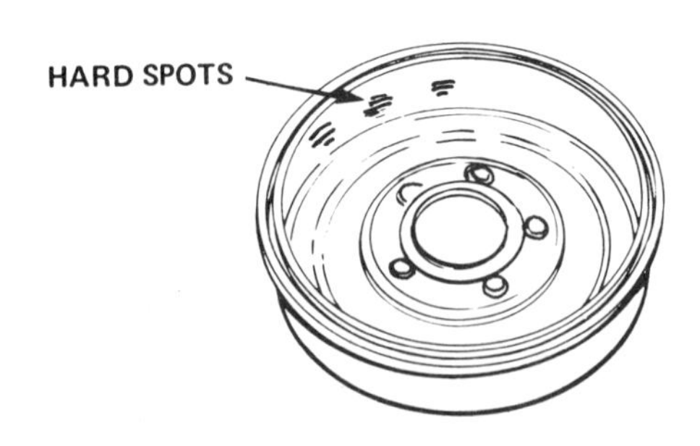

FIGURE 25-13. Hard Spots on Drum Surface. (Courtesy of Wagner Brake Company, Ltd.)

sive heat, the braking surface will have a bluish color, and a number of small cracks (heat cracks) will appear. These tiny cracks normally will be removed during the machining process.

The second type of crack is usually caused by very severe braking or by striking the drum with a heavy tool while removing it. These cracks are often difficult to detect. If the crack is laterally across the face of the drum, it should be noticeable (Figure 25-14A) on many older-type drums, the vertical flange plate (the part that attaches to a wheel hub or axle) is made of steel, and the drum (made of cast iron) is molded around and to the plate. If these two metal parts become unattached (caused by previous reconditioning), a circular fracture will result, and the drum will not maintain its shape, particularly when hot (Figure 25-14B).

One method that may be used to determine a fracture is to wash the drum in hot, soapy water, dry it thoroughly, and then roll a small puddle of gasoline around the inside of the drum. If a crack exists, the gasoline will leak through the crack. If the drum is fractured, it must be replaced. If no crack is detected, once again wash the drum and dry it thoroughly.

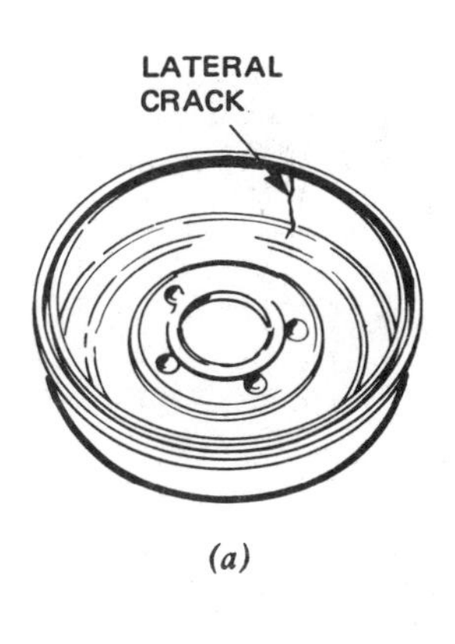

(a)

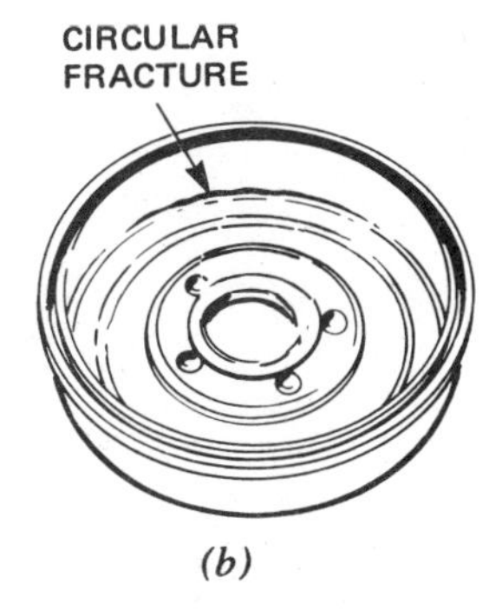

(b)

FIGURE 25-14. Two Types of Drum Cracks. (a) Lateral crack; (b) circular fracture. (Courtesy of Wagner Brake Company, Ltd.)

Out-of-Round. A drum that is out-of-round can best be described as being *elliptical.* This condition is caused by the heating and cooling of the drum during and after brake application. If the parking brake has been applied directly after a severe braking situation, the drum may even become elongated.

A FACT TO REMEMBER

Most brake drum problems are the direct result of excessive heat produced during a severe brake application.

Measuring the Drums

All vehicle manufacturers, without exception, recommend that a drum be measured with a drum micrometer to determine whether it is accurately round and also to assess whether its diameter is oversized. As illustrated in Figure 25-2 manufacturers have now embossed on the drum's exterior, near the flange, the drum's maximum allowable oversize.

To gauge a drum, first determine the drum's diameter by using a ruler. Then set the linear bar of the gauge to the diameter. Insert the gauge into the drum and, holding the gauge at position A in Figure 25-15, move its dial head to obtain a maximum diameter reading. Record the reading. Then measure at positions B and C. If there is a variance of more than 0.127 millimeter (0.005 inch) between the readings, the drum is considered to be out-of-round. Also, if the readings indicate that the drum's diameter

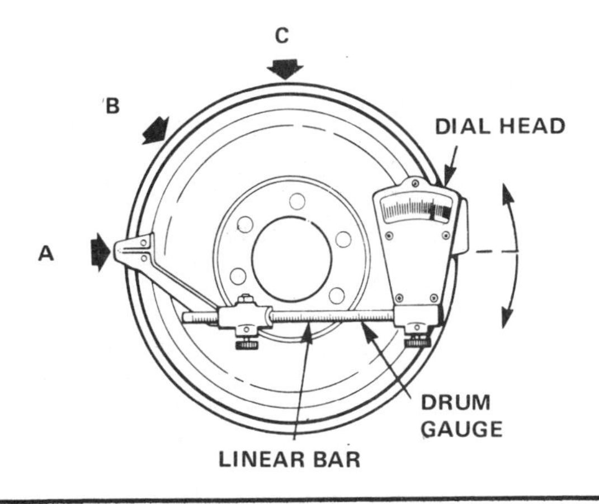

FIGURE 25-15. Measuring Drum Diameter. (Courtesy of Wagner Brake Company, Ltd.)

is greater than the manufacturer's specification, the drum must be replaced.

Machining the Drums

If after a drum has been gauged and found to be within the specified diameter, even though its braking surface appears to be relatively smooth, manufacturers recommend that it be machined to ensure that its surface is indeed true. Also, when a drum is machined, the machining process will remove any contamination on its surface that would have been left from the previous linings. Let's consider the instructions for machining a rear drum.

Step-by-Step Procedure

1. Form a square pad from a sheet of number 80 grit sandpaper. Then, moving the paper in a circular motion, remove any surface rust from the edge of the drum's center opening at the axle flange area.

2. Using a lint-free cloth dampened with varsol, thoroughly clean and remove any chips or contamination from hubless drum adapters and the arbor (shaft) of the lathe.

3. Using the appropriate adapters and spring, mount the drum on the lathe (Figure 25-16).

4. Install the silencer band around the drum to prevent chatter and noise (Figure 25-17).

5. Make certain that the tool bit has a rounded cutting point and is not chipped. Change the bit if necessary.

6. Following the manufacturer's instructions for the lathe that is in your shop, machine (reface) the drum's braking surface (see Figure 25-18A). *Note:* The finishing cut to the drum should be 0.051 millimeter (0.002 inch). The braking surface may also be polished by applying 80 grit sandpaper to the drum while it is revolving in the lathe. If after the initial machining, hard spots are evident, the drum will need to be ground, as illustrated in Figure 25-18B.

7. Remove the drum from the lathe. Using a micrometer, determine the drum's maximum diameter. If the drum is within the specified diameter, wash its interior with denatured alcohol. When the drum's diameter is larger than the manufacturer's specification, the drum must be replaced.

Although we have been discussing vehicles equipped with rear drum brakes, many older vehicles have also been equipped with front drum brake mechanisms. When you are required to machine a front drum, you may find that the drum can be separated from the front hub. If this is the case, the drum must be machined with its hub to maintain the roundness of its circumference. Remove the drum and hub from its axle. Install the wheel nuts to the bolts with the tapered part of the nuts facing outward; tighten the nuts to the correct torque. Remove the seal and the inner bearing from the

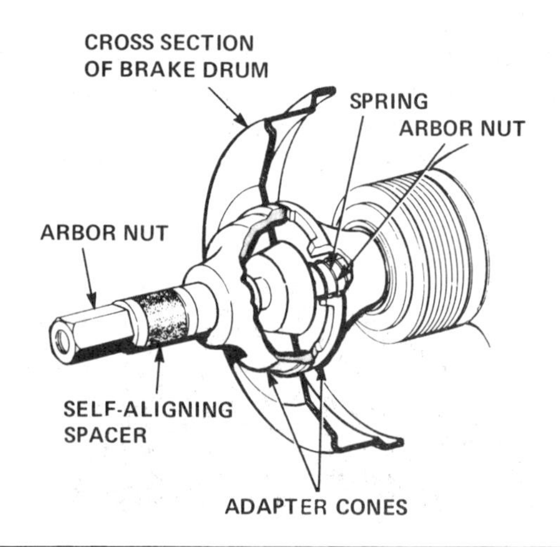

FIGURE 25-16. Mounting Drum on a Lathe for Machining. (Courtesy of Ammco Tools, Inc.)

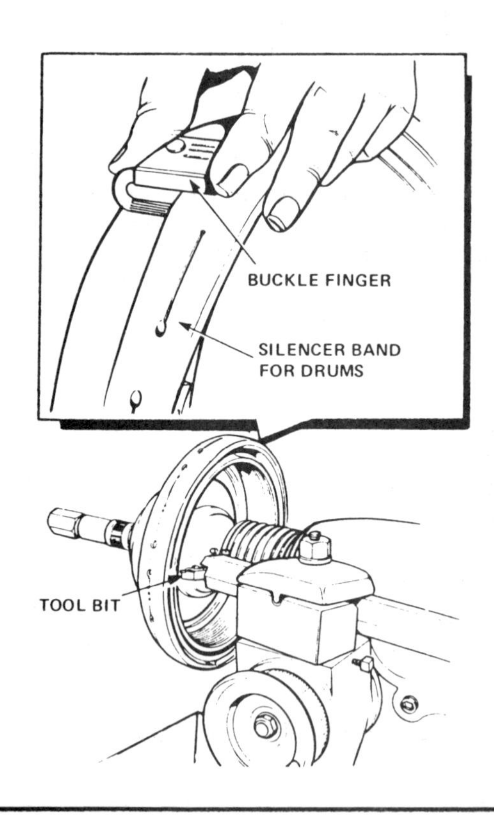

FIGURE 25-17. Installing Drum Silencer Band. (Courtesy of Ammco Tools, Inc.)

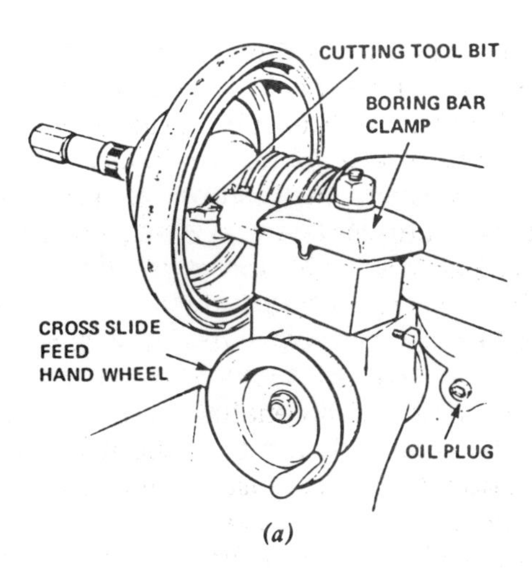

(a)

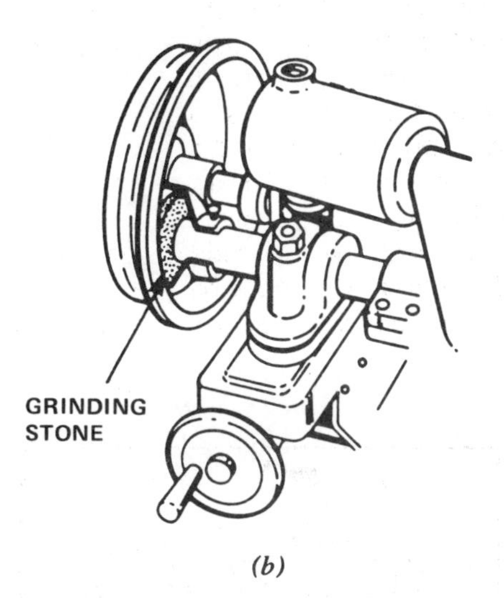

(b)

FIGURE 25-18. Servicing the Drum Brake. (a) Machining (refacing) drum's braking surface; (b) grinding drum's braking surface. (Part a courtesy of Ammco Tools, Inc.; part b courtesy of Wagner Brake Company, Ltd.)

hub as described in Chapter 24, step 2, Machining the Disc.

Some front drums are also riveted to the hub or swaged (a punch peening process) to the wheel studs. If the drum must be replaced, special tools and equipment will be required. It is recommended that the drum and hub assembly be sent to a machine shop that is equipped to perform this service.

FACTS TO REMEMBER

1. Before installing a new drum and/or hub on an axle, always lightly machine cut the drum's braking surface to ensure that it is true.
2. When you are required to machine one front drum, the opposite front drum must also be machined to ensure equal braking performance.
3. Manufacturers recommend that the diameters of the two front drums, or the two rear drums, be within 0.254 millimeter (0.010 inch) of each other after being machined. This will greatly reduce the possibility of brake pull after the brake's shoes have been replaced.

Removing Brake Shoes

Step-by-Step Procedure

1. Using a special tool (Figure 25-19), unhook the secondary brake shoe return spring from the anchor.

2. Unhook the primary brake shoe return spring.

3. Slide the eye of the adjuster cable off the anchor. Unhook the overload spring from the adjuster lever. You may now remove the cable from the cable guide and the anchor plate from the anchor.

4. To disengage the adjuster lever from its spring, use pliers to lift the end of the spring (next to the star wheel). Move the end of the spring over the lever. Remove the lever from its pivot.

5. With the aid of pliers, unhook the long shoe to shoe coil spring from the primary and secondary brake shoes.

6. Using your hands, disengage the brake shoes from the piston pin (if so equipped) by moving the shoes away from the anchor. This step will enable you to remove the parking brake lever strut and its antirattle spring.

7. With the aid of a special tool (Figure 25-20), turn the outer washer (part of the retainer and pin assembly) one-quarter of a revolution. Remove the washers and spring from the pin. At the rear of the

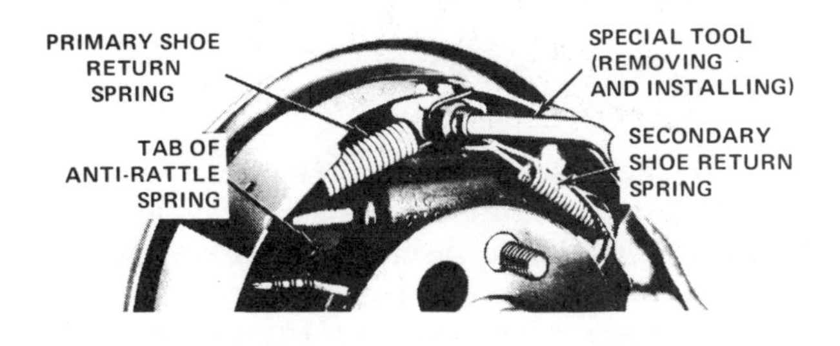

FIGURE 25-19. Using Special Tool to Unhook Secondary Spring. (Courtesy of Chrysler Corporation)

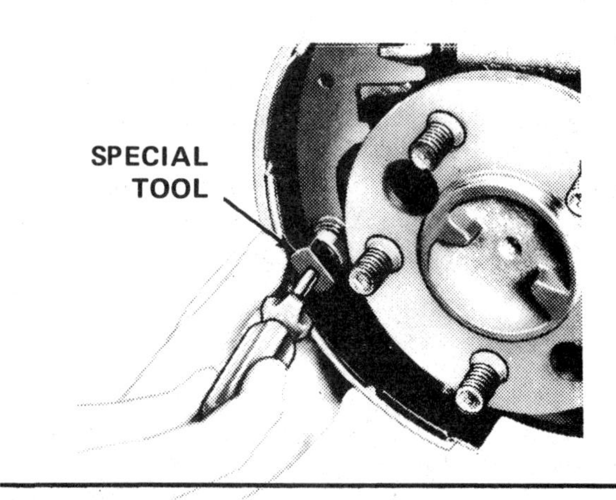

FIGURE 25-20. Removing and Installing Shoe Retainer Spring Assembly. (Courtesy of Chrysler Corporation)

support plate, remove the pin. Since the primary shoe is not attached to the support plate, the shoe may be removed.

8. Follow the same procedure to disconnect the secondary shoe retainer and pin assembly.

9. While holding the secondary shoe in one hand and with the aid of side cutting pliers, compress the parking brake cable spring. This will enable you to disconnect the cable from the parking brake lever.

10. To remove the parking brake lever from the secondary shoe on older model vehicles, place the shoe on the workbench. With the aid of a medium-sized screwdriver, open the U lock, or detach the C clip that retains the brake lever and pivot pin to the shoe. On newer model vehicles, the lever is recessed where it hinges and attaches to the shoe. By slightly twisting the lever, you will be able to remove the part from the shoe.

11. Place all the parts for each wheel in their respective hubcaps. *Remember:* The parts are not interchangeable.

CLEANING AND INSPECTING THE PARTS

Brake Shoes

Once the shoes have been removed from their support plate, examine the surface of the linings to determine if they have been in full contact with the drums. If they have not, the problem may be seized pistons inside a wheel cylinder. If this is the case, the drum's braking surface will be rusty. If the cylinder has been operating correctly, the secondary lining will be worn more than its opposite. This is a normal condition since most of the driving has been forward. You may also see a slight gap that travels down the entire face of the middle of the primary shoe. This is also normal and reduces brake shoe application noise. If the steel webbing that supports the shoe's table and lining is cracked, the shoe has been flexing on application, and the lining will show uneven wear. This condition is usually caused by poor manufacture. Since the removed shoes and their linings will be exchanged with reconditioned shoes and new bonded or riveted linings, do not attempt to clean the removed shoes.

Springs and Adjustment Mechanism

Before inspecting the springs and the adjustment mechanism, brush them dry or clean them with solvent and dry them with compressed air. If the springs are stretched (spaces between the coils) or if the paint at the end of a spring has become burned (darker due to excessive heat), the parts are no longer usable.

Carefully examine the cable to determine if it is frayed where it curves over the cable guide (see Figure 25-1). The other parts that must be examined are the retainer washers, springs, and their pins. The pins tend to become worn where they fit through the shoes. The star wheel must also be inspected to determine if the teeth have become rounded or sheared. Disassemble the adjuster, clean its threads, lubricate with molylube, and reassemble. If any of the parts are worn, they must be replaced.

Wheel Cylinders

When inspecting a wheel cylinder, an inexperienced person may remove (flip back) the lower end of each boot and, if there is no appearance of fluid, assume that the cylinder does not need to be serviced. Let's consider what happens inside a cylinder as the linings wear. As the linings wear, the pistons and cups (seals) within the bore (Figure 25-21) are required to alter their operational position and move outward.

Although modern hydraulic brake systems are sealed, the wall of the bore becomes coated with contamination as the pistons and cups move outward. When new shoes are installed on the support plate, the cups will be required to function within

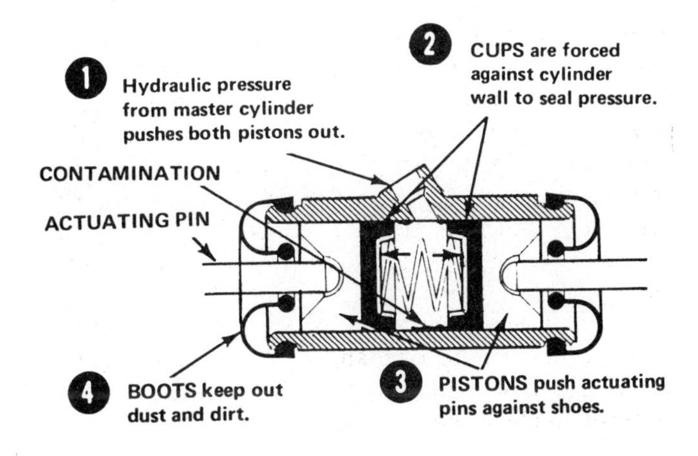

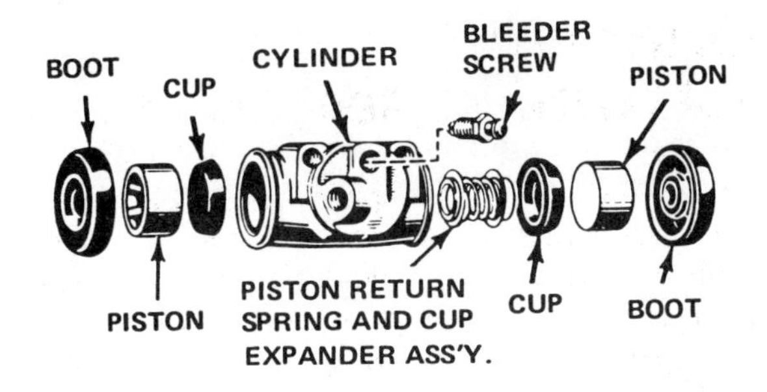

FIGURE 25-21. Sectional View of Wheel Cylinder. (Courtesy of Ford Motor Company)

the contaminated surface of the bore. This condition will induce external leakage of the fluid. An experienced brake repair technician will either overhaul or replace a wheel cylinder when replacing brake shoes.

Most wheel cylinders can be honed and the internal parts replaced without removing the metal casting from the support plate. However, if the plate is designed with piston stops, the casting must be removed from the support.

If you are required to remove the cylinder from the support plate, you will probably find that the inverted line nut that attaches the hydraulic line to the plate will be seized to the line. If this is the case, use side cutting pliers to cut the line next to the nut. Then, replace the steel line.

A PRACTICAL SERVICE TIP

When servicing a wheel cylinder, the first step is to ensure that the bleeder valve is not seized in the cylinder. If the valve is seized, follow step 13 in the Step-by-Step procedure in Chapter 24 in the section entitled Removing and Disassembling the Caliper.

Step-by-Step Procedure

1. Remove the internal parts of the wheel cylinder. *Note:* If the pistons are seized in the bore, the cylinder and its parts must be replaced.

2. Pass a lint-free cloth through the bore several times to remove any loose contamination. *Note:* If the wheel cylinder is aluminum and the surface of the bore must be honed in order to restore its surface, the wheel cylinder must be replaced.

3. Dip the stones of the hone in clean brake fluid or denatured alcohol. With an electric drill, rotate and polish (moving with alternative strokes) the surface of the bore (Figure 25-22).

Caution: If you allow the stones of the hone to slip out of the bore during this step, you will damage the stones.

4. After you have removed the surface contamination and endeavored to polish the bore, wash the inside of the bore with denatured alcohol and dry with compressed air. Use a clean, lint-free cloth to wipe the bore.

5. With the aid of a suitable light, examine the bore's interior for signs of dark areas. A dark area will indicate a low depression in the bore. If, after honing the bore's surface, it is not polished within the operational position of the pistons and cups, the cylinder must be replaced.

6. Clean and inspect the pistons. If they are pitted or show signs of wear, they must be replaced.

7. Once the bore is polished, the next step is to determine if the bore's diameter is oversize. Install the pistons in the cylinder's bore and attempt to insert a 0.137 millimeter (0.005 inch) feeler gauge

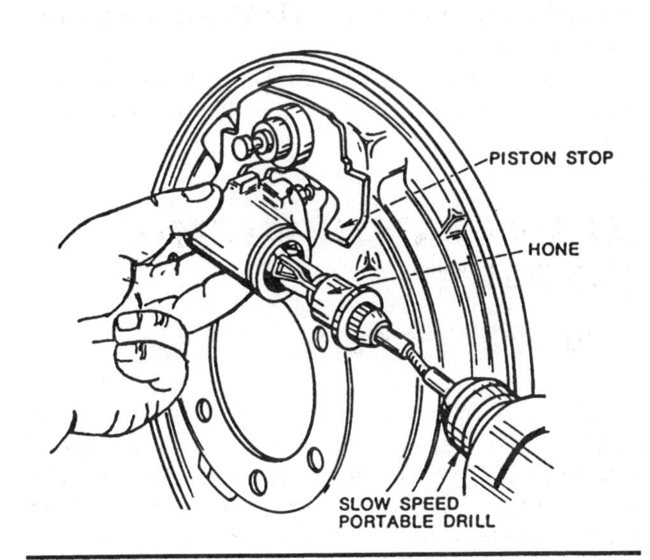

FIGURE 25-22. Honing Cylinder Bore. (Courtesy of Wagner Brake Company, Ltd.)

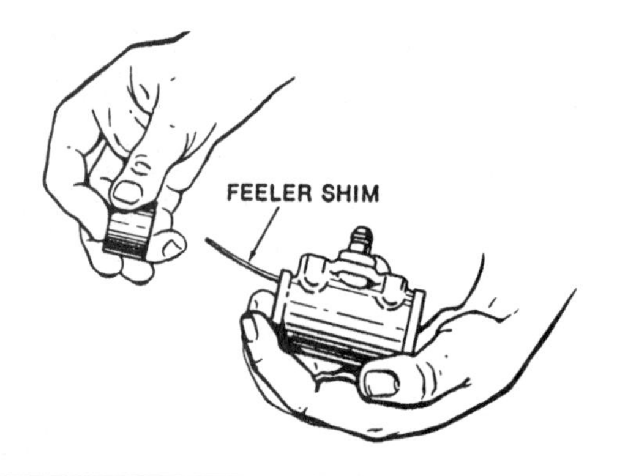

FIGURE 25-23. Using Feeler Gauge to Check Bore Diameter. (Courtesy of Wagner Brake Company, Ltd.)

(Figure 25-23). If the gauge enters, the wheel cylinder is not usable.

If the piston can be inserted with the shim in place, the cylinder is oversize and should be discarded. Depending on the cylinder bore diameter, the shim (or feeler gauge) thickness can vary as follows:

If cylinder bore diameter is:	Shim thickness should be:
19–30 mm (3/4 in.–13/16 in.)	0.15 mm (0.006 in.)
32–37 mm (1 1/4 in.–1 7/16 in.)	0.18 mm (0.007 in.)
38+ mm (1 1/2+ in.)	0.2 mm (0.008 in.)

8. Since the bleeder valve was previously removed from the cylinder, clean its exterior. With compressed air, blow out its internal passage. Coat its threads with molylube. Install and tighten the valve.

9. To assemble the cylinder, apply clean brake fluid to the bore. Dip all the parts of the kit in clean brake fluid to lubricate them and to assist in their installation. Insert the spring, expanders (if equipped), rubber cups, and pistons into the cylinder. *Note:* The flat side of each cup must contact the piston.

10. Install the boots, making sure that the lips are seated in the grooves at the ends of the cylinders.

A FACT TO REMEMBER

When servicing a wheel cylinder, extreme cleanliness is absolutely essential.

Rear Axle Bearing and Seal

Although the rear axle bearing and seal are not considered component parts of a brake mechanism, some mention must be made of them. If the rear axle housing has been overfilled with gear lubricant or if the seal (sometimes part of the bearing) is defective, gear lubricant will leak past the seal onto the support plate, brake shoes, and drum. Depending on the amount of fluid leakage, the brakes may grab or not produce any friction when applied. Another reason a seal will leak is a plugged rear axle breather vent.

To determine if a seal is defective, examine both sides of the support plate where the axle enters the axle housing. If there is any evidence of gear oil, the seal and/or bearing must be replaced. Obtain the manufacturer's shop manual for the vehicle you are servicing and follow the repair instructions.

A PRACTICAL SERVICE TIP

When replacing a rear axle seal, always check to be sure that the vent for the rear axle housing is not restricted.

Support Plate

Although the support plate is not a moving part, it is also subject to wear. Every time the brakes are applied, the shoes move outward and touch the revolving drum. This means that the shoes are moving (floating) against the plate. When and where there is movement, there is the possibility of wear to the contact areas of the plate (Figure 25-24).

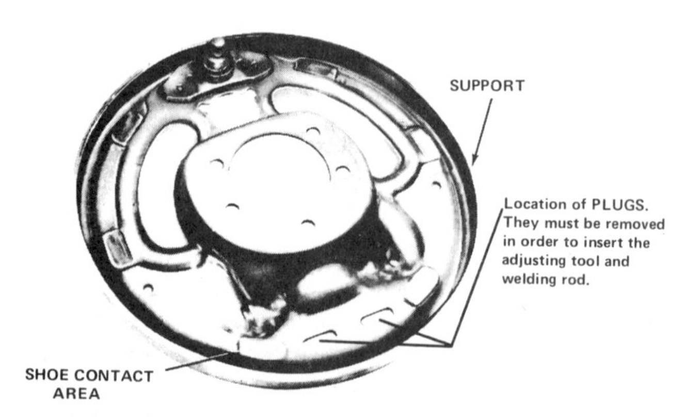

FIGURE 25-24. Shoe Contact Area on Support Plate. (Courtesy of Chrysler Corporation)

Before examining the plate, clean the part by using a whisk broom or a wire brush. If the plate has been covered with brake fluid or rear axle grease, use varsol to remove any residue.

After the plate has been cleaned, carefully examine the plate shoe contact areas. If the areas are grooved from a lack of lubrication, fill in the grooves by brazing them. Then resurface the areas by using a small grinding stone and an electric drill. When the areas have been reconditioned, place a new shoe against the support plate to ensure that all the areas are level and compatible.

If the areas are covered only with surface rust, sand them with number 80 sandpaper until they are free of rust and the metal surfaces are shiny. When the areas have been reconditioned or sanded, apply a light coat of lubricant (molylube) to the support plate shoe contact areas.

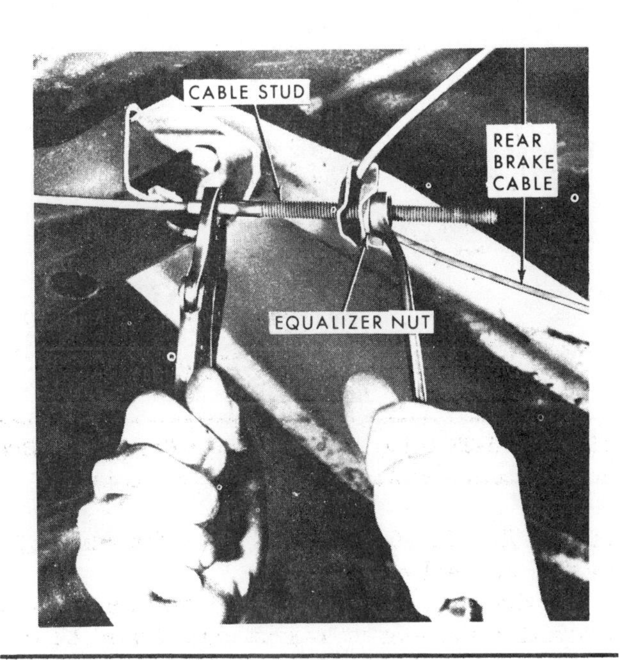

FIGURE 25-25. Unthreading Equalizer Nut from Adjuster. (Courtesy of General Motors of Canada, Limited)

Parking Brake

The parking brake, since it manually operates the rear brakes, must also be examined and serviced if required. Most problems encountered with the operation of a parking brake mechanism are the direct result of moisture seeping through the outer covering of a brake cable. Over time, this moisture will cause the inside cable to seize, not releasing the brake mechanism.

When a brake system receives major service, the cables should be disconnected to allow the repair technician to test the operation of the cables. If the inner cable is considerably longer than its conduit, the cable can be pulled through its covering, cleaned, and lubricated. Some brake cables are not serviceable. If upon inspection they are partially seized, they must be replaced.

A PRACTICAL SERVICE TIP

Before a drum will fit over new brake shoes, it is necessary to release the tension on the rear cables. To release the tension, the adjusting nut will need to be backed off (turned counterclockwise). Since the nut probably will be seized to the threaded end of the front cable, use an oxyacetylene torch to heat the nut to a dull red color. Holding the unthreaded shoulder of the adjuster with vise-grip pliers and using the appropriate wrench, unthread the nut from the adjuster (Figure 25-25).

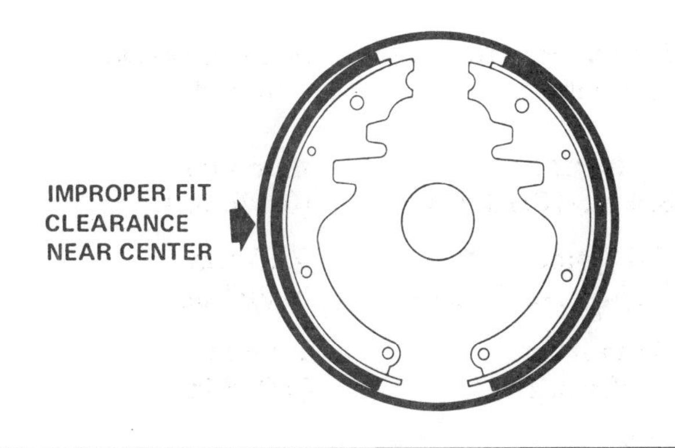

FIGURE 25-26. Example of Improper Fit of Brake Shoes against the Drum. (Courtesy of Wagner Brake Company, Ltd.)

SERVICING PROCEDURES

Grinding (Arcing) a Brake Shoe

When a brake drum has been machined 1.01 to 1.52 millimeters (0.040-0.060 inch) beyond its original size, it is impossible for the brake shoe's lining surface to have the correct contact with the drum's braking surface. Figure 25-26 illustrates an improper fit of two brake shoes. Such a condition will contribute to problems such as squeal, grab, or pull or even cause the brake to lock on application.

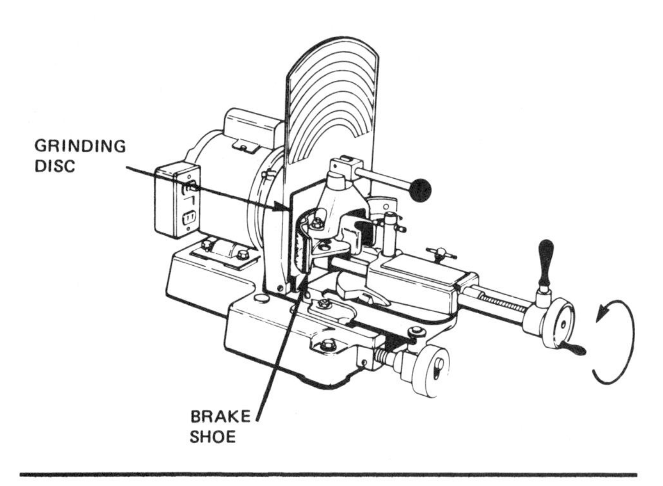

FIGURE 25-27. Machine Used to Grind Brake Shoes. (Courtesy of Wagner Brake Company, Ltd.)

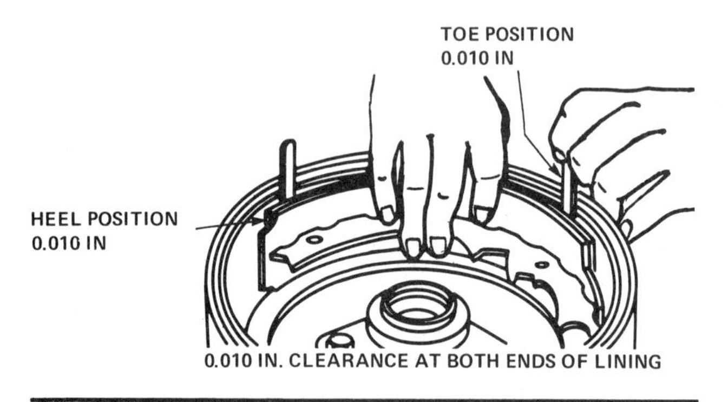

FIGURE 25-28. Using a Feeler Gauge to Measure Clearance at Toe and Heel Positions. (Courtesy of Chrysler Corporation)

To correct this problem, the lining surface of a shoe must be eccentrically ground (on a contour) using a brake shoe grinder to ensure the proper fit. If your repair shop is not equipped to perform this important service, the parts jobber (store) that sells new shoes and linings probably will be equipped to provide the service. Figure 25-27 illustrates a machine designed to grind a brake shoe.

After a brake shoe has been ground, hold the middle of the shoe tightly against the drum (Figure 25-28). A shoe that has been ground correctly will have a clearance of 0.254 millimeter (0.010 inch) at its toe and heel positions. The clearance is measured by using a feeler gauge.

A FACT TO REMEMBER

When operating an unfamiliar piece of shop equipment, always have an instructor or a fellow worker or student provide you with a demonstration. If a demonstration is not possible, follow the manufacturer's operating instructions.

Flushing the Rear Hydraulic System

The last task before assembling the brake shoes to the support plate is flushing the fluid from the rear hydraulic system. Can you remember from the previous chapter why this is necessary? If the fluid remaining in the rear lines contains particles, the lines, when connected to cylinders, will transmit contaminated fluid to the new or reconditioned parts. To prevent this situation from occurring, fill the master cylinder's drum brake reservoir with new brake fluid. Pump the brake pedal until clean fluid is discharged from the steel brake tubes.

Assembling the Brake Mechanism

Let's conclude our study of drum brakes by discussing how to assemble the brake mechanism. For the purpose of our discussion, we'll assume that you have all the necessary parts, tools, and equipment.

Step-by-Step Procedure

1. Holding the wheel cylinder in one hand, with the other hand thread the inverted line nut into the cylinder. This procedure will enable you to connect the line to the cylinder more easily. After the line has been attached, mount the cylinder to the support plate. Tighten the line nut.

2. Using molylube, apply a light coat of lubricant to all the pivotable points, such as: brake shoe to piston pin contact areas, brake shoe to adjuster screw, secondary shoe to parking brake lever, primary shoe and parking brake lever to strut, and cable guide to adjuster cable.

3. Attach the parking brake lever to the secondary shoe.

4. Holding the end of the parking brake cable with pliers, with your other hand and using side cutting pliers, compress the parking brake cable spring. When the spring has been compressed, attach the parking brake lever to the cable.

5. Align the rear piston rod to the secondary shoe. Using one hand to hold the brake shoe against

the support plate and the nail through the plate and shoe, with the aid of the special tool (see Figure 25-20), install the outside washer on the pin. Compress the spring and turn the washer one-quarter of a revolution to lock the washer to the pin.

6. Position the strut behind the axle and into the slot in the parking brake lever. Slide the anti-rattle spring over the free end of the strut.

7. Align the front piston rod to the primary shoe and slide the strut into the shoe. Attach the primary shoe to the support plate following the procedures described in step 5.

8. Install the anchor plate over the anchor and lightly lubricate the outside of the plate. Position the eye of the adjuster cable over the anchor.

9. Engage the primary brake shoe return spring in the web of the shoe. With the special tool used to remove the spring, install the free end over the anchor (Figure 25-29).

10. Position the cable guide against the web of the secondary shoe. While holding the part against the web, engage the secondary shoe return spring through the guide and into the web. Install the free end over the anchor.

11. Using pliers, squeeze the ends of the spring loops (around the anchor) until the ends are parallel.

12. Install the adjusting star wheel assembly between the primary and secondary shoes, with the star wheel next to the secondary shoe (see Figure 25-1). *Note:* Most manufacturers, in order to reduce error, will identify the left star wheel with the letter *L* stamped on the end of the adjusting stud. After the adjusting mechanism has been positioned, install the shoe-to-shoe spring between the shoes.

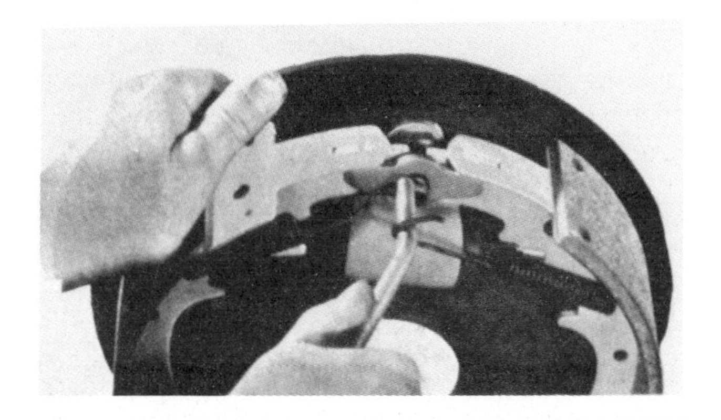

FIGURE 25-29. Using a Special Tool to Install Return Spring over Anchor. (Courtesy Raybestos-Manhattan (Canada), Ltd.)

13. Position the adjuster lever spring over the pivot pin that protrudes through the web of the secondary shoe. Install the adjuster lever to the pin, and position the spring over and around the lever (see Figure 25-1). The spring will retain the lever on the pin.

14. Place the adjusting cable over the cable guide and hook the end of the overload spring in the lever (see Figure 25-1). *Note:* Be sure that the eye of the cable is pulled tight against the anchor plate and in a straight line with the guide.

Adjusting the Shoes to the Drum

Since the brake mechanism has been installed on the support plate, let's discuss how the brake shoes are adjusted to the drum. Figure 25-30A illustrates the use of a brake shoe clearance gauge. When using the gauge, make sure that you are measuring the drum at its maximum diameter. When the diameter has been determined, tighten the gauge's lock screw. Position the opposite side of the gauge over the shoes and rotate the star wheel until the linings just touch the gauge's fingers (see Figure 25-30B). The gauge is designed to provide the correct clearance between the linings and the drums.

Testing the Self-Adjusting Mechanism

Before installing the drum on the axle, one important task that must be completed is to ensure the operational dependability of the self-adjuster mechanism. In Figure 25-31, the repair technician has inserted a screwdriver between the anchor and the top of the secondary shoe. By moving the shoe away from the anchor, the cable should lift the adjuster lever, thereby advancing the star wheel one notch. If the mechanism does not operate before installation of the drum, there is absolutely no point in positioning the drum on the axle. When you are sure that the adjusting mechanism is operating correctly, you may then install the drum on the axle.

Bleeding the Brakes

In the previous chapter, we discussed the procedure for bleeding the brakes manually and/or bleeding the brakes using pressure equipment.

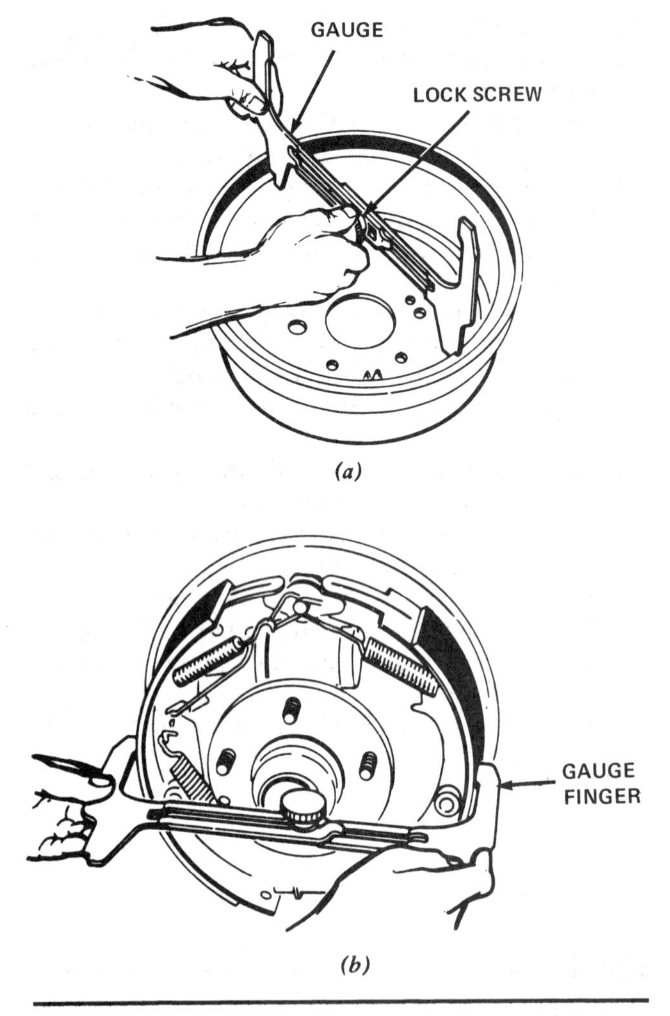

FIGURE 25-30. Adjusting Shoes to the Drum. (a) Tighten the lock screw at the drum's maximum diameter; (b) turn the gauge over and adjust the star wheel so that the linings just touch the gauge's fingers. (Courtesy of General Motors of Canada, Limited)

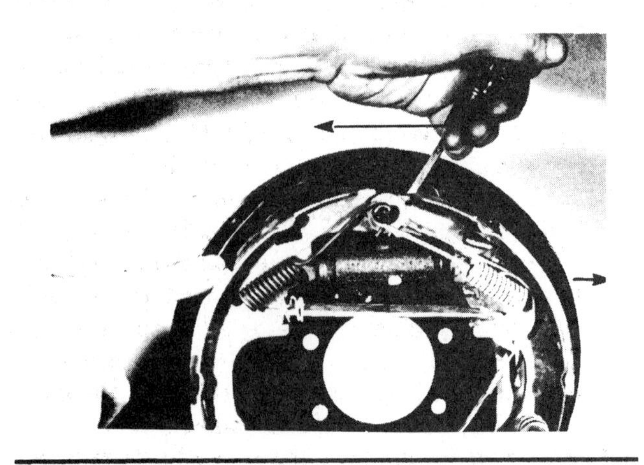

FIGURE 25-31. Testing the Operation of the Self-Adjusting Mechanisms. (Courtesy of Wagner Brake Company, Ltd.)

Either method may be used for bleeding the rear hydraulic system. When bleeding the brakes manually, bleed the wheel cylinder that is located the farthest distance from the master cylinder. This will enable you to expel the air quickly from the system. Finally, when you have bled the lines and cylinder, test the dependability of the vehicle's hydraulic system as described in Chapter 23.

Installing Wheels

Once you have tested the hydraulic systems, install the rear wheels to their indexed axles. Tighten the wheel lug bolts or nuts in the correct sequence to an initial torque specification of 21 newton-meters (15 foot-pounds).

Adjusting the Parking Brake Cable

Because of the design of the parking mechanism, adjusting the parking brake cable is the last task that is performed.

Step-by-Step Procedure

1. Make sure that the parking brake pedal is not applied.
2. If the cables have been disconnected, connect the cables.
3. Tighten the cable adjusting nut until you can feel a slight drag when you turn the rear wheels.
4. Loosen the nut just enough so that both rear wheels turn freely.
5. To ensure that the parking brake mechanism has the necessary clearance, back off the adjusting nut another two full turns.
6. Test the parking brake by applying the pedal several times, then releasing it and making sure no drag is felt at either rear wheel when released.

Now that you have completed the servicing, raise the automobile by using a hydraulic floor jack and remove the safety stands from under the vehicle. Lower the vehicle's rear wheels to the floor. Retighten the lug bolts or nuts in correct sequence and torque to a final specification per the manufacturer's shop manual.

REVIEW TEST

1. List the names of the component parts of a rear drum brake assembly.
2. List three purposes of a brake drum.
3. Briefly describe the operation of a drum brake mechanism.
4. Describe the operation of the self-adjusting assembly.
5. List one distinct advantage of a drum brake mechanism as compared to a disc brake.
6. Briefly describe the operation of the parking brake mechanism.
7. Briefly describe how you would disgnose brake problems associated with the following terms: *pedal pulsation, pull, grab, brake squeal and noise, excessive pedal effort,* and *pedal travel and feel.*
8. Briefly describe how to remove a rear brake drum that had become seized to the axle flange.
9. The brake shoes on a vehicle that you are inspecting are expanded, and you cannot remove the drum. Briefly describe how you would contract the shoes to enable you to remove the drum from the axle.
10. List six brake drum conditions and describe their probable cause.
11. Describe how to measure a drum's diameter.
12. In point form, describe how to machine a rear brake drum.
13. In point form, describe how to remove the brake shoes from the support plate.
14. Assuming that the pistons inside a rear cylinder have become seized, describe how this condition would be evident as you examine the linings and the drum.
15. Explain why pistons change their operational position inside a wheel cylinder when new brake shoes and linings are installed.
16. In point form, describe how to service a wheel cylinder.
17. Briefly explain how a defective rear axle seal will affect a brake mechanism.
18. Describe how to repair the brake shoe contact areas of a support plate.
19. Describe how to clean a support plate.
20. Explain why it is necessary to assess the operation of a parking brake mechanism when servicing a brake.
21. Briefly explain why it is necessary to fit a brake shoe to a drum.

FINAL TEST

This examination is multiple choice. Only one answer will be accepted. Carefully read every statement.

The following questions pertain to a duo-servo drum brake mechanism.

1. Mechanic A says that drum brakes are susceptible to brake fade. Mechanic B says that drum brakes have fewer moving parts than a disc brake. Who is right? (A) Mechanic A, (B) Mechanic B, (C) Both A and B, (D) Neither A nor B.
2. Mechanic A says that the primary and secondary brake shoe linings have the same area size. Mechanic B says that the primary brake shoe lining has the same frictional characteristics as the secondary brake shoe. Who is right? (A) Mechanic A, (B) Mechanic B, (C) Both A and B, (D) Neither A nor B.
3. Mechanic A says that the strut is located above the wheel cylinder. Mechanic B says that the strut is slotted into the secondary shoe. Who is right? (A) Mechanic A, (B) Mechanic B, (C) Both A and B, (D) Neither A nor B.
4. Mechanic A says that the strut's antirattle spring is positioned next to the parking brake lever. Mechanic B says that the strut engages the primary shoe and acts as a pivot point during the engagement of the secondary shoe. Who is right? (A) Mechanic A, (B) Mechanic B, (C) Both A and B, (D) Neither A nor B.
5. Mechanic A says that the bore of the wheel cylinder is straight (not stepped), and there is equal pressure acting on both pistons. Mechanic B says that the star wheel, part of the self-adjusting mechanism, is positioned closer to the secondary shoe than to the primary shoe. Who is right? (A) Mechanic A, (B) Mechanic B, (C) Both A and B, (D) Neither A nor B.
6. Mechanic A says that the brake drum must absorb and dissipate heat. Mechanic B says that a brake drum must protect the braking mechanism

from the elements. Who is right? (A) Mechanic A, (B) Mechanic B, (C) Both A and B, (D) Neither A nor B.

7. Mechanic A says that brake squeal and noise may be caused by a blocked compensating port in the master cylinder. Mechanic B says that insufficient brake pedal reserve height probably will be caused by a defective rear axle seal. Who is right? (A) Mechanic A, (B) Mechanic B, (C) Both A and B, (D) Neither A nor B.
8. Mechanic A says that if the drum is seized onto the axle, you should strike the heat sink of the drum with a hammer. Mechanic B says that if the drum is seized onto the axle, you may apply controlled heat between the wheel studs. Who is right? (A) Mechanic A, (B) Mechanic B, (C) Both A and B, (D) Neither A nor B.
9. Mechanic A says that a new brake drum does not need to be machined prior to installation. Mechanic B says that when a brake drum sustains a bell-mouthed condition, the drum is susceptible to brake fade after machining. Who is right? (A) Mechanic A, (B) Mechanic B, (C) Both A and B, (D) Neither A nor B.
10. Mechanic A says that to determine if a brake drum is cracked, you should test the drum by allowing it to drop on the garage floor. Mechanic B says that to determine if a brake drum is cracked, you should use a small amount of gasoline to detect the crack. Who is right? (A) Mechanic A, (B) Mechanic B, (C) Both A and B, (D) Neither A nor B.
11. Mechanic A says that a brake drum is considered to be out-of-round when there is a variance in the readings of more than 0.178 millimeter (0.007 inch). Mechanic A says that the drum is considered to be out-of-round when there is a variance on the readings of more than 0.229 millimeter (0.009 inch). Who is right? (A) Mechanic A, (B) Mechanic B, (C) Both A and B, (D) Neither A nor B.
12. Mechanic A says that the final finishing cut when machining a drum should be to a depth of 0.051 millimeter (0.002 inch). Mechanic B says that the final finishing cut when machining a drum should be to a depth of 0.152 millimeter (0.006 inch). Who is right? (A) Mechanic A, (B) Mechanic B, (C) Both A and B, (D) Neither A nor B.
13. Mechanic A says that on rear drum brakes, it is normal for the primary shoe lining to wear more than the secondary shoe lining. Mechanic B says that since the adjuster cable is made of steel, the cable does not need to be inspected for signs of wear. Who is right? (A) Mechanic A, (B) Mechanic B, (C) Both A and B, (D) Neither A nor B.
14. Mechanic A says that before removing the brake shoes from the support plate, wash the entire mechanism in varsol. Mechanic B says that as soon as the brake drum is removed from the axle, wash the brake drum and the brake mechanism in water. Who is right? (A) Mechanic A, (B) Mechanic B, (C) Both A and B, (D) Neither A nor B.
15. Mechanic A says that if it is necessary to hone the bore of an aluminum wheel cylinder, the part should be replaced. Mechanic B says that prior to honing the bore of a cast-iron wheel cylinder, dip the stones of the hone in brake fluid. Who is right? (A) Mechanic A, (B) Mechanic B, (C) Both A and B, (D) Neither A nor B.
16. Mechanic A says that when you are assembling a wheel cylinder, coat the inside of the bore with brake fluid. Mechanic B says that when you are assembling a wheel cylinder, coat the inside of the bore with vaseline. Who is right? (A) Mechanic A, (B) Mechanic B, (C) Both A and B, (D) Neither A nor B.
17. Mechanic A says that after the anchor plate has been installed on the anchor, the next part to be installed is the secondary shoe return spring. Mechanic B says that after the anchor plate has been installed on the anchor, the next part to be installed in the anchor is the eye of the adjuster cable. Who is right? (A) Mechanic A, (B) Mechanic B, (C) Both A and B, (D) Neither A nor B.
18. Mechanic A says that before you install the brake drum onto the axle, check the operation of the self-adjusting mechanism. Mechanic B says that before you adjust the brake shoes, release the parking brake cable. Who is right? (A) Mechanic A, (B) Mechanic B, (C) Both A and B, (D) Neither A nor B.

Power Brake Assist Units

LEARNING OBJECTIVES

After studying this chapter, you should be able to:

- List the names of the component parts of a Bendix single diaphragm vacuum power brake assist unit,
- Describe the operational stages of the vacuum power brake assist unit when the brake pedal is not applied, is applied, and is in the holding position,
- Describe how to diagnose operational performance problems related to the vacuum power brake assist unit,
- Describe the design features of a hydro-boost power brake assist unit,
- Describe the operation of the hydro-boost power brake assist unit,
- Describe how to diagnose operational performance problems related to the hydro-boost power brake assist unit.

INTRODUCTION

Power brakes, as they are commonly called, are actually power assist units that enable a driver to produce hydraulic pressure in a brake system with very little braking force. Power assist units greatly increase the efficiency of stopping a moving vehicle, especially in an emergency situation. There are primarily two different design types of power brake assist systems. Let's first discuss the *vacuum power brake assist unit*, its source of power, its design, operation, and diagnosis and then describe the design, operation, and diagnosis of the *hydro-boost power brake assist unit.*

VACUUM SYSTEM

Source of Power

You will recall from Chapter 21 that the vacuum power brake assist unit is positioned between the brake pedal and the dual master cylinder and is attached to the rear wall of the engine compartment (see Figure 21-2). Before we discuss the operation of the unit, we need to ask the question, "Where does its power come from?" The power comes from the force of atmospheric pressure operating against a partial vacuum supplied by the engine.

The air around us, although we cannot feel it, exerts pressure in all directions because of the weight of all the air above it pressing downward. At sea level, this amounts to a pressure of 101 kPa (14.7 psi) and is a pressure that we can put to use (Figure 26-1). All we need is a means to enable it to work for us.

The partial vacuum created by an operating engine provides that means. To find out how strong that vacuum can be, start an automobile's engine and place the palm of your hand over the air intake passage of the carburetor's air filter. You will feel the effect on your hand. If your hand is held tightly over the passage, you will even stop the engine's operation.

What is this so-called vacuum? Technically, it is just *reduced air pressure.* The air inside an operating engine's intake manifold (the part below the carburetor) still exerts some pressure but not nearly so much as the air outside the air filter. At engine idle speed, for example, the air pressure inside the manifold is reduced to approximately 35 to 49 kPa (5 to 7 psi).

Let's discover how the vacuum can be used in the operation of a power assist unit. Figure 26-2 illustrates a cylinder. Inside the container is a movable sliding piston with an attached piston rod. Engine vacuum (low pressure) is present in half the cylinder; at the same time, atmospheric pressure is present in the other half of the cylinder. Since atmospheric pressure is greater than engine vacuum, atmospheric pressure, acting on the surface area of the piston, will cause the piston to move toward the vacuum side (the left of Figure 26-2). The rod,

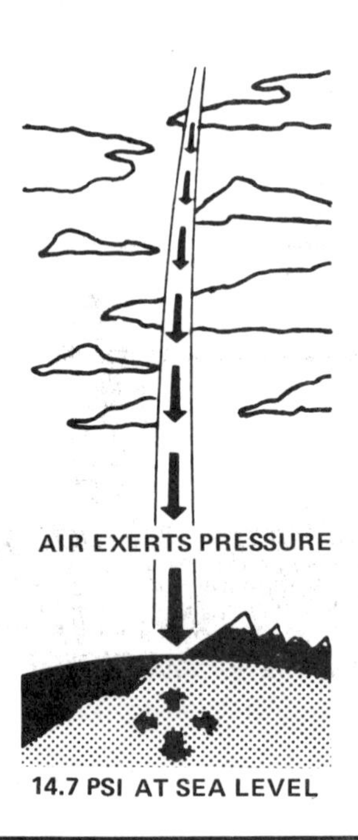

FIGURE 26-1. Constant Atmospheric Pressure. (Courtesy of Chrysler Corporation)

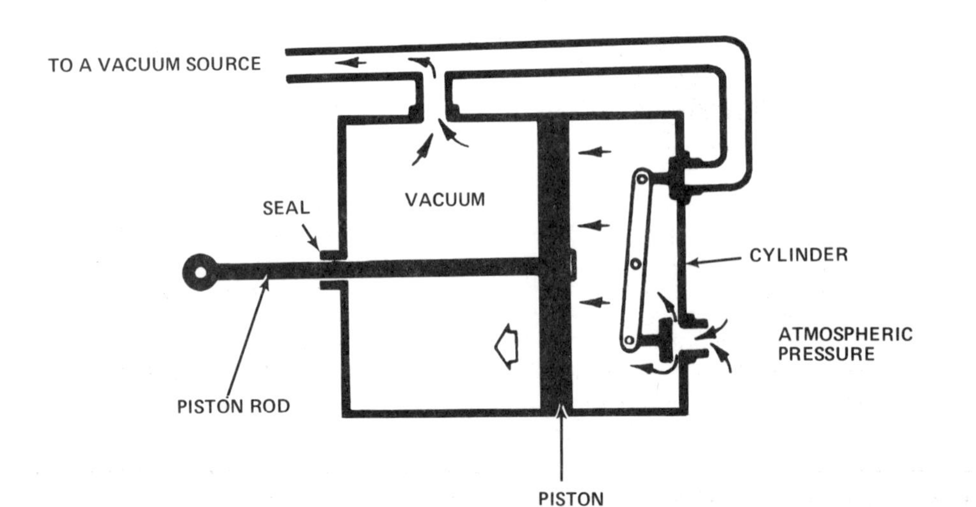

FIGURE 26-2. Cylinder Operation. Atmospheric pressure causes the piston in the cylinder to move toward the vacuum and thus move the piston rod. (Courtesy of Wagner Brake Company, Ltd.)

since it is attached to the piston, will move outward. The movement and force transmitted through the piston rod may then be used to apply the pistons within the bore of a dual master cylinder.

Types of Design

Bendix Single Diaphragm Unit. Figure 26-3 illustrates a Bendix single diaphragm power brake unit. The outside housing, made of steel, is a two-part shell joined together at its middle by an interlocking device. A vacuum check valve, mounted at the front of the unit, provides an attachment for a vacuum hose (not shown) that connects the unit to the intake manifold. The purpose of the vacuum check valve is to provide sufficient vacuum inside the booster for two brake applications in the event that the vehicle's engine ceases to operate. Fitted over and around the rear hub shell and the brake pedal push rod is a dust boot. The dust boot protects that part of the unit since the hub (part of the piston) must move laterally during the unit's operation. Internally, the unit is divided into two cham-

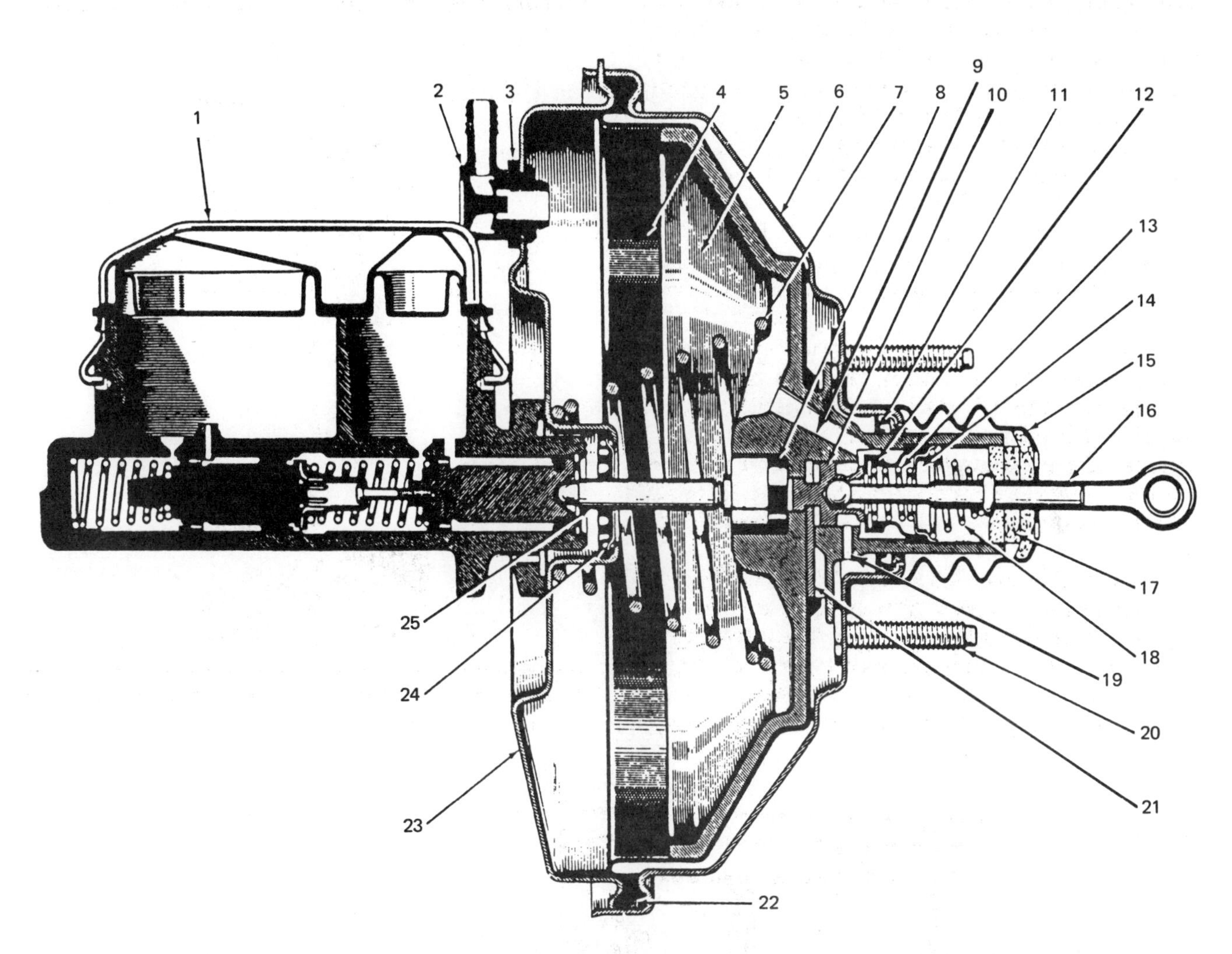

1. MASTER CYLINDER
2. VACUUM CHECK VALVE
3. GROMMET
4. FLEXIBLE RUBBER DIAPHRAGM
5. PISTON
6. REAR SHELL
7. PISTON RETURN SPRING
8. RUBBER REACTION DISC
9. VACUUM PASSAGE
10. VALVE PLUNGER
11. REAR HUB SHELL
12. POPPET VALVE
13. POPPET VALVE SPRING
14. POPPET RETAINER
15. DUST BOOT
16. BRAKE PEDAL PUSH ROD
17. AIR FILTER AND SILENCER
18. VALVE RETURN SPRING
19. PASSAGE
20. REAR MOUNTING STUD
21. AIR VALVE LOCK PLATE
22. INTERLOCKING DEVICE
23. FRONT SHELL
24. FRONT SEAL
25. MASTER CYLINDER PUSH ROD

FIGURE 26-3. Bendix Single Diaphragm Power Brake Assist Unit. (Courtesy of General Motors of Canada, Limited)

bers (sections) by a flexible rubber diaphragm and is supported by a piston, usually made of a plastic Bakelite material.

The piston is designed so that it contains the master cylinder push rod, a rubber reaction disc, a valve plunger, a poppet valve, a poppet valve spring, a valve return spring, and two passages. These passages are used to transmit vacuum and atmospheric pressure during various stages of operation. A piston return spring, located at the front of the piston, returns the piston to its nonoperational position. To prevent dust and dirt from entering the unit when the brakes are applied, an air filter, which also acts as a silencer, is positioned at the hub's opening. Since the master cylinder push rod and hub are external as well as internal moving parts, a front seal and a rear hub seal prevent the entry of atmospheric pressure at their indicated locations.

Bendix Double Diaphragm Unit. Not all power brake assist units are the same size or shape. Some units are designed with a single small piston and others with a large piston. In cases where additional braking assist is required but space inside the engine compartment is limited, an assist unit will be designed with two pistons (front and rear). By designing a unit with dual pistons, total piston area is doubled, thereby multiplying the force on the master cylinder push rod (Figure 26-4).

Pressure Booster Unit. The pressure booster unit is connected into the hydraulic system downstream from the master cylinder, and the valve mechanism is controlled with low hydraulic pressure originated in the master cylinder by manual effort. The power section strokes a piston in another main hydraulic cylinder, which is part of the unit. The

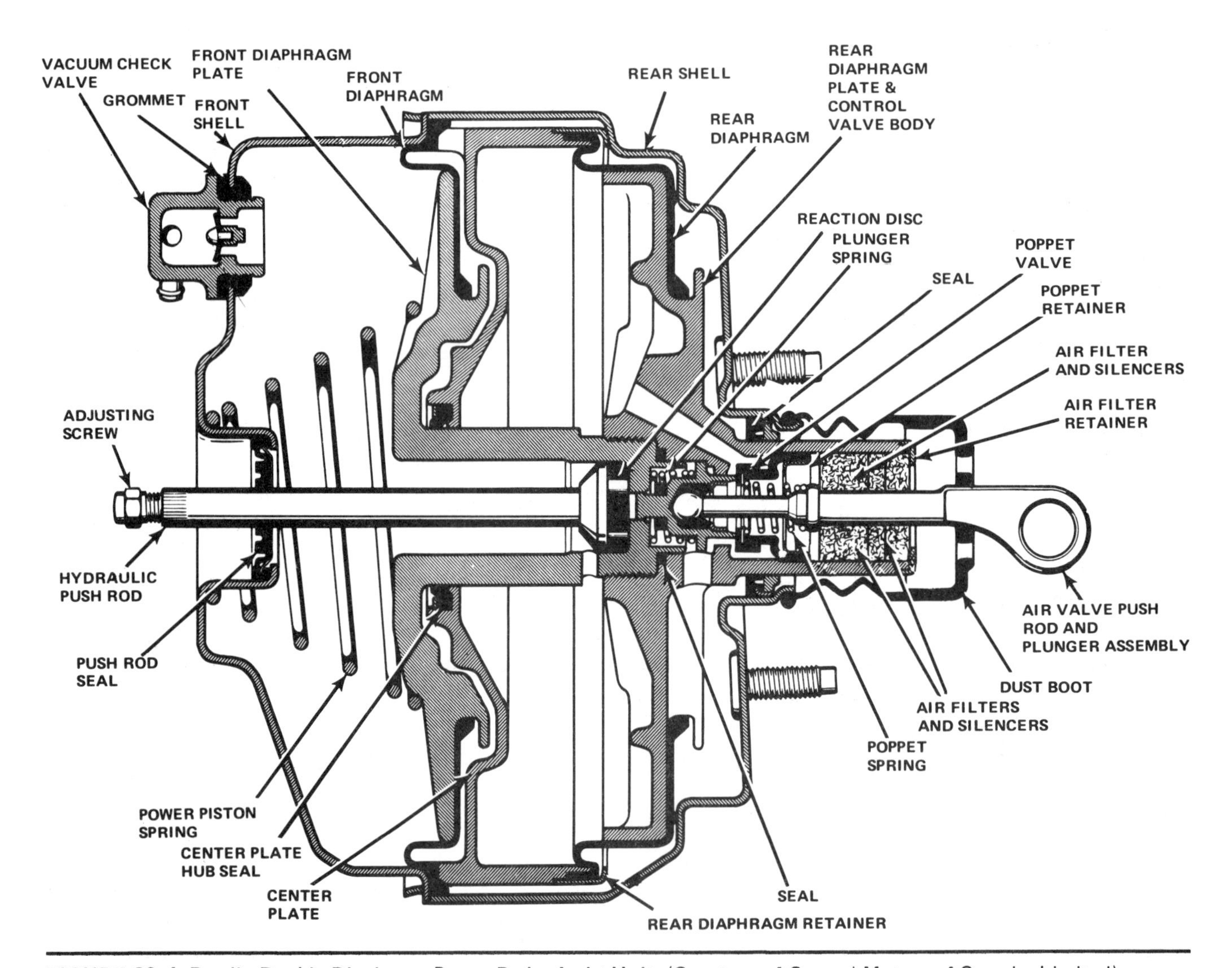

FIGURE 26-4. Bendix Double Diaphragm Power Brake Assist Unit. (Courtesy of General Motors of Canada, Limited)

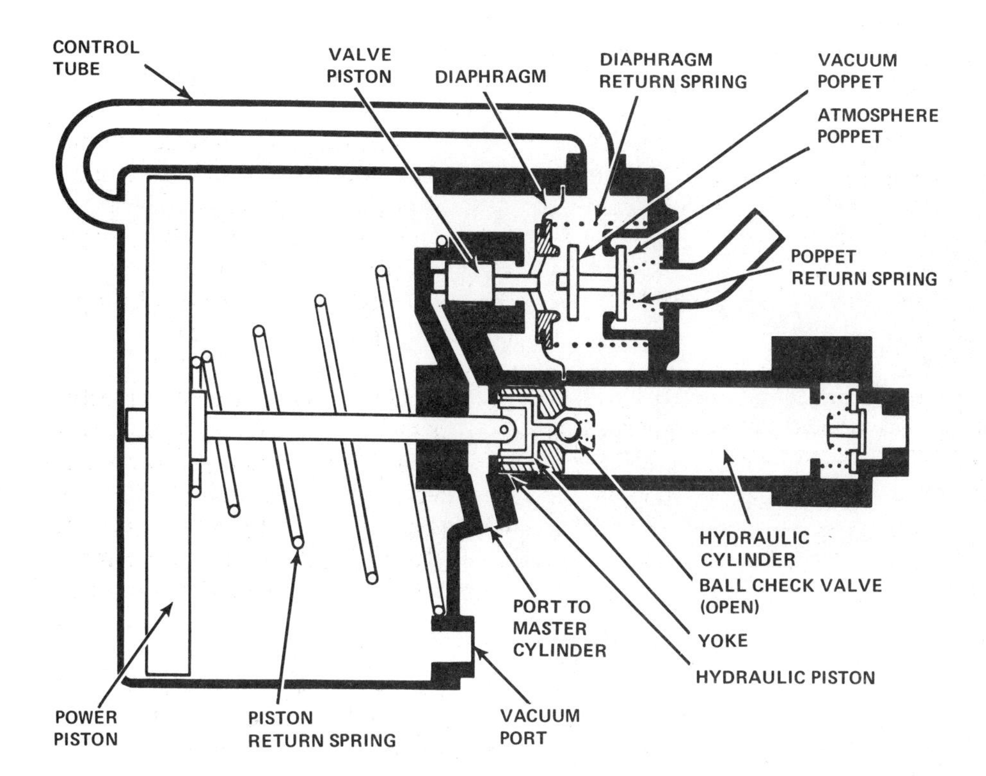

FIGURE 26-5. Cross-Section of Pressure Booster Unit. (Courtesy of Raybestos-Manhattan, Ltd.)

beginning of brake operation is manual, power cutting in at 172 to 551 kPa (25 to 80 psi) hydraulic pressure, depending on the model. Then the main cylinder begins to increase hydraulic pressure to the wheel cylinders above that supplied from the master cylinder. On some systems, the hydraulic residual pressure check valve is located in this main cylinder rather than in the brake master cylinder. Hydrovac (Bendix trade name) pressure boosters of this design are now generally confined to commercial vehicles and some later model European passenger cars (Figure 26-5).

Bendix Single Diaphragm Power Brake Operation

Released Stage of Operation. When the engine is operating and the brakes are released, vacuum from the intake manifold is admitted through a vacuum hose and a vacuum check valve to the front chamber of the unit (Figure 26-6). In the released position, the leading face of the popper valve, influenced by the tension of the valve spring, is held tightly against therearward face of the valve plunger. In this stage of operation, vacuum is also present in the rear chamber of the unit; vacuum has entered this section through the vacuum passage. Since vacuum is present on both sides of the diaphragm, the diaphragm is suspended in a partial vacuum, and the piston return spring is not compressed.

Applied Stage of Operation. As the brakes are applied by the driver, the brake pedal push rod and the valve plunger move to the left (Figure 26-7) in the piston assembly. This movement compresses the poppet valve return spring and brings the poppet valve into contact with the vacuum port seat in the valve housing (part of the piston), closing the vacuum port.

Additional movement of the pedal push rod in the applied direction moves the valve plunger away from the poppet valve to open the atmospheric port. At this time, atmospheric pressure is admitted through the air filter and passages to the rear chamber of the brake assist unit. With the front side of the diaphragm subject to vacuum and atmospheric pressure acting on the rear side of the diaphragm, a force is developed to move the diaphragm and piston assembly toward the front of the unit. Since the master cylinder push rod is supported by the piston, the rod will cause the primary and secondary pistons within the master cylinder to move into the bore of the cylinder. As hydraulic pressure is developed within the master cylinder, a counter (opposite) force, acting through the master cylinder push rod and the rubber reaction disc, sets up a reaction

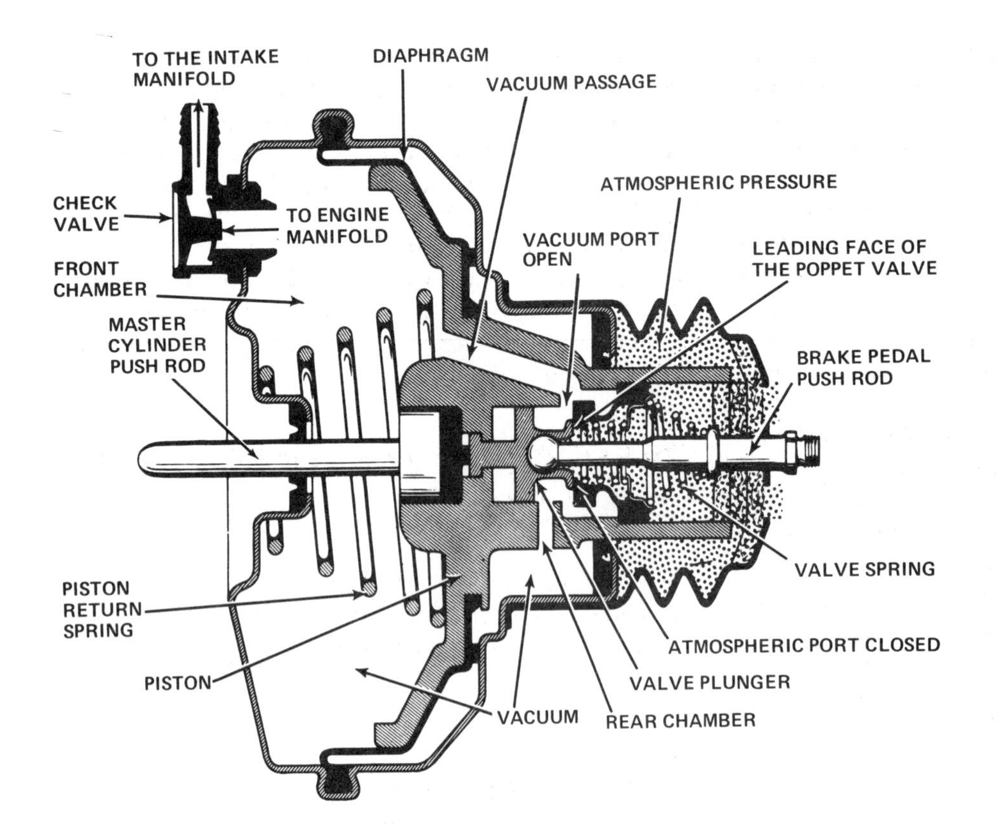

FIGURE 26-6. Bendix Single Diaphragm Power Brake: Released Stage of Operation. (Courtesy of General Motors of Canada, Limited)

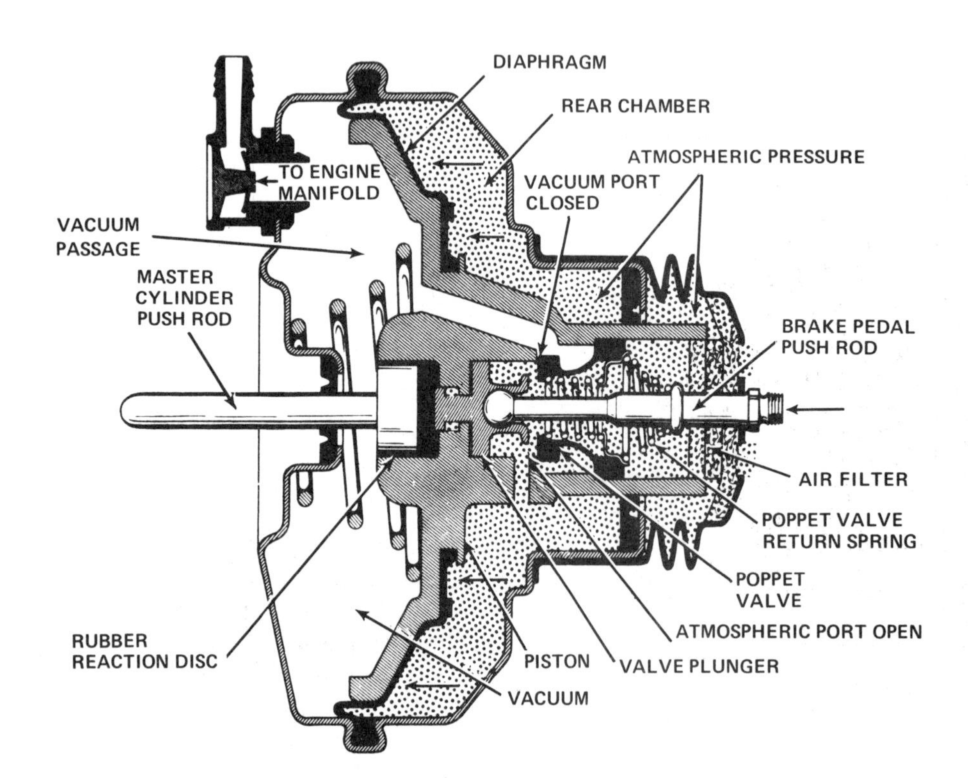

FIGURE 26-7. Bendix Single Diaphragm Power Brake: Applied Stage of Operation. (Courtesy of General Motors of Canada, Limited)

force against the piston and diaphragm and the valve plunger. The rubber reaction disc distributes the pressure between the piston assembly and the valve plunger in proportion to their respective contact areas.

The pressure acting against the valve plunger and the brake pedal push rod tends to move the valve plunger slightly rearward in relation to the piston and valve housing assembly to close off the atmospheric port. Since this counter force, or reaction force, is in direct proportion to the hydraulic pressure developed within the brake systems, the driver, with a foot on the brake pedal, is able to maintain a feel for the degree of brake application.

During an emergency braking situation, the valve plunger holds the poppet valve away from the atmospheric port seal to admit the maximum of atmospheric pressure to the rear side of the diaphragm and piston assembly. With vacuum present in the front chamber of the unit, full power application is attained. Any increase in hydraulic pressure beyond this point must be supplied by the physical effort of the driver.

Holding Stage of Operation. During normal application of the brakes, the reaction against the valve plunger is working against the driver to close the atmospheric port. With both atmospheric and vacuum ports closed, the power brake assist unit is said to be in the *holding position* (Figure 26-8). When both valves are closed, any degree of braking application attained will be held until the atmospheric port is reopened either by an increase in brake pedal force or by a decrease in pedal force to reopen the vacuum port. Whenever the force applied to the brake pedal is held constant for a moment, the valve returns to its holding position. When the brake pedal is released, vacuum is present in both chambers of the unit.

DIAGNOSING VACUUM SYSTEM PROBLEMS

Modern vacuum power brake assist units seldom provide problems. If the unit is suspected of being faulty, the service repair technician can quickly test the unit's operation by following proper diagnostic procedures.

Step-by-Step Procedure

1. Since the operation of a power brake unit is dependent on the vacuum produced by the operation of the vehicle's engine, first determine the condition of the vacuum hose that attaches to the

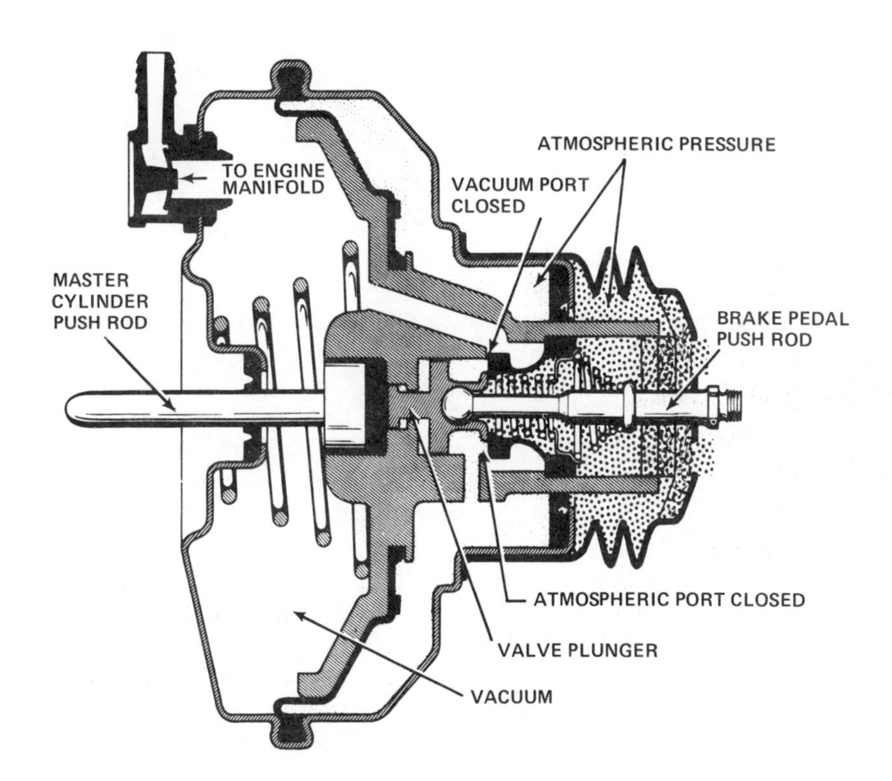

FIGURE 26-8. Bendix Single Diaphragm Power Brake: Holding Stage of Operation. (Courtesy of General Motors of Canada, Limited)

vacuum check valve. To determine the condition of the hose, disconnect the hose from the valve. Placing a finger over the opening of the hose or attaching a vacuum gauge to the hose, have a fellow student or worker start and operate the engine. If the hose is not restricted, you will feel the effect of the vacuum on your finger, or you will see the vacuum gauge register a reading. If the hose is restricted, replace the part.

An alternative way of checking a hose is to remove the part and, using your mouth, blow through the hose. If a great deal of effort is expended blowing through the hose, the hose is restricted and must be replaced. If the hose is not defective, attach the hose to the manifold and to the vacuum check valve.

2. To perform the second check, start and operate the engine. Without applying the brake pedal, determine if the engine is idling smoothly or roughly. If the engine is idling roughly, gently apply the brake pedal and determine if the engine's operation is smooth. If this is the case, the cause of the engine's roughness is located inside the assist unit. Verify your diagnosis by disconnecting the hose from the check valve and blocking the opening of the hose. If the engine's operation is smooth, you have confirmed your diagnosis. The power assist unit must be replaced or repaired.

3. This test is also conducted while the engine is operating. Gently apply the brake pedal and determine if the engine's idle is continuously rough. If this situation exists, listen for a slight hissing sound. If the engine is rough and the sound is noticeable, the assist unit must be serviced.

4. The fourth test is conducted after the brake pedal has been pumped several times with the vehicle's engine not operating. This is done to deplete any existing vacuum in the unit's front chamber. Apply the brake pedal with a moderate force. Then start the engine. If the unit is functioning correctly, the brake pedal will fall slightly (sink lower), and less force will be required to apply the pedal as soon as the engine becomes operational. If the pedal does not sink when the engine is operating, the assist unit is defective.

5. The last test is conducted after the engine has been allowed to idle without the application of the brake pedal. After the engine has idled, turn the ignition key to the off position. If the vacuum check valve is operating correctly, there should be sufficient vacuum retained in the unit's front chamber for two brake pedal applications, assisted by the power assist unit. If the pedal's application requires additional force when depressed, the vacuum check valve is defective. You may confirm your diagnosis by removing the valve from the unit and attempting to blow air through the valve with your mouth. If air is passed through the valve in both directions, the valve is defective and must be replaced.

SERVICING POWER BRAKE ASSIST UNITS

Most vehicle manufacturers recommend that a vacuum power brake assist unit be serviced by replacing the unit. The reason for this policy is that special tools are required to dismantle and reassemble the unit. Factory-remanufactured units are available for some later model vehicles from the dealer.

When you are replacing an assist unit, first determine if the part has failed because gasoline has entered the unit through the vacuum hose. If gasoline has entered the unit, the flexible rubber diaphragm will become porous. To determine if gasoline is the cause of the failure, check whether a strong odor of fumes is emitted from the unit after you remove the master cylinder. If gasoline fumes are detected, the engine's carburetor mixture may be too rich, or the vacuum hose, leading to the check valve, may be positioned lower than the top of the carburetor. If the carburetor is the cause of the problem, it must be serviced before another assist unit is installed. If the hose is lower than the top of the carburetor, the hose must be repositioned against the fire wall so that the hose is at least 2.54 centimeters (1 inch) above the top of the carburetor. The reason for relocating the hose is to allow the part to act as an inverted reverse trap, thereby reducing the possibility of fuel vapors entering the assist unit. Before attaching the master cylinder to the assist unit, check the length of the master cylinder push rod (Figure 26-9).

After the power assist unit has been assembled, test the unit's operation to confirm that the unit is functioning correctly. If the unit is designed and operated by an additional lever and linkage mechanism between its rear chamber and the rear wall of the engine compartment, examine the mechanism to ensure that it is not seized and that it is operating freely.

Caution: Manufacturers state emphatically that when you are required to bleed the brakes, the vehicle's engine must not be operating. If you bleed the brakes with the engine operating, you may cause the diaphragm to flex beyond its design limitation and rupture.

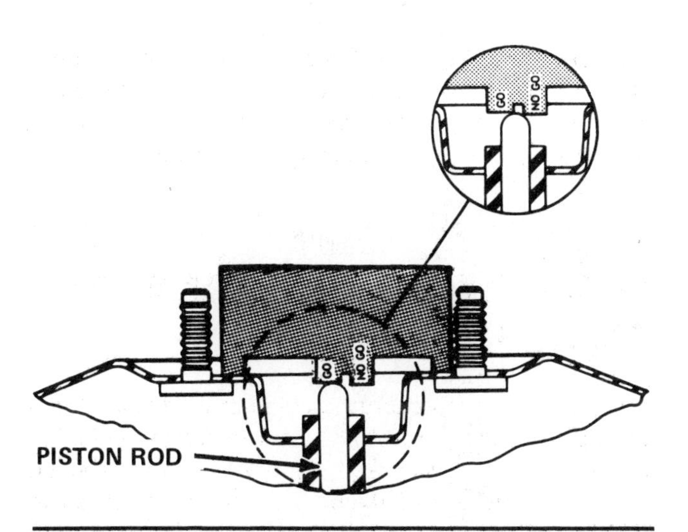

FIGURE 26-9. Checking Push Rod Adjustment. (Courtesy of General Motors of Canada, Limited)

FACTS TO REMEMBER

1. Do not attempt to overhaul a vacuum power brake assist unit unless you have the necessary tools.
2. When you have the necessary tools and are required to overhaul a unit, follow the instructions as described in the manufacturer's shop manual.

HYDRO-BOOST SYSTEM

The hydro-boost power brake assist unit is now being used by some manufacturers as an alternative means for providing brake assist for some larger automobiles and commercial vehicles. In this system, the power steering pump provides the main source for fluid pressure to operate both the hydro-boost unit and the power steering gearbox mechanism. Figure 26-10 illustrates the routing of the lines between the power steering pump, hydro-boost unit, and power steering gearbox. The figure also illustrates an external view of a Bendix hydro-boost assist unit.

Unit Design

The hydro-boost unit contains an internal open center spool valve and sleeve assembly (Figure 26-11), which controls the power steering pump pressure magnitude during brake application. A lever mechanism, operated by the input rod, controls the position of the valve. A boost piston, which contains an output rod, provides the force necessary to operate the pistons of a conventional dual master cylinder. The master cylinder is attached to the front of the booster unit (see Figure 26-10). The hydro-boost unit also has a reserve system, which is designed to store sufficient fluid under pressure to provide at least two brake applications in the event that fluid flow from the power steering pump is not available. When the reserve system has been depleted, the hydraulic brake systems can then be applied manually; however, more effort will be required.

Unit Operation

Released Stage. When the brake pedal is in the released position, the spool valve return spring holds the spool valve open (Figure 26-11). In the open position, the spool valve provides an unrestricted flow of fluid between the power steering pump and the power steering gearbox. Fluid pressure is blocked from entering the boost pressure chamber by the *lands* (the raised portions of the spool) on the spool valve. Therefore, as fluid pressure increases with steering demand, the pressure has no effect on the boost pressure chamber. The boost pressure chamber is vented through the spool valve to the pump return port, thereby allowing the power steering fluid to return to the power steering pump.

Applied Stage. As the brake pedal is depressed, the pedal moves the input rod and initiates the movement of the spool valve. This preliminary movement of the valve closes the fluid return port to the power steering pump from the boost chamber and admits fluid into the boost chamber from the pressure port. Additional valve movement restricts fluid flow between the power steering pump and the power steering gear, causing the power steering pump to increase fluid pressure to maintain essentially the same flow rate to the power steering gear. As the fluid pressure increases in the boost chamber, the pressure forces the boost piston and output rod forward, actuating the pistons inside the dual master cylinder. The amount of forward travel is in proportion to the force applied to the brake pedal and the hydraulic assist that is attained on brake application (Figure 26-12).

Brake Reserve Stage. The hydro-boost unit also has a reserve system (compressed gas accumulator), which is designed to store sufficient fluid

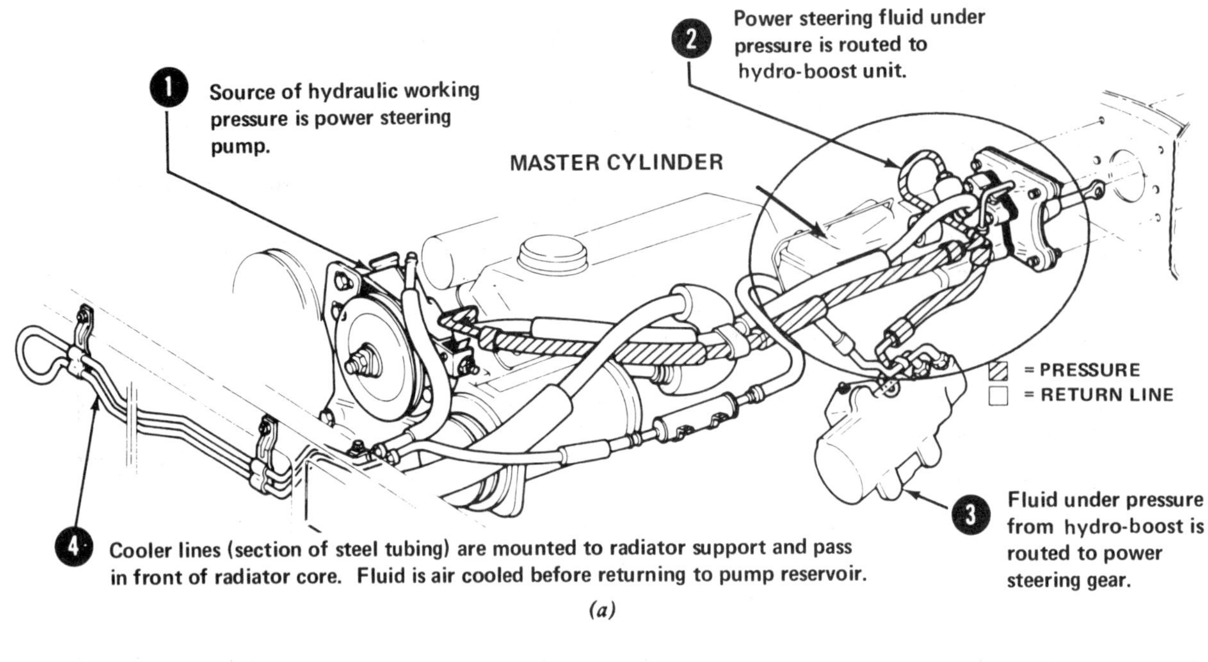

(a)

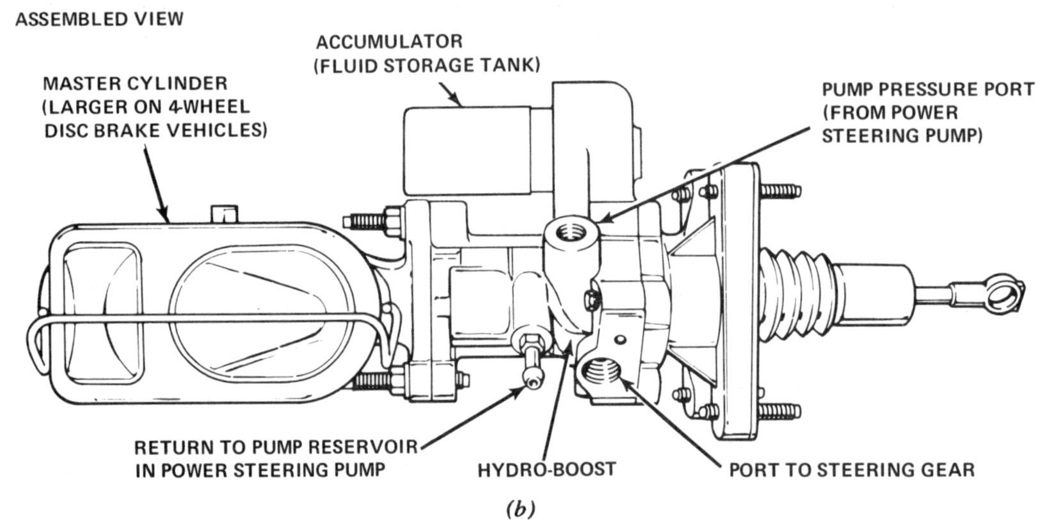

(b)

FIGURE 26-10. Hydro-Boost System. (a) Fluid flow through hydraulic system; (b) external view of Bendix hydro-boost power brake assist unit. (Courtesy of Ford Motor Company)

under pressure to provide at least two brake applications in the event that fluid flow from the power steering pump is not available. On some larger vehicles, an accessory hydraulic electrically driven pump, which takes the place of the accumulator, is incorporated into the hydro-boost unit. If for any reason the reserve system is not operational, the unit is designed so that the hydraulic brake system also can be applied manually. Figure 26-13 illustrates a typical schematic layout of a hydro-boost system.

DIAGNOSING AND SERVICING HYDRO-BOOST UNITS

The hydro-boost unit is only one component in a vehicle's braking systems. Complaints concerning

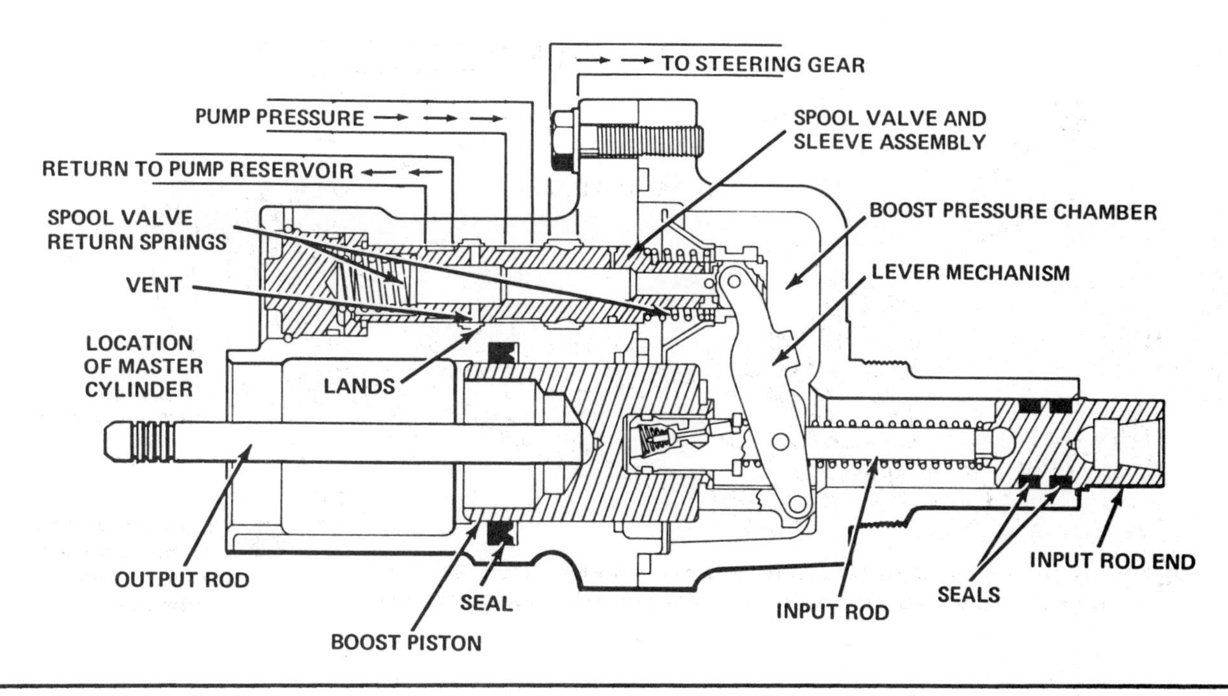

FIGURE 26-11. **Hydro-Boost Power Brake: Released Stage of Operation.** (Courtesy of Ford Motor Company)

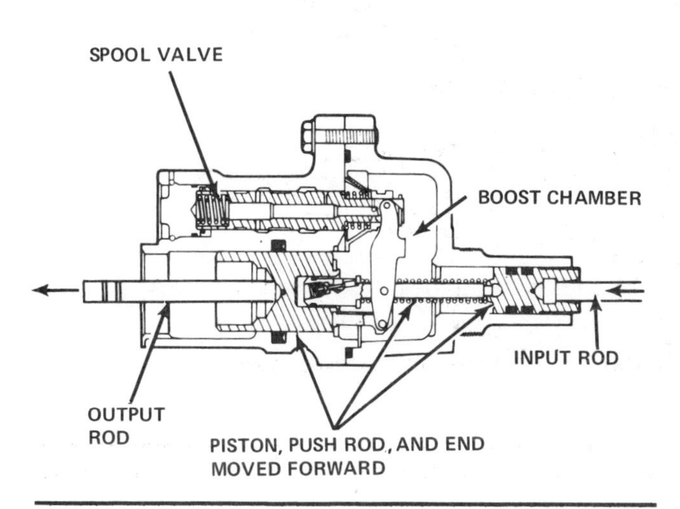

FIGURE 26-12. **Hydro-Boost Power Brake: Applied Stage of Operation.** In the applied position, the piston and input rod are moved forward. (Courtesy of Ford Motor Company)

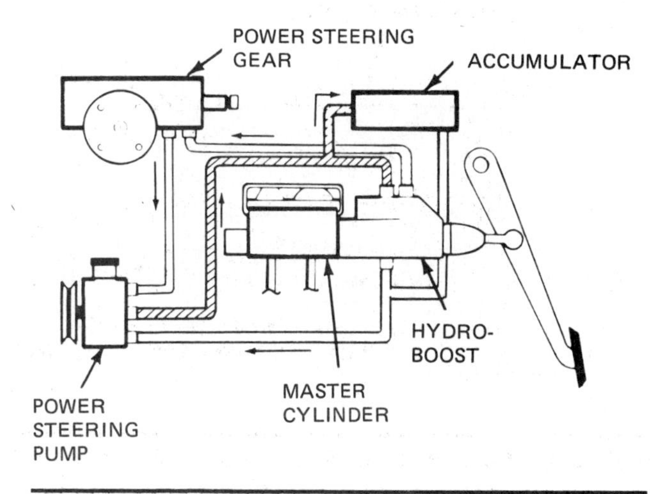

FIGURE 26-13. **Schematic Layout of Hydro-Boost System.** (Courtesy of Ford Motor Company)

the hydro-boost may be related to problems in either the power steering system or the service brakes.

Preliminary Checks

Before performing any diagnostic tests, first check and eliminate mechanical problem sources that could affect the proper operation of the unit.

Step-by-Step Procedure

1. Talk to the customer and obtain a description of the complaint. Verify the difficulty by driving the vehicle. If necessary, compare the operation of the customer's brakes to the operation of the brakes in a vehicle you know steers and brakes well. *Remember:* A hydro-boost unit cannot cause a noisy brake, a fading brake pedal, or a pulling brake. If one of these conditions exists, components other

than the hydro-boost are the cause of the problem.

2. Check the fluid level in the master cylinder, and fill the cylinder to the proper level if necessary.

3. Allow the engine to operate until the power steering fluid is at operating temperature of 74°C to 80°C (165°F to 175°F). Turn off the ignition and check the level of the fluid. Insufficient steering fluid can cause a loss of assist to both the steering and the braking systems.

4. If the level of the fluid is down, carefully examine the power steering system and hose connections for possible leaks.

5. If while inspecting the fluid level on the pump's dipstick, you notice signs of aeration (air bubbles in the power steering oil), the power steering system will need to be purged of air following instructions in the manufacturer's shop manual.

6. Check the power steering pump belt tension and adjust if necessary. If the belt is cracked or glazed, the belt must be replaced.

Functional Tests

If you have completed all the preliminary checks without finding the source of the problem, you can check the operation of the hydro-boost by performing a hydro-boost functional test and an accumulator leakdown test.

Note: If the hydro-boost is not operating correctly, verify that the power steering pump is functioning correctly by performing a pump flow test and a pump pressure relief test, as described in the manufacturer's shop manual. If the pump is not capable of producing the required operating pressure, the pump must be serviced and repaired before proceeding to the tests.

Hydro-Boost Functional Test. If the operation of the power steering pump conforms with the manufacturer's specifications, you may then proceed to hydro-boost functional test.

Step-by-Step Procedure

1. Check the hydraulic brake system for leaks or insufficient fluid in the master cylinder reservoirs.

2. With the parking brake applied and the transmission in neutral, apply the brake pedal several times to deplete all accumulator reserve pressure.

3. Hold the pedal depressed with medium applied force 11 to 15 kilograms (25 to 35 pounds), and start and operate the engine. If the hydro-boost unit is operating correctly, the brake pedal will fall slightly and then push back against the foot as the force against the pedal is maintained. If the pedal does not fall slightly and if there is no reaction, the hydro-boost unit is not functioning.

Accumulator Leakdown Test. The second test is the accumulator leakdown test.

Step-by-Step Procedure

1. Start and operate the engine at the correct idle speed as specified in the manufacturer's shop manual. Turn the steering wheel to the full right or full left position (against the stop) for no more than 5 seconds. Then turn the steering wheel so that the front wheels are positioned in the straight-ahead direction. Turn the ignition key to the off position.

2. Checking the time with a watch, wait for 90 seconds to elapse. Then, apply the brake pedal twice, using medium force. If the brake applications are not power assisted, the hydro-boost unit is defective.

A FACT TO REMEMBER

Changes in brake feel or travel are usually the first indicators that something is wrong with the brake systems.

Service Procedure Policy

Not all manufacturers recommend that the hydro-boost unit be disassembled and repaired. Before attempting any repair procedures, always follow the manufacturer's shop manual instructions pertaining to hydro-boost service procedures.

REVIEW TEST

1. List the names of the component parts of a Bendix single diaphragm vacuum power brake assist unit.
2. Describe the operation of a vacuum power brake assist unit during the released stage, applied stage, and holding stage.
3. Briefly describe the five tests you would use to diagnose the performance of a vacuum power brake assist unit.
4. Briefly describe the design features of a hydro-boost power brake assist unit.
5. Briefly describe the operation of the hydro-boost power brake assist unit during the applied stage of operation.

6. Briefly describe mechanical problems that could affect the proper operation of a hydro-boost unit.
7. Explain what can be determined by conducting a pump flow test and a pump pressure relief valve test.
8. Describe in detail how to conduct a hydro-boost functional test.
9. Describe in detail how to conduct an accumulator leakdown test.

FINAL TEST

This examination is multiple choice. Only one answer will be accepted. Carefully read every statement.

1. Mechanic A says that assuming that the vehicle's engine is operating correctly, pressure inside the intake manifold is approximately 35 to 49 kPa (5 to 7 psi). Mechanic B says that assuming that the vehicle's engine is operating correctly, as pressure inside the intake manifold decreases, vacuum increases. Who is right? (A) Mechanic A, (B) Mechanic B, (C) Both A and B, (D) Neither A nor B.
2. Mechanic A says that during the released stage of operation, vacuum is present on both sides of the piston. Mechanic B says that during the released stage of operation, atmospheric pressure is present on both sides of the piston. Who is right? (A) Mechanic A, (B) Mechanic B, (C) Both A and B, (D) Neither A nor B.
3. Mechanic A says that during the applied stage of operation, the poppet valve return spring is compressed. Mechanic B says that the valve plunger moves toward the brake pedal push rod. Who is right? (A) Mechanic A, (B) Mechanic B, (C) Both A and B, (D) Neither A nor B.
4. Mechanic A says that during the applied stage of operation, the hub or the piston moves inward toward the center of the brake assist unit. Mechanic B says that during the applied stage of operation, atmospheric pressure is present in the front chamber. Who is right? (A) Mechanic A, (B) Mechanic B, (C) Both A and B, (D) Neither A nor B.
5. Mechanic A says that during the applied stage of operation, the reaction disc is not compressed. Mechanic B says that during the applied stage of operation, the poppet valve admits a vacuum to the front chamber. Who is right? (A) Mechanic A, (B) Mechanic B, (C) Both A and B, (D) Neither A nor B.
6. Mechanic A says that during the holding stage of operation, the piston return spring is not compressed. Mechanic B says that during the holding stage of operation, both the atmospheric and vacuum ports are closed. Who is right? (A) Mechanic A, (B) Mechanic B, (C) Both A and B, (D) Neither A nor B.
7. Mechanic A says that the piston is an internal and external moving part. Mechanic B says that the reaction disc is located between the valve plunger and the master cylinder push rod. Who is right? (A) Mechanic A, (B) Mechanic B, (C) Both A and B, (D) Neither A nor B.
8. Mechanic A says that a defective hydro-boost unit will cause the brakes to be noisy upon application. Mechanic B says that a defective hydro-boost unit will cause the brakes to pull upon application. Who is right? (A) Mechanic A, (B) Mechanic B, (C) Both A and B, (D) Neither A nor B.
9. Mechanic A says that a loss of braking assist in a hydro-boost unit can be caused by a defective vacuum check valve. Mechanic B says that the loss of braking assist in a hydro-boost unit can be caused by the slipping power steering brake belt. Who is right? (A) Mechanic A, (B) Mechanic B, (C) Both A and B, (D) Neither A nor B.
10. Mechanic A says that the prime purpose of the power brake assist unit is to multiply hydraulic pressure. Mechanic B says that the prime purpose of the power brake assist unit is to multiply force. Who is right? (A) Mechanic A, (B) Mechanic B, (C) Both A and B, (D) Neither A nor B.
11. Mechanic A says that when a vacuum power brake assist unit is designed with two pistons, hydraulic pressure produced by the master cylinder is doubled. Mechanic B says that when a vacuum power brake assist unit is designed with two pistons, the assist unit has two brake reaction mechanisms. Who is right? (A) Mechanic A, (B) Mechanic B, (C) Both A and B, (D) Neither A nor B.
12. Mechanic A says that to determine the dependability of the vacuum check valve, the engine

must be operating during the brake application. Mechanic B says that to determine if the vacuum assist unit is operating correctly, you must pump the pedal several times, apply the pedal, and then start and operate the engine. Who is right? (A) Mechanic A, (B) Mechanic B, (C) Both A and B, (D) Neither A nor B.

13. Mechanic A says that the hydro-boost is operated by brake fluid transmitted from the power steering pump. Mechanic B says that the operation of the hydro-boost will be affected if the flow control valve in the power steering pump is stuck in the open position. Who is right? (A) Mechanic A, (B) Mechanic B, (C) Both A and B, (D) Neither A nor B.

Troubleshooting Guide for Brakes

INTRODUCTION

Proper diagnosis of a vehicle's brake system is very important. A proper road test and reference to this troubleshooting guide will enable you to locate and repair the problem area.

A road test is very important to evaluate the brake performance and to determine the extent of the required service. Whenever practical, a road test should be given before beginning any work on the brake system. After the brake work is completed, a road test should be made in every case as a measure of safety and to check the work performed.

Depressing the brake pedal before test driving the car is essential to make sure that there is adequate pedal reserve. Then, a series of stops at low speed should be made to ensure that the brakes are safe for road testing. After the condition of the brakes is determined, either by road testing the car or by listening to the customer's description of the problem, this troubleshooting guide should be referred to for possible causes and corrective procedures.

This troubleshooting guide is divided into five parts. The first part covers disc brakes, the second covers drum brakes, the third part covers vacuum power brake assist units, the fourth part covers hydro-boost power brake assist units, and the last part is a facsimile of a brake systems inspection report. The charts were made available courtesy of Wagner Electric Corporation.

TROUBLESHOOTING CHART FOR DISC BRAKES

Condition	Possible Cause	Corrective Action
Depressing brake pedal produces no braking effect	Reservoir fluid level low	Check for cause of fluid leak, repair as required, refill the reservoir
	Air in the hydraulic system	Bleed the system
	Bleeder screw open	Close the bleeder screw and bleed the system
	Improperly positioned pads (while servicing the calipers, pistons may have been pushed back in cylinder bores)	Reposition pads. (Depress the pedal a second time)
	Maladjusted rear brakes	Adjust properly
	Leak in rear brake cylinder	Hone the cylinder bore. Replace with new piston seals or replace caliper
	Piston seal damaged in one or more of the cylinders	Disassemble the caliper and replace the piston seals
	Leak past piston cups in master cylinder	Recondition or replace the master cylinder
Excessive pedal effort	Pads worn below minimum thickness	Install new pads
	Master cylinder or power brake malfunction	Check and correct the affected parts
	"Faded" overheated condition, glazed pads, "blued" or heat-checked rotors	Remachine the rotor
	Grease, oil, brake fluid on linings	Install new pads
	Seized or frozen pistons	Disassemble caliper and free up pistons
Excessive pedal travel	Air in the system	Bleed the system, replenish the fluid
	Insufficient or improper brake fluid (boil)	Drain, flush, and refill with correct fluid
	Master cylinder or power brake not functioning properly	Check and repair the affected parts
	Excessive disc runout	Check with dial indicator. Install new disc if runout exceeds maximum specified
	Excessively tapered or warped pads	Install new pads
	Damaged caliper piston seal	Install new piston seal
	Rear brakes out of adjustment	Check and adjust as necessary
	Wheel bearings loosely adjusted	Readjust to specified torque
Pedal pulsation (Brake roughness or chatter)	Excessive lateral runout of brake disc	Check with dial indicator. Install new disc if runout exceeds the maximum specified
	Excessive out-of-parallelism of brake disc	Check the parallelism (disc thickness variation) with micrometer, and remachine the disc or install a new one if the parallelism exceeds the maximum allowed
	Steering or suspension deflections	Front end alignment
	Excessive front bearing clearance	Readjust the bearings to specified torque

Condition	Possible Cause	Corrective Action
Car pulls to one side	Brake fluid, oil, grease, water, mud, or dirt on linings	Install new pads
	Unmatched linings	Install matched sets of pads
	Brakes grab; power brake malfunction	Check and correct power unit
	Seized or frozen pistons	Disassemble caliper and repair
	On rear brakes—maladjustment, sticking pistons or broken spring	Adjust the rear brakes, free up the pistons, install new spring
	Incorrect tire pressure	Inflate (or deflate) tires to the recommended pressure
	Distorted disc brake shoes	Install new brake pads
	Restricted hose or line	Examine the hoses and lines, and replace as necessary
	Front end out of alignment	Make proper wheel alignment
Leaky wheel cylinder	Cylinder bore surface scored or corroded	Disassemble caliper, hone cylinder bore. Install new seal
	Caliper piston seal damaged or worn	Disassemble caliper and install new seal
Front brakes heat up during driving and fail to release	Power brake malfunction	Check and repair the power brake unit
	Residual check valve in the master cylinder front brake outlet	Remove that valve from the master cylinder at front brake line outlet
	Sticking pedal linkage	Free up and lubricate as necessary
	Frozen or seized pistons	Disassemble the caliper and free up the pistons
	The driver in the habit of riding the brake pedal	Instruct the driver to give up that habit
Brake noises		
Chatter	Excessive lateral runout of rotor	Check the runout with a dial indicator. Install new disc if the runout exceeds the maximum specified
	Rotor parallelism (lack of)	Check the parallelism with a micrometer Remachine the disc, or install a new one, as required
	Loose wheel bearing	Readjust the bearings to specified torque
Scraping	Rust or mud build-up on edges of rotor and on caliper housing	Clean or replace, as necessary
	Worn pad, metal tabs or backing exposed to rotor	Replace the pad. Repair or replace faulty parts
	Faulty caliper alignment permitting rotor to scrape on housing	Adjust the caliper alignment properly
	Loose wheel bearing	Readjust wheel bearings to correct specifications
Groan or creep-groan	Slowly releasing the brakes	Slightly increase or decrease the pedal effort to eliminate the noise
Rattle	Excessive clearance between shoe and caliper	Install new pads
	Shoe anti-rattle spring missing or not properly positioned (Kelsey-Hayes or Budd types only)	Install new anti-rattle spring or position properly

TROUBLESHOOTING CHART FOR DRUM BRAKES

Condition	Possible Cause	Corrective Action
Low pedal	Low fluid level permits air to enter the system	Refill and bleed the master cylinder
	Leak in the hydraulic system	Check the master cylinder, wheel cylinders, tubes, and hoses for leakage. Repair or replace faulty parts
	Poor quality brake fluid (low boiling point)	Drain the hydraulic system and fill with the approved brake fluid
	Improperly adjusted manual master cylinder push rod	If the car has an adjustable push rod, adjust it
	Improperly adjusted power brake hydraulic push rod	Adjust the rod
	Excessive clearance between drums and linings	Adjust the brakes
	Automatic adjusters not working	Make several forward and reverse stops; if pedal doesn't come up, repair the automatic adjusters. On the star-wheel type, check the pawl advancement
	Brakes out of adjustment	Adjust the brakes
	Manual adjusters don't work	Repair or replace manual adjusters
	Bent or distorted brake shoes	Replace the shoes and linings in axle sets
	Residual pressure check valves don't hold pressure in lines	Repair or replace the master cylinder
Spongy or springy pedal	Air in the brake system	Refill and bleed. Check for possible fluid leaks at: • Master cylinder secondary cup • Wheel cylinders or disc calipers • Hose and line connections • Power brake vacuum line. Disconnect the vacuum line and check for the presence of fluid
	Fluid boil, vaporized by over-heating	Check by opening the bleeder screw. **Caution:** Cooling will restore braking, but condition will return when brakes are heated. Drain and replace the fluid. Check for cause of boiling: • Fluid—contaminated or wrong type for the system • Over-heated brakes—thin drums or rotors, spotty lining contacts, severe overloading of the vehicle Correct accordingly
	Weak hoses expanding (ballooning) under pressure	Replace the weak and faulty hoses
	Loss of system residual pressure may permit air to seep into wheel cylinders	Repair the valve leak or replace the residual pressure check valve
	Improper use of parking brake (after the pedal is released) causes a pressure loss	Check and correct the parking brake
	Shoes bent, distorted, not centered in the drum	Replace shoes and linings in axle sets. Adjust anchor pins (if provided). Check lining seating and drum eccentricity
	Brake drum too thin	Replace drum

Condition	Possible Cause	Corrective Action
Pedal drops or fades	Plugged master cylinder compensating port(s), and/or swollen master cylinder cups	Rebuild or replace the master cylinder. Flush the hydraulic system and fill with approved brake fluid
	Fluid seepage at hose joints and line connections	Replace faulty parts
	Fluid seepage at (or sticking) wheel cylinder piston cups	Rebuild or replace wheel cylinder
	Weak or broken shoe return springs	Replace shoe return springs
	Excessive drum expansion	Check for re-bored thin drum, cracked drum, over-heated conditions. Correct accordingly
Pedal pulsation	Worn or damaged front wheel bearings	Replace bearings
	Out-of-round drums	Recondition or replace the drums in axle sets
	Bent rear axle	Replace the axle
	Excessive drum runout due to faulty installation of new drums	Remove and properly install the drums
Brakes drag, lock, or overheat	Rough or corroded master cylinder bore. Master cylinder by-pass port blocked by dirt or swollen primary cup	Rebuild or replace the master cylinder
	Binding brake pedal linkage; bind prevents the return of pedal against its stop	Free up and lubricate
	Weak pedal return spring (if used)	Replace the spring
	Valve leaks, weak reaction springs, or bind in the power mechanism	Repair or replace the power section of the power brake
	Loose or worn front wheel bearings	Adjust to specifications or replace
	Defective brake hose or hydraulic tubing, preventing the return of brake fluid	Replace the faulty tubes or hoses, as necessary
	Soft or swollen rubber parts caused by contaminated or incorrect brake fluid	Replace all rubber parts, flush the hydraulic system, and fill with approved brake fluid
	Incorrect, bent, or distorted shoes	Replace shoes and linings in axle sets
	Incorrect linings, or linings loose on shoes, or grease or brake fluid on linings	Replace with correct shoes and linings. Repair grease seal or wheel cylinder
	Weak or broken shoe return springs	Replace the shoe return springs
	Improper brake shoe adjustment	Adjust shoes, and repair the automatic adjuster, if necessary
	Shoes not centered in drums	Adjust anchor pins (if provided)
	Loose or distorted brake backing plate	Tighten or replace backing plate
	Scored, hard spotted, or out of round drums	Refinish or replace the drums in axle sets
	Frozen parking brake cables	Free up, and lubricate cables, or replace
	Piston stuck in wheel cylinder	Rebuild or replace wheel cylinder
Severe stops, dive	Unbalanced "pull," any condition causing a brake to "grab"	Refer to the next condition, "car pulls to one side "
	Abnormal front end "dive" • Front suspension wear or alignment causing abnormal deflection, worn shocks, etc.	Front end wheel alignment needed. Repair or replace the affected parts
	• Temporary "morning sickness" due to moisture and rust accumulation on cooled friction surfaces	No cause for worry, if the conditions are temporary. If the condition persists, repair the affected parts

Condition	Possible Cause	Corrective Action
Car pulls to one side	Faulty suspension parts	Repair suspension system
	Incorrect linings on shoes	Replace with correct shoes and linings.
	Brake fluid, oil, grease, water, mud, dirt on linings	Repair grease seal or wheel cylinder, as necessary, clean the affected parts, replace shoes and linings in axle sets
	Distorted or incorrect shoes	Replace with correct shoes and linings
	Shoes not centered in drums	Adjust anchor pins (if provided)
	Loose or distorted brake backing plate	Tighten or replace the plate
	Weak shoe return springs or hold-downs	Replace with new parts
	Drum bell-mouthed, barrel shaped, threaded, cracked or high hard spots on drum surface	Recondition or replace drum
	Sticking wheel cylinder piston	Repair or replace wheel cylinder
	Abnormal deflection, wear and improper alignment including caster, camber, weak or broken springs, broken spring center bolt, loose U-bolts, loose steering gear or tie rods, worn shock absorbers	Check and correct front end alignment. Repair or replace faulty parts
	Tire improperly inflated	Inflate to correct pressure
	Faulty brake hose or obstruction in brake line	Replace hose or line
	Charred or badly worn brake linings	Replace linings and other worn parts
Brake noises		
Squeal	With new linings: Improper arc grind or eccentric drum causes the lining to "pinch" against the drum	Reface or grind the drums
	Dust or other particles imbedded in the lining	Remove foreign matter. Replace the lining if necessary
	Loose lining rivets	Reset the rivets or reline the shoes if the rivet holes in the linings are elongated
	Loose backing plate, anchor, drum or wheel cylinder	Tighten all loose parts and replace them if necessary
	Shoes scrape on support pads	Clean and lubricate
	Weak or broken shoe hold-down springs	Replace springs
	Support plate bent, shoes twisted	Replace the plate and/or shoes
	Drums out-of-round or out-of-square	Recondition or replace the drum
	Drum silencer (screen door) spring missing or weak	Replace the spring
Grind	Lining worn. Rivets or shoe table rub on drum	Replace shoes and linings.
	Foreign material imbedded in lining	Replace shoes and linings
	Rough drum surface	Recondition or replace the drum

Condition	Possible Cause	Corrective Action
Chatter	Lining clearance too tight or too loose	Adjust the linings to proper clearance
	Loose backing plate	Tighten or replace the plate
	Grease, fluid or dust on lining surface	Clean or replace the linings
	Twisted brake shoe	Replace the shoes
	Drum out-of-round, bell-mouthed, or barrel-shaped, cracked	Recondition or replace the drum
	Weak or broken return spring	Replace the springs
	Star wheel adjuster threads dry	Lubricate the threads
	Loose wheel bearing	Readjust wheel bearing
Click	Shoes lift off the support pads and snap-back:	
	Threaded drum surface	Replace the drum, or recondition
	Weak hold-downs	Replace the hold-downs
	Grooves in brake shoe support pads	Stone smooth and lubricate
Thumps	Excessive lining clearance. Applying, linings thump drums; releasing, shoes thump anchors	Adjust the brakes to proper clearance
Groans	Unlubricated metal contacts	Lubricate metal-to-metal contacts
Snapping	Dry metal to metal contacts	Lubricate
	Grooved shoe support pads	Stone smooth
	Loose backing plate or loose drum	Tighten or replace the affected parts
	Loose or worn front end components	Tighten or replace the components
Excessive stopping distance and/or pedal effort	After extended use, linings worn out or, poor quality brake linings	Replace with approved shoes and linings in axle sets
	"Faded" over-heated condition, glazed linings, "blued" or heat-checked drums	Sand the lining surface or replace shoes and linings in axle sets. Recondition or replace drums
	Excessive lining clearance	Check and correct pawl advancement on the automatic adjusters—star-wheel type. Adjust the manual adjusters
	Grease—or brake fluid—soaked linings	Repair grease seal or wheel cylinder, as necessary, and replace shoes and linings in axle sets
	Low engine vacuum supply to power brake booster or defective power brake vacuum check valve	Tune or repair the engine to obtain correct vacuum, or replace the check valve
	Faulty power brake booster	Replace the power booster
	Loose or leaking vacuum hose to power brake	Tighten clamps, or replace hose, as required
	Wheel cylinder or master cylinder piston stuck	Rebuild or replace appropriate cylinder
	Brake drum bell-mouthed or barrel shaped	Refinish drum to proper limits, or replace it (if beyond refinishing limits)

TROUBLESHOOTING CHART FOR VACUUM POWER BRAKE ASSIST UNITS

Condition	Possible Cause	Corrective Action
Hard pedal	Faulty vacuum check valve	Do not repair or oil the valve, replace it
	Engine intake manifold doesn't have sufficient (17% - 20%/432 - 508 mm) vacuum	Tune the engine, repair and adjust to obtain proper vacuum
	Plugged breather in power unit	Clean or replace
	Vacuum hoses and line fittings plugged	Clean the passages, or replace the faulty parts
	External vacuum leaks in reserve tank and loose vacuum connections	Replace with new tank, tighten all vacuum connections
	Wrong type of lining	Reline with new pads
	Restricted power brake air filter	Wash filter in non-oil base cleaning fluid. Allow to dry before replacing, or install a new filter
	Brake fluid, oil, grease, or foreign matter on lining surface	Correct the leaks. Reline the brakes with new shoe and lining sets
	Rusted or dry vacuum piston	Lubricate vacuum piston, or replace the power brake unit
	Frozen brake pedal linkage	Free up, lubricate, and replace the parts as necessary
	Collapsed vacuum hose or dented tubing from the intake manifold to the power brake unit	Replace with new hose or tubing
Pedal goes to floor	Worn out brakes	Reline with new pads
	Air in the system	Bleed the hydraulic system
	Low fluid level, or external hydraulic fluid leaks	Check for the leaks, and repair as necessary; replenish the fluid level
	Faulty residual check valve, primary cup, or contamination or rust	Replace defective parts. Flush the system. Replace the power brake unit, if defective
Pedal fails (or slow) to release	Pedal linkage frozen	Free up, lubricate, replace necessary parts
	Plugged breather in power brake unit	Clean or replace
	Plugged compensator port in the power brake unit	Adjust the pedal linkage, flush the system, or replace the power brake unit
	Piston stroke interference, or faulty residual check valve, or broken piston return spring, or sticky slide valve (do not oil) in the power brake unit	Replace the power brake unit
	Excessive hydraulic seal friction in the wheel cylinders and the power brake unit	Replace the wheel cylinders or the power brake unit

Condition	Possible Cause	Corrective Action
Sensitive brakes	Brake fluid, oil, or grease on the lining surface	Correct the leaks, install new pads
	Lining surface burnt (over-heated)	Replace the lining
	Improper adjustment of the stop light switch	Adjust or replace the switch
	Binding pedal linkage	Check for freeness, relieve, and lubricate
	Reaction diaphragm leakage, or sticky slide valve action, or closed or restricted atmospheric passage in the vacuum piston assembly, or binding rubber seals in the power brake unit	Replace the power brake unit
Brake pedal chatter	Air in the system	Bleed the system
Pedal vibrates and power brakes chatter	Air in the system	Bleed the system
	Improperly adjusted brake pedal free play	Readjust brake pedal free play

TROUBLESHOOTING CHART FOR HYDRO-BOOST POWER BRAKE ASSIST UNITS

Condition	Possible Cause	Corrective Action
Slow brake pedal return	Excessive seal friction in booster unit	Overhaul with new seal kit
	Spool return and/or piston return spring broken	Install appropriate new spring(s)
	Faulty spool action	Flush steering system while pumping brake pedal
	Clogged return line from booster to pump reservoir	Replace the line
Excessive brake pedal effort	Power steering pump belt loose or broken	Tighten or replace the belt
	Leaks in power steering, booster or accumulator connections, tube fittings, or hoses	Tighten fittings and/or replace tube seats and faulty parts
	External leakage at accumulator	Replace retainer and "O" ring
	Power steering reservoir empty	Fill reservoir, and check for external leaks
	Leakage at booster flange vent due to faulty booster piston seal, or leakage between housing and cover due to faulty booster cover seal	Overhaul with appropriate new seal kit
	Faulty booster spool plug seal	Overhaul with new spool plug seal kit
	Faulty booster input rod seal with leakage at input rod end	Replace booster
System does not hold charge	Accumulator leak down Steering hydro-boost system contaminated	Flush steering system while pumping brake pedal
	Internal leakage in accumulator system	Overhaul the system using accumulator re-build kit and seal kit
Pedal vibrates and booster chatters	Loose power steering pump belt	Tighten the belt
	Faulty spool operation due to contamination in system	Flush steering system while pumping brake pedal
	Low fluid level in power steering pump reservoir	Fill reservoir adequately, and check for external leaks
Brakes grab	Spool return spring damaged	Replace the spring
	Faulty spool operation caused by contamination in system	Flush steering system while pumping brake pedal

Caution: Power steering fluid and brake fluid should not be mixed. If brake seals contact steering fluid, or steering seals contact brake fluid, seal damage will result.

INSPECTION REPORT

After completing the inspection of various components of the brake system, a service mechanic is encouraged to fill out the inspection report chart that follows.

The completion of this report is a safeguard against missing any component of the brake system. Also it is very helpful to the customer in maintaining the brake system on his or her car in the best condition, which is very important from a safety standpoint.

BRAKE SYSTEM INSPECTION REPORT

Vehicle Make ____________ Year ________ Odometer ________ mi/km

Inspected by ____________________ Date ____________

The condition of the brake system on this vehicle is as follows:

Brake System Components	OK as is	Required Services	$ Estimate	
			Parts	Labor
Lines and Hoses		☐ Replace		
Stop Light		☐ Replace Bulb, ☐ Replace Switch ☐ Adjust		
Master Cylinder		☐ Needs Brake Fluid, ☐ Repair ☐ Replace		
Wheel Cylinder(s)		☐ Repair, ☐ Replace		
Brake Fluid		☐ Discard old, and add new fluid		
Main Brakes		☐ Front, ☐ Rear, ☐ Adjust ☐ Replace		
Power Brakes		☐ Adjust, ☐ Replace		
Parking Brake		☐ Adjust, ☐ Replace, ☐ Lubricate		
Wheel Bearings		☐ Adjust, ☐ Replace, ☐ Repack		
Grease Seals		☐ Replace		
Brake Drums		☐ Recondition, ☐ Replace		
Disc Rotor		☐ Recondition, ☐ Replace		
Disc Caliper		☐ Repair, ☐ Replace		
Miscellaneous				
Miscellaneous				
Miscellaneous				
		TOTAL		

Approximate time for services required (if any) ____________________

Customer Name and Address ____________________

____

Appendix: Introduction to Metrics

Most threaded fasteners are covered by specifications that define required mechanical properties, such as tensile strength, yield strength, proof load, and hardness. These specifications are carefully considered in initial selection of fasteners for a given application. To assure continued satisfactory vehicle performance, replacement fasteners used should be of the correct strength, as well as the correct nominal diameter, thread pitch, length, and finish.

Most original equipment fasteners (English system or metric) are identified with markings or numbers indicating the strength of the fastener. These markings are described in the pages that follow. Attention to these markings is important in assuring that the proper replacement fasteners are used. Further, some metric fasteners, especially nuts, are colored blue. This metric blue identification is in most cases a temporary aid for production start-up, and color will generally revert to normal black or bright after start-up.

Appendix reproduced courtesy of Ford Motor Company.

NOMENCLATURE FOR BOLTS

(ENGLISH) INCH SYSTEM
Bolt, 1/2-13x1

METRIC SYSTEM
Bolt M12-1.75x25

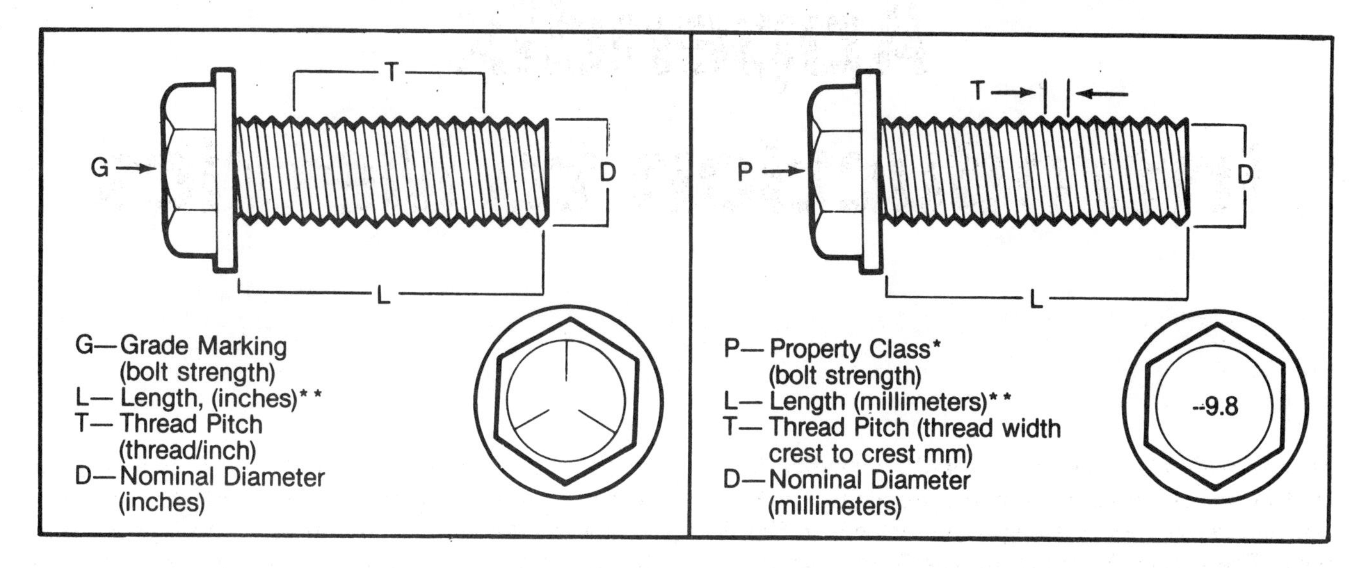

*The property class is an Arabic numeral distinguishable from the slash SAE English grade system.
**The length of all bolts is measured from the underside of the head to the end.

BOLT STRENGTH IDENTIFICATION

(ENGLISH) INCH SYSTEM

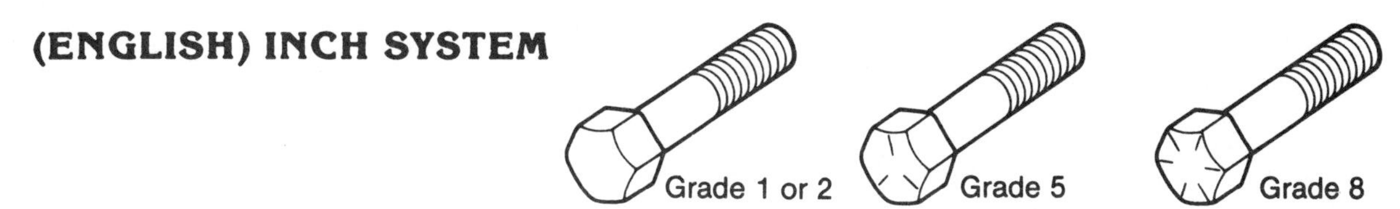

English (Inch) bolts—Identification marks correspond to bolt strength—increasing number of slashes represent increasing strength.

METRIC SYSTEM

Metric bolts—Identification class numbers correspond to bolt strength—increasing numbers represent increasing strength. Common metric fastener bolt strength property are 9.8 and 10.9 with the class identification embossed on the bolt head.

HEX NUT STRENGTH IDENTIFICATION

(ENGLISH) INCH SYSTEM

Grade	Hex Nut Grade 5	Hex Nut Grade 8
Identification	3 Dots	6 Dots

Increasing dots represent increasing strength.

METRIC SYSTEM

Class	Hex Nut Property Class 9	Hex Nut Property Class 10
Identification	9 Arabic 9	10 Arabic 10

May also have blue finish or paint daub on hex flat. Increasing numbers represent increasing strength.

OTHER TYPES OF PARTS

Metric identification schemes vary by type of part, most often a variation of that used of bolts and nuts. Note that many types of English and metric fasteners carry no special identification if they are otherwise unique.

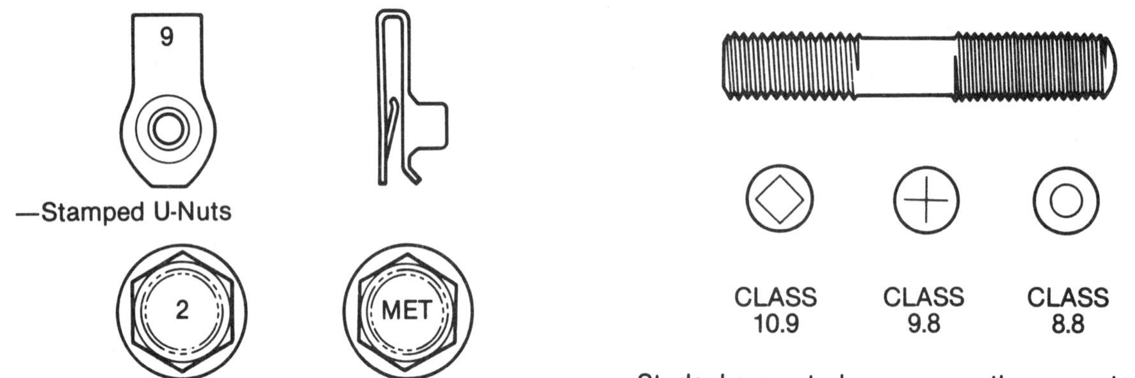

—Stamped U-Nuts

—Tapping, thread forming and certain other case hardened screws

—Studs, Large studs may carry the property class number. Smaller studs use a geometric code on the end.

ENGLISH METRIC CONVERSION

Description	Multiply	By	For Metric Equivalent
ACCELERATION	Foot/sec²	0.304 8	metre/sec² (m/s²)
	Inch/sec²	0.025 4	metre/sec²
TORQUE	Pound-inch	0.112 98	newton-metres (N·m)
	Pound-foot	1.355 8	newton-metres
POWER	horsepower	0.746	kilowatts (kw)
PRESSURE or STRESS	inches of water	0.2488	kilopascals (kPa)
	pounds/sq. in.	6.895	kilopascals (kPa)
ENERGY or WORK	BTU	1 055.	joules (J)
	foot-pound	1.355 8	joules (J)
	kilowatt-hour	3 600 000. or 3.6×10^6	joules (J = one W's)
LIGHT	foot candle	10.76	lumens/metre² (lm/m²)
FUEL PERFORMANCE	miles/gal	0.425 1	kilometres/litre (km/l)
	gal/mile	2.352 7	litres/kilometre (l/km)
VELOCITY	miles/hour	1.609 3	kilometres/hr. (km/h)
LENGTH	inch	25.4	millimetres (mm)
	foot	0.304 8	metres (m)
	yard	0.914 4	metres (m)
	mile	1.609	kilometres (km)
AREA	inch²	645.2	millimetres² (mm²)
		6.45	centimetres² (cm²)
	foot²	0.092 9	metres² (m²)
	yard²	0.836 1	metres²
VOLUME	inch³	16 387.	mm³
	inch³	16.387	cm³
	quart	0.016 4	litres (1)
	quart	0.946 4	litres
	gallon	3.785 4	litres
	yard³	0.764 6	metres³ (m³)
MASS	pound	0.453 6	kilograms (kg)
	ton	907.18	kilograms (kg)
	ton	0.90718	tonne
FORCE	kilogram	9.807	newtons (N)
	ounce	0.278 0	newtons
	pound	4.448	newtons
TEMPERATURE	degree fahrenheit	0.556 (°F -32)	degree Celsius (°C)

DECIMAL AND METRIC EQUIVALENTS

Fractions	Decimal Inch	Metric mm
1/64	.015625	.397
1/32	.03125	.794
3/64	.046875	1.191
1/16	.0625	1.588
5/64	.078125	1.984
3/32	.09375	2.381
7/64	.109375	2.778
1/8	.125	3.175
9/64	.140625	3.572
5/32	.15625	3.969
11/64	.171875	4.366
3/16	.1875	4.763
13/64	.203125	5.159
7/32	.21875	5.556
15/64	.234375	5.953
1/4	.250	6.35
17/64	.265625	6.747
9/32	.28125	7.144
19/64	.296875	7.54
5/16	.3125	7.938
21/64	.328125	8.334
11/32	.34375	8.731
23/64	.359375	9.128
3/8	.375	9.525
25/64	.390625	9.922
13/32	.40625	10.319
27/64	.421875	10.716
7/16	.4375	11.113
29/64	.453125	11.509
15/32	.46875	11.906
31/64	.484375	12.303
1/2	.500	12.7

Fractions	Decimal Inch	Metric mm
33/64	.515625	13.097
17/32	.53125	13.494
35/64	.546875	13.891
9/16	.5625	14.288
37/64	.578125	14.684
19/32	.59375	15.081
39/64	.609375	15.478
5/8	.625	15.875
41/64	.640625	16.272
21/32	.65625	16.669
43/64	.671875	17.066
11/16	.6875	17.463
45/64	.703125	17.859
23/32	.71875	18.256
47/64	.734375	18.653
3/4	.750	19.05
49/64	.765625	19.447
25/32	.78125	19.844
51/64	.796875	20.241
13/16	.8125	20.638
53/64	.828125	21.034
27/32	.84375	21.431
55/64	.859375	21.828
7/8	.875	22.225
57/64	.890625	22.622
29/32	.90625	23.019
59/64	.921875	23.416
15/16	.9375	23.813
61/64	.953125	24.209
31/32	.96875	24.606
63/64	.984375	25.003
1	1.00	25.4

TORQUE CONVERSION

NEWTON METRES (N·m)	POUND-FEET (LB-FT)
1	0.7376
2	1.5
3	2.2
4	3.0
5	3.7
6	4.4
7	5.2
8	5.9
9	6.6
10	7.4
15	11.1
20	14.8
25	18.4
30	22.1
35	25.8
40	29.5
50	36.9
60	44.3
70	51.6
80	59.0
90	66.4
100	73.8
110	81.1
120	88.5
130	95.9
140	103.3
150	110.6
160	118.0
170	125.4
180	132.8
190	140.1
200	147.5
225	166.0
250	184.4

POUND-FEET (LB-FT)	NEWTON METRES (N·m)
1	1.356
2	2.7
3	4.0
4	5.4
5	6.8
6	8.1
7	9.5
8	10.8
9	12.2
10	13.6
15	20.3
20	27.1
25	33.9
30	40.7
35	47.5
40	54.2
45	61.0
50	67.8
55	74.6
60	81.4
65	88.1
70	94.9
75	101.7
80	108.5
90	122.0
100	135.6
110	149.1
120	162.7
130	176.3
140	189.8
150	203.4
160	216.9
170	230.5
180	244.0

Glossary

Ackerman principle: geometric principle used to provide toe-out on turns. (The ends of the steering arms are angled so that the inside wheel turns more than the outside wheel when a car is making a turn.)

alignment: adjustment of components to bring them into a predetermined position; usually considered a combination of camber, caster, and toe-in adjustments.

anchor: point on a drum brake support plate where the forces are transmitted to the chassis.

asbestos: fiber that is heat resistant and nonburning; used for brake linings.

axial play: movement that is parallel to the axis of rotation.

ball joint: joint of connection where a ball moves within a socket so as to allow rotary motion while the angle of the axis of rotation changes.

ball joint inclination: inward tilt of the top of the steering axis centerline through the ball joints as viewed from the front of the vehicle.

bead: part of a tire that contacts the rim of a wheel.

bellows: movable cover or seal, usually of a rubber-like material, that is pleated or folded like an accordian to allow for expansion and contraction.

bias ply belted tire: bias ply tire with reinforcing strips of belts placed over the plies at the thread section.

bias ply tire: tire constructed of alternate plies positioned so that the cords cross the centerline of the tire at an angle of about 35°.

boiling point: temperature at which a liquid begins to boil.

brake: energy-conversion device used to slow, stop, or hold a vehicle or mechanism; a device that changes the kinetic energy of motion into useless and wasted heat energy.

brake drag: constant, relatively light contact between brake linings and drums or disks when the brakes are not applied.

brake drum: rotating cylindrical member of the drum brake assembly that is acted upon by the brake shoes to slow or stop a vehicle.

brake fade: temporary reduction, or fading out, of braking effectiveness; caused by overheating from excessively long and hard brake application or by water reducing the friction between braking surfaces.

brake feel: reaction of the brake pedal against the driver's foot; tells the driver how heavily the brakes are being applied.

brake fluid: special fluid used in hydraulic braking systems that is incompressible and transmits hydraulic force from the master cylinder to the wheel cylinders or calipers.

brake pad: friction lining and shoe assembly that is forces against the rotor to cause braking action in a disk brake.

brake shoe: curved metal part, faced with the brake lining, that is forced against the brake drum to produce braking action.

bushing: liner or separator between parts. Usually made of soft metal, plastic, or rubber, it is used to reduce friction and wear between parts.

caliper: C-shaped housing that fits over the rotor, holds the pads, and contains the hydraulic components that force the pads against the rotors when braking.

cam bolt: bolt fitted with an eccentric that will cause parts to change position when the bolt is turned.

camber: inward or outward tilt of the top of a wheel; angle formed by the centerline of the wheel and true vertical.

camber roll: inherent characteristic of independent suspension systems to change camber angles when cornering.

camber wear: wear on one side of a tire tread caused by the angle at which the tire tread contacts the road surface.

castellated nut: nut that has slots through which a cotter pin may be passed to secure the nut to its bolt or stud.

caster: forward or backward tilt of the top of the steering axis from the true vertical when viewed from the side.

center bolt: bolt that maintains the alignment of the leaves in a leaf spring; also maintains the position of the axle housing on the spring.

center of gravity: point about which the vehicle weight is evenly distributed or balanced.

centrifugal force: outward force from the center of a rotating object.

chassis: frame, suspension systems, engine, and drive train of a vehicle. Assembled parts of an automobile without the body.

check valve: valve that opens to permit the passage of air or fluid in one direction only or operates to prevent (check) some undesirable action.

coil convolutions: height measurement between two circular coils of a coil spring.

coil spring: length of spring-steel wire wound in the shape of a spiral.

combination valve: brake warning lamp valve in combination with a proportioning and/or metering valve.

compression: loading or storing of energy in a spring.

compression loaded: force that pushes the round ball of the tapered ball joint stud against the top of the ball joint housing; caused by the weight of the vehicle.

concentric: having the same center.

constant displacement pump: rotary hydraulic pump where the internal parts are closely fit. (The volume of oil produced by this type of pump will vary relatively to its revolutions per minute.)

contact area: portion of a tire that contacts the road at any given moment.

control arms, front: horizontal arms that connect the front wheels to the car and that support the weight of the front of the car.

control arms, rear: horizontal arms that connect the rear axle housing to the frame when coil springs are used in the rear suspension system. (The arms maintain axle alignment and handle the driving and torque loads.)

conventional differential: type of rear axle that allows the inside rear wheel to turn fewer revolutions than the outside rear wheel—for example, when the vehicle is cornering.

crocus cloth: cloth material with iron oxide. (It contains no grit material and is used to remove slight blemishes from machined surfaces.)

curb riding height: distance from the vehicle's frame or lower points of the front and rear suspension to a level surface. (To measure, tires must be properly inflated, fuel tank full, and no passenger or luggage compartment load. The vehicle must also be positioned on a level floor or alignment machine rack.)

curb weight: weight of a vehicle with a full supply of fuel, oil, and coolant but with no driver, passenger, or luggage.

cylinder hone: expandable rotating tool with abrasive stones turned by an electric motor; used to clean and smooth the inside surface of a cylinder.

datum line: an imaginary reference line or plane, established by vehicle manufacturers, located a fixed distance below the vehicle from which vertical measurements can be made.

degree: unit used to measure angles; 1/360th of a circle; usually abbreviated by the symbol placed behind a number.

diagonal brake system: dual brake system with separate hydraulic circuits connecting diagonal wheels together (right front to left rear and left front to right rear).

dial indicator: precision instrument that indicates linear measurement on a dial face.

directional control and stability: ability of a car to travel in a straight line with a minimum of driver control.

disc brake: brake in which the frictional forces act upon the faces of a revolving disk to slow or stop it.

double lap flare: type of flare used on the ends of brake lines for extra strength. (The flared end of the tubing is doubled over.)

drag link (steering connecting rod): tube or rod connecting the steering gear pitman arm to the tie-rods or steering knuckle arms.

dual master cylinder: master cylinder with two separate pressure chambers; used in dual brake systems.

dynamic wheel balance: balance in motion; the balance of a wheel while it is rotating; the total weight distributed evenly over both the axis of rotation and the centerline of the wheel.

dynamic wheel unbalance: unbalance of a wheel while it is rotating; the total weight distributed unevenly over both the axis of rotation and the centerline of the wheel.

energy-absorbing steering column: steering column designed to collapse, or telescope, at a controlled rate in the event of a frontal collision.

feeler gauge: series of various thicknesses of shim stock used to measure width distance between two allied parts.

friction: resistance to motion between two bodies in contact with each other.

front end wheel alignment factor: interrelationship of camber, caster, toe kingpin inclination, and toe-out on turns. (These factors combine to provide directional control and stability. The end result should reduce driver fatigue, provide maximum vehicle control, and increase tire life.)

hub: central part of a wheel; housing for the bearings on which the wheel rotates around the spindle.

hydraulics: science of water and other liquids in motion, their uses in engineering, and the laws of their actions.

hydraulic pressure: force per unit area exerted in all parts of a hydraulic system by a liquid.

hydrostatics: branch of physics concerned with the potential (static) energy of fluid under pressure and its use in performing work.

idler arm: arm or lever that can rotate about its support and is used to support one end of a relay rod; usually duplicates the motion of the pitman arm.

included angle: sum of the angles of camber and steering axis inclination.

independent suspension: suspension systems by which a wheel on one side of a car can move vertically without affecting the wheel on the other side of the car.

inertia: tendency of matter (weight) to remain at rest or continue in a fixed direction unless affected by an outside force.

jounce travel: upward movement of a wheel from its normal position; usually caused by a road bump or a weight transfer onto a wheel.

kinetic balance: balance of the radial forces on a spinning tire; determined by an electronic wheel balancer.

kingpin: pin or shaft on which the steering spindle assembly rotates.

kingpin inclination: inward tilt of the top of the steering axis centerline through the kingpin as viewed from the front of the vehicle.

lash: movement or play between parts; clearance between moving parts, such as meshing gear teeth.

lateral tire/rim runout: wobble, or side-to-side movement, of a rotating wheel or of a rotating wheel and tire assembly.

lead: slight pull to one side.

leaf spring: spring made of several long, thin strips of steel of graduated lengths used to support vehicle weight and absorb road shock.

limited slip differential: differential that utilizes a clutch device to deliver power to either rear wheel when the opposite wheel is spinning.

linkage: system of rods and levers used to transmit motion or force.

load range: alphabetic system used to identify the service limitations of a tire.

mechanical advantage (M.A.): gear ratio of the steering gearbox.

millimeter: metric unit of measurement equal to 0.039370 inch. Usually abbreviated as mm, as in 1 mm.

molylube: grease containing molybdenum; a special plating ingredient with excellent lubricating characteristics that is not affected by moisture.

NIASE: National Institute for Automotive Service Excellence.

nonslip differential: prevents the slippage of one or both rear wheels when the tread is on a slippery surface. When balancing a rear wheel on a vehicle with a nonslip differential, the speed of the wheel being balanced is the same as the speedomoter. (Some vehicle manufacturers use the term *limited slip on positive track* to describe this rear axle design.)

offset: lateral movement of a wheel in relationship to the vehicle centerline.

oil aeration: excessive retention of air or gas in oil. (When oil is subject to this condition, the oil tends to become compressible producing a groaning noise in the pump when it produces flow and pressure.)

oscillate: to move back and forth.

out-of-round: wheel or tire defect in which the wheel or tire is not round.

overall steering ratio (O.S.R.): total mechanical advantage of the total steering system.

overinflation: condition of a tire that is inflated to more than the recommended pressure.

oversteer: tendency of a car to turn more sharply than the driver intends while negotiating a turn.

parallelogram steering linkage: commonly used steering linkage system that utilizes a relay rod or center link to connect the pitman arm to an idler arm that duplicates the pitman arm's length and motion. Separate tie-rods connect the steering arms to the relay rod. The assembled linkage resembles a parallelogram in shape, and the centerline of the pivot points are parallel.

pitman arm: arm connected to the steering gear sector shaft that transforms the rotating motion of the shaft into lateral motion at the relay rod.

plane of rotation: established centerline through a wheel that is perpendicular to the axis line of the wheel.

play: movement between parts.

plumb bob: weight hung on the end of a line to establish a true vertical position.

ply rating: method of indicating relative tire strength. The ply rating usually does not indicate the actual number of plies.

point of intersection: apex of the extended centerline of the pivot axis and the extended centerline of a front wheel (see Figure 18-12).

power steering: steering system that utilizes hydraulic pressure to boost the steering effort of the driver.

preload: thrust load applied to bearings that support a rotating part to eliminate axial play or movement.

Prussian blue dye: dye used to distinguish mating surfaces between contact surfaces.

pull: tendency of a car to veer to one side.

rack and pinion steering gear: steering gear design that utilizes a small pinion gear attached to the steering shaft to move a long, toothed bar, called a rack gear. (The ends of the rack gear are attached to the steering arms by means of of tie-rods.)

radial (vertical support) load: influence of the vertical support load carried by a bearing.

radial force variation: difference in stiffness at two or more points on a tire.

radial play: movement at 90° to an axis of rotation.

radial ply tire: tire constructed of alternate plies positioned so that the cords cross the tire centerline at an angle of 90°.

radial (elliptical) tire/rim runout: variation in the radius of a wheel or a wheel and tire assembly; out-of-round.

rag joint: flexible coupling that contains a rubberized fabric disc or water; usually used in steering systems.

ratios: when discussing shock absorbers, a term that applies to the total resistance or control of a shock absorber. (A shock absorber valued at 70/30 would mean that 70 percent of the total control would go on the extension cycle and 30 percent of the total control would be on the compression cycle.)

rebound travel: downward movement of a wheel from its normal position; examples are the sudden drop of a wheel into a depression and a weight transfer away from the wheel.

rigid axle suspension. *See* solid axle suspension

road crown: slope or pitch of a road from its center to the curbs or shoulders.

road feel: feeling transmitted back to the steering wheel by the wheels of the vehicle.

road shock: shock or movement transmitted from the road surface to the steering wheel through the steering gear and linkage.

SAE: Society of Automotive Engineers.

sawtooth wear pattern: tire wear pattern in which the tread ribs wear more on one side than on the other side; usually caused by incorrect toe-in.

scrub radius area: distance between the extended centerline of the steering axis and the centerline of the tire at the point where the tire contacts the road.

sector (gear): segment of a gear with two or more teeth.

sector shaft: shaft to which the sector (gear) is attached; output shaft of a steering gear.

self-energizing brakes: drum brake in which braking action pulls the show lining tighter against the drum.

serrations: grooves or teeth formed in parts so that they do not shift position when tightened together.

setback: situation in which one wheel is rearward of the opposite wheel.

sheering: vertical breakage of a metal part.

shim: spacer used to adjust the distance between two parts.

shimmy: rapid oscillation, or wobble, of a wheel and tire assembly about the steering axis.

shock absorber: device used to dampen spring oscillations.

slip angle: angle between the true centerline of the tire and the actual path followed by the tire while rounding a turn.

solid axle suspension: suspension system in which the wheels are mounted at each end of a solid, or undivided, axle or axle housing.

space-saver spare tire: inflatable spare tire. When the deflated spare tire has been removed from the trunk, it is inflated by means of an aerosol-type container.

spindle: part of the front suspension system about which the front wheel rotates; shaft or pin about which another part rotates.

spinner: electrically driven drum or roller used to spin a wheel and tire assembly on the car to check for imbalance.

sprung weight: weight supported by the spring of a vehicle—that is, frame, engine, transmission, and body.

stability: property of a body that causes it, when disturbed from a condition of equilibrium or steady motion, to develop forces or tendencies to restore it to its original condition.

stabilizer: device that uses the torsional resistance of a steel bar to reduce the roll of a car and to prevent too great a difference in the spring action at the two front wheels.

static wheel balance: balance at rest; a distribution of weight around the axis of rotation so that a wheel has no tendency to rotate by itself, regardless of its position.

static wheel unbalance: distribution of uneven weight around the axis of rotation.

steering arms: arms that transmit the steering motion from the rods to the steering knuckles. (In some instances, they are forged in one piece with the steering knuckles.)

steering axis inclination: inward tilt of the top of the steering axis centerline through the kingpin or ball joints as viewed from the front of the vehicle.

steering column: support for the steering wheel. (It includes the mast jacket, steering shaft, and shift tube and also serves as a mounting for other controls.)

steering gear: device made of gears to transmit steering effort to the steering linkage for the purpose of directing the car. (A high mechanical advantage is usually designed into the steering gear.)

steering geometry: term used to describe the relationship of the various measurements and angles in the steering and suspension systems.

steering knuckle: forging consisting of a spindle and its mounting, mounted between the upper and lower ball joints and pivots for steering.

steering linkage: system of links, rods, and levers used to transmit motion from the steering gear to the steering knuckles or spindle assemblies.

steering shaft: steel rod that connects the steering wheel to the steering gear.

steering system: combination of the steering gear, steering wheel, and steering linkage that enables the driver to turn the front wheels and guide the car.

steering wheel play: any movement of the steering wheel that does not produce movement of the front wheels.

stick-on lead weight: adhesive lead weight made for balancing magnesium rims, and sold in 12 inch strips.

strut rod: brace used between a control arm and the frame.

suspension height: distance from a specified point on a car to the road surface when the car is at curb weight.

sway bar. *See* stabilizer.

tensile bolts and nuts: fasteners made of extra-strength steel that are used to attach parts where there is additional stress and torque.

tension loaded: force that pulls the tapered stud within the ball joint away from the top of the ball joint housing.

thrust (lateral, side to side) load: lateral load or force on a bearing.

tie-rod: rod used to connect the relay rod to a steering arm.

tie-rod end: ball and socket joint at the end of a tie-rod.

tie-rod sleeves: tubes or pipes with internal threads into which the tie-rods and tie-rod ends are threaded. (Turning the sleeves extends or retracts the tie-rod, changing its length.)

tire print: pattern made by the tire at the point of road contact.

tire wear pattern: characteristics of the wear shown by the tread of a tire.

toe: difference in the linear distance between the front of the front wheels of a vehicle at axle height and the distance at the rear of the front wheels at axle height. All distances are measured from the centerline of the tires' tread.

toe change: in and out (lateral) movement of the front wheels as the vehicle is driven; normally caused by incorrect riding height, incorrect camber and/or caster front end alignment settings, and the unparallel relationship of the tie-rods and lower control arms of the vehicle.

toe-in: condition in which the fronts of the wheels on a common axle are closer together than the rears of the wheels.

toe-out: condition in which the fronts of the wheels on a common axle are farther apart than the rears of the wheels.

toe-out on turns: number of degrees the front inside wheel (on a corner) turns out as compared to number of degrees the opposite wheel turns in.

torque: force that produces a twisting effect.

torsion bar: spring steel bar supported and anchored at one end, whereas the other end is supported but allowed to twist. (The bar's resistance to any torque or twisting effort provides spring action.)

torsional stress: strain placed on the frame caused by the lateral twisting forces created by the reaction of the wheels and suspension in response to the road surface.

track width: linear distance from the outside edge of the tread of the right front wheel to the outside tread edge of the left front wheel.

tracking: relationship of the paths taken by the front wheels and the rear wheels; alignment of the center of the tread distance of the front wheels with the center of the tread distance of the rear wheels.

traction: adhesion (static friction) between two bodies. (When a person walks on a surface, the bottom of one foot must momentarily adhere to the surface to allow movement; without friction, no movement would be possible.)

transverse torsion bars: torsion bars mounted so that they extend across the frame.

tread: distance between the centerlines of the wheels on a common axle.

tread wear indicators: ridges molded into the grooves between the ribs of a tire tread. (The indicators become visible when the tread is worn to a depth of less than 1.6 millimeters or 1/16 inch.)

turning resistance and road feel: steering and braking control of the vehicle by the driver that occurs through the traction of the front wheels on the road surface. (Without road feel, the driver would not realize the amount of effort required to turn the steering wheel or the applied force on the brake pedal to stop the vehicle.)

underinflation: condition of a tire that is inflated to less than the recommended pressure.

understeer: tendency of a car to turn less sharply than the driver intends while negotiating a turn.

unitized body: design that does not use a frame. The body of the car, reinforced at appropriate points, provides the mounting for the suspension system.

unsprung weight: weight of that portion of a vehicle not supported by the springs, such as wheels, tires, axles, and control arms.

vacuum suspended power brake: type of power brake in which both sides of the piston are subjected to vacuum; therefore, the piston is suspended in vacuum.

variable spring rate: load required to move a spring a given distance. Rate is expressed in pounds per inch and is an indicator of the softness or firmness of a given spring or suspension.

waddle: side-to-side movement of a car; usually caused by a tire that has a belt that has been installed crookedly.

wander: tendency of a car to veer, or drift, to either side from a straight path.

wheel alignment service: adjusting the front end steering factors to provide maximum tire tread life and steering control.

wheel cylinder: drum brake device for converting hydraulic fluid pressure to mechanical force for brake application.

wheel hop: vertical movement of a wheel assembly that is statically unbalanced.

wheel spindle: short, tapered axle shaft.

wheel tramp: vertical movement of a revolving wheel due to the effect of centrifugal force.

wheel base: distance between the centers of the front and rear wheels.

Final Test Answers

Chapter 1: 1. A, 2. B, 3. C, 4. A, 5. D, 6. C, 7. B, 8. A, 9. C, 10. D, 11. C, 12. B, 13. B, 14. C

Chapter 2: 1. C, 2. C, 3. A, 4. A, 5. C, 6. A, 7. B, 8. D

Chapter 3: 1. A, 2. D, 3. D, 4. C, 5. A, 6. B, 7. B, 8. A, 9. C, 10. D, 11. D, 12. B, 13. C, 14. D, 15. B, 16. C

Chapter 4: 1. C, 2. C, 3. A, 4. D, 5. B, 6. A, 7. A, 8. B, 9. D, 10. D, 11. A, 12. B, 13. C, 14. C, 15. A

Chapter 5: 1. A, 2. C, 3. C, 4. D, 5. D, 6. B, 7. A, 8. B, 9. B, 10. D

Chapter 6: 1. C, 2. A, 3. C, 4. B, 5. D, 6. A, 7. B, 8. D, 9. C

Chapter 7: 1. A, 2. C, 3. C, 4. B, 5. B, 6. C, 7. D, 8. D, 9. A, 10. C, 11. D, 12. D

Chapter 8: 1. C, 2. B, 3. C, 4. A, 5. A, 6. D, 7. C, 8. C, 9. A, 10. B

Chapter 9: 1. B, 2. D, 3. C, 4. C, 5. A, 6. D, 7. A, 8. B, 9. B, 10. C, 11. C, 12. C, 13. C, 14. B, 15. D, 16. A

Chapter 10: 1. A, 2. B, 3. C, 4. C, 5. A, 6. A, 7. D, 8. C, 9. A, 10. B, 11. D, 12. A, 13. C, 14. D

Chapter 11: 1. C, 2. C, 3. B, 4. D, 5. C, 6. A, 7. C, 8. B, 9. A

Chapter 12: 1. C, 2. B, 3. C, 4. A, 5. D, 6. B, 7. B, 8. C, 9. C, 10. A, 11. C, 12. C, 13. C, 14. B

Chapter 13: 1. C, 2. C, 3. C, 4. B, 5. C, 6. A, 7. A, 8. A, 9. A

Chapter 14: 1. B, 2. B, 3. A, 4. D, 5. B, 6. C, 7. C, 8. C, 9. A, 10. A

Chapter 15: 1. A, 2. D, 3. D, 4. C, 5. B, 6. C, 7. D, 8. B, 9. C

Chapter 16: 1. B, 2. D, 3. A, 4. C, 5. C, 6. C, 7. B, 8. C, 9. A, 10. A, 11. D, 12. D, 13. B

Chapter 17: 1. D, 2. A, 3. B, 4. C, 5. C, 6. B

Chapter 18: 1. B, 2. B, 3. C, 4. D, 5. C, 6. A, 7. A, 8. A, 9. A

Chapter 19: 1. D, 2. A, 3. C, 4. A, 5. B, 6. C, 7. B, 8. C, 9. D, 10. D

Chapter 20: 1. D, 2. A, 3. B, 4. C, 5. C, 6. A, 7. B, 8. C, 9. D, 10. A, 11. C, 12. D, 13. C, 14. B, 15. C

Chapter 21: 1. C, 2. A, 3. D, 4. D, 5. C, 6. C, 7. B, 8. A, 9. C, 10. B, 11. A

Chapter 22: 1. C, 2. B, 3. D, 4. A, 5. B, 6. A, 7. C, 8. D, 9. C, 10. B, 11. B, 12. B, 13. D, 14. A

Chapter 23: 1. B, 2. A, 3. D, 4. C, 5. D, 6. A, 7. B, 8. D

Chapter 24: 1. C, 2. D, 3. C, 4. A, 5. B, 6. B, 7. B, 8. D, 9. A, 10. C, 11. A, 12. B, 13. C

Chapter 25: 1. C, 2. D, 3. D, 4. B, 5. C, 6. C, 7. D, 8. B, 9. B, 10. B, 11. D, 12. A, 13. D, 14. B, 15. B, 16. A, 17. B, 18. C

Chapter 26: 1. C, 2. A, 3. A, 4. A, 5. D, 6. B, 7. C, 8. D, 9. B, 10. B, 11. D, 12. B, 13. D

Index